Veterinary Public Health

Veterinary Public Health

Edited by:

Patricia Marques

Delve
PUBLISHING

www.delvepublishing.com

Veterinary Public Health

Patricia Marques

Delve Publishing
224 Shoreacres Road
Burlington, ON L7L 2H2
Canada
www.delvepublishing.com
Email: orders@arclereducation.com

ISBN: 978-1-77469-171-7 (Hardcover)

DECLARATION

ABOUT THE EDITOR

Patricia obtained her PhD from University College Dublin in 2010. Her interests are on Microbiology and Parasitology. She is currently working as a Postdoc at University of Maryland Baltimore, USA on Chlamydia infections

TABLE OF CONTENTS

Section 2: "One Health" Concept

LIST OF CONTRIBUTORS

Victoria Ng
Centre for Public Health and Zoonoses, Department of Population Medicine, Ontario Veterinary College, University of Guelph, Guelph, Canada

Jan M. Sargeant
Centre for Public Health and Zoonoses, Department of Population Medicine, Ontario Veterinary College, University of Guelph, Guelph, Canada

Catherine Kansiime
Department of Health Policy, Planning and Management, Makerere University School of Public Health, College of Health Sciences, Kampala, Uganda

Lynn M. Atuyambe
Department of Community Health and Behavioral Sciences, Makerere University School of Public Health, College of Health Sciences, Kampala, Uganda

Benon B. Asiimwe
Department of Medical Microbiology, College of Health Sciences, Makerere University, Kampala, Uganda

Anthony Mugisha
Department of Veterinary Medicine, College of Veterinary Medicine, Animal Resources and Bio-security, Makerere University, Kampala, Uganda

Samuel Mugisha
Department of Biological Sciences, College of Natural Sciences, Makerere University, Kampala, Uganda

Victor Guma
Department of Mental Health and Community Psychology, Makerere University College of Humanities and Social Sciences, Kampala, Uganda

Innocent B. Rwego
Department of Biological Sciences, College of Natural Sciences, Makerere University, Kampala, Uganda
Ecosystem Health Initiative, College of Veterinary Medicine, University of Minnesota, St. Paul, Minnesota, United States of America

Elizeus Rutebemberwa
Department of Health Policy, Planning and Management, Makerere University School of Public Health, College of Health Sciences, Kampala, Uganda

Angela Song-En Huang
Office of Preventive Medicine, Centers for Disease Control, Taipei, Taiwan

Wan-Chin Chen
Office of Preventive Medicine, Centers for Disease Control, Taipei, Taiwan

Wan-Ting Huang
Office of Preventive Medicine, Centers for Disease Control, Taipei, Taiwan

Shih-Tse Huang
Office of Preventive Medicine, Centers for Disease Control, Taipei, Taiwan

YiChun Lo
Office of Preventive Medicine, Centers for Disease Control, Taipei, Taiwan

Sung-Hsi Wei
Office of Preventive Medicine, Centers for Disease Control, Taipei, Taiwan

Hung-Wei Kuo
Epidemic Intelligence Center, Centers for Disease Control, Taipei, Taiwan

Pei-Chun Chan
Office of Preventive Medicine, Centers for Disease Control, Taipei, Taiwan

Min-Nan Hung
Office of Preventive Medicine, Centers for Disease Control, Taipei, Taiwan

Yu-Lun Liu
Office of Preventive Medicine, Centers for Disease Control, Taipei, Taiwan

Jung-Jung Mu
Center for Research and Diagnostics, Centers for Disease Control, Taipei, Taiwan

Jyh-Yuan Yang
Center for Research and Diagnostics, Centers for Disease Control, Taipei, Taiwan

Ding-Ping Liu
Epidemic Intelligence Center, Centers for Disease Control, Taipei, Taiwan

Jih-Haw Chou
Office of Deputy Director, Centers for Disease Control, Taipei, Taiwan

Jen-Hsiang Chuang
Office of Deputy Director, Centers for Disease Control, Taipei, Taiwan

Feng-Yee Chang
National Defense Medical Center, Tri-Service General Hospital, Taipei, Taiwan

Valerie Hongoh
The Research Group on Epidemiology of Zoonoses and Public Health (GREZOSP), Faculty of Veterinary Medicine, Universite´ de Montre´al, Saint-Hyacinthe, Canada
Department of Pathology and Microbiology, Faculty of Veterinary Medicine, Universite´ de Montre´al, Saint-Hyacinthe, Canada

Pierre Gosselin
Institut national de sante´ publique Que´bec (INSPQ), Que´bec, Canada
Ouranos, Consortium on Regional Climatology and Adaptation to Climate Change, Montreal, Canada

Pascal Michel
The Research Group on Epidemiology of Zoonoses and Public Health (GREZOSP), Faculty of Veterinary Medicine, Universite´ de Montre´al, Saint-Hyacinthe, Canada
National Microbiology Laboratory at Saint-Hyacinthe, Public Health Agency of Canada, Saint-Hyacinthe, Canada

Andre´ Ravel
The Research Group on Epidemiology of Zoonoses and Public Health (GREZOSP), Faculty of Veterinary Medicine, Universite´ de Montre´al, Saint-Hyacinthe, Canada
Department of Pathology and Microbiology, Faculty of Veterinary Medicine, Universite´ de Montre´al, Saint-Hyacinthe, Canada

JeanPhilippe Waaub
Group for Research in Decision Analysis (GERAD, HEC Montre´al, Polytechnique Montre´al, McGill, UQAM), Montreal, Canada

Ce´line Campagna
Institut national de sante´ publique Que´bec (INSPQ), Que´bec, Canada
Department of Social and Preventive Medicine, Universite´ Laval, Que´bec, Canada

Karim Samoura
Universite´ Aube Nouvelle, Ouagadougou, Burkina Faso

Laura Contalbrigo
Istituto Zooprofilattico Sperimentale delle Venezie, Viale Dell'Università 10, 35020 Legnaro, (PD), Italy

Stefano Borgo
Laboratory for Applied Ontology, ISTC-CNR, Via alla Cascata, 56/C, 38123 Trento, Italy

Giandomenico Pozza
Istituto Zooprofilattico Sperimentale delle Venezie, Viale Dell'Università 10, 35020 Legnaro, (PD), Italy

Stefano Marangon
Istituto Zooprofilattico Sperimentale delle Venezie, Viale Dell'Università 10, 35020 Legnaro, (PD), Italy

M. Minara Khatun
Department of Microbiology and Hygiene, Faculty of Veterinary Science, Bangladesh Agricultural University, Mymensingh 2202, Bangladesh

M. Ariful Islam
Department of Microbiology and Hygiene, Faculty of Veterinary Science, Bangladesh Agricultural University, Mymensingh 2202, Bangladesh

M. Mufizur Rahman
Department of Microbiology and Hygiene, Faculty of Veterinary Science, Bangladesh Agricultural University, Mymensingh 2202, Bangladesh

S. M. Thumbi
Paul G. Allen School for Global Animal Health, Washington State University, Pullman, Washington, United States of America
Center for Global Health Research, Kenya Medical Research Institute, Kisumu, Kenya

M. Kariuki Njenga
Centers for Disease Control and Prevention, Nairobi, Kenya

Thomas L. Marsh
Paul G. Allen School for Global Animal Health, Washington State University, Pullman, Washington, United States of America
School of Economic Sciences, Washington State University, Pullman, Washington, United States of America

Susan Noh
Paul G. Allen School for Global Animal Health, Washington State University, Pullman, Washington, United States of America
Animal Diseases Research Unit, Agricultural Research Service, United States Department of Agriculture, Pullman, Washington, United States of America

Elkanah Otiang
Center for Global Health Research, Kenya Medical Research Institute, Kisumu, Kenya

Peninah Munyua
Centers for Disease Control and Prevention, Nairobi, Kenya

Linus Ochieng
Center for Global Health Research, Kenya Medical Research Institute, Kisumu, Kenya

Eric Ogola
Center for Global Health Research, Kenya Medical Research Institute, Kisumu, Kenya

Jonathan Yoder
Paul G. Allen School for Global Animal Health, Washington State University, Pullman, Washington, United States of America
School of Economic Sciences, Washington State University, Pullman, Washington, United States of America

Allan Audi
Center for Global Health Research, Kenya Medical Research Institute, Kisumu, Kenya

Joel M. Montgomery
Centers for Disease Control and Prevention, Nairobi, Kenya

Godfrey Bigogo
Center for Global Health Research, Kenya Medical Research Institute, Kisumu, Kenya

Robert F. Breiman
Emory University, Atlanta, Georgia, United States of America

Guy H. Palmer
Paul G. Allen School for Global Animal Health, Washington State University, Pullman, Washington, United States of America

Terry F. McElwain
Paul G. Allen School for Global Animal Health, Washington State University, Pullman, Washington, United States of America

Yvonne Regier
Institute for Medical Microbiology and Infection Control, University Hospital, Goethe-University, Frankfurt am Main, Germany

Fiona O'Rourke
Institute for Medical Microbiology and Infection Control, University Hospital, Goethe-University, Frankfurt am Main, Germany

Volkhard A. J. Kempf
Institute for Medical Microbiology and Infection Control, University Hospital, Goethe-University, Frankfurt am Main, Germany

Tai Xie
Faculty of Health Service, Second Military Medical University, Shanghai, China
Faculty of Naval Medicine, Second Military Medical University, Shanghai, China

Wenbao Liu
Faculty of Naval Medicine, Second Military Medical University, Shanghai, China

Benjamin D. Anderson
Division of Infectious Diseases and Duke Global Health Institute, Duke University, Durham, North Carolina, United States of America

Xiaorong Liu
Faculty of Health Service, Second Military Medical University, Shanghai, China

Gregory C. Gray
Division of Infectious Diseases and Duke Global Health Institute, Duke University, Durham, North Carolina, United States of America

Christian Sonne
Department of Bioscience, Arctic Research Centre (ARC), Aarhus Univer sity, Faculty of Science and Technology, Frederiksborgvej 399, 4000 Roskilde, Denmark

Robert James Letcher
Ecotoxicology and Wildlife Health Division, Environment and Climate Change Canada, National Wildlife Research Centre, Carleton University, Ottawa, ON K1A 0H3, Canada

Bjørn Munro Jenssen
Department of Bioscience, Arctic Research Centre (ARC), Aarhus Univer sity, Faculty of Science and Technology, Frederiksborgvej 399, 4000 Roskilde, Denmark
Department of Biology, Norwegian University of Science and Technology, 7491 Trondheim, Norway
Department of Arctic Technology, The University Centre in Svalbard, 9171 Longyearbyen, Norway

JeanPierre Desforges
Department of Bioscience, Arctic Research Centre (ARC), Aarhus Univer sity, Faculty of Science and Technology, Frederiksborgvej 399, 4000 Roskilde, Denmark

Igor Eulaers
Department of Bioscience, Arctic Research Centre (ARC), Aarhus Univer sity, Faculty of Science and Technology, Frederiksborgvej 399, 4000 Roskilde, Denmark

Emilie AndersenRanberg
Department of Bioscience, Arctic Research Centre (ARC), Aarhus Univer sity, Faculty of Science and Technology, Frederiksborgvej 399, 4000 Roskilde, Denmark

Kim Gustavson
Department of Bioscience, Arctic Research Centre (ARC), Aarhus Univer sity, Faculty of Science and Technology, Frederiksborgvej 399, 4000 Roskilde, Denmark

Bjarne Styrishave
Toxicology Laboratory, Department of Pharmacy, Faculty of Health and Medical Sciences, University of Copenha gen, Universitetsparken 2, 2100 Copenhagen, Denmark

Rune Dietz
Department of Bioscience, Arctic Research Centre (ARC), Aarhus Univer sity, Faculty of Science and Technology, Frederiksborgvej 399, 4000 Roskilde, Denmark

Mercedes Arguello-Casteleiro
School of Computer Science, University of Manchester, Manchester, UK

Robert Stevens
School of Computer Science, University of Manchester, Manchester, UK

Julio Des-Diz
Hospital do Salnés, Villagarcía de Arousa, Pontevedra, Spain

Chris Wroe
BMJ, Tavistock Square, London, UK

Maria Jesus Fernandez-Prieto
Salford Languages, University of Salford, Salford, UK

Nava Maroto
Departamento de Lingüística Aplicada a la Ciencia y a la Tecnología, Universidad Politécnica de Madrid, Madrid, Spain

Diego Maseda-Fernandez
Midcheshire Hospital Foundation Trust, NHS England, Crewe, UK
School of Medical Sciences, University of Manchester, Manchester, UK

George Demetriou
School of Computer Science, University of Manchester, Manchester, UK

Simon Peters
School of Social Sciences, University of Manchester, Manchester, UK

Peter-John M. Noble
Small Animal Veterinary Surveillance Network, University of Liverpool, Liverpool, UK

Phil H. Jones
Small Animal Veterinary Surveillance Network, University of Liverpool, Liverpool, UK

Jo Dukes-McEwan
Small Animal Teaching Hospital, University of Liverpool, Liverpool, UK

Alan D. Radford
Small Animal Veterinary Surveillance Network, University of Liverpool, Liverpool, UK

John Keane
School of Computer Science, University of Manchester, Manchester, UK
Manchester Institute of Biotechnology, University of Manchester, Manchester, UK

Goran Nenadic
School of Computer Science, University of Manchester, Manchester, UK
Manchester Institute of Biotechnology, University of Manchester, Manchester, UK
Health eResearch Centre, University of Manchester, Manchester, UK

Stefan Berg
TB Research Group, Veterinary Laboratories Agency, Weybridge, New Haw, Addlestone, Surrey, United Kingdom

Rebuma Firdessa
Armauer Hansen Research Institute, Addis Ababa, Ethiopia

Meseret Habtamu
Armauer Hansen Research Institute, Addis Ababa, Ethiopia

Endalamaw Gadisa
Armauer Hansen Research Institute, Addis Ababa, Ethiopia

Araya Mengistu
Armauer Hansen Research Institute, Addis Ababa, Ethiopia

Lawrence Yamuah
Armauer Hansen Research Institute, Addis Ababa, Ethiopia

Gobena Ameni
Armauer Hansen Research Institute, Addis Ababa, Ethiopia
Aklilu Lemma Institute of Pathobiology, Addis Ababa University, Addis Ababa, Ethiopia

Martin Vordermeier
TB Research Group, Veterinary Laboratories Agency, Weybridge, New Haw, Addlestone, Surrey, United Kingdom

Brian D. Robertson
Department of Infectious Disease and Microbiology, Imperial College, South Kensington Campus, London, United Kingdom

Noel H. Smith
TB Research Group, Veterinary Laboratories Agency, Weybridge, New Haw, Addlestone, Surrey, United Kingdom

Howard Engers
Armauer Hansen Research Institute, Addis Ababa, Ethiopia

Douglas Young
Department of Infectious Disease and Microbiology, Imperial College, South Kensington Campus, London, United Kingdom

R. Glyn Hewinson
TB Research Group, Veterinary Laboratories Agency, Weybridge, New Haw, Addlestone, Surrey, United Kingdom

Abraham Aseffa
Armauer Hansen Research Institute, Addis Ababa, Ethiopia

Stephen V. Gordon
TB Research Group, Veterinary Laboratories Agency, Weybridge, New Haw, Addlestone, Surrey, United Kingdom

School of Agriculture, Food Science and Veterinary Medicine, University College Dublin, Dublin, Ireland
School of Medicine and Medical Science, University College Dublin, Dublin, Ireland
School of Biomolecular and Biomedical Science, College of Life Sciences, University College Dublin, Dublin, Ireland
UCD Conway Institute of Biomolecular and Biomedical Research, University College Dublin, Dublin, Ireland

Yasser M. Sanad
Food Animal Health Research Program, Ohio Agricultural Research and Development Center, The Ohio State University, Wooster, Ohio, United States of America
Department of Veterinary Preventive Medicine, College of Veterinary Medicine, The Ohio State University, Columbus, Ohio, United States of America

Issmat I. Kassem
Food Animal Health Research Program, Ohio Agricultural Research and Development Center, The Ohio State University, Wooster, Ohio, United States of America

Melanie Abley
Department of Veterinary Preventive Medicine, College of Veterinary Medicine, The Ohio State University, Columbus, Ohio, United States of America

Wondwossen Gebreyes
Department of Veterinary Preventive Medicine, College of Veterinary Medicine, The Ohio State University, Columbus, Ohio, United States of America

Jeffrey T. LeJeune
Food Animal Health Research Program, Ohio Agricultural Research and Development Center, The Ohio State University, Wooster, Ohio, United States of America
Department of Veterinary Preventive Medicine, College of Veterinary Medicine, The Ohio State University, Columbus, Ohio, United States of America

Gireesh Rajashekara
Food Animal Health Research Program, Ohio Agricultural Research and Development Center, The Ohio State University, Wooster, Ohio, United States of America
Department of Veterinary Preventive Medicine, College of Veterinary Medicine, The Ohio State University, Columbus, Ohio, United States of America

Rebuma Firdessa
Armauer Hansen Research Institute, Addis Ababa, Ethiopia

Rea Tschopp
Armauer Hansen Research Institute, Addis Ababa, Ethiopia
Centre for Molecular Microbiology and Infection, Imperial College London, London,

United Kingdom
Swiss Tropical and Public Health, Basel, Switzerland

Alehegne Wubete
National Animal Health Diagnostic and Investigation Center, Sebeta, Addis Ababa, Ethiopia

Melaku Sombo
National Animal Health Diagnostic and Investigation Center, Sebeta, Addis Ababa, Ethiopia

Elena Hailu
Armauer Hansen Research Institute, Addis Ababa, Ethiopia

Girume Erenso
Armauer Hansen Research Institute, Addis Ababa, Ethiopia

Teklu Kiros
Armauer Hansen Research Institute, Addis Ababa, Ethiopia

Lawrence Yamuah
Armauer Hansen Research Institute, Addis Ababa, Ethiopia

Martin Vordermeier
Department for Bovine Tuberculosis, Animal Health and Veterinary Laboratories Agency, Weybridge, Surrey, United Kingdom

R. Glyn Hewinson
Department for Bovine Tuberculosis, Animal Health and Veterinary Laboratories Agency, Weybridge, Surrey, United Kingdom

Douglas Young
Centre for Molecular Microbiology and Infection, Imperial College London, London, United Kingdom

Stephen V. Gordon
School of Veterinary Medicine, University College Dublin, Dublin, Republic of Ireland

Mesfin Sahile
National Animal Health Diagnostic and Investigation Center, Sebeta, Addis Ababa, Ethiopia

Abraham Aseffa
Armauer Hansen Research Institute, Addis Ababa, Ethiopia

Stefan Berg
Department for Bovine Tuberculosis, Animal Health and Veterinary Laboratories Agency, Weybridge, Surrey, United Kingdom

Nma Bida Alhaji
Department of Public Health and Epidemiology, Niger State Ministry of Livestock and Fisheries, Minna, Nigeria

Mohammed Baba Aliyu
Department of Public Health and Epidemiology, Niger State Ministry of Livestock and Fisheries, Minna, Nigeria

Ibrahim Ghali-Mohammed
Department of Veterinary Public Health and Preventive Medicine, University of Ilorin, Ilorin, Nigeria

Ismail Ayoade Odetokun
Department of Veterinary Public Health and Preventive Medicine, University of Ilorin, Ilorin, Nigeria

Mateus de Souza Ribeiro Mioni
Departamento de Higiene Veterina´ria e Sau´de Pu´blica, Universidade Estadual Paulista "Ju´lio de Mesquita Filho", Botucatu, São Paulo, Brazil

Francisco Borges Costa
Departamento de Patologia, Universidade Estadual do Maranhão, São Luı´s, Maranhão, Brazil

Bruna Letı´cia Devide´ Ribeiro
Departamento de Higiene Veterina´ria e Sau´de Pu´blica, Universidade Estadual Paulista "Ju´lio de Mesquita Filho", Botucatu, São Paulo, Brazil

Wanderson Sirley Reis Teixeira
Departamento de Higiene Veterina´ria e Sau´de Pu´blica, Universidade Estadual Paulista "Ju´lio de Mesquita Filho", Botucatu, São Paulo, Brazil
Instituto de Quı´mica, Universidade de São Paulo, São Paulo, São Paulo, Brazil

Vanessa Cristina Pelicia
Departamento de Higiene Veterina´ria e Sau´de Pu´blica, Universidade Estadual Paulista "Ju´lio de Mesquita Filho", Botucatu, São Paulo, Brazil

Marcelo Bahia Labruna
Departamento de Medicina Veterina´ria Preventiva e Sau´de Animal, Universidade de São Paulo, São Paulo, São Paulo, Brazil

E´lodie Rousset
Anses, French Agency for Food, Environmental and Occupational Health Safety, Sophia Antipolis laboratory, Animal Q Fever Unit, Sophia Antipolis, France

Karim Sidi-Boumedine
Anses, French Agency for Food, Environmental and Occupational Health Safety, Sophia Antipolis laboratory, Animal Q Fever Unit, Sophia Antipolis, France

Richard Thie´ry
Anses, French Agency for Food, Environmental and Occupational Health Safety, Sophia Antipolis laboratory, Animal Q Fever Unit, Sophia Antipolis, France

Jane Megid
Departamento de Higiene Veterina´ria e Sau´de Pu´blica, Universidade Estadual Paulista "Ju´lio de Mesquita Filho", Botucatu, São Paulo, Brazil

Yanlong Cong
College of Animal Science and Veterinary Medicine, Jilin University, Changchun, Jilin, China

Guangmei Wang
College of Animal Science and Veterinary Medicine, Jilin University, Changchun, Jilin, China

Zhenhong Guan
Institute of Zoonoses, Jilin University, Changchun, Jilin, China

Shuang Chang
College of Animal Science and Veterinary Medicine, Jilin University, Changchun, Jilin, China

Quanpeng Zhang
College of Animal Science and Veterinary Medicine, Jilin University, Changchun, Jilin, China

Guilian Yang
College of Animal Science and Technology, Jilin Agricultural University, Changchun, Jilin, China

Weili Wang
Jilin Entry-Exit Inspection and Quarantine Bureau, Changchun, Jilin, China

Qingfeng Meng
Jilin Entry-Exit Inspection and Quarantine Bureau, Changchun, Jilin, China

Weiming Ren
College of Animal Science and Technology, Jilin Agricultural University, Changchun, Jilin, China

Chunfeng Wang
College of Animal Science and Technology, Jilin Agricultural University, Changchun, Jilin, China

Zhuang Ding
College of Animal Science and Veterinary Medicine, Jilin University, Changchun, Jilin, China

Santosh Dhakal
National Zoonoses and Food Hygiene Research Centre, Kathmandu, Nepal

Durga Datt Joshi
National Zoonoses and Food Hygiene Research Centre, Kathmandu, Nepal

Anita Ale
National Zoonoses and Food Hygiene Research Centre, Kathmandu, Nepal

Minu Sharma
National Zoonoses and Food Hygiene Research Centre, Kathmandu, Nepal

Meena Dahal
National Zoonoses and Food Hygiene Research Centre, Kathmandu, Nepal

Yogendra Shah
National Zoonoses and Food Hygiene Research Centre, Kathmandu, Nepal

Dhan Kumar Pant
National Zoonoses and Food Hygiene Research Centre, Kathmandu, Nepal

Craig Stephen
Centre for Coastal Health, Nanaimo British Columbia, Canada and Faculty of Veterinary Medicine and Faculty of Medicine, University of Calgary, Calgary, Alberta, Canada

Puteri Azaziah Megat Abd Rani
School of Veterinary Science, The University of Queensland, Queensland 4072, Australia

Peter J Irwin
School of Veterinary and Biomedical Science, Murdoch University, Western Australia 6150, Australia

Mukulesh Gatne
Bombay Veterinary College, Maharastra Animal and Fisheries Sciences University, Parel, Mumbai 400012, India

Glen T Coleman
School of Veterinary Science, The University of Queensland, Queensland 4072, Australia

Linda M McInnes
School of Veterinary and Biomedical Science, Murdoch University, Western Australia 6150, Australia

Rebecca J Traub
School of Veterinary Science, The University of Queensland, Queensland 4072, Australia

Yipeng Sun
Key Laboratory of Zoonosis of Ministry of Agriculture, College of Veterinary Medicine, China Agricultural University, Beijing, China

Yuhai Bi
Key Laboratory of Zoonosis of Ministry of Agriculture, College of Veterinary Medicine, China Agricultural University, Beijing, China
CAS Key Laboratory of Pathogenic Microbiology and Immunology Institute of Microbiology, Chinese Academy of Sciences, Beijing, China

Juan Pu
Key Laboratory of Zoonosis of Ministry of Agriculture, College of Veterinary Medicine, China Agricultural University, Beijing, China

Yanxin Hu
Key Laboratory of Zoonosis of Ministry of Agriculture, College of Veterinary Medicine, China Agricultural University, Beijing, China

Jingjing Wang
Key Laboratory of Zoonosis of Ministry of Agriculture, College of Veterinary Medicine, China Agricultural University, Beijing, China

Huijie Gao
Key Laboratory of Zoonosis of Ministry of Agriculture, College of Veterinary Medicine, China Agricultural University, Beijing, China

Linqing Liu
Key Laboratory of Zoonosis of Ministry of Agriculture, College of Veterinary Medicine, China Agricultural University, Beijing, China

Qi Xu
Key Laboratory of Zoonosis of Ministry of Agriculture, College of Veterinary Medicine, China Agricultural University, Beijing, China

Yuanyuan Tan
Key Laboratory of Zoonosis of Ministry of Agriculture, College of Veterinary Medicine, China Agricultural University, Beijing, China

Mengda Liu
Key Laboratory of Zoonosis of Ministry of Agriculture, College of Veterinary Medicine, China Agricultural University, Beijing, China

Xin Guo
Key Laboratory of Zoonosis of Ministry of Agriculture, College of Veterinary Medicine, China Agricultural University, Beijing, China

Hanchun Yang
Key Laboratory of Zoonosis of Ministry of Agriculture, College of Veterinary Medicine, China Agricultural University, Beijing, China

Jinhua Liu
Key Laboratory of Zoonosis of Ministry of Agriculture, College of Veterinary Medicine, China Agricultural University, Beijing, China
The Shandong Animal Disease Control Center, Jinan, Shandong, China

Hang Lu
Department of Communication, Cornell University, Ithaca, New York, United States of America

Katherine A. McComas
Department of Communication, Cornell University, Ithaca, New York, United States of America

Danielle E. Buttke
Wildlife Health Branch, Biological Resources Division, National Park Service, Fort Collins, Colorado, United States of America

Sungjong Roh
Lee Kong Chian School of Business, Singapore Management University, Singapore, Singapore

Margaret A. Wild
Wildlife Health Branch, Biological Resources Division, National Park Service, Fort Collins, Colorado, United States of America

Sonia M. Hernandez
Warnell School of Forestry and Natural Resources, University of Georgia, Athens, Georgia, United States of America
Southeastern Cooperative Wildlife Disease Study, Department of Population Health, College of Veterinary Medicine, University of Georgia, Athens, Georgia, United States of America

Catharine N. Welch
Warnell School of Forestry and Natural Resources, University of Georgia, Athens, Georgia, United States of America

Valerie E. Peters
Institute for Environment and Sustainability, Department of Zoology, Miami University, Columbia, Ohio, United States of America

Erin K. Lipp
Department of Environmental Health Science, University of Georgia, Athens, Georgia, United States of America

Shannon Curry
Warnell School of Forestry and Natural Resources, University of Georgia, Athens, Georgia, United States of America

Michael J. Yabsley
Warnell School of Forestry and Natural Resources, University of Georgia, Athens, Georgia, United States of America
Southeastern Cooperative Wildlife Disease Study, Department of Population Health, College of Veterinary Medicine, University of Georgia, Athens, Georgia, United States of America

Susan Sanchez
Department of Infectious Diseases, College of Veterinary Medicine, University of Georgia, Athens, Georgia, United States of America

Andrea Presotto
Department of Geography, University of Georgia, Athens, Georgia, United States of America

Peter GernerSmidt
Centers for Disease Control and Prevention, U.S. Department of Health and Human Services, Atlanta, Georgia, United States of America

Kelley B. Hise
Centers for Disease Control and Prevention, U.S. Department of Health and Human Services, Atlanta, Georgia, United States of America

Elizabeth Hammond
Lion Country Safari Park, Loxahatchee, Florida, United States of America

Whitney M. Kistler
Southeastern Cooperative Wildlife Disease Study, Department of Population Health, College of Veterinary Medicine, University of Georgia, Athens, Georgia, United States of America

Marguerite Madden
Department of Geography, University of Georgia, Athens, Georgia, United States of America

April L. Conway
Warnell School of Forestry and Natural Resources, University of Georgia, Athens, Georgia, United States of America

Tiffany Kwan
Poultry Diagnostic and Research Center, Department of Population Health, College of Veterinary Medicine, University of Georgia, Athens, Georgia, United States of America

John J. Maurer
Poultry Diagnostic and Research Center, Department of Population Health, College of Veterinary Medicine, University of Georgia, Athens, Georgia, United States of America

Annika van Roon
Centre for Infectious Disease Control, National Institute for Public Health and the Environment (RIVM), Bilthoven, the Netherlands

Miriam Maas
Centre for Infectious Disease Control, National Institute for Public Health and the Environment (RIVM), Bilthoven, the Netherlands

Daniela Toale
Centre for Infectious Disease Control, National Institute for Public Health and the Environment (RIVM), Bilthoven, the Netherlands

Nedzib Tafro
Border inspection, Netherlands Food and Consumer Product Safety Authority (NVWA), Amsterdam, The Netherlands

Joke van der Giessen
Centre for Infectious Disease Control, National Institute for Public Health and the Environment (RIVM), Bilthoven, the Netherlands

LIST OF ABBREVIATIONS

AEFI	Adverse events following immunization
ACIP	Advisory Committee on Immunization Practices
ALERT	Africa Leprosy, Tuberculosis and Rehabilitation Training Centre
ANOVA	Analysis of variance
ASFs	Animal source foods
AMU	Annual antimicrobial usage
AMR	Antimicrobial resistance
AMAP	Arctic Monitoring and Assessment Programme
AHRI	Armauer Hansen Research Institute
AZI	Azithromycin
BIBO	Bibliographic Ontology Specification ontology
bTB	Bovine tuberculosis
BFRs	Brominated flame retardants
CFIA	Canadian Food Inspection Agency
CBOW	Continuous Bag-of-Words
CDC	Centers for Disease Control and Prevention
CSF	Cerebrospinal fluids
CBC	Choice-based conjoint
CLSI	Clinical and Laboratory Standards Institute
CC	Clonal complexes
CFU	Colony forming units
CIs	Community interviewers
CIDT	Comparative intradermal tuberculin test
CITES	Convention on International Trade in Endangered Species of wild flora and fauna
COA	Council of Agriculture
CLB	Coxiella-like bacteria
CYP	Cytochrome
DL	Description Logic

DFA	Direct fluorescent antibody
DOTS	Directly observed treatment short course
DLS	Directorate of Livestock Services
EEG	Electroencephalography
FGDs	Focus Group Discussions
FAO	Food and Agriculture Organization
GAIA	Geometrical Analysis for Interactive Aid
GLMM	Generalized linear mixed models
GE	Genome equivalent
GIS	Geographical information systems
GAPDH	Glyceraldehyde-3-phosphate dehydrogenase
GDP	Gross domestic product
HDSS	Health and Demographic Surveillance System
HPAI	Highly pathogenic avian influenza
IFA	Immunofluorescence assay
IgG	Immunoglobulin G
MRPs	Macro restriction profiles
MDCK	Madin-Darby canine kidney
MaxDiff	Maximum Difference Scaling
MedDRA	Medical Dictionary for Regulatory Activities
MIC	Minimal inhibitory concentrations
MoARD	Ministry of Agriculture and Rural Development
MAPA	Ministry of Agriculture, Livestock, and Food Supply
mCCDA	modified Cefoperazone Charcoal Deoxycholate Agar
MCMC	Monte Carlo Markov Chain
MLST	Multilocus Sequence Typing
NER	Named Entity Recognition
NMDA	N-methyl-D-aspartic acid
NTM	Non-tuberculous mycobacteria
OCPs	Organochlorine pesticides
PSP	Paralytic Shellfish Poisoning
PHAC	Public Health Agency of Canada
SNA	Sambucus nigra agglutinin
SAVSNET	Small Animal Veterinary Surveillance Network

SD	System dynamics
TCDC	Taiwan Centers for Disease Control
TFDA	Taiwan Food and Drug Administration
TDI	Tolerable daily intake
UMLS	Unified Medical Language System
VPH	Veterinary Public Health
WHO	World Health Organization

PREFACE

Veterinary public health is a component of public health that focuses on the application of veterinary science to protect and improve the physical, mental and social well-being of humans. The World Health Organization defines veterinary public health as, "The sum of all contributions to the complete physical, mental and social well-being of humans through an understanding and application of veterinary medical science." With this definition every veterinary surgeon contributes to public health, whether through provision of health care for pets, protection of animal welfare, biomedical research, or ensuring adequate food animal production and food safety.

A zoonosis may be defined as any disease and/or infection which is naturally transmissible between animals and man. They are a major public concern. Headlines on issues like avian influenza, BSE (Mad cow disease) and salmonella of eggs have dominated the UK newspaper headlines for the last thirty years. Veterinary public health concerns the surveillance and control of zoonoses at many different levels be it via disease control programs at farm level or wild animals or in the abattoir.

One Health is "the collaborative efforts of multiple disciplines working locally, nationally, and globally, to attain optimal health for people, animals and our environment", as defined by the One Health Initiative Task Force (OHITF).

A major zoonosis is Rabies. Approximately 5,000 animal rabies cases are reported annually to CDC, and more than 90% of those cases occur in wildlife. This marks a dramatic change in the types of animals reported as rabid since 1960, when the majority of cases were in domestic animal species, primarily dogs. The principal rabies reservoir hosts in the United States today include bats, raccoons, skunks, and foxes. The number of rabies-related human deaths in the United States declined during the twentieth century, from more than 100 annually in the early 1900's to just one or two per year since 1960. This decline can be attributed to successful pet vaccination and animal control programs, public health surveillance and testing, and availability of post-exposure prophylaxis (PEP) for rabies.

This book is divided in 6 sections: Introduction, "One Health" Concept, Livestock, Companion Animals, Animal Models and Wild Animals. The Livestock sections in sub-divided on Cows and Pigs. On the introduction section the chapters cover topics of quantification of Zoonotic diseases, animal vaccination, animal rabies, vector-borne diseases and data distribution. Section 2, the "One Health" approach on diseases is analyzed. Section 3 covers public health concern diseases in cows and pigs. Section 4 discusses Filariasis in dogs. Section 5, talks about the use of gunea pigs has animal models for infection with influenza virus. The last section of this book (section 7) is dedicated to exotci and wild animals in Public Health.

SECTION 1: INTRODUCTION

CHAPTER 1

A Quantitative and Novel Approach to the Prioritization of Zoonotic Diseases in North America: A Public Perspective

Victoria Ng, Jan M. Sargeant

Centre for Public Health and Zoonoses, Department of Population Medicine, Ontario Veterinary College, University of Guelph, Guelph, Canada

ABSTRACT

Background

Zoonoses account for over half of all communicable diseases causing illness in humans. As there are limited resources available for the control and prevention of zoonotic diseases, a framework for their prioritization is necessary to ensure resources are directed into those of highest importance. Although zoonotic outbreaks are a significant burden of disease in North

Citation: Ng V, Sargeant JM (2012) A Quantitative and Novel Approach to the Prioritization of Zoonotic Diseases in North America: A Public Perspective. PLoS ONE 7(11): e48519. https://doi.org/10.1371/journal.pone.0048519

America, the systematic prioritization of zoonoses in this region has not been previously evaluated.

Methodology/Principal Findings

This study describes the novel use of a well-established quantitative method, conjoint analysis (CA), to identify the relative importance of 21 key characteristics of zoonotic diseases that can be used for their prioritization in Canada and the US. Relative importance weights from the CA were used to develop a point-scoring system to derive a recommended list of zoonoses for prioritization in Canada and the US. Over 1,500 participants from the general public were recruited to complete the online survey (761 from Canada and 778 from the US). Hierarchical Bayes models were fitted to the survey data to derive CA-weighted scores. Scores were applied to 62 zoonotic diseases of public health importance in Canada and the US to rank diseases in order of priority.

Conclusions/Significance

This was the first study to describe a systematic and quantitative approach to the prioritization of zoonoses in North America involving public participants. We found individuals with no prior knowledge or experience in prioritizing zoonoses were capable of producing meaningful results using CA as a novel quantitative approach to prioritization. More similarities than differences were observed between countries suggesting general agreement in disease prioritization between Canadians and Americans. We demonstrate CA as a potential tool for the prioritization of zoonoses; other prioritization exercises may also consider this approach.

INTRODUCTION

Zoonotic diseases are defined as those that are naturally transmitted between vertebrate animals and humans [1]. Zoonoses account for over half of all communicable diseases causing illness in humans [2], [3]. As there are limited resources available for research, surveillance, control and prevention of zoonoses, it is necessary to prioritize diseases in order to direct resources into those with the greatest needs. While there is general consensus amongst medical and veterinary professionals for the need to prioritize zoonoses, there are numerous challenges. Firstly, zoonotic diseases vary greatly in their occurrence and in their health impact on the human and animal populations, making it difficult to compare their overall public health importance [2].

Second, there is no universal agreement on the measurable criteria by which to quantify and prioritize zoonoses, nor is there agreement on the methodologies to elicit such information [4]–[6]. Lastly, there are a number of stakeholders involved, each with their own objectives [7]. It is therefore difficult to establish a universally accepted priority list for zoonotic diseases.

A number of studies have attempted to methodically prioritize communicable diseases of public health concern [6], [8]–[13]. With respect to zoonoses, a framework for their prioritization has been developed in France (non-foodborne zoonoses) [14], [15], Belgium (foodborne zoonoses) [16], The Netherlands (emerging zoonoses) [17] and Europe (food-producing animal diseases and zoonoses) [18]. Krause and colleagues [4] compared prioritization methodologies published between 1995 and 2005 and noted there was no uniformity in the objectives, methodological approaches, criteria for prioritization, number of criteria considered, level of standardization and weighting of criteria. Despite methodological differences, it is agreed that risk-based priority should be systematic, empirical and quantitative, easy to implement, based on good science, transparent, flexible, reproducible and informative to public policy [4], [19]. In recent studies, there has been a shift towards the use of novel quantitative approaches to overcome limitations in traditional methods and to address the complexity of disease prioritization [4], [6], [13], [16], [17].

Conjoint analysis (CA) has been used in market research of consumer preferences over the last 40 years [20]. More recently, this method has gained widespread use in the health and medical setting in eliciting preferences for healthcare [21]–[23]. The theory behind CA is that any product (goods or services) can be described by a set of characteristics and the extent to which an individual places value on a product is determined by the level of those separate characteristics (part-worth of the product) and the combination of those characteristics (overall worth of the product) [21], [22], [24]. CA forces individuals to make decisions by presenting competing products with both desirable and undesirable characteristics and asking individuals to state a preference. In expressing preference, individuals make a trade-off between the desirable and undesirable characteristics in those products allowing researchers to determine the true value of each characteristic relative to all other characteristics. A relative weighted score for each characteristic is derived as well as an overall score for each competing product as a whole. In the context of zoonoses, each disease can be treated as a product described by a set of disease criteria (characteristics), and the value of the disease is determined by the level of each criterion and the combination of the value of those levels. The overall score for each disease can be used to rank diseases in order of priority. Although there are similar methods to CA including Maximum Difference Scaling (MaxDiff) that has an emphasis on prioritizing a list of characteristics together [24], the complexity of zoonoses requires an understanding of preferences in the face of multiple levels of multiple characteristics combined together. Conjoint Analysis allows for the exploration of inter-relationships between various levels of disease characteristics together and was considered a more appropriate method for exploring preferences for disease characteristics and levels.

Current prioritization methods are limited by numerous challenges [4]–[6], [8]–[18], many of these can be overcome using CA, these include: (1) generating relative weighted scores; a better representation of criteria importance compared to arbitrary scores and subjective weights, (2) eliciting preferences through choice; an intuitive approach as opposed to ranking or rating diseases, (3) considering disease criteria jointly rather than separately; a more realistic approach to decision-making that recognizes disease criteria are not equal nor independent from each other, (4) presenting zoonoses as a set of disease characteristics without identifying diseases; this forces individuals to prioritize based on science thereby reducing biases associated with disease names, these can include the potential fear of a disease name compared to a lesser-known disease or prioritizing diseases based on professional interest and/or personal gain), and (5) presenting respondents with all the information necessary to prioritize diseases; this allows for prioritization to be based on wide social participation including public participation and experts who may not be familiar with the full range of diseases. We note in particular point number 4, by choosing to conduct an unlabeled prioritization study.

Although zoonotic outbreaks are a significant burden of disease in North America [25]–[27], the systematic prioritization of zoonoses in this region has not been previously evaluated. The primary objective of this study was to describe the novel use of a quantitative method, CA, to develop a point scoring system for disease criteria considered important in determining zoonotic disease priority. The second objective was to use the CA-derived scores to develop and compare a priority list of zoonoses in Canada and the US. This paper will focus on the results of individuals from the general public; this is the first study to describe a systematic and quantitative approach to the prioritization of zoonotic diseases from a public perspective.

METHODS

Focus Groups and Criteria Identification

The authors previously described the six focus groups using the nominal group technique (NGT) to inform this study [7]. Individuals from the public and medical and veterinary professionals identified 59 unique disease criteria to prioritize zoonoses. Of these, 21 were used to inform the CA experimental design (Tables 1, 2, 3).

Table 1: Disease criteria and standardized part-worth utility values (β) for disease criteria levels by country

https://doi.org/10.1371/journal.pone.0048519.t001

Disease criteria[1] and corresponding levels	Canada		US		*t*[4]
	MUV[2]	*SD*[3]	*MUV*[2]	*SD*[3]	
Case-fatality in humans					
No deaths or deaths are rarely reported	−68.97	30.90	−89.60	29.97	13.30***
Case-fatality is low (6%)	−53.69	21.49	−45.59	20.23	7.62***
Case-fatality is moderate (35%)	20.59	22.78	22.02	26.43	1.14^
Case-fatality is high (80%)	102.07	32.25	113.17	28.60	7.14***^
Incidence of the disease in the Canadian/US human population in the last five years					
0 cases	−72.23	27.16	−85.10	31.02	8.66***^
5 cases	−39.50	23.14	−37.25	22.11	1.95
100 cases	17.79	18.97	23.00	16.45	5.76***^
10,000 cases	93.94	37.71	99.34	33.69	2.96**^
Case-fatality in animals					
No deaths or deaths are rarely reported	−61.85	27.48	−49.87	19.72	9.81***^
Case-fatality is low (6%)	−27.10	17.61	−33.58	15.82	7.59***^
Case-fatality is moderate (35%)	10.72	23.28	8.52	19.12	2.02*^
Case-fatality is high (80%)	78.23	29.56	74.93	22.65	2.46*^
Incidence of the disease in the Canadian/US animal population in the last five years					
0 cases	−57.44	30.36	−63.48	21.73	4.48***^
5 cases	−27.41	21.62	−33.30	22.75	5.20***
100 cases	11.75	21.22	17.59	20.46	5.50***
10,000 cases	73.11	34.79	79.18	24.75	3.94***^
Severity of illness in humans					
No clinical symptoms or illness that is not noticeable	−54.46	22.61	−54.64	26.54	0.15^
Mild clinical symptoms (time off work, some medical assistance and personal care at home)	−28.07	23.11	−26.93	14.68	1.16^
Moderate clinical symptoms (urgent medical care and hospital admission)	8.29	19.19	8.28	18.40	0.02
Severe clinical symptoms (failure of major organ system/s necessitating long-term hospital admission)	74.24	29.72	73.29	29.71	0.63
Disease trend in Canada/US in the last five years in humans					
Decline over the last five years	−58.29	24.03	−58.83	18.49	0.50^
Stable over the last five years	−34.30	20.40	−26.89	18.53	7.45***^
Increase over the last five years	29.29	19.26	26.43	17.85	3.02**^
New emerging disease, rapid increase over the last five years	63.30	26.80	59.29	24.37	3.07**^
Transmission potential between humans					
No transmission between humans	−52.93	22.50	−47.07	18.19	5.61***^
Low transmission between humans	−30.86	19.95	−34.42	21.40	3.37**
Moderate transmission between humans	12.71	17.35	17.57	18.87	5.28***^
High transmission between humans	71.08	18.90	63.92	21.28	6.98***^

[1]Presented in order of importance to Canadian participants.
[2]Mean part-worth utility values (β) across respondents.
[3]Standard deviation of mean part-worth utility values (β) across respondents.
[4]*t*-statistic; *d.f.* = 1,537.
*$p<0.05$,
**$p<0.01$,
***$p<0.001$.
^Adjusted for unequal variance using the Welch *t*-test; Satterthwaite's *d.f.* = 1,282.61 to 1531.22.
doi:10.1371/journal.pone.0048519.t001

These criteria were selected on the basis of having a high mean score (derived from the NGT) across groups, identified by three of more groups, or corresponding to human-related criteria that were deemed important (e.g. incidence in humans scored high, therefore, incidence in animals was included despite a lower score). A number of criteria were merged together, for example, *pathogenicity*, *immunogenicity*, *incubation period*, *communicability* and *mode of transmission* were integrated into *transmission potential*. This process allowed for a more efficient number of the most relatively important disease characteristics to be included in the study.

Table 2: Disease criteria and standardized part-worth utility values (β) for disease criteria levels by country

https://doi.org/10.1371/journal.pone.0048519.t002

Disease criteria[1] and corresponding levels	Canada		US		t^4
	MUV[2]	*SD*[3]	*MUV*[2]	*SD*[3]	
Duration of illness in humans					
No illness observed or only a few days of illness	−47.58	22.40	−46.58	31.05	0.72^
Short-term illness (weeks)	−26.00	23.78	−22.08	16.16	3.78***^
Medium-term illness (months)	6.26	21.34	3.24	23.55	2.64**^
Chronic illness (years) or illness with permanent deficits	67.32	30.94	65.42	30.09	1.23
Transmission potential from animals to humans					
No transmission from animals to humans	−38.81	18.86	−41.89	17.39	3.33**^
Low transmission from animals to humans	−28.33	18.40	−33.07	17.43	5.19***
Moderate transmission from animals to humans	5.23	17.15	13.03	16.83	9.00***
High	61.91	21.12	61.93	21.09	0.02
Disease trend in Canada/US in the last five years in animals					
Decline over the last five years	−47.50	21.98	−48.38	21.66	0.80
Stable over the last five years	−24.50	18.50	−26.24	19.01	1.82
Increase over the last five years	25.03	18.23	28.13	17.74	3.38**
New emerging disease, rapid increase over the last five years	46.97	22.54	46.49	24.71	0.40^
Economic burden in humans					
No cost to the health care system and individuals	−35.66	26.38	−34.54	24.26	0.86^
Low cost ($100 per sick individual)	−14.78	22.27	−22.16	18.03	7.14***^
Moderate cost ($1,000 per sick individual)	6.50	16.63	8.61	16.44	2.50*
High cost ($10,000 per sick individual)	43.94	36.21	48.10	31.32	2.41*^
Transmission potential from humans to animals					
No transmission from humans to animals	−26.88	17.37	−27.19	21.58	0.31^
Low transmission from humans to animals	−25.98	18.96	−27.03	19.44	1.07
Moderate transmission from humans to animals	6.74	15.47	13.44	15.48	8.49***
High transmission from humans to animals	46.13	26.65	40.79	22.95	4.21***^
Duration of illness in animals					
No illness observed or only a few days of illness	−28.32	19.05	−33.47	18.99	5.31***
Short-term illness (weeks)	−9.97	20.26	−10.85	17.58	0.91^
Medium-term illness (months)	1.21	18.71	6.68	17.37	5.93***^
Chronic illness (years) or illness with permanent deficits	37.08	24.28	37.63	22.53	0.47^
Transmission potential between animals					
No transmission between animals	−27.27	21.02	−23.24	18.21	4.01***^
Low transmission between animals	−13.53	15.78	−18.35	13.60	6.42***^
Moderate transmission between animals	4.76	15.33	3.52	18.29	1.45^
High transmission between animals	36.03	18.62	38.08	18.44	2.17*

[1]Presented in order of importance to Canadian participants.
[2]Mean part-worth utility values (β) across respondents.
[3]Standard deviation of mean part-worth utility values (β) across respondents.
[4]*t*-statistic; *d.f.* = 1,537.
**p*<0.05,
***p*<0.01,
****p*<0.001.
^Adjusted for unequal variance using the Welch *t*-test; Satterthwaite's *d.f.* = 1,282.61 to 1531.22.
doi:10.1371/journal.pone.0048519.t002

Criteria that were difficult to quantify in measureable units (e.g. *ability of the disease pathogen to mutate*) or lacked scientific information for quantification (e.g. *bioterrorism potential*) were excluded. Criteria pertaining to public disruption, awareness and perception were also excluded. Although the professional groups acknowledged disease prioritization was often driven by public and political pressure, they agreed it should not steer the process [7]. The final 21 selected criteria could all be quantitatively measured with scientific information available in the literature.

Disease Selection and Literature Search

We identified 62 existing and emerging zoonotic and enteric diseases of public health importance in North America (Tables 4, 5). These diseases were selected using the following criteria: a) nationally notifiable to the Public Health Agency of Canada (PHAC), Canadian Food Inspection Agency (CFIA), Centers for Disease Control and Prevention (CDC) or the United States Department of Agriculture (USDA), b) internationally notifiable to the World Health Organization (WHO), the World Organization for Animal Health (OIE) or the Food and Agriculture Organization (FAO) or c) identified as a priority by PHAC at a national meeting on non-enteric zoonoses in 2009. We note that the majority of enteric diseases in this study are zoonotic; hence, the group of diseases in this paper will be referred to as zoonotic diseases. However, there are three diseases in this study that are strictly enteric diseases with no zoonotic involvement, these are, Cholera, Hepatitis A and Paralytic Shellfish Poisoning (PSP). As enteric diseases typically fall under the responsibility of the Zoonotic Division in public health (for example, PHAC and the CDC), these diseases were also included in our study.

A literature search was conducted for each zoonosis in the study. Diseases exhibiting multiple forms (e.g. acute/chronic, latent/active, encephalitic/non-encephalitic) were divided into separate syndromes and assigned approximate proportions that were informed by the literature. There were 117 separate disease syndromes identified from the 62 diseases (Tables 4, 5). The literature search obtained information for each criterion for each disease syndrome. The literature search included: a) website searches of human and animal health organizations involved in zoonotic disease prevention and control including national organizations, intergovernmental organizations, provincial organizations and academic institutions; b) reference textbooks [2], [28], [29] and c) PubMed cataloged peer-reviewed publications. Key search terms used included the disease criterion, the scientific and/or common name of diseases, and a combination of the two (e.g. *case-fatality rate* and/or *rabies*).

Table 3: Disease criteria and standardized part-worth utility values (β) for disease criteria levels by country

https://doi.org/10.1371/journal.pone.0048519.t003

Disease criteria[1] and corresponding levels	**Canada**		**US**		t^4
	MUV[2]	*SD*[3]	*MUV*[2]	*SD*[3]	
Economic and social burden on trade in animals					
No cost to trade in animals	−25.21	16.44	−15.51	19.10	10.69***^
Low cost to trade in animals (vaccination of herds)	−11.08	17.47	−16.85	19.54	6.10***^
Moderate cost to trade in animals (restriction of movement and trade)	1.26	12.90	2.33	17.34	1.38^
High cost to trade in animals (culling of herds or destroying infected crops/produce)	35.03	23.65	30.02	21.60	4.34***^
Severity of illness in animals					
No apparent clinical signs or the animal-source of infection is non-living (e.g. food-source)	−20.92	18.56	−24.17	15.23	3.75***^
Mild clinical signs (minor distress in animals such as fever, lethargy, shivering, constipation, loose feces)	−15.40	16.98	−13.87	16.75	1.78
Moderate clinical signs (moderate distress in animals such as difficult breathing, bleeding from openings, aborted fetuses)	4.75	18.39	4.41	20.16	0.35^
Severe clinical signs (severe distress in animals such as convulsion, organ failure, neurological involvement)	31.57	19.80	33.64	18.07	2.14*^
High risk groups in humans					
No	−25.66	18.68	−24.05	14.53	1.89^
Yes	30.39	20.31	29.32	17.19	1.12^
Unknown	−4.73	16.74	−5.28	14.27	0.69^
Control measures in humans					
Highly effective in reducing disease burden	5.53	33.46	3.75	32.58	1.06
Moderately effective in reducing disease burden	−0.52	18.55	−7.72	17.88	7.76***
Minimally effective in reducing disease burden	−5.00	17.71	0.52	19.35	5.84***^
Not effective at all in reducing disease burden	−0.01	31.90	3.46	27.23	2.29*^
Control measures in animals					
Highly effective in reducing disease burden	9.02	27.81	5.17	23.03	2.96**^
Moderately effective in reducing disease burden	1.34	18.40	−2.47	15.26	4.42***^
Minimally effective in reducing disease burden	−6.21	19.71	−0.66	15.35	6.16***^
Not effective at all in reducing disease burden	−4.15	26.50	−2.04	22.38	1.68^
How much is known scientifically about the disease					
Knowledge of the disease is well known and scientifically valid	−10.17	27.16	−3.96	30.17	4.26***^
Knowledge of the disease exists but the validity of the information is uncertain	−2.97	23.19	4.23	16.08	7.06***^
Knowledge of the disease is currently insufficient	6.91	17.24	3.39	23.08	3.39**^
There is no scientific knowledge of the disease	6.22	18.96	−3.66	18.58	10.33***
High risk groups in animals					
No	−13.79	17.69	−13.37	13.23	0.53^
Yes	14.40	16.49	13.73	18.63	0.75^
Unknown	−0.61	14.64	−0.36	15.22	0.33

[1]Presented in order of importance to Canadian participants.
[2]Mean part-worth utility values (β) across respondents.
[3]Standard deviation of mean part-worth utility values (β) across respondents.
[4]*t*-statistic; *d.f.* = 1,537.
*$p<0.05$,
**$p<0.01$,
***$p<0.001$.
^Adjusted for unequal variance using the Welch *t*-test; Satterthwaite's *d.f.* = 1,282.61 to 1531.22.
doi:10.1371/journal.pone.0048519.t003

Table 4: List of 62 diseases and their separate disease syndromes and proportions

https://doi.org/10.1371/journal.pone.0048519.t004

#	Diseases or disease syndromes	Proportion	#	Diseases or disease syndromes	Proportion
1	1. American Trypanosomiasis/Chagas' disease (acute)	70%	31	17. Cyclosporiasis (healthy individuals)	98%
2	1. American Trypanosomiasis/Chagas' disease (chronic)	30%	32	17. Cyclosporiasis (immunocompromised individuals)	2%
3	2. Anaplasmosis/Canine granulocytic anaplasmosis (acute)	50%	33	18. Cysticercosis/Taeniasis (non-neurological involvement)	99.8%
4	2. Anaplasmosis/Canine granulocytic anaplasmosis (chronic)	50%	34	18. Cysticercosis/Taeniasis (neurocysticercosis)	0.2%
5	3. Anthrax (cutaneous)	95%	35	19. Dengue fever	99%
6	3. Anthrax (inhalational)	5%	36	19. Dengue haemorrhagic fever	1%
7	4. Babesiosis (mild)	95%	37	20. Eastern equine encephalitis (non-neurological involvement)	66.7%
8	4. Babesiosis (shock and renal failure)	5%	38	20. Eastern equine encephalitis (encephalitic/neurological involvement)	33.3%
9	5. Bartonellosis (Cat-scratch disease) (mild)	90%	39	21. Ebola virus haemorrhagic fever (*Reston* strain)	0.4%
10	5. Bartonellosis (Cat-scratch disease) (bacteremia/systemic disease)	10%	40	21. Ebola virus haemorrhagic fever (other Ebola strains)	99.6%
11	6. Baylisascariasis (visceral larval migrans)	95%	41	22. Echinococcosis (cystic)	70%
12	6. Baylisascariasis (neural or ocular larval migrans)	5%	42	22. Echinococcosis (alveolar)	30%
13	7. Botulism (mild)	50%	43	23. *Escherichia coli* infection	85%
14	7. Botulism (paralysis)	50%	44	23. *Escherichia coli* infection (hemolytic-uremic syndrome)	15%
15	8. Bovine Tuberculosis (latent)	92.5%	45	24. Giardiasis (healthy individuals)	98%
16	8. Bovine Tuberculosis (active)	7.5%	46	24. Giardiasis (immunocompromised individuals)	2%
17	9. Brucellosis (mild)	75%	47	25. Hantavirus pulmonary syndrome (moderate)	50%
18	9. Brucellosis (undulant)	25%	48	25. Hantavirus pulmonary syndrome (respiratory failure)	50%
19	10. Campylobacteriosis (healthy individuals)	98%	49	26. Hendra virus (rare, consistent prognosis suspected)	100%
20	10. Campylobacteriosis (immunocompromised individuals)	2%	50	27. Hepatitis A (mild)	85%
21	11. Chlamydiosis (*C. abortus* spp.)	50%	51	27. Hepatitis A (prolonged relapse)	15%
22	11. Chlamydiosis (*C. felis* spp.)	50%	52	28. H1N1 Influenza (mild)	97.5%
23	12. Cholera (mild)	80%	53	28. H1N1 Influenza (respiratory failure)	2.5%
24	12. Cholera (severe dehydration, kidney failure, hypovolemic shock)	20%	54	29. HPAI H5N1 Influenza (moderate)	40%
25	13. Coccidioidomycosis (acute)	99%	55	29. HPAI H5N1 Influenza (respiratory failure)	60%
26	13. Coccidioidomycosis (disseminated)	1%	56	30. Japanese encephalitis (mild)	99.5%
27	14. Crimean-Congo haemorrhagic fever (consistent prognosis)	100%	57	30. Japanese encephalitis (neurological involvement)	0.5%
28	15. Cryptosporidiosis (healthy individuals)	98%	58	31. La Cross virus (non-encephalitic)	99%
29	15. Cryptosporidiosis (immunocompromised individuals)	2%	59	31. La Cross virus (encephalitic)	1%
30	16. Cutaneous larva migrans/Ancylostomiasis (dose-dependent)	100%			

doi:10.1371/journal.pone.0048519.t004

When there was conflicting information between multiple sources, the most consistent sources were used to define the disease criteria; where only two sources were available, the most recent source was used. When information was not available for North America, information available for other developed countries was used. Country-specific information for Canada and the US were collected for four criteria – *disease incidence in the last five years* (humans and animals) and *disease trend in the last five years* (humans and animals). An assumption was made that the remaining criteria were consistent between countries. When there was a gap in the literature, particularly pertaining to animal-related criteria [30], information from corresponding human-related criteria were used to define animal-related criteria on the assumption that disease criteria were similar between humans and animals. Animal-related criteria, however, were not used to inform human-related criteria due to existing data for humans.

Table 5: List of 62 diseases and their separate disease syndromes and proportions

https://doi.org/10.1371/journal.pone.0048519.t005

#	Diseases or disease syndromes	Proportion	#	Diseases or disease syndromes	Proportion
60	32. Lassa fever (mild)	80%	**89**	47. Rift Valley fever (mild)	99%
61	32. Lassa fever (severe multi-system failure)	20%	**90**	47. Rift Valley fever (haemorrhagic fever)	1%
62	33. Leishmaniasis (mild cutaneous)	30%	**91**	48. Rocky Mountain spotted fever (US only, prognosis unknown)	100%
63	33. Leishmaniasis (moderate cutaneous)	33.7%	**92**	49. Salmonellosis (enteric)	98%
64	33. Leishmaniasis (visceral (kala-azar))	33.3%	**93**	49. Salmonellosis (septicemic/enteric in immunocompromised)	2%
65	34. Leptospirosis (aniceteric)	93%	**94**	50. Severe Acquired Respiratory Syndrome (mild)	90%
66	34. Leptospirosis (icteric)	7%	**95**	50. Severe Acquired Respiratory Syndrome (respiratory failure)	10%
67	35. Listeriosis (healthy individuals)	50%	**96**	51. Shigellosis (mild)	95%
63	35. Listeriosis (pregnant, newborns, elderly, immunocompromised)	50%	**97**	51. Shigellosis Reiter's (syndrome and chronic arthritis)	5%
69	36. Lyme disease (early localized)	50%	**98**	52. St. Louis encephalitis (non-encephalitic)	99%
70	36. Lyme disease (early dissemination)	49%	**99**	52. St. Louis encephalitis (encephalitic)	1%
71	36. Lyme disease (chronic dissemination)	1%	**100**	53. Toxocariasis (visceral larval migrans)	98%
72	37. Marburg haemorrhagic fever (acute)	80%	**101**	53. Toxocariasis (ocular larval migrans)	2%
73	37. Marburg haemorrhagic fever (chronic)	20%	**102**	54. Toxoplasmosis (healthy individuals)	95%
74	38. Monkeypox (consistent prognosis)	100%	**103**	54. Toxoplasmosis (immunocompromised individuals)	5%
75	39. Nipah virus encephalitis (acute)	80%	**104**	55. Trichinosis (dose-dependent)	100%
76	39. Nipah virus encephalitis (residual neurological deficits)	20%	**105**	56. Tularemia (ulceroglandular and glandular)	90%
77	40. Old/New World Screwworm (consistent prognosis)	100%	**106**	56. Tularemia (typhoidal)	10%
78	41. Paralytic shellfish poisoning (dose-dependent)	100%	**107**	57. Typhus fever (epidemic louse-borne)	50%
79	42. Plague (bubonic)	90%	**108**	57. Typhus fever (endemic flea-borne)	50%
80	42. Plague (septicemic)	7.5%	**109**	58. variant Creutzfeldt-Jakob disease (CJD)/BSE (consistent prognosis)	100%
81	42. Plague (pneumonic)	2.5%	**110**	59. Venezuelan equine encephalitis (non-encephalitic)	97.5%
82	43. Powassan virus (moderate leading to recovery)	66.7%	**111**	59. Venezuelan equine encephalitis (encephalitic)	2.5%
83	43. Powassan virus (severe leading to death)	33.3%	**112**	60. West Nile virus (non-neurological involvement)	80%
84	44. Psittacosis/Avian Chlamydiosis (mild)	99%	**113**	60. West Nile virus (neurological involvement)	20%
85	44. Psittacosis/Avian Chlamydiosis (severe multi-system failure)	1%	**114**	61. Western equine encephalitis (systemic, non-neurological involvement)	96.5%
86	45. Q fever (acute)	90%	**115**	61. Western equine encephalitis (encephalitic, neurological involvement)	3.5%
87	45. Q fever (chronic)	10%	**116**	62. Yellow fever (mild)	85%
88	46. Rabies (consistent prognosis)	100%	**117**	62. Yellow fever (hepato-renal failure)	15%

doi:10.1371/journal.pone.0048519.t005

Defining Levels for Disease Criteria

Three or four levels for each criterion were defined according to the range exhibited in the literature (Tables 1, 2, 3). For example, the duration of illness in humans among diseases ranged from no duration (asymptomatic) to chronic disability (permanent deficits). For the criteria, *case-fatality* and *disease incidence* (both humans and animals), levels were defined according to the 25th, 50th, 75th and 90th percentile values.

Survey Development and Administration

We used 21 criteria with criteria levels ranging from three to four levels that reflected the full range of each criterion across the 62 diseases to inform our survey instrument. Due to the large number of criteria, a partial-profile choice-based conjoint (CBC) survey was developed comprising of 14 choice tasks [31], [32]; each choice task presented participants with five disease combinations (zoonoses 1–5) described by criteria levels for 5 of the 21 criteria. Disease criteria and levels varied between choice tasks. Participants were asked to select one zoonosis to prioritize in either Canada or the US with the objective being for policy implementation for control and prevention (Figure 1). Surveys were administered electronically to allow for flexibility in the study design and wide participation across North America. Surveys were offered in English, French and Spanish. We used Sawtooth Software CBC module v7 to create 300 survey versions to generate an efficient experimental design (*D*-efficiency of 908.13326 relative to a full-orthogonal design with a standard error of <0.05 for each criterion level). The selected design ensured a number of important features of an efficient experimental design were included, these were: level balance (each criteria level appeared approximately an equal number of times across the 14 choice tasks), orthogonality (criteria levels were selected independently of other criteria levels) and a balanced overlap approach (the design permitted levels to appear more than once per choice task for a more precise measurement of interaction terms) [33]. Disease combinations were designed to be easily understood, to mimic the presentation of zoonoses in nature and to enhance informant and statistical efficiency [24], [31], [32], [34].

In your opinion, which of the following diseases should be prioritized for policy implementation for their control and prevention in Canada/US, assuming all other characteristics not presented here are the same between diseases:

Disease criteria	Zoonoses 1	Zoonoses 2	Zoonoses 3	Zoonoses 4	Zoonoses 5
Transmission potential from animals to humans	Moderate	Low	High	High	Moderate
Transmission potential between humans	Low	Moderate	High	No transmission between humans	No transmission between humans
Severity of illness in humans	Mild clinical symptoms	No clinical symptoms or illness that is not noticeable	Moderate clinical symptoms	Moderate clinical symptoms	Severe clinical symptoms
Duration of illness in animals	Medium-term illness (months)	Chronic illness or illness with permanent deficits	No illness observed or only a few days of illness	Short-term illness (weeks)	Medium-term illness (months)
High-risk groups in humans	No	No	Yes	Yes	Unknown

Figure 1: Example of one choice task set completed by each study participant.

As multiple survey versions were administered randomly to each person, a different combination of disease criteria and levels was presented to study participants. The ordering of the presentation of disease criteria within each choice task was randomized to reduce ordering bias.

https://doi.org/10.1371/journal.pone.0048519.g001

Two additional fixed choice tasks were included in all survey versions to determine the reliability of responses. In these choice tasks, one zoonosis was more severe as described by the set of criteria compared to the remaining four, giving participants an incentive to select that zoonosis to prioritize. Fixed choice task 1 presented one zoonosis with the highest incidence in humans (10,000 cases), most severe illness in humans (severe clinical symptoms), highest transmission potential between humans (high), highest case-fatality in humans (80%) and the most costly economic burden in humans ($10,000 per sick individual). In comparison, the remaining four zoonoses contained a combination of lower and less severe criteria levels. Fixed choice task 2 presented one zoonosis with the most severe illness in animals (severe clinical symptoms), highest case-fatality in animals (80%), most costly socioeconomic burden in trade in animals (high cost such as culling of herds or destroying infected crops/produce), longest duration of illness in animals (chronic illness or permanent deficits) and rapid change in disease trend in the human population (new emerging disease, rapid increase over the last five years). In comparison, the remaining four zoonoses contained a combination of lower and less severe criteria levels. The fixed choice tasks tested the reliability of responses by identifying respondents who did not understand the choice task process and/or fatigue responders. The ordering of the presentation of disease criteria within each choice task was randomized to reduce ordering bias. The fixed choice tasks were also randomized to reduce ordering bias and as an additional measure of reliability given the tasks were designed to test the reliability of respondents to understand the choice task process rather than the reliability of respondents to make the same choices consistently across the two fixed choice task sets. Sawtooth Software SSI Web v7 was used to randomly assign a survey version to each study participant.

Study Participants

The study was approved by the Research Ethics Board at the University of Guelph. The target study participants were individuals from the public in Canada and the US with no knowledge or experience in the prioritization of zoonotic diseases. Participants from both countries were recruited using an online panel through Research Now™; these are groups of pre-screened individuals who have expressed a willingness to participate in online surveys. Surveys were collected anonymously. All participants acknowledged an informed consent assuring confidentiality and the option to withdraw from participation without penalty. Sawtooth Software SSI Web v7 was used to screen participants through a series of demographic questions prior to survey commencement. Participants were disqualified if they did not reside in North America or were employed in the following fields: epidemiology, public health, medical sciences, veterinary sciences, infectious disease research, laboratory technician, nursing or dentistry. Quotas were used to ensure that no particular age group, gender or geographic region dominated the survey responses and that study populations were representative of the national populations in their respective countries (Table 6).

Table 6: Demographic characteristics of Canadian and US study participants in comparison to their respective national population characteristics

https://doi.org/10.1371/journal.pone.0048519.t006

	Canada (n = 761)			US (n = 778)			
	Study Participant	**National Population**[1]	χ^2		**Study Participant**	**National Population**[2]	χ^2
Gender *Male*	48.0%	48.5%	0.09	**Gender**	48.2%	49.2%	0.31
Female	52.0%	51.5%			51.8%	50.8%	
Age group *18 to 34*	27.3%	27.9%	14.92*	**Age group**	29.7%	30.6%	0.30
35 to 50	35.1%	29.1%			27.5%	27.2%	
50+	37.5%	43.0%			42.8%	42.2%	
Unknown	0.1%	–			–	–	
Province *Alberta*	10.6%	10.6%	12.9	**Region**[3]			0.64
British Columbia	13.1%	13.4%		*Midwest*	22.6%	21.7%	
Manitoba	3.8%	3.5%		*Northeast*	18.1%	18.3%	
New Brunswick	2.1%	2.3%		*South*	35.9%	37.0%	
Newfoundland and Labrador	1.4%	1.6%		*West*	23.4%	23.0%	
Nova Scotia	2.8%	2.8%					
Northwest Territories	0.1%	0.1%					
Nunavut	0.0%	0.1%					
Ontario	38.9%	38.2%					
Prince Edward Island	0.9%	0.4%					
Quebec	22.7%	23.9%					
Saskatchewan	3.0%	3.0%					
Yukon	0.4%	0.1%					
Educational attainment[4]				**Educational attainment**[5]			
High school graduate or less	34.8%	45.1%	262.39*		42.9%	44.5%	307.07*
Diploma, trade or college degree	25.4%	35.1%			4.5%	27.0%	
Bachelor's degree	27.1%	12.7%			35.0%	18.7%	
Master's degree	7.4%	5.8%			13.0%	7.1%	
Professional degree (MD, DVM)	3.3%	0.6%			2.8%	1.4%	
Doctorate degree	1.5%	0.8%			1.8%	1.3%	
Unknown	0.7%	–			–	–	

[1]2011 population data for individuals 18 years and older in Canada was obtained from Statistics Canada [36].
[2]2010 population data for individuals 18 years and older in the US was obtained from the US Census Bureau [38].
[3]Regions were: **Midwest** (Illinois, Indiana, Iowa, Kansas, Michigan, Minnesota, Missouri, Nebraska, North Dakota, Ohio, South Dakota, Wisconsin); **Northeast** (Connecticut, Maine, Massachusetts, New Hampshire, New Jersey, New York, Pennsylvania, Rhode Island, Vermont); **South** (Alabama, Arkansas, Delaware, District of Columbia, Florida, Georgia, Kentucky, Louisiana, Maryland, Mississippi, North Carolina, Oklahoma, South Carolina, Tennessee, Texas, Virginia, West Virginia); **West** (Alaska, Arizona, California, Colorado, Hawaii, Idaho, Montana, Nevada, New Mexico, Oregon, Utah, Washington, Wyoming).
[4]2006 education data for individuals 20 years and over in Canada (most current and available data) [35].
[5]2010 education data for individuals 18 years and over in the US [37].
*$p<0.001$.
doi:10.1371/journal.pone.0048519.t006

Data Analysis

We used Chi Square and Mann-Whitney tests to compare the demographic and survey characteristics of study participants to their respective national populations and to make comparisons by country. National population data for age, gender, education and region was obtained from Statistics Canada [35], [36] and the US Census Bureau [37], [38]. Unpaired t-tests and *F*-tests were used to explore differences in standardized importance scores and part-worth utility values between Canada and the US. Twenty-one part-worth utility values, one for each disease criterion, were assigned to the 117 separate disease syndromes by matching the level of each disease criterion to those of disease syndromes. Part-worth utility values were summed up in proportion to the relative frequency of each syndrome within a disease (Tables 4, 5) to derive an overall score for each of the 62 diseases. The overall scores were used to rank-order diseases; the higher the score, the higher the ranking on the priority list. We chose to use summed part-worth utility values over other approaches such as market simulations because we wanted to apply CA-derived scores to a set of disease using a method that is comparable to current traditional prioritization methods [4], [11]–[13], [16]–[18]. We also selected this approach because our goal was to prioritize the list of diseases in ranked order, which can be achieved using the simple summed part-worth utility values approach.

Hierarchical Bayes (HB) was used to compute parameter estimates (weighted scores) from survey choice data [39]. We used Sawtooth Software CBC/HB v5.2.8 to estimate individual-level parameter estimates (β). The software uses a combination of Bayes theorem, a Monte Carlo Markov Chain (MCMC) procedure and the Metropolis/Hastings algorithm to derive parameter estimates from two distributions: an upper-level model (prior) drawn from a multivariate normal distribution representing parameter estimates at the population level and a lower-level model (posterior) described by a multinomial logit model representing parameter estimates at the individual level. Bayesian updating of probabilities using MCMC and the Metropolis/Hastings algorithm provided an iterative process to update parameter estimates drawing on the upper-level and lower-level model [39]. The final individual-level parameter estimates reflected an optimal mix of the population average and individual choices [40]. We computed 30,000 preliminary iterations before convergence was assumed (and observed) and an additional 30,000 iterations per respondent to estimate parameters.

Zero-centered part-worth utility values (β estimates) were standardized by setting the average range of the parameter values of all disease criteria to 100. Part-worth utility values represent the relative influence each criterion level had on respondent choices with higher values indicating a stronger influence on choice [24]. To estimate the influence of each criterion collectively, importance scores were calculated as a percentage by dividing the difference in range between the highest and lowest criterion level part-worth utility value by the sum of all part-worth utility value ranges across all criteria. The more variation between the levels in a criterion, the higher the importance score and the stronger the influence the criterion had on the decision to prioritize [24]. Part-worth utility values and importance scores were calculated directly using Sawtooth Software SMRT v4.22.0. We computed *t*-statistics by dividing the mean difference in range in part-worth utility values across each criterion by the standard error of the differences to test each disease criterion for statistical significance in the final model. The goodness of fit of the individual-level HB models were estimated using Sawtooth Software CBC/HB v5.2.8 and are presented as a percent certainty fit and a root likelihood (RLH); both of these measures are calculated as the probability of each respondent choosing as he/she did on each choice task using a logit model fitted with the current estimates of each respondent's part-worth utility values [39]. The percent certainty and RLH both indicate how much better the model is than a chance model, as compared to a perfect model.

RESULTS

Demographic and Survey Characteristics

In total, 46,547 and 8,298 email invitations were sent to Canadian and US participants, respectively. Of these, 1,313 (2.8%) and 1,309 (15.8%) completed surveys were returned in 8 and 7 business days, respectively. The majority of completed surveys were in English (77.1% in Canada, 99.4% in the US) while remaining surveys were completed in French (22.9% in Canada) and Spanish (0.6% in US). Participants passed the survey if all 14 choice task sets were completed and the correct diseases were selected for both fixed choice tasks[2]. The pass rate was 58.0% in Canada (761) and 59.4% in the US (778); there was no significant difference in the pass rate between countries ($\chi^2=0.59$, $p=0.44$). The median completion time for passed surveys was 26.9 minutes in Canada and 28.1 minutes in the US. There were 1,539 completed and passed surveys in this study. The study population was generally representative of the national population by gender, age and geography (Table 6). There was a significantly higher response rate in the 35 to 50 years age group in the Canadian study population compared to the Canadian national population ($p<0.001$). A higher than expected educated population was observed in both study populations compared to their respective national populations ($p<0.001$ for both countries).

Model Fit

Both the Canadian and the US models had a percent certainty fit of 79.4% and a root likelihood (RLH) of 0.72. The expected percent certainty for a chance model is 0% and a perfect model is 100% while the expected RLH for a chance model is 0.2 (one divided by five disease combinations per task) and a perfect model is 1.0 [39].

Disease Criteria Importance Scores and Part-worth Utility Values

The importance scores for disease criteria indicate the degree to which each criterion contributed to the decision to prioritize (Table 7). Human-related criteria were preferred over corresponding animal-related criteria with each of the eight matching criteria exhibiting this trend in both countries. The four transmission potential criteria were ranked in the following order of preference in both countries: human-human, animal-human, humans-animals and animal-animal; thus also revealing a stronger preference for human-related criteria over animal-related criteria. Although the contribution of each disease criterion in the decision to prioritize zoonoses varied, each criterion was statistically significant (P-value <0.05) in the final model for both countries validating the choice of appropriate criteria for assessing disease prioritization and highlighting their varying degree of importance in the overall decision.

The part-worth utility values (β) indicate the relative influence each level had on respondent choices with higher values representing a stronger degree of influence on choice (Tables 1, 2, 3). The part-worth utility value trends for the ten disease criteria with significant difference in importance score between countries (Table 7) can be broadly summarized as follows:

Both groups considered *case-fatality in humans* and *incidence of the disease in the last five years in humans* to be the most influential criteria in the decision to prioritize zoonoses (Table 7). Although the US group was influenced more strongly by these criteria (P-value <0.001 for both, Table 7), expressed by a wider range in part-worth utility values between the lowest and highest levels (Tables 1, 2, 3), there was agreement between countries on the importance placed on the levels within these criteria. The same trend was observed for the criterion *incidence in the last five years in animals*.

Table 7: Disease criteria importance scores by country

https://doi.org/10.1371/journal.pone.0048519.t007

Disease criteria	Canada (n = 761)			US (n = 778)			*t*[6]
	R[3]	*MS*[4]	*SD*[5]	*R*	*MS*	*SD*	
Case-fatality (H)[1]	1	8.57	2.20	1	**9.84**	2.04	11.81***^
Incidence of the disease in the last five years (H)	2	8.18	2.44	2	**9.06**	2.31	7.29***
Case-fatality (A)[2]	3	**6.91**	2.06	5	6.16	1.48	8.17***^
Incidence of the disease in the last five years (A)	4	6.63	2.20	3	**7.03**	1.70	3.98***^
Severity of disease (H)	5	6.41	1.88	4	6.39	1.95	0.19
Disease trend in the last five years (H)	6	**6.16**	1.70	7	5.81	1.69	4.14***
Transmission potential between humans	7	**6.16**	1.44	8	5.59	1.50	7.59***
Duration of illness (H)	8	5.82	1.95	6	5.85	1.83	0.29
Transmission potential from animals to humans	9	5.15	1.41	9	5.27	1.38	1.81
Disease trend in the last five years (A)	10	4.89	1.53	10	4.99	1.50	1.35
Economic burden (H)	11	4.43	2.08	11	4.51	1.85	0.79^
Transmission potential from humans to animals	12	4.01	1.75	12	4.01	1.40	0.50^
Duration of illness (A)	13	3.72	1.14	13	3.76	1.34	0.64^
Transmission potential between animals	14	3.48	1.24	14	3.39	1.23	1.37
Economic and social burden on trade (A)	15	**3.29**	1.33	16	3.10	1.23	2.93**^
Severity of disease (A)	16	3.15	1.19	15	3.24	1.10	1.62^
High risk groups (H)	17	**3.05**	1.32	17	2.76	1.16	4.54***^
Control measures (H)	18	**2.87**	1.63	18	2.71	1.46	1.98*^
Control measures (A)	19	**2.61**	1.37	20	2.12	1.14	7.54***^
How much is known scientifically about the disease	20	2.57	1.20	19	2.51	1.32	0.92^
High risk groups (A)	21	1.95	1.07	21	1.87	1.03	1.37

[1](H) = human-related characteristic, for example, case-fatality in *humans*.
[2](A) = animal-related characteristic, for example, case-fatality in *animals*.
[3]Relative rank of disease criteria by importance scores; presented in order of importance to Canadian participants.
[4]Mean importance score across respondents.
[5]Standard deviation of importance scores across respondents.
[6]*t*-statistic; *d.f.* = 1,537.
*$p<0.05$.
**$p<0.01$.
***$p<0.001$.
^Adjusted for unequal variance using the Welch *t*-test; Satterthwaite's *d.f.* = 1,377.25 to 1529.68.
Scores in **bold** indicate disease criteria with statistically significant difference in importance scores between Canada and the US; scores for the country with the highest score (i.e. placed more importance on) are in **bold**.
doi:10.1371/journal.pone.0048519.t007

The Canadian group was more strongly influenced by case-fatality in animals, disease trend in the last five years in humans, transmission potential between humans, economic and social burden on trade in animals and high risk groups in humans (P-value <0.01 for all, Table 7), nonetheless, there was agreement between countries on the levels of least importance (lowest part-worth utility values) and levels of highest importance (highest part-worth utility values) with incremental increases in part-worth utility values for the levels in between (Tables 1, 2, 3).

The criteria *control measures in humans* and *control measures in animal* did not exhibit this sequential pattern in level preference in either country, however, these criteria were statistically significant (*P*-value <0.05) in the final models. Although preference was given to 'highly effective in reducing disease burden' (highest part-worth utility values) in both humans and animals and in both countries, there was no clear order of preference in the remaining levels (Tables 1, 2, 3). Both groups expressed next preference for the level 'not effective at all in reducing disease burden' in humans (second highest part-worth utility values).

Table 8: Disease priority list by country

https://doi.org/10.1371/journal.pone.0048519.t008

Canada	Score	rank	US	score	rank	Difference in rank (relative to Canada)
Nipah virus encephalitis	284.01	1	variant Creutzfeldt-Jakob disease (CJD)	368.89	1	5
Rabies	280.02	2	Rabies	295.44	2	0
Ebola virus haemorrhagic fever	260.24	3	Nipah virus encephalitis	286.10	3	−2
Marburg haemorrhagic fever	225.13	4	Ebola virus haemorrhagic fever	276.87	4	−1
Influenza (H1N1)	208.70	5	Marburg haemorrhagic fever	250.86	5	−1
variant Creutzfeldt-Jakob disease (CJD)	194.02	6	Influenza (H1N1)	207.22	6	−1
Listeriosis	177.78	7	Listeriosis	200.75	7	0
Hendra virus	64.79	8	Tularemia	164.88	8	4
Influenza (H5N1)	64.69	9	Anaplasmosis/Canine granulocytic anaplasmosis	137.19	9	36*
Salmonellosis	37.65	10	Hantavirus pulmonary syndrome	106.09	10	10*
Leishmaniasis	23.44	11	Paralytic shellfish poisoning	104.85	11	25*
Tularemia	10.33	12	Babesiosis	90.74	12	38*
Escherichia coli infection	−8.46	13	American trypanosomiasis (Chagas' disease)	81.17	13	5
Cryptosporidiosis	−11.29	14	Plague	79.65	14	9
Eastern Equine Encephalitis	−26.50	15	Hendra virus	65.12	15	−7
Botulism	−33.51	16	Influenza (H5N1)	62.25	16	−7
Shigellosis	−36.76	17	Shigellosis	55.89	17	0
American trypanosomiasis (Chagas' disease)	−52.78	18	Eastern Equine Encephalitis	54.28	18	−3
Giardiasis	−54.12	19	Leishmaniasis	53.60	19	−8
Hantavirus pulmonary syndrome	−59.94	20	Salmonellosis	47.74	20	−10*
Campylobacteriosis	−60.02	21	Escherichia coli infection	38.07	21	−8
Toxoplasmosis	−60.58	22	Q fever	19.95	22	5
Plague	−62.54	23	Cryptosporidiosis	10.44	23	−9
Psittacosis/Avian chlamydiosis	−74.75	24	Rocky Mountain spotted fever	7.94	24	9
Leptospirosis	−79.55	25	Botulism	−26.23	25	−9
Chlamydiosis	−79.67	26	Campylobacteriosis	−27.72	26	−5
Q fever	−94.88	27	Leptospirosis	−32.95	27	−2
West Nile virus	−109.20	28	Lyme Disease	−45.26	28	2
Bartonellosis	−114.43	29	Brucellosis	−47.38	29	5
Lyme Disease	−124.52	30	Chlamydiosis	−52.67	30	−4
Crimean-Congo haemorrhagic fever	−130.11	31	Psittacosis/Avian chlamydiosis	−53.33	31	−7
Powassan virus	−142.24	32	Toxoplasmosis	−58.94	32	−10*
Rocky Mountain spotted fever	−145.83	33	Giardiasis	−70.07	33	−14*
Brucellosis	−149.39	34	Powassan virus	−84.47	34	−2
Anthrax	−167.66	35	West Nile virus	−85.51	35	−7
Paralytic shellfish poisoning	−170.06	36	Bartonellosis	−94.16	36	−7
Echinococcosis	−180.88	37	Typhus	−103.02	37	9
Toxocariasis	−183.97	38	Coccidioidomycosis	−109.79	38	20*
Cutaneous larva migrans	−199.71	39	Crimean-Congo haemorrhagic fever	−132.98	39	−8
Lassa fever	−203.02	40	Anthrax	−144.16	40	−5
Baylisascariasis	−219.41	41	Echinococcosis	−147.03	41	−4
Severe Acquired Respiratory Syndrome	−222.05	42	Baylisascariasis	−155.26	42	−1
Old/New World Screwworm	−245.17	43	Toxocariasis	−157.14	43	−5
Western Equine Encephalitis	−250.06	44	Cutaneous larva migrans	−158.61	44	−5
Anaplasmosis/Canine granulocytic anaplasmosis	−256.71	45	Cysticercosis/Taeniasis	−168.25	45	12*
Typhus	−272.70	46	Western Equine Encephalitis	−185.49	46	−2
Japanese encephalitis	−273.33	47	Severe Acquired Respiratory Syndrome	−194.66	47	−5
Monkeypox	−279.78	48	Hepatitis A	−205.98	48	6
Trichinosis	−316.26	49	Japanese encephalitis	−230.44	49	−2
Babesiosis	−316.48	50	Lassa fever	−231.04	50	−10*
Venezuelan Equine Encephalitis	−329.70	51	Old/New World Screwworm	−251.11	51	−8
Yellow Fever	−330.35	52	Monkeypox	−274.35	52	−4
Cholera	−342.29	53	Venezuelan Equine Encephalitis	−279.00	53	−2
Hepatitis A	−359.51	54	Yellow Fever	−303.53	54	−2
Bovine Tuberculosis	−370.50	55	Trichinosis	−338.97	55	−6
Rift Valley fever	−372.81	56	St. Louis encephalitis	−363.37	56	6
Cysticercosis/Taeniasis	−443.42	57	Cyclosporiasis	−363.46	57	2
Coccidioidomycosis	−459.90	58	La Crosse encephalitis	−394.32	58	3
Cyclosporiasis	−490.94	59	Bovine Tuberculosis	−397.95	59	−4
Dengue fever	−520.64	60	Cholera	−416.70	60	−7
La Crosse encephalitis	−589.41	61	Dengue fever	−422.66	61	−1
St. Louis encephalitis	−597.52	62	Rift Valley fever	−425.87	62	−6

*Diseases that deviated by more than 10 ranked positions between countries.
doi:10.1371/journal.pone.0048519.t008

While preferences appear contradictory, this is due to the difference in preferences between individuals who choose to prioritize when the opportunity to control is attainable (highly effective control measures) and those who choose to prioritize when the opportunity to control is not viable (no control measures).

Although the strength of preference in disease criteria importance scores differed between countries, there was general agreement in the contribution of disease criteria in the decision to prioritize zoonoses (Table 7). However, marginal differences in the part-worth utility values within disease criteria (Tables 1, 2, 3) and in disease incidence and trend contributed to a unique disease priority list by country (Table 8).

Disease Priority Lists

The final ranking of diseases by their overall CA-derived scores is presented in Table 8. The higher the score, the higher the ranking on the priority list. The range in the overall scores by diseases differed between Canada and the US and correlates with the part-worth utility values derived by country (Tables 1, 2, 3). As part-worth utility values are interval data, the overall scores cannot be directly compared both within and between countries [24]. We can, however, compare the difference in disease ranking as an overall measure of proximity of diseases both within and between countries. Although differences in disease ranking were observed between countries, the majority of diseases (77%) were within ten ranked positions of each other indicating a general consensus in ranks between countries. Diseases of high priority generally exhibited high incidence, high case-fatality, severe symptoms, prolonged duration of illness, high transmission potential and are increasing or emerging in humans and animals, though it was not necessary to exhibit all of these characteristics to be prioritized (e.g., *Nipah virus encephalitis*, *Ebola* and *Marburg* do not occur naturally in North America while *H1N1 Influenza* has a low case-fatality rate). Diseases of low priority generally comprised rare diseases or diseases with a large proportion of asymptomatic cases, low case-fatality, short duration of illness, mild symptoms, stable disease trend and low transmission potential in humans and animals.

Canadians considered *Giardiasis*, *Salmonellosis*, *Lassa fever*, *Cryptosporidiosis*, *Toxoplasmosis* and *Botulism* of higher priority than Americans. Conversely, Americans considered *Babesiosis, Anaplasmosis, Paralytic shellfish poisoning, Coccidioidomycosis, Cysticercosis,*

Hantavirus pulmonary syndrome, Rocky Mountain spotted fever, Typhus, and the *plague* of higher priority than Canadians. This was due to regional differences in disease incidence (*Anaplasmosis*, *Rocky Mountain spotted fever* and *Babesiosis* do not occur naturally in Canada but are endemic in the US), differences in disease trend (*Paralytic shellfish poisoning* has been increasing in the US but is stable in Canada) or differences in the part-worth utility values within disease criteria by country (Tables 1, 2, 3).

DISCUSSION

Disease prioritization has engaged the interest of numerous research groups recently [4], [6], [12], [13], [15]–[18]. As zoonotic outbreaks are a significant public health burden in North America [25]–[27], and with decreasing resources available for their prevention and control, a strategic framework for the prioritization of zoonoses would be valuable. We present on the novel use of a quantitative methodology, CA, for the prioritization of zoonoses in North America. There are a number of unique features in our study that overcome constraints in traditional prioritization methodologies. Our method fits a statistical model to a robust experimental design to generate relative weighted scores for disease criteria and their levels thus overcoming the limitation of a subjective weighting process for disease criteria by expert panels [4], [6], [13], [16], [18] or an arbitrary scoring system for levels [4], [6], [8]–[13], [16], [18]. Because CA generates weighted scores by assessing disease combinations rather than disease criterion separately, our scores are relative weighted scores that do not assume disease criteria are independent or of equal importance [4], [6], [8]–[13], [16], [18]. The study explored preference for disease criteria, rather than actual diseases; thus, individuals were forced to prioritize on the basis of scientific information, eliminating biases associated with disease names. We elicited preferences by choice rather than by ranking or rating diseases allowing for a more intuitive approach to expressing preferences, particularly when choices are similar [17].These features of the CA trade-off methodology brings additional value to the disease prioritization process compared to traditional methods.

Perhaps the most distinct feature of this study was the ability to engage wide social participation from the public. This has not been previously conducted and understanding the perception of the public may offer healthcare professionals the opportunity to improve public education and risk communication. Our CA models indicate individuals with no prior knowledge or experience in prioritizing zoonoses were capable of producing meaningful results with a satisfactory model fit. Part-worth utility values exhibited face validity with a stronger preference for salient disease criteria levels than non-salient alternatives. The degree to which each disease criterion contributed to the overall decision to prioritize also demonstrated face validity and are consistent with findings from similar studies [4], [6], [13], [16]–[18]. The priority lists derived from applying part-worth utility values from the CA models to actual diseases produced a list of diseases that were reasonable, from a public perspective, for prioritization. More similarities than differences were observed in the strength of preference in disease criteria importance scores between countries suggesting general agreement in disease prioritization between Canadians and Americans.

There are some limitations associated with this study. The response rate was low given the length and technical content of the survey, we also used a pre-screened panel to recruit participants; the study populations were therefore highly educated compared to the national populations and the results may not be truly representative of the general public. However, public education and risk communication has the highest impact on those who are educated and actively involved in the community, thus, our study populations likely represent the target population for public health intervention. There was a marked difference in the response rate between the Canadians and Americans, although the cause for this difference is unknown as recruitment strategies and survey design were identical between countries. It is important to note that while response rates differed dramatically, the study populations did not deviate considerably from their respective national populations or from each other in terms of gender, age and education. The difference in response rate is therefore unlikely to have resulted in substantial differences in the results between countries.

Our percent certainty model fit was 79.4% for both countries; an opportunity therefore exists to improve the models further to estimate more precise part-worth utility values. Marginal changes in part-worth utility values may have a substantial impact on the relative ranking of diseases on the priority list; thus, our current priority lists are only acceptable if we accept the current model fit. Nonetheless, this is the first disease prioritization study to validate disease criteria scores against respondent choices and may serve as a standard for prioritization methodologies. We also acknowledge that not all of the 59 disease criteria identified from focus groups was considered in this study [7], the inclusion of additional criteria would have an impact on the relative ranking of diseases. The current priority lists are therefore only applicable to the 21 disease criteria in the study, in spite of this; we believe these criteria capture some of the most relatively important aspects for priority setting.

Finally, there are multiple objectives for prioritizing zoonoses (e.g. prioritizing for research, regulation, control, prevention, management, vaccination, diagnosis, cost-effective and surveillance), although we specified to participants to prioritize "for policy implementation for the control and prevention of zoonoses", they may have prioritized with another objective in mind. There is no way to measure this type of error and we assume that participants were consistent in their objective to prioritize. Further, we asked participants "which of the following diseases should be prioritized" with each choice task set, which may be ambiguous in the face of multiple competing diseases. Although we did state in the instructions (not presented) to "select one of the five disease profiles that characterizes a disease that, in your opinion, should be prioritized above the others", we could have reinforced this by explicitly specifying, "which of the following diseases should receive the highest priority" with each choice task set. We assume that participants understood the research question and were consistent in their interpretation.

In conclusion, we describe the first zoonotic disease prioritization exercise, involving public participation, in North America. Given the established quantitative approach, robust experimental design, satisfactory model fit and reasonable disease criteria level scores [24], [34], we have demonstrated CA as a potential tool for the prioritization of zoonoses. Periodical updating of criteria levels to match current disease trends will allow for the revision of disease priority lists that will reflect the most current state. Finally, the application of this methodology is versatile and not limited to zoonoses or North America; thus, other prioritization exercises may consider this approach with different criteria inputs.

ACKNOWLEDGMENTS

We acknowledge the following individuals for helping us with literature searches: Karren Lemay, Mai Pham, Amber Reed and Ashley Whiteman. Dr. David Fisman is acknowledged for his early contributions to the conception of this project.

AUTHOR CONTRIBUTIONS

Conceived and designed the experiments: VN JMS. Performed the experiments: VN. Analyzed the data: VN. Contributed reagents/materials/analysis tools: VN JMS. Wrote the paper: VN.

REFERENCES

1. WHO (2012) World Health Organization website.Neglected Zoonotic Diseases.Available: http://www. who. int/neglected_diseases/diseases/zoonoses/en/index. html.Accessed 2012 May 24.
2. Heymann DL (2008) Control of Communicable Diseases Manual; Heymann DL, editor.Washington D. C: American Public Health Association.746 p.
3. Taylor LH, Latham SM, Woolhouse ME (2001) Risk factors for human disease emergence.Phil Trans R Soc Lond B 356: 983–989.
4. Krause G, Alpers K, Benzler J, Bremer V, Claus H, et al.(2008) Prioritisation of infectious diseases in public health - call for comments. Eurosurveillance 13: 1–6.
5. Gilsdorf A, Krause G (2011) Prioritisation of infectious diseases in public health: feedback on the prioritisation methodology, 15 July 2008 to 15 January 2009.Euro Surveillance 16: pii: 19861.
6. Krause G, Alpers K, Benzler J, Bremer V, Claus H, et al.(2008) How can infectious diseases be prioritized in public health.EMBO Reports. European Molecular Biology Organization 9: S22–27.
7. Ng V, Sargeant J (2012) A Stakeholder-Informed Approach to the Identification of Criteria for the Prioritization of Zoonoses in Canada. PLoS ONE 7: e29752 .
8. Horby P, Rushdy A, Graham C, O'Mahony M (2001) PHLS overview of Communicable Diseases 1999.Communicable Disease and Public Health 4: 8–17.
9. Rushdy A, O'Mahony M (1998) PHLS overview of communicable diseases 1997: results of a priority setting exercise.Communicable Disease Report Supplement 8: S1–12.
10. WHO (2003) The Dubrovnik pledge on surveillance and prioritization of infectious diseases.Report on a WHO meeting in Bucharest, Romania 21–23 November, 2003.Copenhagen, WHO Regional Office for Europe: World Health Organisation.27 p.
11. Doherty J (2000) Establishing priorities for national communicable disease surveillance.Canadian Journal of Infectious Diseases 11: 21–24.
12. Doherty J (2006) Final Report and Recommendations from the National Notifiable Diseases Working Group.Canadian Communicable Disease Report 32: 211–225.

13. Balabanova Y, Gilsdorf A, Buda S, Burger R, Eckmanns T, et al.(2011) Communicable Diseases Prioritized for Surveillance and Epidemiological Research: Results of a Standardized Prioritization Procedure in Germany, 2011.PLoS ONE 6: e25691 .
14. InVS (2002) Définition des priorités dans le domaine des zoonoses non alimentaires 2000–2002.Paris, France: Institut de Veille Sanitaire (InVS).40 p.
15. InVS (2010) Définition des priorités dans le domaine des zoonoses non alimentaires 2008–2009.Paris, France: Institut de Veille Sanitaire (InVS).31 p.
16. Cardoen S, van Huffel X, Berkvens D, Quoilin S, Ducoffre G, et al.(2009) Evidence-based semiquantitative methodology for prioritization of foodborne zoonoses.Foodborne pathogens and disease 6: 1083–1096.
17. Havelarr AH, van Rosse F, Bucura C, Toetenel MA, Haagsma JA, et al.(2010) Prioritizing Emerging Zoonoses in The Netherlands.PLoS ONE 5: e13965 .
18. Humblet M, Vandeputte S, Albert A, Gosset C, Kirschvink N, et al. .(2012) Multidisciplinary and Evidence-based Method for Prioritizing Diseases of Food-producing Animals and Zoonoses.Emerging Infectious Diseases 18: doi: https://doi. org/10. 3201/eid1804. 111151.
19. Mangen M, Batz M, Käsbohrer A, Hald T, Morris J, et al.(2009) Integrated approaches for the public health prioritization of foodborne and zoonotic pathogens.Risk Analysis 30: 782–797.
20. Green P, Srinivasan V (1978) Conjoint analysis in consumer research: Issues and outlook.Journal of Consumer Research 5: 103–123.
21. Ryan M, Farrar S (2000) Using conjoint analysis to elicit preferences for health care.British Medical Journal 320: 1530–1533.
22. Mele N (2008) Conjoint analysis: using a market-based research model for healthcare decision making.Nursing Research 57: 220–224.
23. Sampietro-Colom L, Espallargues M, Rodríguez E, Comas M, Alonso J, et al.(2008) Wide social participation in prioritizing patients on waiting list for joint replacement: a conjoint analysis.Medical Decision Making 28: 554–566.
24. Orme BK (2010) Getting started with conjoint analysis: Strategies for product design and pricing research.Madison, Wisconsin: Research Publishers, LLC.210 p.

25. Lindsey N, Staples J, Lehman J, Fischer M (2010) CDC (2010) Surveillance for human West Nile virus disease - United States, 1999–2008.MMWR Surveillance Summary 59: 1–17.
26. Kermode-Scott B (2009) Canada has world's highest rate of confirmed cases of A/H1N1, with Aboriginal people hardest hit.British Medical Journal 339: b2746.
27. Bacon R, Kugeler K, Mead P (2008) CDC (2008) Surveillance for Lyme Disease - United States, 1992–2006.MMWR Surveillance Summary 57: 1–9.
28. PAHO (2003) Zoonoses and communicable diseases common to man and animals; Acha P, Szyfres B, editors.Washington DC: Pan American Health Organization.Three volume set, 408, 378 and 395 pages.
29. Merck (2010) The Merck Veterinary manual; Kahn C, Line S, editors.Whitehouse Station, NJ, USA: Merck Sharp & Dohme Corporation.2945 p.
30. Sargeant J, Ng V (2011) The current state of knowledge of zoonoses and the implications for research outcomes in disease prioritization in Canada.Ecohealth 7: 34.
31. Patterson M, Chrzan K (2003) Partial Profile Discrete Choice: What's the Optimal Number of Attributes? Sawtooth Software Conference Proceedings, San Antonio, Texas, April 15–17 2003: 173–185.
32. Chrazn K (2010) Using partial profile choice experiments to handle large numbers of attributes.International Journal of Market Research 52: 827–840.
33. Sawtooth (2008) CBC v6. 0 Technical Paper: The CBC System for Choice-Based Conjoint Analysis.Sequim, Washington: Sawtooth Software.26 p.
34. Bridges J, Hauber A, Marshall D, Lloyd A, Prosser L, et al.(2011) Conjoint Analysis Applications in Health - a Checklist: A Report of the ISPOR Good Research Practices for Conjoint Analysis Task Force. Value in Health 14: 403–413.
35. StatsCan (2006) Statistics Canada website.Population 15 years and over by highest certificate, diploma or degree, by age groups, 2006 Census.Available: http://www. statcan. gc. ca/tables-tableaux/sum-som/l01/cst01/educ43a-eng. htm.Accessed 2012 June 20.
36. StatsCan (2011) Statistics Canada website.Age and Sex Highlight Tables, 2011 Census.Available: http://www12. statcan. gc. ca/census-

recensement/2011/dp-pd/hlt-fst/as-sa/?Lang=E.Accessed 20 June 2012.

37. USCB (2010) United States Census Bureau website.Educational Attainment in the United States: 2010 - Detailed Tables.Available: http:// www. census. gov/hhes/socdemo/education/data/cps/2010/tables. html.Accessed 20 June 2012.
38. USCB (2010) United States Census Bureau website.2010 Census Data.Available: http://2010. census. gov/2010census/data/.Accessed 20 June 2012.
39. Sawtooth (2009) CBC/HB Version 5. 0 Technical Paper: The CBC/ HB System for Hierarchical Bayes Estimation.Sequim, Washington: Sawtooth Software.31 p.
40. Howell J (2009) Sawtooth Software Research Papers Series: CBC/HB for Beginners.Sequim, Washington: Sawtooth Software.5 p.

CHAPTER 2

Community Perceptions on Integrating Animal Vaccination and Health Education by Veterinary and Public Health Workers in the Prevention of Brucellosis among Pastoral Communities of South Western Uganda

Catherine Kansiime[1], Lynn M. Atuyambe[2] , Benon B. Asiimwe[3] , Anthony Mugisha[4] , Samuel Mugisha[5] , Victor Guma[6] , Innocent B. Rwego[5,7], Elizeus Rutebemberwa[1]

[1] Department of Health Policy, Planning and Management, Makerere University School of Public Health, College of Health Sciences, Kampala, Uganda

Citation: Kansiime C, Atuyambe LM, Asiimwe BB, Mugisha A, Mugisha S, Guma V, et al. (2015) Community Perceptions on Integrating Animal Vaccination and Health Education by Veterinary and Public Health Workers in the Prevention of Brucellosis among Pastoral Communities of South Western Uganda. PLoS ONE 10(7): e0132206. https://doi.org/10.1371/journal.pone.0132206.

[2] Department of Community Health and Behavioral Sciences, Makerere University School of Public Health, College of Health Sciences, Kampala, Uganda

[3] Department of Medical Microbiology, College of Health Sciences, Makerere University, Kampala, Uganda

[4] Department of Veterinary Medicine, College of Veterinary Medicine, Animal Resources and Bio-security, Makerere University, Kampala, Uganda

[5] Department of Biological Sciences, College of Natural Sciences, Makerere University, Kampala, Uganda

[6] Department of Mental Health and Community Psychology, Makerere University College of Humanities and Social Sciences, Kampala, Uganda

[7] Ecosystem Health Initiative, College of Veterinary Medicine, University of Minnesota, St. Paul, Minnesota, United States of America

ABSTRACT

Background

Brucellosis is a zoonotic disease of veterinary, public health, and economic significance in most developing countries, yet there are few studies that show integrated human and veterinary health care intervention focusing on integration at both activity and actors levels. The aim of our study, therefore, was to explore community perceptions on integration of animal vaccination and health education by veterinary and public health workers in the management of brucellosis in Uganda.

Methods

This study used a qualitative design where six Focus Group Discussions (FGDs) that were homogenous in nature were conducted, two from each sub-county, one with the local leaders, and another with pastoralists and farmers. Five Key Informant Interviews (KIIs) with two public health workers and three veterinary extension workers from three sub-counties in Kiruhura district, Uganda were conducted. All FGDs were conducted in the local language and tape recorded with consent from the participants. KIIs were in English and later transcribed and analyzed using latent content data analysis method.

Results

All the groups mentioned that they lacked awareness on brucellosis commonly known as Brucella and its vaccination in animals. Respondents perceived improvement in human resources in terms of training and recruiting more health personnel, facilitation of the necessary activities such as sensitization of the communities about brucellosis, and provision of vaccines and diagnostic tests as very important in the integration process in the communities. The FGD participants also believed that community participation was crucial for sustainability and ownership of the integration process.

Conclusions

The respondents reported limited knowledge of brucellosis and its vaccination in animals. The community members believed that mass animal vaccination in combination with health education about the disease is important and possible if it involves government and all other stakeholders such as wildlife authorities, community members, local to national political leaders, as well as the technical personnel from veterinary, medical and public health sectors since it affects both humans and animals.

INTRODUCTION

Brucellosis is a zoonosis of veterinary, public health and economic significance in most developing countries [1]. It affects human health, as well as the health and reproductive performance of livestock, thus reducing their productivity, leads to abortions and weak offspring and is a major impediment for trade and export [2]. Studies on animal brucellosis in Uganda have reported a herd prevalence of 55.6% and animal prevalence of 15.8% in the pastoral dairy system in Mbarara district [3], while higher figures of up to 100% at herd level and 30% at animal level have been reported in the Central district of Nakasongola [4]. Recently, a cross sectional study of the disease in two sub counties of Kazo and Buremba in Ibanda district, Western Uganda, found a prevalence of 38.4% and 26% in cows and goats respectively [5]. Human brucellosis has been reported to be prevalent in both rural and urban settings [6,7] and a more recent study revealed that 12.6% of informally marketed milk in urban Kampala was contaminated with *B. abortus* at purchase; and the annual human incidence rate in Kampala was estimated to be 5.8 per 10,000 people [8].

The explosive human population growth and environmental changes have resulted in increased numbers of human-animal interactions which presents a potential zoonotic threat [9] that may cause spillover of infections from wildlife to cattle in areas around national parks. Brucellosis is also prevalent among the Ugandan wildlife populations [10], and other studies have reported prevalence of brucellosis in buffaloes at 10% in Egypt and 5.05% in Pakistan [11] with evidence of spill over to outdoor reared domestic pigs and cattle.

With increasing human-animal interface, integration of services is critical to encourage working together of both public health and veterinary professionals. Integration is at the heart of systems theory and, therefore, central to organizational design and performance [12]. Despite the fact that systems are comprised of separate hierarchical structures with interconnected components; these components are supposed to play complementary roles in order to accomplish their joint tasks [13]. However, the division, decentralization, and specialization found in many organizations usually interfere with efficiency and quality goals [14]. Therefore, the fulfillment of system aims necessitates co-operation and collaboration among and between the various parts of the system. In this sense, integration is the "glue" that bonds the entity together, thus enabling it to achieve common goals and optimal results [15].

There is no universal definition of integration, but various researchers have defined it differently. Gröne and Garcia-Barbero [16] sums up health care integration as the "bringing together of inputs, delivery, management and organization of services as a means of improving access, quality, user satisfaction and efficiency." Brown & McCool [17] suggest that integration allows for greater efficiency and effectiveness, less duplication and waste, more flexible service provision, and better co-ordination and continuity. On the other hand, the Program for Appropriate Technology in Health [18] defines health services integration as the organization, management, and delivery of a continuum of preventive and curative health services.

While great successes have been achieved through vertical programming notably in childhood immunizations, HIV/AIDS and tuberculosis diagnosis and care in Tanzania, as well as integrated care for HIV/AIDS and maternal and child health in Kenya [18], it is recognized that integration of health services can derive greater overall impact from health resources and systems.

Traditionally, prevention of human brucellosis has mainly depended on control of the disease in domestic animals [19], with scholars arguing that these are the main reservoir of the disease [20]. A study in Greece [21] found that vaccination of animals against brucellosis combined with health education on preventive measures among agricultural and pastoral population led to a reduction in the incidence of human brucellosis. However, in areas bordering conservation areas, vaccination is complicated by the domestic animals—wildlife interaction since there are no vaccination efforts for the wildlife species [22]. In Uganda, *Brucella* S19 vaccine is used against animal brucellosis with emphasis on cattle as a response to a reported outbreak. The focus is mainly on *B. abortus*. Vaccination is a voluntary exercise, but government also provides the vaccine through Ministry of Agriculture, Animal Industry and Fisheries (MAAIF) and it is also done by District Local Government headed by the District Veterinary Officer during disease outbreak. There is no test and slaughter policy in place, due to the related costs of farmer compensation. Despite the significant role of vaccination in the prevention of brucellosis, there is a need to introduce other measures such as health education on the nature and mode of transmission of the disease. It is important to note that good knowledge of the disease is essential for the success of a control program as documented elsewhere [23]. In a study of non-sedentary pastoralist communities, Ward et al [24] recommended health education, especially on zoonoses, as one branch of collaboration between veterinary and public health services in such populations.

In communities where livelihoods depend on domestic animals and their products, such as the pastoral rangelands of South Western Uganda, human-animal interaction is intimate. There is therefore a need to greatly recognize that animal and human health are closely linked and that veterinary and public health sectors share the common goal of protecting, promoting and improving the health and wellbeing of human populations [25]. Closer co-operation between human and animal healthcare providers can also lead to financial savings in the two sectors [26]. In resource poor countries, control measures targeting zoonotic diseases at the human-domestic animal-wildlife interface could work provided that they are designed and adopted by the local population [27]. There are few studies of integrated human and veterinary health care intervention focusing on integration at both activity and actors levels. Those commonly described often deal with delivery of health care under special circumstances such as human disease outbreak necessitating both local and international funding [24]. The aim of our study therefore was to explore community perceptions on integration at two levels: the activity level which focused on animal vaccination and health education, and the actor level which involved veterinary and public health workers in the management of brucellosis among pastoralist communities living in close proximity with wildlife in Kiruhura district, South Western Uganda.

Conceptual framework on Approach to Integrated Health Services PATH 2011[18]

This framework encompasses four levels of a country's health structure—client-centered services at the community level, health operations planning at the organization level, health system coordination at the national level, and inter-sectoral initiatives across development sectors.

Client-centered services at the facility and community levels involves integrated programming that fit the needs of clients, including individuals and families, as well as the broader community. It may involve increased awareness of brucellosis in terms of symptoms, improved access to treatment services, drugs, preventive methods both in animals and humans.

Operational elements at the health organization levels, often requires changes in how services are delivered by ministries of health, nongovernmental or local organizations, and private-sector agencies. Existing or new health system inputs (such as resources, time, money, or expertise) may need to be allocated differently to support planning, management, staffing, interpersonal communication, or the measurement of integrated services.

Broader governance and capacity issues at the health system level such as; new levels of coordination or joint planning of the policies, processes, and infrastructure that make up a health system may be needed to deliver integrated services. Integration at this level often requires significant involvement and support from all stakeholders, including donors, ministries of health, politicians, advocacy groups, the private sector, and non-governmental organizations.

Inter-sectoral coordination such as cross-sector integration is when public health personnel jointly work with veterinary extension staff to provide health education on brucellosis during animal vaccination. Integration across sectors requires engagement and commitment at multiple levels of the organization as discussed above.

The strength of this framework is that it looks at integration of services in a wider perspective to include all sectors as well as client needs. However, it fails to describe challenges likely to be encountered by different sectors in integration of services. Awareness of likely challenges is an important issue for successful integration.

METHODS

Study Area and Population

The study was conducted in the pastoralist communities of Nyabushozi County, Kiruhura district, South Western Uganda. Three sub-counties of Kanyaryeru to the West, Nyakashashara to the East, and Sanga to the North of Lake Mburo National Park (LMNP) were purposively selected. This was because of their close proximity to wildlife which poses a potential risk for zoonotic spill overs. The communities in this region derive their livelihoods from consumption and selling of cattle and their products but have recently started growing food crops at a subsistence level. The economic livelihoods in this area include: the farmers who are settled and solely grow crops (mostly the recent immigrants to the area); the agro-pastoralists who rear cattle as well as practice crop farming, and lastly the pure pastoralists/ semi-nomads who only rear cattle but have permanent shelters, only moving their animals in dry seasons in search of water and pasture. The majority of cattle keepers in this area, however, are agro-pastoralists. In this area, animals are mainly vaccinated against Foot and Mouth Disease, and treated against Trypanosomiasis, and Lumpy Skin Disease and this is only done during suspected disease outbreaks.

Study Design and Sampling

This was a qualitative study that utilized two data collection methods namely: Focus Group Discussions (FGDs) and Key Informant Interviews (KIIs). Six FGDs of 10 participants each and five KIIs were conducted. These were purposively selected. The FGDs included local leaders (local council chairpersons, councilors, opinion leaders, religious leaders, as well as youth and women representatives), pastoralists and farmers. The KIIs comprised of veterinary and public health practitioners.

Data Collection

Focus Group Discussions (FGDs) refer to a qualitative methodology where people with similar background or experiences gather to discuss a specific topic of interest with a researcher [28]. The researcher gains understanding of a particular issue from the perspective of the group participants [29]. This technique allows a small group of participants to discuss issues led by a moderator using a discussion guide [30]. On average, the number of participants in each FGD is recommended to be at least eight [31]. In the current study, 57 of the 60 expected FGD participants turned up. The discussions were held at a venue, in each of the three sub-counties, identified by and convenient to the participants. Prior to data collection, study tools were pre-tested so as to help identify emerging themes. Highlighted below are the key topics identified and later explored during the FGDs and KIIs: i) community perception on integration of animal vaccination and health education on brucellosis (at activity level), ii) community perceptions on integration of the two services by veterinary and public health workers (at actor level), iii) how the integration can be conducted (at both levels), iv) challenges likely to be faced as well as recommendations to overcome them (at both levels).

The FGD guide was translated from the original English version into the local language (Runyankole) by Makerere University Institute of Languages. All FGDs were conducted in the local language and tape recorded with verbal consent from the participants. The moderator introduced the topic and aims of the study, and guided the discussions which lasted for one and a half hours until saturation where participants repeated the same information when they were asked the questions again. Recordings were later transcribed into English by the note taker and the principal investigator.

A total of five key informants were purposively selected on the basis of their professions and experience in order to provide an overview on the integration process. Two Clinical Officers (these are in-charge of Health Centre IIIs at sub-county level) and three Veterinary Officers (who are in-charge of sub-counties and responsible for veterinary extension services to the farmers) were interviewed to supplement the FGDs. The KIIs were conducted at the venue that was convenient to them (their place of work). Semi-structured interviews were conducted with Key Informant (KIs) because of their position or experience, and in-depth knowledge [32,33] on animal vaccination and health education on brucellosis. When respondents are well informed people, they provide an overall view of the community [34]. We also used KIIs to discover the subjective meanings and interpretations of peoples' experiences. In this case, people's responses are less influenced by the presence of their peers [35].

All KIIs were conducted in English and lasted an average of 45 minutes each. The first author (CK) conducted the interviews which were tape recorded with consent from the participants. Later the interviews were transcribed verbatim by the note taker and the first author. Tape recording was used in this study because it allows more time for eye contact between the moderator and the respondent during the interview [36] and the interviewer was able to get details and accuracy that would not be ascertained from field notes alone.

Data Management and Analysis

The unit of analysis was the transcripts from the FGDs and the KIIs. Transcripts were first read through several times while making notes on them. All transcripts were cross-checked to ensure completeness of data and were kept under lock and key. Content analysis using both Latent and manifest analysis were used; Latent analysis technique refers to what the text talks about and involves in-depth interpretation of the underlying meanings of the text while manifest content analysis refers to the analysis of the obvious visible components [37,38]. Data were condensed without losing quality as described above.

All transcripts were read though and data were discussed by the authors who came up with codes. Responses with similar codes were re-categorized under a unifying sub-theme or theme. A matrix was created and individual matrices were discussed by the authors (CK and VG) until an agreement was reached. The categories were then interpreted for their descriptive meaning. Descriptive quotes representing key themes were identified by the authors and included in the final report writing.

Ethical Statement

The study protocol was approved by Makerere University School of Public Health Higher Degrees, Research and Ethics Committee as well as Uganda National Council for Science and Technology. The study objective was explained to participants in their local language (Runyankole), written consent to participate in the study was obtained from all individuals before the discussions and interviews begun. Permission was also asked verbally from participants to have the sessions recorded and they were informed that the recordings were for the researchers only. For anonymity and confidentiality in the information provided, participants were assured that their names would not be used in analysis. Participants were informed that their participation was voluntary and they were free to withdraw anytime. Each focus group discussion was held within the village where the participants came from and at a time and place of their choice. The key informant interviews were conducted at a time and place convenient to each person.

RESULTS

This study set out to explore community perceptions on the integration of animal vaccination and health education by veterinary and public health workers in the management of brucellosis among pastoralists of Kanyaryeru, Sanga and Nyakasharara sub-counties in Kiruhura district, South Western Uganda.

Socio-demographic Characteristics of the Participants

A total of 57 participants from three sub-counties of Nyakashahara, Sanga and Kanyaryeru were recruited into the Focus Group Discussions (FGDs) of which 51(89.5%) were males. The mean age of the participants was 46 years with a standard deviation of 10.13. There were five key informants (including two females).

The key themes that the research targeted were: i) Community perceptions on integration of animal vaccination and health education on brucellosis (the activity level), ii) community perceptions on integration of the two services by veterinary and public health workers (the actor level), iii) how to conduct the integration process (at both levels) and, iv) challenges likely to be faced as well as recommendations to overcome them at both levels. These themes are discussed below:

Community Perceptions on Integration of Animal Vaccination and Health Education on Brucellosis (the activity level)

In the study area, brucellosis is commonly known as *Brucella*. From the discussions held with the community members, lack of awareness on brucellosis such as signs, symptoms, prevention, mode of transmission and vaccination in animals was highlighted and perceived as a limitation to the integration process. Five out of the six FGDs believed the integration process will work if the community is provided with adequate information on brucellosis.

"*We have never heard about animal vaccination against Brucella. We would like to know more about it and about the signs and symptoms because when our people and cattle are suffering from Brucella, we cannot tell unless the doctor or veterinary doctor tests and tells us. We therefore want to know more about Brucella in humans and animals, the cause, symptoms and how we can prevent it. This is important for integration because you cannot integrate what the people do not know*" (FGD with farmers and pastoralist, Kanyaryeru sub-county).

Respondents expressed their concerns about the free movements of wild animals in the communities. They believed that integration of animal vaccination and health education alone would not be effective in the prevention of brucellosis unless the wild animals are restricted. This is illustrated below:

"*First, government and park management should work towards restricting the wild animals from our communities by fencing them in the park because without doing that, the disease will remain in the communities*" (KII veterinary officers, Sanga and Kanyaryeru sub-counties).

"*Before you think of reducing Brucella in humans you must first reduce it in wild animals and livestock because they are the source of the disease. Only when this is done should we proceed with health education and vaccination of our animals*" (FGD local leaders, farmers and pastoralists, Nyakashashara and Kanyayeru sub-county).

Four out of five FGDs and two key informants believed that animal vaccination against brucellosis is more important than health education as illustrated below:

"*When you tell us to stop drinking unboiled milk and other practices, it is difficult to stop. For instance, I cannot burn all the cow dung to kill all bacteria nor will I fail to help my cow to deliver because you say I should not. The important thing to do is to vaccinate our animals against Brucella and later teach us about the disease since animals are the source of the disease. Otherwise, teaching us about the disease will not help us if our animals remain sick*" (FGD local leader, Nyakashashara sub-county).

"*The main focus should be on mass vaccination of animals against brucellosis because they are the main source of disease. When this is done, it will contribute to the reduction of the disease in humans and then health education comes in as an additional informative action to further reduce brucellosis in the communities*" (KII veterinary officer, Kanyaryeru sub-county).

Community Perceptions on Integration of the Two Services by Veterinary and Public Health Workers (the actor level)

During the discussions and interviews, participants were given an opportunity to express their opinions regarding the integration of the two services by the veterinary and health workers. Participants believed that veterinary and health personnel could work together but they emphasized the need to also train local people especially in disseminating information about brucellosis in churches, at parties and on radios:

"*The health providers and veterinary doctors will need to train at least two people in every parish about Brucella in humans and animals. These could spread the information across the villages, in churches, and at parties. Information should also be broadcasted on the radios especially Radio Five, Radio West and Radio Uganda; these are the most listened to in this locality*" (FGD local leaders, Kanyaryeru sub-county).

All the FGDs and key informants mentioned shortage of veterinary officers as a limitation in the integration process. The participants believed that there is need for facilitation in terms of more health providers and veterinarians if the integration process is to succeed.

"*I think the veterinary officers should be facilitated by training and recruiting more of them because there is only one veterinary doctor in each sub-county. This would help them work with public health workers to provide joint services to the communities. I think if government increased their numbers and extended them to the communities they would be able to teach and treat animals*" (FGD farmers and pastoralists, Sanga sub-county and KII veterinary officer, Kanyaryeru sub-county).

Focus group participants and KII respondents were engaged and allowed to discuss their opinions and perceptions on how to integrate the two services by the veterinary and public health workers in the management of brucellosis. Joint provision of the services by both veterinary and public health workers was mentioned as good for the management of brucellosis in the community. It was suggested that this could be achieved through joint programming of activities by veterinary and health providers.

"*I think if the two services are integrated in the providers' different programs, they will be able to work together, otherwise this would be the biggest challenges. For example, the public health workers could incorporate health education of Brucella in their outreach programs and also work with veterinary officers jointly during animal vaccination to teach people about Brucella in humans*" (KII veterinary, Kanyaryeru sub-county).

Gender issues were raised as important in the integration at activity level. Participants mentioned gender roles of women in handling animal products and care seeking for children as important opportunities to teach them about brucellosis while for men, animal vaccination would be the best way to capture and educate them about the disease at vaccination sites.

"Integrated services are important for instance, when people go for outreach programs like immunization it would be good if veterinary officers would come along and teach the women who come for antenatal or bring children for immunization about boiling milk, water, and proper cooking of animal-derived foods. The women are involved in cooking, handling of most of the animal products and hygiene issues. If veterinary officers could organize activities at the sub-county such as testing and vaccinating animals and government intervenes by providing more veterinary officers, the integration process would work" (FGD local leaders, Sanga and Kanyaryeru sub-counties).

"Provide continuous free animal vaccination against brucellosis so that more men can be captured during vaccination and given health education about the disease since most men do not attend outreach programs and this would be one of the ways to involve them in the integration process" (KII veterinary officer, Sanga sub-county).

All KII respondents believed that government and stakeholder involvement in the integration at actor level was important since veterinary and public health workers would not manage on their own. They highlighted the role of government in sensitization, training and improving human resources.

"Involve Ministry of Health, district, sub-county, local council members and the private sector in the planning, management and implementing of these services in the communities" (KII veterinary officer, Sanga sub-county).

"Government should be involved in the vaccination process by providing free Brucella vaccines like it does during outbreaks for foot and mouth disease or provide the vaccine at subsidized prices because it is very expensive and pastoralists cannot afford it" (KII veterinary officer, Kanyaryeru and Nyakashashara sub-counties).

Four FGDs and all Key Informants believed that animal vaccination and health education about brucellosis is not enough unless it is accompanied by infrastructural development. They highlighted the need for laboratory facilities for rapid diagnosis and treatment of humans; and vaccination or test and slaughter for their livestock in the integration process as illustrated below:

"Integration of the two services to be successful needs establishment of equipped laboratories centered at the sub-counties to facilitate testing for both the livestock and the humans. This enables us prevent the disease

and increases access to diagnosis in humans and treatment (vaccination in animals because Brucella is rampart in these areas as seen from the numerous animal abortions, infertility and weak calves" (FGD local leaders, Kanyaryeru sub-county).

"Veterinarians can closely work with health providers to provide the two services but this should be well planned with more diagnostic laboratories for both animals and humans and equipping these facilities is essential since the disease is often misdiagnosed" (KIs senior clinical officer, and Veterinary officer Sanga and Kanyaryeru sub-counties).

Challenges Likely to be Faced During the Integration Process as well as Recommendations for Successful Integration

During the discussions and interviews, a number of challenges likely to be encountered during the integration process at both the actor and activity levels were believed to be; workload, human resource shortage and facilitation problems.

"The problems the veterinary and health personnel may face include overload of work, for instance, the public health workers have other programs they conduct during their outreaches, and veterinary officers also have their different schedules to conduct in the communities and this may not fit into the other's schedule unless they are supported by government to incorporate animal vaccination and health education of Brucella into their programs and then, the two professionals agree on when, and where to conduct jointly these two services" (FGD leaders, farmers and pastoralists, Kanyaryeru sub-county).

Inadequate human resources, especially fewer veterinary personnel, were believed to be a big hindrance to vaccination and unless this is addressed, the integration at actor level may not be possible.

"In this sub-county, we have about 500,000 cattle but only one veterinary doctor who is also the District Veterinary Officer. You do not expect him to handle all the animals in the communities. It is important therefore that government increases the number of veterinary officers at sub-county level if the integration is to work" (FGD local leaders, farmers and pastoralists, Nyakashashara sub-county).

"More manpower especially the veterinary officers and more training for both providers would be needed to facilitate the integration program" (KII Senior Clinical Officer, Nyakashashara sub-county).

Four out of five key informant respondents emphasized the need to facilitate the veterinary and health personnel in form of allowance and transport as important for the integration process.

"Transport costs are one major problem since our areas have a poor road network and the people are sparsely populated. For the integration process to work, facilitation will be needed in form of allowances and transport" (KII Senior Clinical Officer, Nyakashashara and Sanga sub-counties).

All the Key Informants mentioned that brucellosis vaccines in animals were very expensive and not readily available making it unaffordable by most of the pastoralists with the exception of a few who request for it.

"We carry out animal vaccination against brucellosis only at the request of the farmers. This service is private and not free, and only about five people request for it in a month. The challenges are failure to control brucellosis because the vaccine is expensive and usually unavailable. There are no vaccine programs for brucellosis yet there is a high population of susceptible animals in this sub county: about 500,000 cattle, 40,000 goats and 10,000 sheep plus wildlife" (KII veterinary officer, Sanga sub-county).

In regard to human brucellosis, the participants acknowledged that brucellosis services such as health education and treatment are more readily available as compared to information and vaccination against the disease in animals. For instance, five FGDS were not aware of the symptoms, transmission and availability of animal vaccination against brucellosis. However, all the participants mentioned unavailability of both animal and human diagnostics tests at the community level.

DISCUSSION

Community and provider perceptions on the integration of health education and animal vaccination against brucellosis by both public health workers and veterinarians were explored in our study. These focused on integration of animal vaccination and health education on brucellosis (the activity level), integration of the two services by veterinary and public health workers (the actor level), as well as how to conduct the integration process, and challenges likely to be faced. It was noted that integration of animal vaccination and health education would be feasible if government and other stakeholders were involved in facilitation of the programs. Key areas of improvement mentioned included; restricting wildlife movements in the communities, facilitating the necessary activities such as sensitization of the communities about brucellosis, provision of brucellosis vaccines and diagnostic tests as well as collaborations between all stakeholders to improve human resources.

All the Focus Group participants were not aware of vaccination against brucellosis, neither had they administered, nor done it for their animals. Five out six of the groups in our study believed that lack of knowledge about brucellosis was a hindering factor to integration; as a result, the community needs to first understand the importance of the integration before it is implemented. An earlier study done in the same area [39] reported moderate and low community knowledge on human and animal brucellosis respectively. Elsewhere, a study in Saudi Arabia found that awareness of the disease and its routes of transmission among livestock keepers may, on occasions, be poor [40]. It is important to note that good knowledge of the disease is essential to the success of a control program [23] as well as mass vaccination of livestock [2]. Ward et al, [24] recommended health education, especially on zoonoses as one branch of collaboration between veterinary and public health services. Therefore, efforts should be made to equip the veterinary and public health workers and the general community with adequate knowledge of brucellosis on; symptoms, improved access to treatment services, drugs, preventive methods both in animals and humans in order for integrated programming to fit the needs of clients, including individuals and families, as well as the broader community.

All the participants as well as the key informants believed that integration of animal vaccination would be effective if free movement of wildlife in the community was restricted to reduce the spread of brucellosis. This was attributed to close proximity and increased interaction with livestock. Increased interaction at the wildlife-human—livestock interface could be because of prolonged droughts that force wild animals to move to the communities as well as pastoralists moving their animals to the park in search of water and pasture. A study done by Lake Mburo National Park Authority in 2009 reported that farmers, mostly from Sanga, Kanyaryeru, and Nyakashashara sub-counties, moved over 15,000 cattle to the park that competed with wild animals for food and water and increased the risk for the transmission of animal diseases [41]. Although, wildlife brucellosis represents a potential zoonotic threat [9] which may cause spillover of infections to cattle in areas around national parks, no vaccine has proven to be safe and to provide significant degree of protection in wildlife species [22]. Therefore, there is need for community and park authorities to work together to allow temporary and safe access to Lake Mburo National Park during dry seasons, because it is not just wildlife transgressing boundaries but also the humans pushing into the wildlife space.

Despite the significant role of vaccination in the prevention of brucellosis, there is a need to combine vaccination with other measures such as health education on the nature and mode of transmission of the disease. This could be done through joint collaborations on policies, processes, and infrastructure which may be needed to deliver integrated services. Integration at this level, requires existing or new health system inputs (such as resources, time, money, or expertise) to be allocated differently to support planning, management, staffing, interpersonal communication, or the measurement of integrated services. And within the broader health system, new levels of coordination or joint planning of the policies, processes, and infrastructure that make up a health system may be needed in the delivery of the integrated services [18]. This often requires significant involvement and support from all stakeholders, including donors, ministries of health, politicians, advocacy groups, the private sector, and non-governmental organizations to effectively integrate these services and manage brucellosis in the communities.

Respondents in our study believed that veterinary and health personnel could work together but they emphasized the need to also train local people especially in disseminating information about brucellosis. Training of local community members such as village health teams to assist in mobilization and dissemination of information was highlighted as important for community participation and sustainability of the integration process. Elsewhere [2], it has been noted that community participation and joint collaboration between public health workers and veterinary officers is crucial for the acceptability and sustainability of effective control of brucellosis.

Respondents and participants believed that better general education and information on brucellosis and animal vaccination would help in the integration process and this could best be disseminated using radio broadcasting, in church announcement, parties and training village teams. A study done in nomadic pastoralists communities in Chad [42] noted that communities preferred that information on future vaccination campaigns be directed to their representatives and the complementary mode of information transmission mentioned was radio broadcasting. Therefore, animal vaccinations should be accompanied by continuous health education on the disease especially to those directly involved in the animal and food processing sector in order to guarantee sustainability of the campaigns [43]. This could be done through the media, advertisement and social gathering.

Integration of animal vaccination and health education by the veterinary and public health workers is believed to be important in the management of brucellosis in communities if it is jointly provided. However, the key informants and group participants believed that adequate funding and joint planning by the two professions in terms of time and resources would be key to the success of the effort. A study done in Greece [44] among agricultural and pastoral populations found that vaccination of animals against brucellosis combined with health education led to a reduction in the incidence of human brucellosis from 13.2 per 1,000 population in 1997 to 0.7 per 1,000 population in 2002, a drop of 12.5% in five years.

Joint programming of animal vaccination and health education is believed to be important for successful integration in communities. For example, three respondents mentioned that outreach programs such as child immunization days, antenatal care and HIV counseling and testing would be useful to capture more people especially women because they are the ones mainly involved in handling animal products at household level. A study done in Southern Sudan [24] found that increased coverage for childhood vaccination under the Expanded Programs for Immunization was achieved when livestock owners brought their cattle for vaccination against rinderpest at the same time and in the same location. Additionally, other authors [45] have proposed joint veterinary and health services in order to reach pastoralists in remote zones, so as to reduce costs and increase acceptance. Therefore, there is need to encourage joint service provision for both animals and humans by both veterinary and public health workers for successful provision of services in pastoral communities of Uganda.

Major challenges to the integration process, according to the respondents and group participants in our study, would be the lack of diagnostic facilities and tests for both humans and animals and inadequate veterinary staffs. They reported having only one government veterinary officer at each sub-county yet the animal population is large. All the key informants highlighted the high costs of brucellosis vaccine and its unavailability as a challenge in animal vaccination. In most developing countries, control of brucellosis in cattle is not straight forward as mass vaccination and stumping out programs are not feasible or sustainable in terms of cost [46]. Significant joint efforts, both in terms of human and financial resources as well as time, therefore needs to be allocated differently to support planning, management, staffing, interpersonal communication, or the measurement of integrated services in the management of brucellosis. This should be accompanied by an increase in the number of human resource especially veterinary officers as well as providing affordable, simple and rapid screening tests for brucellosis that can be used outside the established laboratory.

In our study, we modified the "Approach to Integrated Health Services" adapted from PATH 2011 [18] to discuss our findings. The strength of this framework is that it looks at integration of services in a wider perspective to include all sectors as well as client needs. However, it fails to describe challenges likely to be encountered by different sectors in integration of services. Awareness of likely challenges is an important issue for successful integration. In our study, the community members were able to articulate likely challenges as seen in the discussion section above. There were no major differences in opinion between the farmers and the local leaders group, as they were able to agree on a common approach.

Methodological Considerations

The strength of our study is that it was conducted among pastoralists who are exposed to zoonotic diseases such as brucellosis, and their experiences contributed to their perceptions on the integration of animal vaccination and health education in the management of the disease. The use of triangulation of methods (FGDs, KIIs and a previous knowledge study in the same area [39]) brought about a more accurate assessment of the phenomena studied by testing the consistency in the diversity of perspectives on integration of services both at the activity and actors level as recommended elsewhere [35]. Possible limitations were; we did not specifically conduct KII with policy makers to enable us capture the logistical barriers to integration of activities across sectors from the policy perspective and there were also few key informants especially the veterinarians to give a wider overview of the communities on the integration process. This was due to few veterinary officers in the area (one government veterinary extension worker per sub-county) and unavailability of health providers during the study period.

For further studies, we recommend that future epidemiological surveys in this setting should the assess role of wildlife in the epidemiology of brucellosis and other zoonotic diseases.

CONCLUSION

The respondents reported limited knowledge of brucellosis and its vaccination in animals. The community members believed that mass animal vaccination in combination with health education about the disease is important and possible if it involves government and all other stakeholders such as wildlife authorities, community members, local to national political leaders, as well as the technical personnel from veterinary, medical and public health sectors since it affects both humans and animals.

ACKNOWLEDGMENTS

The authors are grateful for the financial support from International Development Research Centre (Canada), all the interview respondents who participated in the study, the community leaders and to all the research assistants for their contribution to the success of this research.

AUTHOR CONTRIBUTIONS

Conceived and designed the experiments: CK AM SM IBR BBA ER. Performed the experiments: CK SM ER. Analyzed the data: CK VG LMA. Contributed reagents/materials/analysis tools: ER AM SM BBA. Wrote the paper: CK AM VG IBR SM BBA ER. Read and approved the final version of the manuscript: CK LMA BBA AM SM VG IBR ER.

REFERENCES

1. World Health Organisation (1997) World Health Organisation. Fact Sheet: N173.
2. Smits HL, Cutler SJ (2004) Contributions of biotechnology to the control and prevention of brucelliosis in Africa. African Journal of Biotechnology 12: 631–636.
3. Faye B, Vincent C, Matthieu L, David R, James D (2005) Tuberculosis and brucellosis prevalence survey on dairy cattle in Mbarara milk basin (Uganda). Prev Vet Med 67: 267–281. pmid:15748756
4. Magona JW, Walubengo J, Galiwango T, Etoori A (2009) Seroprevalence and potential risk of bovine brucellosis in zerograzing and pastoral dairy systems in Uganda. Kampala: Makerere. pp. 1765–1771.
5. Mugabi R (2012) Brucellosis Epidemiology, virulence factors, control and molecular targets to prevent bacterial infectious diseases. Makerere University, Kampala. Department Veterinary and Microbiological Sciences. (Master's degree thesis.)
6. Makita K, Fevre EM, Waiswa C, Kaboyo W, De Clare Bronsvoort BM, et al. (2008) Human brucellosis in urban and peri-urban areas of Kampala, Uganda. Ann N Y Acad Sci 1149: 309–311. pmid:19120236
7. Makita K, Fevre EM, Waiswa C, Kaboyo W, Eisler MC, et al. (2011) Spatial epidemiology of hospital-diagnosed brucellosis in Kampala, Uganda. Int J Health Geogr 10: 52. pmid:21962176
8. Makita K, Fèvre EM, Waiswa C, Eisler MC, Welburn SC (2010) How Human Brucellosis Incidence in Urban Kampala Can Be Reduced Most Efficiently? A Stochastic Risk Assessment of Informally-Marketed Milk. PLOS ONE 5(12): e14188. pmid:21152072
9. Rhyan JC (2000) Brucellosis in terrestrial wildlife and marine mammals. In: Brown K, Bolin C, editors. Emerging diseases of animals. Washington, DC: ASM Press. pp. 161–184.
10. Kalema-Zikusoka G, Bengis RG, Michel AL, Woodford MH (2005) A preliminary investigation of tuberculosis and other diseases in African buffalo (Synceruscaffer) in Queen Elizabeth National Park, Uganda. Onderstepoort J Vet Res 72: 145–151. pmid:16137132
11. Gul ST, Khan A (2007) The Epidemiology and Epizootiology of Brucellosis: A Review Pakistan. Department of Veterinary Pathology, University of Agriculture, Faisalabad, Pakistan. Vet J 27: 145–151.

12. Scott R (1992) Organizations: Rational, natural and open systems. 3. Englewood Cliffs, NJ: Prentice Hall. p. 199.
13. Pfeffer J (1982) Organizations and organizational theory. Boston, MA: Pitman.
14. Lawrence P, Lorsch J (1967) Organization and environment: managing differentiation and integration. Boston, MA: Division of Research, Harvard Business School.
15. Kodner D, Spreeuwenberg C (2002) Beyond care management: the logic and promise of vertically integrated systems of care for the frail elderly. In: Knapp M, et al. , editors. Towards equitable and efficient social care: Festschrift for Bleddyn Davies. London: PSSRU, London School of Economics and Political Science.
16. Garcia-Barbero M (2001) Integrated care: a position paper of the WHO European Office for Integrated Health Care Services. International Journal of Integrated Care. Available: http://www. ijic. org/.
17. Brown M, McCool B (1992) Vertical integration: exploration of a popular strategic concept. In: Brown M, editor. Health care management: Strategy, structure and process. Gaithersburg, MD: Aspen Publishers.
18. Program for Appropriate Technology in Health (PATH) (2011) Harvard Global health Review. Integration of Health Services: Theory and Practices. Available: http://www. hcs. harvard. edu/hghr/print/spring-2011/health-services-integration/.
19. Thakur SD, Kumar R, Thapliyal DC (2002) Human brucellosis:review of an under-diagnosed animal transmitted diseases. J Commun Dis 34: 287–301. pmid:14710861
20. Ibironke AA, McCrindle CM (2008) Evaluation of problems and possible solutions linked to the surveillance and control of bovine brucellosis in sub-Saharan Africa, with special emphasis on Nigeria. Vet Ital 44: 549–556. pmid:20405451
21. Jelastopulu E, Bikas C, Petropoulos C, Leotsinidis M (2008) Incidence of human brucellosis in a rural area in Western Greece after the implementation of a vaccination programme against animal brucellosis. BMC Public Health 8: 241. pmid:18637172
22. Godfroid J, Käsbohrer A (2002) Brucellosis in the European Union and Norway at the turn of the twenty-first century. Vet Microbiol 90: 135–145. pmid:12414139

23. Reviriego FJ, Moreno MA, Dominguez L (2000) Risk factors for brucellosis seroprevalence of sheep and goat flocks in Spain. Prev Vet Med 44: 167–173. pmid:10760400

24. Ward DE, Ruppanner R, Marchot PJ, Hansen JW (1993) One medicine—practical application for non-sedentary pastoral populations. Nomadic Peoples 32: 55–63.

25. Godfroid J, Cloeckaert A, Liautard JP, Kohler S, Fretin D, et al. (2005) From the discovery of the Malta fever's agent to the discovery of a marine mammal reservoir, brucellosis has continuously been a re-emerging zoonosis. Vet Res 36: 313–326. pmid:15845228

26. Zinsstag J, Schelling E, Wyss K, Mahamat MB (2005) Potential of cooperation between human and animal health to strengthen health systems. Lancet 366: 2142–2145. pmid:16360795

27. Roth F, Zinsstag J, Orkhon D, Chimid-Ochir G, Hutton G, et al. (2003) Human health benefits from livestock vaccination for brucellosis: case study. Bulletin of the World Health Organization 81: 867–876. pmid:14997239

28. Dawson S, Manderson L, Tallo VL (1992) A Manual for the Use of Focus Groups. Methods for Social Research in Disease. Boston, MA: International Nutrition Foundation for Developing Countries. Available: http://whqlibdoc. who. int/publications/1993/0963552228. pdf.

29. Khan ME, Manderson L (1992) Focus groups in rapid assessment procedures for tropical disease research. Health Policy and Planning 7, 1: 56–66.

30. Lakshman M, Charles M, Biswas M, Sinha L, Arora NK (2000) Focus group discussions in medical research. Indian J Pediatr 67: 358–362. pmid:10885209

31. Khan ME (1991) The use of focus group discussions in social and behavioural research: some methodological issues. World Health Stat Q 44: 145–149. pmid:1949882

32. Goering PN, Streiner DL (1996) Reconcilable differences: The marriage of qualitative and quantitative methods. Canadian Journal of Psychiatry 41: 491–497.

33. Newman WL (2003) Social Research Methods: Qualitative and Quantitative. Approaches, Fifth Edition. Boston, MA: Pearson Education Inc. 584 p.

34. Marshall C, Rossman GB (2006) Designing Qualitative Research. Thousand Oaks, CA: Sage Publications.
35. Rice PL, Ezzy D (1999) Qualitative Research Methods: A Health Focus. New York: Oxford University Press.
36. Holstein JA, Gubrium JF (1995) The active interview. Thousand Oaks, CA: Sage Publications.
37. Downe-Wamboldt B (1992) Content analysis: method, applications, and issues. Health Care Women Int 13: 313–312. pmid:1399871
38. Kondracki NL, Wellman NS, Amundson DR (2002) Content analysis: review of methods and their applications in nutrition education. J Nutr Educ Behav 34: 224–230. pmid:12217266
39. Kansiime C, Mugisha A, Makumbi F, Mugisha S, Sempa J, et al. (2014) Knowledge and perceptions of brucellosis in the pastoral communities adjacent to Lake Mburo National Park, Uganda. BMC Public Health 14: 242. pmid:24612845
40. Kang'ethe EK, Ekuttan CE, Kimani VN, Kiragu MW (2007) Investigations into the prevalence of bovine brucellosis and the risk factors that predispose humans to infection among urban dairy and non-dairy farming households in Dagoretti Division, Nairobi, Kenya. East Afr Med J 84: 96–100.
41. Choudhary V, van Engelen A, Sebadduka S, Valdivia P (2011) Uganda Dairy Supply Chain Risk Assessment. Washington, DC: World Bank. Available: http://www-wds. worldbank. org/external/default/WDSContentServer/WDSP/IB/2013/05/10/000333037_20130510095815/Rendered/PDF/775940WP0Ugand0Box0342041B00PUBLIC0.pdf.
42. Schelling E, Krönke F, Bechir M, Wiese M, Daoud S, et al. (2002) Nomadic pastoralists's perception on future joint human and animal vaccination campaigns. Available: http://edoc. unibas. ch/57/1/DissB_6478. pdf.
43. Nichter M (1995) Vaccinations in the third world: a consideration of community demand. Social Science and Medicine 41: 617–632. pmid:7502096
44. Jelastopulu E, Christos B, Chrysanthos P, Leotsinidis M (2008) Incidence of human brucellosis in a rural area in Western Greece after the implementation of a vaccination programme against animal brucellosis. BMC Public Health 8: 241. pmid:18637172

45. Meslin FX (1996) Global changes: perspectives on future roles of veterinarians in public health. Proceedings of the VIII International Conference of Institutions of Tropical Veterinary Medicine, Berlin, Germany. p. 101.
46. Makita K, Fèvre EM, Waiswa C, Eisler MC, Welburn SC (2010) How Human Brucellosis Incidence in Urban Kampala Can Be Reduced Most Efficiently? A Stochastic Risk Assessment of Informally-Marketed Milk. PLOS ONE 5(12): e14188. pmid:21152072

CHAPTER 3

Public Health Responses to Reemergence of Animal Rabies, Taiwan, July 16–December 28, 2013

Angela Song-En Huang[1] , Wan-Chin Chen[1] , Wan-Ting Huang[1] , Shih-Tse Huang[1] , YiChun Lo[1] , Sung-Hsi Wei[1] , Hung-Wei Kuo[2] , Pei-Chun Chan[1] , Min-Nan Hung[1] , Yu-Lun Liu[1] , Jung-Jung Mu[3] , Jyh-Yuan Yang[3] , Ding-Ping Liu[2] , Jih-Haw Chou[4] , Jen-Hsiang Chuang[4], Feng-Yee Chang[5]

[1] Office of Preventive Medicine, Centers for Disease Control, Taipei, Taiwan

[2] Epidemic Intelligence Center, Centers for Disease Control, Taipei, Taiwan

[3] Center for Research and Diagnostics, Centers for Disease Control, Taipei, Taiwan

[4] Office of Deputy Director, Centers for Disease Control, Taipei, Taiwan

[5] National Defense Medical Center, Tri-Service General Hospital, Taipei, Taiwan

Citation: Huang AS-E, Chen W-C, Huang W-T, Huang S-T, Lo Y-C, Wei S-H, et al. (2015) Public Health Responses to Reemergence of Animal Rabies, Taiwan, July 16–December 28, 2013. PLoS ONE 10(7): e0132160. https://doi.org/10.1371/journal.pone.0132160.

ABSTRACT

Taiwan had been free of indigenous human and animal rabies case since canine rabies was eliminated in 1961. In July 2013, rabies was confirmed among three wild ferret-badgers, prompting public health response to prevent human rabies cases. This descriptive study reports the immediate response to the reemergence of rabies in Taiwan. Response included enhanced surveillance for human rabies cases by testing stored cerebrospinal fluids (CSF) from patients with encephalitides of unknown cause by RT-PCR, prioritizing vaccine use for postexposure prophylaxis (PEP) during periods of vaccine shortage and subsequent expansion of PEP, surveillance of animal bites using information obtained from vaccine application, roll out of preexposure prophylaxis (PrEP) with vaccine stock restoration, surveillance for adverse events following immunization (AEFI), and ensuring surge capacity to respond to general public inquiries by phone and training for healthcare professionals. Enhanced surveillance for human rabies found no cases after testing 205 stored CSF specimens collected during January 2010–July 2013. During July 16 to December 28, 2013, we received 8,241 rabies PEP application; 6,634 (80.5%) were consistent with recommendations. Among the 6,501persons who received at least one dose of rabies vaccine postexposure, 4,953 (76.2%) persons who were bitten by dogs; only 59 (0.9%) persons were bitten by ferret-badgers. During the study period, 6,247 persons received preexposure prophylaxis. There were 23 reports of AEFI; but no anaphylaxis, Guillain-Barré syndrome, or acute disseminated encephalomyelitis were found. During the study period, there were 40,312 calls to the Taiwan Centers for Disease Control hotline, of which, 8,692 (22%) were related to rabies. Recent identification of rabies among ferret-badgers in a previously rabies-free country prompted rapid response. To date, no human rabies has been identified. Continued multifaceted surveillance and interministerial collaboration are crucial to achieve the goal of rabies-free status in Taiwan.

INTRODUCTION

Taiwan had been free of indigenous human and animal rabies case since canine rabies was eliminated in 1961 [1, 2], despite the lower than recommended 70% rabies vaccine coverage among dogs and cats. The successful maintenance of rabies-free status was attributed to Taiwan's relative geographic isolation as an island and strict animal importation control [2].

However, human rabies remains nationally notifiable in Taiwan. Since the last locally-acquired human rabies case was confirmed in 1959, three imported cases of human rabies have been identified, including a female traveler from China (July 2002), a businessman returning from China (July 2012), and a male worker from the Philippines (May 2013), all of which occurred after dog bites outside Taiwan without receiving rabies postexposure prophylaxis (PEP) [3–5].

The Taiwan Council of Agriculture (COA) conducts surveillance of animal rabies by histopathologic examination of animal brain tissues using direct fluorescent antibody (DFA) testing and immunohistochemistry (IHC) [6–9]. No rabies virus were found among 6,841 dog specimens tested during 1999–2012 and 322 bat specimens tested during 2008–2012 [10].

In 2012, COA contracted two universities to conduct surveillance of emerging infectious diseases among wild animals; rabies real-time reverse transcription-polymerase chain reaction (RT-PCR) testing was added in 2013. Three dead ferret-badgers with unexplained encephalitides found in central Taiwan during May to December 2012 underwent a series of diagnostic tests, eventually tested positive for rabies virus by RT-PCR on June 24, 2013. The specimens were submitted to COA on June 26 for confirmatory testing. On July 16, COA confirmed rabies in these three ferret-badgers based on positive RT-PCR, DFA, and IHC results [1]. Of the 827 dead or behaviorally abnormal ferret-badgers collected during July 16 to December 27, 2013, rabies was confirmed by DFA in 273 animals found in central, southern and eastern Taiwan (Fig 1) [10]. In addition, COA confirmed rabies in a house shrew on July 30 and in a dog bitten by a rabid ferret-badger on September 10 [11]. Furthermore, three laboratory-confirmed rabies cases were found after testing 11 stored ferret-badger brain specimens collected during July 2010 to August 2012 [12].

In response to the reemergence of animal rabies, the Ministry of Health and Welfare and COA established an interministerial rabies working group on July 24 for integrated public health and agricultural response [13]. The Central Epidemic Command Center, chaired by the Vice Premier, was established on August 1 to coordinate interministerial efforts against rabies [14]. This report summarizes surveillance activities and immediate public health response to rabies reemergence in Taiwan.

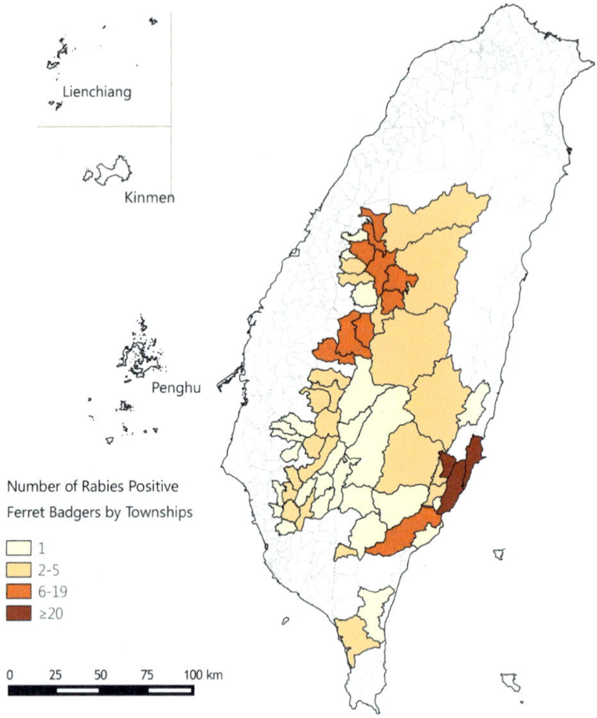

Figure 1: Distribution of rabies positive ferret-badgers, July 16–December 27, 2013 (n = 273).

Rabies positive ferret-badgers found in colored areas.

https://doi.org/10.1371/journal.pone.0132160.g001

METHODS

After the first rabid ferret-badger was identified, initial response focused on providing adequate supply of rabies vaccine and rabies immunoglobulin (RIG) for PEP [15]. Surveillance of rabid animals was established to respond to this reemerging rabies by COA [13]. Pathology and RT-PCR was performed to screen rabies virus among dead mammals and mammals with aggressive behavior, collected by forestry workers, wildlife rescue centers and local animal health agencies. Positive results were confirmed using DFA. Priorities for testing were given to animals with aggressive behavior and those which had bitten humans. Rabies test results were provided to Taiwan Centers for Disease Control (TCDC) daily so public health practitioners could follow-up on bite victims, ensuring the completeness of PEP [14]. Control measures implemented include expanded animal rabies surveillance, animal bite surveillance, revision of human rabies vaccine and RIG recommendations, expanded human rabies PEP service points, health education for the public and mammal bite care training for healthcare professionals.

Enhanced Surveillance of Human Rabies

Human rabies has been a notifiable disease in Taiwan since 1952. Rabies surveillance mandates physicians to notify clinically suspected cases to the health authority within 24 hours of diagnosis and submit clinical specimens including saliva, serum and cerebrospinal fluid (CSF) to TCDC laboratories. Rabies is confirmed if any of the specimens tested positive for rabies virus by diagnostic RT-PCR targeting both L and N genes of Rabies virus [16]. PCR products were sequenced in both directions to characterize the rabies amplicon. In addition, specific anti-rabies antibody testing by enzyme-linked immunosorbent assay [17] was performed in unvaccinated individuals. In addition, TCDC has conducted sentinel surveillance of human encephalitis of undetermined etiology since 2010. Physicians from 46 sentinel hospitals were requested to submit serum, throat swabs and CSF from patients with either encephalopathy or ataxia of unknown etiologies, plus any one of the following: fever, seizure, focal neurological signs, abnormal CSF profile, abnormal electroencephalography (EEG) and brain images. Submitted specimens were tested for viruses using a multiplex PCR panel containing probes targeting viruses causing encephalitis. Retrospective testing to detect rabies virus was conducted on stored CSF taken during January 1, 2010 to July 15, 2013, by rabies RT-PCR. The multiplex real-time PCR detected the following pathogens using primers designed based on previously published sequences: Herpes simplex virus 1 and 2 [18], VZV [18], CMV [18], Human Herpesvirus 6 and 7 [18], Hendra virus [19], Nipah virus [20], Dengue virus [21], chikungunya virus [22], Japanese encephalitis virus [23], West Nile virus [23], adenovirus [24], bocavirus [24], coronavirus (229E, HKU1, OC43 and NL63) [25], enterovirus [24], human metapneumovirus [26], human parechovirus [27], influenza virus (types A [28] and B[29]), parainfluenza virus (types 1–3, 4A and 4B) [24] [25], parvovirus B19 [30], polyomavirus (JC and BK) [31], respiratory syncytial virus (RSV) [29] and rhinovirus [24].

Surveillance of Mammal Bites

Surveillance of mammal bites is conducted through a newly established PEP application system. Because of the initial shortage of RIG and rabies vaccines, distribution was tightly controlled by TCDC. Physicians needing to provide patients with RIG or rabies vaccines must send in their requests directly to TCDC by fax or email for review. Physicians must indicate biting animals, whether the animal was wild or owned, bite site on body of the bitten, bite category, according to WHO classification, circumstances of the incident, and the disposition of the animal that bit. Medical officers in TCDC would then review the application, and decide on whether to approve the application for PEP. Approval to provide treatment is given by phone within 30 minutes. After one month, as physicians have become familiar with rabies PEP recommendations, physicians were asked to provide rabies PEP immediately, and have their requests reviews retrospectively, to ensure that all patients needing rabies PEP were given appropriate PEP.

After the reemergence of animal rabies, the Advisory Committee on Immunization Practices (ACIP) in Taiwan established new recommendations of rabies PEP. During July 18–29, ACIP recommended PEP for individuals with category II or III exposure, as defined by the World Health Organization [32, 33], to wild mammals nationwide and stray dogs and cats in affected mountainous townships. PEP recommended include five doses of rabies vaccine given on days 0, 3, 7, 14, and 28, plus 20 IU/kg of human RIG given within 7 days of the first rabies vaccine dose at the wound site. On July 31, ACIP expanded PEP recommendations to individuals exposed to stray dogs and cats nationwide. RIG was initially prioritized for individuals with category III [33] exposure to ferret-badgers nationwide and to house shrews in Taitung. On September 6, with increased availability of RIG, individuals with category II exposure to ferret badgers in which rabies was confirmed by DFA were also recommended for RIG, to ensure that no human rabies occurred, even though this is contrary to WHO recommendation.

Stockpile and Distribution of Vaccines

Government-funded PEP based on ACIP recommendations has been available through TCDC-contracted hospitals to individuals bitten by mammals. Before the reemergence of rabies in Taiwan, rabies preexposure prophylaxis (PrEP) and PEP were provided through 12 designated travel clinics to travelers. During 2010 to 2012, approximately 250 doses of rabies vaccines and 10 doses of RIG (HyperRAB S/D, [Talecris Biotherapeutics]) were administered annually. To provide easy access for all persons exposed, rabies PEP service points were expanded to 28 hospitals by the end of July 2013, and further expanded to 60 hospitals by early August. International procurement of vaccines was expedited with the support of Taiwan Food and Drug Administration (TFDA). Because of possible human RIG shortage, 2000 vials of 5 ml equine RIG (RabAvert [Novartis]) was also imported in August. PrEP programs began on July 27, targeting high risk groups, which included dog catchers, game wardens, animal shelter workers and veterinarians conducting rabies surveillance. Characteristics of all rabies immunization recipients, including demographics of the recipients, animal types and classification of animal-bite wound, were registered in a web-based system.

Retrospective Identification of Ferret-badger Exposures

From August 1 to September 30, 2013, persons exposed to ferret-badgers during July 2010 to July 2013 were urged to contact TCDC. Patients were asked about their age; sex; date, place, body part and extent of exposure and injury; history of vaccination against rabies; and phone number. Information was used for individual risk assessment of potential rabies exposure and issuance of PEP. Because no patients had fresh wounds, categorizing exposure based on WHO classification were based only on history.

Monitoring Adverse Events following Administration of Rabies Biologics

The national passive vaccine safety surveillance system, collaboratively operated by TCDC and TFDA, has served as a tool for adverse event following immunization (AEFIs) collection and vaccine safety signal detection since the 2009 influenza pandemic [34]. All reports of adverse events following RIG or rabies vaccines received during July 16–December 28, 2013 were reviewed and AEFIs coded with Medical Dictionary for Regulatory Activities (MedDRA) terms [35]. A report was categorized as serious when the patient outcome involved death, life-threatening illness, hospitalization, prolongation of an existing hospitalization, permanent disability, or congenital anomaly [36]; all other reports were considered nonserious. TCDC provided advice and assistance on the management of potentially serious adverse events (SAEs) for subsequent rabies PrEP or PEP vaccine doses.

Health Education, Public Health Inquiries, and Healthcare Professional Training

Multiple press releases and conferences were conducted educating the public on the dangers of animal bites and the management following bites. For additional information, anyone can call "1922", a toll-free hotline set up by TCDC, using any landline or mobile phones. All inquiries were recorded automatically. We collected rabies-related inquiries from the record. The inquiries from July 16 to December 28, 2013 were included for analysis.

Healthcare professional also received training on animal bite wound management, evaluation of animal exposures for PEP using rabies vaccine and RIG, and evaluation of high risk groups for PrEP. Rapid development and roll out of training sessions were conducted by TCDC, providing information to healthcare workers both in hospitals and in the public health sector.

Ethics

Specimen submitted by sentinel hospitals were collected previously for encephalitis surveillance and re-tested for rabies virus. Information collected from persons requiring rabies postexposure prophylaxis was for public health use. All specimen and information collection were done by the pronouncement of the Central Epidemic Command Center, and, in accordance with regulations of Taiwan, therefore did not require institutional review board approval. Only oral consent was obtained for specimens collected for surveillance purposes. The data was de-identified prior to analysis.

RESULTS

Enhanced Surveillance of Human Rabies

During July 16–December 28, 2013, there were no human rabies cases reported. During January 2010–July 2013, TCDC was notified of 365 cases with encephalitis of undetermined etiologies. Of the 205 available stored CSF specimens, none tested positive for rabies virus.

Surveillance of Mammal Bites

From July 17 to December 28, 2013, TCDC received requests for rabies PEP from 8,241 individuals. The peak of requests occurred on the week of July 28, two days after rabies was confirmed in a house shrew (Fig 2), and the median number of PEP requests per week was 289 (range, 107–985). PEP was denied in 1,607 (20%). The most common reasons for denial included category II or III wounds inflicted by pet dogs or cats (n = 1,245) and exposure as category I (n = 42). Persons sustained bites requiring PEP concentrated in coastal urban areas. Persons exposed to ferret-badger bites were distributed all over the island, and none in the offshore islands (Fig 3).

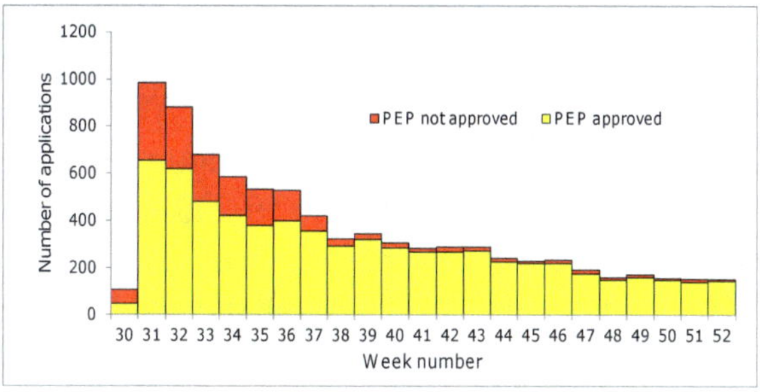

Figure 2: Rabies postexposure prophylaxis applications (n = 8,241) and public inquiries to Taiwan Centers for Disease Control hotline for information on rabies, July 14–December 28, 2013 (n = 8,692).

https://doi.org/10.1371/journal.pone.0132160.g002

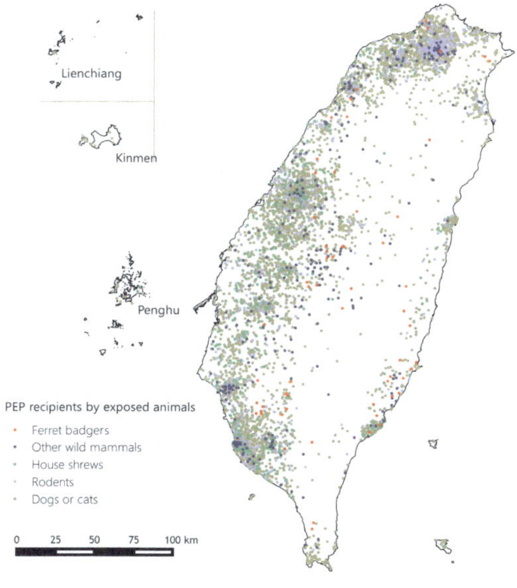

Figure 3: Geographic distribution of animal bites exposure location, July 16–December 28, 2013.

https://doi.org/10.1371/journal.pone.0132160.g003

Of the 6,634 requests approved for PEP, 6,501 (98%) received at least one dose of vaccination. The median age of the PEP recipients was 42 years (range, 0–94), and 3,642 (56%) were male. The earliest date of animal exposure was February 16, 2011, as determined by retrospective identification of ferret-badger exposures, and the median interval from exposure to PEP application was 0 days (range, 0–896). Among those who received PEP, the most commonly exposed animals were dogs or cats (n = 4,953, 76%), followed by rodents (n = 893, 14%), house shrews (n = 406, 6%) and other wild mammals (n = 223, 3.4%). Category II exposures consisted of 2,855 applicants (43.9%) and category III in 3,613 applicants (55.6%). Ferret-badgers accounted for 59/223 (26%) of the wildlife exposures; among them, 37 (63%) persons had category III exposure, and 32 received RIG. Eight persons, even though they had category III exposure and RIG requests consistent with ACIP recommendations, did not receive RIG following physician evaluation. Physicians who chose to not give RIG were not asked for their reasons for doing so.

By February 24, 2014, there were 6501 persons who received at least one dose of PEP vaccination. Among them, 5,692 persons received at least three doses; 4,861 persons completed the first three doses within 7 days of starting the first dose. There were 4,867 persons who completed 5 doses of PEP vaccination; 3,051 persons completed PEP within 28 days, the recommended schedule for PEP vaccination.

Of the 6,501 PEP recipients, the disposition of animals that bit humans were not available in 5,955 (92%) at the time of vaccine request; among these, 4,534 (76%) were dogs or cats. There were 399 (8%) dogs or cats reportedly healthy at initial evaluation. Rabies testing was conducted in 149 (2%) animals, including 45 dogs or cats, 17 ferret-badgers, and 42 house shrews. Among them, 12 ferret-badgers and 1 house shrew tested positive for rabies.

As of December 28, 2013, no patients who received PEP had developed symptoms suspected to be associated with rabies.

Stockpile and Distribution of Vaccine

With limited rabies vaccine and RIG available at the start of animal rabies reemergence, vaccines were PEP prioritized to persons who sustained bites from wild mammals or stray domestic animals. Vaccine uptake for PEP were low, therefore, vaccines were made available for PrEP. After vaccines were made available for PrEP, vaccine use increased. Vaccine use peaked during week 33, and weekly volume leveled after week 41 (Fig 4). During the study period, there were 6,247 persons who received PrEP, with 18,353 doses used.

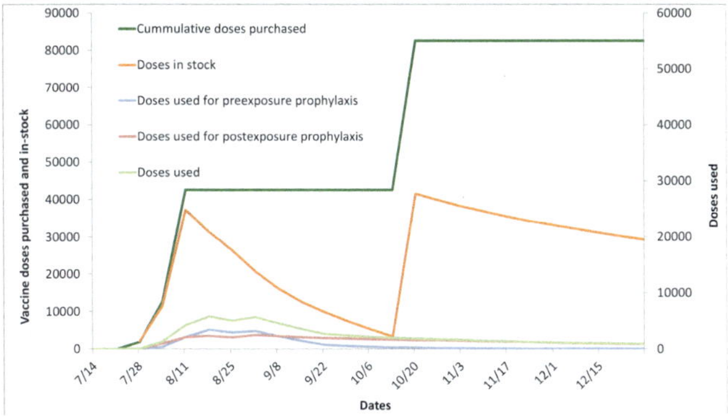

Figure 4: Rabies vaccine and immunoglobulin procurement and use, July 14–December 28, 2013.

https://doi.org/10.1371/journal.pone.0132160.g004

Retrospective Identification of Ferret-badger Exposures

TCDC received 13 reports of ferret-badger exposures that occurred during October 2012 to June 2013. The median age of the 13 patients was 53 years (range 31–71); 10 were male. All the ferret-badger bites were determined to be category III exposures; 10 received RIG and rabies vaccines and 3 received vaccines only.

AEFI Monitoring

As of December 28, 2013, there were 23 reports of AEFI after 48,612 rabies vaccine doses administered (reporting rate 4.7 per 10,000 doses). Of the 23 reports, 4 occurred following PrEP and 19 following PEP; none were classified as anaphylaxis, Guillain-Barré syndrome, or acute disseminated encephalomyelitis. The two SAE reports involved a child, aged 1 year and 11 months, who was hospitalized for acute bronchiolitis 2 days following receipt of the third rabies PEP vaccine dose; and an adult, aged 66 years, who had significant weight loss 6 days following the first rabies PEP vaccine dose and was diagnosed with pulmonary tuberculosis. Among the 21 nonserious reports, the most frequently reported adverse events were rash (n = 8), dizziness (n = 5), and pruritus (n = 4). Seven (30%) patients, including four PEP recipients who reported nonserious adverse events, did not complete the required vaccination series.

Health Education, Public Health Inquiries, and Healthcare Professional Training

As soon as the outbreak of rabies among ferret-badgers was confirmed, TCDC conducted a press conference and posted press release urging the public not to approach unknown animals, especially wild animals. During July 16 to December 28, there were 40,312 calls to the TCDC hotline, of which, 8,692 (22%) were related to rabies (Fig 2). The most common questions concerned PEP and PrEP (n = 5,155, 59%). Others included risk of animal exposure, clinical manifestations of rabies, transmission routes, and where to receive medical care for rabies. Daily inquiries about rabies increased to 115 after the first ferret-badger bit human on July 24, and the peak of calls occurred on July 29. After August 20, daily calls never reached above 100 (mean 21.0).

Healthcare professional training rolled out on July 26. Nationwide, during July 26 to September 1, 2013, there were 21 training sessions conducted, providing healthcare professional updated information on the ferret-badger epidemic, animal bite wound evaluation and management, PEP, and PrEP. These training sessions were hosted by TCDC along with local health departments and medical societies. Over 5,100 persons attended these training sessions, including physicians and nurses in hospitals, central and local health departments, and schools. The decrease in the numbers of call inquiries and PEP not being approved by September may be attributed to intensive public education and training.

DISCUSSION

As of December 28, 2013, no human rabies cases have been identified in Taiwan through enhanced surveillance despite the continued rabies epizootic among ferret-badgers. However, the newly established mammal bite surveillance demonstrated that bites from stray dogs or cats and wildlife including ferret-badgers, are common and pose increased risk for rabies if PEP is not promptly administered. Swift public health response to rabies reemergence in Taiwan has ensured public awareness of the importance of and access to PEP to prevent occurrence of locally-acquired human rabies cases.

Immediate public health challenges after the first wildlife rabies was confirmed included shortage of human rabies vaccines and RIG as PEP. Prior to confirming Taiwan as a rabies endemic country, few vaccines and RIGs were required countrywide. Therefore, all vaccines and RIGs were procured by TCDC, and then distributed to hospitals as needed. Having the experience of centralized procurement and supply of rabies PEP meant that the emergent procurement process could be pushed forward without having to seek new pharmaceutical firm collaboration, and that TCDC and hospitals already have set standard operating procedure for the distribution of these supplies. Through rapid mobilization of government fund and prompt procurement, TCDC guaranteed the availability of PEP to all who requires vaccines and RIG following animal bites. However, rabies prevention through PEP is costly. A course of vaccination and RIG costs more than USD800, and the US Centers for Disease Control and Prevention estimates that it costs USD10,000 to 100 million for each life saved [37]. Intradermal vaccination protocol has been endorsed by WHO to reduce vaccine use. This protocol has not been endorsed by the national health authorities because intradermal administration of vaccines requires additional training of healthcare personnel, and is not available in all vaccine administration sites. In addition, the vaccine, once opened, must be used up within a few hours, and Taiwan does not have the volume of patients that make intradermal administration cost-saving. Therefore, intradermal PEP protocols were not adopted in Taiwan. The government-funded PEP program has become a financial burden on public funding and sustainability should be taken into account in long-term preparedness plans. As of January 1, 2014, rabies PEP is being offered free of charge through Taiwan's National Health Insurance scheme, ensuring the sustainable medical care to bite patients, ending the role played by TCDC during the first six month of the animal rabies epidemic.

Because the last case of locally-acquired human and animal rabies cases occurred more than half a century ago, there exists a knowledge gap among the public and medical personnel in Taiwan. Having medical officers trained in multiple disciplines of medicine within TCDC significantly eased the tension during the height of public rabies panic. These medical officers updated the human rabies prevention manual using the latest information, assisted in providing accurate information for public health education, and developed healthcare professionals training materials.

Educating the public on animal-human interaction may reduce animal bites. Furthermore, "catch, vaccinate, sterilize and release" activities and encouraging responsible pet ownership may also decrease stray animals and opportunities for humans to come into contact with unfamiliar animals [38]. Rodent bites consist of 14% of animal bite injuries during our surveillance period. According to the World Health Organization, rodents are usually spillover animals for rabies and do not transmit the disease [15]. Furthermore, rodent bites were often provoked by men trying to rid of rodents from human habitat. Therefore, we must remind citizens to avoid injury from rodent bites when attempting to catch animals.

The peak of PEP requests came shortly after a house shrew was found to be infected by rabies and the revision of the ACIP recommendation for rabies PEP recommendation on the following day. The subsequent panic-driven vaccine-seeking behavior from the public was expected. Because physicians in Taiwan had not been educated about the use of rabies PEP, blanket ACIP recommendation to cover all rodent and shrew bites were unavoidable, even though this is contrary to WHO recommendation. In the ensuing months, physicians treating animal bites must be educated to evaluate bites more thoroughly, and provide vaccines to those who truly require vaccination and RIG. Increased postexposure vaccine-seeking behavior and increased physician delivery of vaccines have been shown to be associated with episodic importation of rabies by Lardon, et al [39]. Media alerts regarding the presence of rabies in the community is shown to be associated with enhanced sense of risk. Following the peak of vaccine-seeking behavior in the wake of the media attention on rabies in Taiwan, PEP use decreased. The reason for the decrease of PEP use in the last two months of our surveillance period is currently unknown. This may have been the result of increased physician awareness of the rabies transmission risk among different animals, decreased medical seeking behavior among patients with animal bites without media stimulation, decreased risky behavior leading to animal bites through public education, or a combination of these causes. High risk personnel receiving PrEP has been advised before rabies was identified in animal population, but uptake of vaccine was low, with approximately 250 doses administered annually. The low uptake is suspected to be because PrEP must be paid by individual patient, while PEP was provided free of charge. With animal rabies reemergence in Taiwan, there is increased perceived risk of illness, thus increased PrEP uptake.

Despite efforts to prevent rabies in humans, rabies control among animals is even more important. Mass canine vaccination and population control is paramount to minimizing the risk transmission of rabies to humans [40]. Following the identification of rabies among ferret-badgers, the Taiwan Council of Agriculture began expanded efforts to provide free rabies vaccination to all owned dogs and cats, especially in rural areas. Stray dogs and cats were also vaccinated and tagged to prevent rabies among animals with frequent human contact. In addition, pets were banned from entering national parks and forested areas where ferret-badgers are known to inhabit, to prevent rabies transmission from ferret-badger to pets. According to a press release from the Bureau of Animal and Plant Health Inspection and Quarantine, Taiwan Council of Agriculture, by March 2014, the 93% of the dogs and cats in areas with rabid ferret-badgers identified were vaccinated, and 79% in areas with ferret-badger activities were vaccinated, and 63% in all other areas were vaccinated [41]. However, on December 29, 2014, a sick civet found in southern Taiwan tested positive for rabies in the same lineage as the rabies virus group from ferret-badger in Taitung [42–44]. Because, to date, only one civet out of the 215 civets tested in 2013 has been found to have rabies [45], it is not yet known if this represent spillover disease from ferret-badger or the virus is now established in the civet population in Taiwan.

There are limitations to this study. Non-medically-attended mammal bites and bites which require no rabies PEP would not have been counted, resulting in underestimation. The true incidence of animal bites is likely to be much higher than currently reported. In addition, PCR testing of CSF is not sensitive for detecting rabies infection [46]. Current strategy in reporting patients with encephalitis of unknown etiologies may also not be sensitive in selecting patients with furious rabies. A more sensitive strategy may be needed to increase the sensitivity of identifying unrecognized human rabies. Because sentinel hospitals submitting specimens for encephalitis of unknown etiology are not geographically representative of Taiwan, human rabies cases may also not have been detected as a result. To ameliorate geographic misrepresentation, on October 28, 2013, TCDC has requested that all hospitals to submit CSF and saliva specimens for rabies testing, if patients have encephalitides of unidentified etiology requiring intensive care or resulting in death. Starting November 2013, rabies virus has been added to the multiplex PCR panel to identify encephalitides of unknown cause.

Because rabies was identified in the ferret-badger population in Taiwan, a previously rabies-free country, the establishment of animal bite surveillance, increased animal testing, emergent RIG and vaccine procurement, and establishment of public education were all important aspects of response. There are still much to be done for the control of animal rabies and prevent human rabies, including studies to characterize interactions between humans and ferret-badgers, establishing improved animal rabies reporting for human disease control, and control of rabies among ferret-badgers. Continued sharing of surveillance, epidemiologic and virologic data between public health and agricultural authorities may facilitate rabies control in both humans and animals. In the long run, continued multifaceted surveillance of animal bites and clinical encephalitides, encompassing both animal and human surveillance, and interministerial collaboration are crucial to achieve the goal of rabies-free status in Taiwan. Continued surveillance is required for countries with and without rabies, to ensure early identification of disease in both humans and animals.

ACKNOWLEDGMENTS

The authors would like to thank the coworkers at the Taiwan Centers for Disease Control and Council of Agriculture for providing the data for surveillance that culminated to this manuscript.

AUTHOR CONTRIBUTIONS

Conceived and designed the experiments: ASH WCC YCL PCC JJM JYY J.H. Chuang J.H. Chou FYC. Performed the experiments: ASH WCC WTH STH YCL JJM JYY HWK SHW MNH. Analyzed the data: ASH WCC WTH STH YLL JJM HWK SHW. Contributed reagents/materials/analysis tools: WCC WTH JJM HWK. Wrote the paper: ASH WCC WTH STH YCL SHW HWK PCC MNH YLL JJM JYY DPL J.H. Chuang J.H. Chou FYC.

REFERENCES

1. World Organization for Animal Health. Rabies; Chinese Taipei. Weekly Disease Information; July 18 2013. Available from: http://www. oie. int/wahis_2/public/wahid. php/Reviewreport/Review?page_refer=MapFullEventReport&reportid=13775.
2. Weng HY, Wu PI, Yang PC, Tsai YL, Chang CC. A quantitative risk assessment model to evaluate effective border control measures for rabies prevention. Veterinary research. 2010;41(1):11. pmid:19822125; PubMed Central PMCID: PMC2775170.
3. Chou CY WH, Yang JY, Tung HP, Wang HF. The first case report of imported rabies in 2012. Taiwan Epidemiology Bulletin. 2013;29:30–1. Epub 2013/2/5.
4. Hsu YH, Cho LC, Wang LS, Chen LK, Lee JJ, Yang HH. Acute respiratory distress syndrome associated with rabies: a case report. The Kaohsiung journal of medical sciences. 2006;22(2):94–8. pmid:16568727.
5. Tuan YC HM, Chuang MF, Lin HC, Tsai YY, Chen MJ, You CY, Chang CC. Investigation of the first reported rabies case in 2013. Taiwan Epidemiology Bulletin. 2013;29:201.
6. KR T. Surveillance of bat rabies in Taiwan 2008. Available from: http:// m3. nvri. gov. tw/documents/huan0051-2. doc.
7. Lee SH CK, Tsai KR. Surveillance on rabies of dogs in Taiwan in 2003. Animal Health Research Institute Report. 2003;39:19–23.
8. Lee SH CK, Tsai KR, Chang JC, Hung CT. Surveillance and monitoring on transmissible spongiform encephalopathy and rabies in Taiwan in 2005. Animal Health Research Institute Report. 2006;41:11–24.
9. Lee SH TK, Chang KH, Chang JC, Hung CT, Ting LJ, Sung W. Surveillance of transmissible spongiform encephalopathy and rabies in 2004. Animal Health Research Institute Report [Internet]. 2004; 40:[51–63 pp.]. Available from: http://m5. nvri. gov. tw/Documents/ rep040006. pdf.
10. Taiwan Animal Health Research Institute. Surveillance of rabies on animal brain tissue, 1999–2013 2013. Available from: http://www. baphiq. gov. tw/newsview. php?typeid=1948&typeid2=&news_ id=7971
11. World Organization for Animal Health. Rabies; Chinese Taipei. Weekly Disease Information; August 1 2013. Available from: http://www. oie.

int/wahis_2/public/wahid. php/Reviewreport/Review?page_refer=MapFullEventReport&reportid=13835.

12. World Organization for Animal Health. Rabies; Chinese Taipei. Weekly Disease Information; September 6 2013. Available from: http://www. oie. int/wahis_2/public/wahid. php/Reviewreport/Review?page_refer=MapFullEventReport&reportid=14032.
13. Taiwan Centers for Disease Control. Ministry of Health and Welfare and Council of Agriculture establishes rabies prevention inter-ministerial working group on July 24, 2013 to coordinate rabies management and ensure health and wellbeing of public 2013 [updated 07/26/2013]. Available from: http://www. cdc. gov. tw/english/info. aspx?treeid=bc2d4e89b154059b&nowtreeid=ee0a2987cfba3222&tid=1A2A88F1001C56AE.
14. Taiwan Centers for Disease Control. In response to the ongoing rabies situation, Executive Yuan activates Central Epidemic Command Center for Rabies and convenes first meeting held by CECC for Rabies to coordinate cross-ministry efforts and ensure health of people in Taiwan 2013 [updated August 1, 2013]. 2013/8/1:[Available from: http://www. cdc. gov. tw/english/info. aspx?treeid=bc2d4e89b154059b&nowtreeid=ee0a2987cfba3222&tid=BBA29C5119882713.
15. World Health Organization. WHO Expert Consultation on Rabies: second report. WHO technical report series. 2013;(982).
16. Heaton PR, Johnstone P, McElhinney LM, Cowley R, O'Sullivan E, Whitby JE. Heminested PCR assay for detection of six genotypes of rabies and rabies-related viruses. J Clin Microbiol. 1997;35(11):2762–6. D—NLM: PMC230057 EDAT- 1997/11/14 MHDA- 1997/11/14 00:01 CRDT- 1997/11/14 00:00 PST—ppublish. 9350729; PubMed Central PMCID: PMC230057. pmid:9350729
17. Feyssaguet M, Dacheux L, Audry L, Compoint A, Morize JL, Blanchard I, et al. Multicenter comparative study of a new ELISA, PLATELIA RABIES II, for the detection and titration of anti-rabies glycoprotein antibodies and comparison with the rapid fluorescent focus inhibition test (RFFIT) on human samples from vaccinated and non-vaccinated people. Vaccine. 2007;25(12):2244–51. pmid:17224214.
18. Sugita S, Shimizu N, Watanabe K, Mizukami M, Morio T, Sugamoto Y, et al. Use of multiplex PCR and real-time PCR to detect human herpes virus genome in ocular fluids of patients with uveitis. The

British journal of ophthalmology. 2008;92(7):928–32. pmid:18408082; PubMed Central PMCID: PMC2564807.

19. Smith IL, Halpin K, Warrilow D, Smith GA. Development of a fluorogenic RT-PCR assay (TaqMan) for the detection of Hendra virus. J Virol Methods. 2001;98(1):33–40. pmid:11543882.
20. Guillaume V, Contamin H, Loth P, Georges-Courbot MC, Lefeuvre A, Marianneau P, et al. Nipah virus: vaccination and passive protection studies in a hamster model. Journal of virology. 2004;78(2):834–40. D—NLM: PMC368848 EDAT- 2003/12/25 05:00 MHDA- 2004/02/12 05:00 CRDT- 2003/12/25 05:00 PST—ppublish. 14694115; PubMed Central PMCID: PMC368848. pmid:14694115
21. Callahan JD, Wu SJ, Dion-Schultz A, Mangold BE, Peruski LF, Watts DM, et al. Development and evaluation of serotype- and group-specific fluorogenic reverse transcriptase PCR (TaqMan) assays for dengue virus. J Clin Microbiol. 2001;39(11):4119–24. pmid:11682539; PubMed Central PMCID: PMC88496.
22. Edwards CJ, Welch SR, Chamberlain J, Hewson R, Tolley H, Cane PA, et al. Molecular diagnosis and analysis of Chikungunya virus. Journal of clinical virology: the official publication of the Pan American Society for Clinical Virology. 2007;39(4):271–5. pmid:17627877.
23. Shirato K, Miyoshi H, Kariwa H, Takashima I. Detection of West Nile virus and Japanese encephalitis virus using real-time PCR with a probe common to both viruses. J Virol Methods. 2005;126(1–2):119–25. WOS:000229003000015. pmid:15847927
24. Bonzel L, Tenenbaum T, Schroten H, Schildgen O, Schweitzer-Krantz S, Adams O. Frequent detection of viral coinfection in children hospitalized with acute respiratory tract infection using a real-time polymerase chain reaction. Pediatr Infect Dis J. 2008;27(7):589–94. WOS:000257176600002. pmid:18520973
25. Chidlow G, Harnett G, Shellam G, Smith D. An Economical Tandem Multiplex Real-Time PCR Technique for the Detection of a Comprehensive Range of Respiratory Pathogens. Viruses. 2009;1(1):42–56. pmid:21994537
26. Maertzdorf J, Wang CK, Brown JB, Quinto JD, Chu M, de Graaf M, et al. Real-time reverse transcriptase PCR assay for detection of human metapneumoviruses from all known genetic lineages. J Clin Microbiol. 2004;42(3):981–6. WOS:000220376900006. pmid:15004041

27. Noordhoek GT, Weel JF, Poelstra E, Hooghiemstra M, Brandenburg AH. Clinical validation of a new real-time PCR assay for detection of enteroviruses and parechoviruses, and implications for diagnostic procedures. Journal of clinical virology: the official publication of the Pan American Society for Clinical Virology. 2008;41(2):75–80. pmid:18354820.
28. Ward CL, Dempsey MH, Ring CJ, Kempson RE, Zhang L, Gor D, et al. Design and performance testing of quantitative real time PCR assays for influenza A and B viral load measurement. Journal of clinical virology: the official publication of the Pan American Society for Clinical Virology. 2004;29(3):179–88. pmid:14962787.
29. Chen Y, Cui D, Zheng S, Yang S, Tong J, Yang D, et al. Simultaneous detection of influenza A, influenza B, and respiratory syncytial viruses and subtyping of influenza A H3N2 virus and H1N1 (2009) virus by multiplex real-time PCR. J Clin Microbiol. 2011;49(4):1653–6. pmid:21270233; PubMed Central PMCID: PMC3122825.
30. Knoll A, Louwen F, Kochanowski B, Plentz A, Stussel J, Beckenlehner K, et al. Parvovirus B19 infection in pregnancy: Quantitative viral DNA analysis using a kinetic fluorescence detection system (TaqMan PCR). J Med Virol. 2002;67(2):259–66. WOS:000175415900015. pmid:11992588
31. Funahashi Y, Iwata S, Ito Y, Kojima S, Yoshikawa T, Hattori R, et al. Multiplex real-time PCR assay for simultaneous quantification of BK polyomavirus, JC polyomavirus, and adenovirus DNA. J Clin Microbiol. 2010;48(3):825–30. pmid:20053854; PubMed Central PMCID: PMC2832414.
32. Taiwan Centers for Disease Control. Rabies postexposure prophylaxis recommendations 2013 [updated 2013/7/24; cited 2014 2014/1/10]. Available from: http://webcache. googleusercontent. com/search?q=cache:PJyGcizuSioJ:www. cdc. gov. tw/downloadfile. aspx%3Ffid%3D37E93B9D9F6F16FA+&cd=1&hl=en&ct=clnk&gl=tw.
33. World Health Organization. Rabies: Guide for post-exposure prophylaxis [2014/1/10]. Available from: http://www. who. int/rabies/human/postexp/en/.
34. Huang WT, Chen WW, Yang HW, Chen WC, Chao YN, Huang YW, et al. Design of a robust infrastructure to monitor the safety of the pandemic A(H1N1) 2009 vaccination program in Taiwan. Vaccine. 2010;28(44):7161–6. pmid:20804804.

35. Medical Dictionary for Regulatory Activities. 2013.
36. Ministry of Health and Welfare EY, Taiwan. Pharmacovigilance in Taiwan. Available from: http://medwatch. fda. gov. tw/adr-eng/index. htm.
37. U. S. Centers for Disease Control and Prevention. Cost of Rabies Prevention 2011. Available from: http://www. cdc. gov/rabies/location/ usa/cost. html.
38. World Health Organization. Vaccinate dogs to save human lives—World Rabies Day 2012 2012. Available from: http://www. who. int/ features/2012/world_rabies_day/en/.
39. Lardon Z, Watier L, Brunet A, Bernede C, Goudal M, Dacheux L, et al. Imported episodic rabies increases patient demand for and physician delivery of antirabies prophylaxis. PLoS neglected tropical diseases. 2010;4(6):e723. pmid:20582307; PubMed Central PMCID: PMC2889823.
40. Fooks AR, Banyard AC, Horton DL, Johnson N, McElhinney LM, Jackson AC. Current status of rabies and prospects for elimination. Lancet. 2014;384(9951):1389–99. pmid:24828901.
41. Quarantine BoAaPHIa. Get dogs and cats vaccinated for rabies [Press release]. 2014 [updated 03/04/2014]. Available from: http://www. baphiq. gov. tw/newsview. php?typeid=1939&typeid2=&news_ id=9010.
42. Quarantine BoAaPHIa. Test confirmes rabies in civet found by resident [Press release]. 2014 [updated 12/29/2014]. Available from: http:// www. baphiq. gov. tw/newsview. php?menu=1054&typeid=1056&N ewsMainSeq=7367.
43. Quarantine BoAaPHIa. Rabies virus from civet found in Kenting is of the same genetic lineage as rabies virus from ferret-badgers in Taitung [Press release]. 2014. Available from: http://www. baphiq. gov. tw/ newsview. php?menu=1054&typeid=1056&NewsMainSeq=7374.
44. Chiou H-Y, Hsieh C-H, Jeng C-R, Chan F-T, Wang H-Y, Pang VF. Molecular Characterization of Cryptically Circulating Rabies Virus from Ferret Badgers, Taiwan. Emerging Infectious Diseases. 2014;20(5):790–8. PMC4012806.
45. Quarantine BoAaPHIa. Animal rabies surveillance results 12/31/2014 2014 [updated 12/31/2014]. Available from: http://www. baphiq. gov. tw/userfiles/%E6%AF%8F%E6%97%A5%E6%9B%B4%E6%96%B

0PPT1031231-%E6%94%BE%E7%B6%B2%E7%AB%99%E7%94%A8. pdf.

46. Crepin P, Audry L, Rotivel Y, Gacoin A, Caroff C, Bourhy H. Intravitam diagnosis of human rabies by PCR using saliva and cerebrospinal fluid. J Clin Microbiol. 1998;36(4):1117–21. WOS:000072565800053. pmid:9542950

CHAPTER 4

Criteria for the Prioritization of Public Health Interventions for Climate-Sensitive Vector-Borne Diseases in Quebec

Valerie Hongoh[1,2], Pierre Gosselin[3,4], Pascal Michel[1,5], Andre´ Ravel[1,2], JeanPhilippe Waaub[6] , Ce´line Campagna[3,7], Karim Samoura[8]

[1] The Research Group on Epidemiology of Zoonoses and Public Health (GREZOSP), Faculty of Veterinary Medicine, Universite´ de Montre´al, Saint-Hyacinthe, Canada

[2] Department of Pathology and Microbiology, Faculty of Veterinary Medicine, Universite´ de Montre´al, Saint-Hyacinthe, Canada

[3] Institut national de sante´ publique Que´bec (INSPQ), Que´bec, Canada

Citation: Hongoh V, Gosselin P, Michel P, Ravel A, Waaub J-P, Campagna C, et al. (2017) Criteria for the prioritization of public health interventions for climate-sensitive vector-borne diseases in Quebec. PLoS ONE 12(12): e0190049. https://doi.org/10.1371/journal.pone.0190049.

[4] Ouranos, Consortium on Regional Climatology and Adaptation to Climate Change, Montreal, Canada

[5] National Microbiology Laboratory at Saint-Hyacinthe, Public Health Agency of Canada, Saint-Hyacinthe, Canada

[6] Group for Research in Decision Analysis (GERAD, HEC Montre´al, Polytechnique Montre´al, McGill, UQAM), Montreal, Canada

[7] Department of Social and Preventive Medicine, Universite´ Laval, Que´bec, Canada, 8 Universite´ Aube Nouvelle, Ouagadougou, Burkina Faso

ABSTRACT

Prioritizing resources for optimal responses to an ever growing list of existing and emerging infectious diseases represents an important challenge to public health. In the context of climate change, there is increasing anticipated variability in the occurrence of infectious diseases, notably climate-sensitive vector-borne diseases. An essential step in prioritizing efforts is to identify what considerations and concerns to take into account to guide decisions and thus set disease priorities. This study was designed to perform a comprehensive review of criteria for vector-borne disease prioritization, assess their applicability in a context of climate change with a diverse cross-section of stakeholders in order to produce a baseline list of considerations to use in this decision-making context. Differences in stakeholder choices were examined with regards to prioritization of these criteria for research, surveillance and disease prevention and control objectives. A preliminary list of criteria was identified following a review of the literature. Discussions with stakeholders were held to consolidate and validate this list of criteria and examine their effects on disease prioritization. After this validation phase, a total of 21 criteria were retained. A pilot vector-borne disease prioritization exercise was conducted using PROMETHEE to examine the effects of the retained criteria on prioritization in different intervention domains. Overall, concerns expressed by stakeholders for prioritization were well aligned with categories of criteria identified in previous prioritization studies. Weighting by category was consistent between stakeholders overall, though some significant differences were found between public health and non-public health stakeholders. From this exercise, a general model for climate-sensitive vector-borne disease prioritization has been developed that can be used as a starting point for further public health prioritization exercises relating to research, surveillance, and prevention and control interventions in a context of climate change. Multi-stakeholder engagement in prioritization can help broaden the range of criteria taken into account, offer opportunities for early identification of potential challenges and may facilitate acceptability of any resulting decisions.

INTRODUCTION

Prioritizing resources for optimal response to an ever growing list of existing and emerging infectious disease risks presents an important challenge to public health administrations and their core intervention domains [1,2]. Ongoing global changes such as climate change, large scale land use transformations, increasing global travel and political instability in various regions of the world, contribute to variations in the patterns and occurrence of a number of infectious diseases, notably vector-borne diseases, which are known to be sensitive to weather and climate [3]. Changes in terms of the season of occurrence and the geographical distribution of these diseases are anticipated to increase as weather and climate are known to be drivers of the transmission and distribution of vector-borne diseases [4]. Prioritizing resources between existing and climate sensitive vector-borne diseases is complex but a necessary reality as potentially difficult trade-offs need to be made while taking into account a diversity of viewpoints [5]. For example, how much should be invested in surveillance activities, including vector surveillance and laboratory testing of novel pathogens, for diseases that are not yet endemic in Canada when resources for surveillance of existing diseases are limited? In other words, what non-endemic diseases should we be conducting surveillance for? Adaptation to vector-borne diseases will require both incremental and transformational approaches [6]. This study examines prioritization of climate-sensitive vector-borne diseases as an incremental step in improving management of these diseases within the primary domains of public health activity: research, surveillance, and prevention & control. Prioritization often serves as an initial step in aligning efforts and guiding public health decisions. In Quebec, MCDA methods have been used by the National Public health Institute and the Ministry of health for the prioritisation of Lyme disease surveillance and West nile virus interventions. The prioritization exercises referred to in this study pertain to planning of activities over a 1-5-year time scale while recognizing that these exercises may need to be updated and revisited within that time frame as knowledge evolves. The interventions included in these three domains range from the search for new treatments and diagnostic methods to education and outreach interventions as well as field interventions such as vector control or modifications to the built environment to help reduce the human health impact of infectious diseases. Several disease prioritization exercises have been undertaken in public health and veterinary public health contexts over the last few decades [7–21]. Traditionally, where stakeholders have been involved in these processes, with the exception of the studies by Ng and Sargeant (2012a) and Brookes et al (2014b), experts in public health have been the primary stakeholders included in the process. Prioritization exercises assist in explicitly bringing into focus the issues of concern around the decision problem and systematically evaluating available evidence on these issues. In this way, these exercises help structure reflection and guide decisions around resource allocation in order to ensure effectiveness within organisations and across various levels of government for effective public health delivery [22]. In other words, what are the concerns, what is the state of knowledge around these concerns, what are options going forward and how do these options perform in light of stated concerns. Prioritization exercises have been carried out at various scales from institutional-level [10], country-level [16,23–29] to continent-level [30,31]. Depending on the context, these exercises have been carried out either by institutional representative groups, a cross-section of the expert community concerned, advisory groups, national officials or researchers. Concern over how to prioritize infectious diseases in a context of climate change is more recent and at the the time of the initial review, had been documented only in the prioritization exercise by Cox and colleagues (2013). An evolution in the way prioritization exercises have been carried out over the last few decades can be seen in the examination of the literature [5] and reveals common goals and concerns that have persisted in their undertaking; notably, a push towards a systematic and transparent process and growing awareness of the increasing viewpoints that should be included in such exercises [13,22,25,30]. Refinements to the disease prioritization process over time have sought to separate information on the diseases (criteria measurement) from values pertaining to prioritization concerns (criteria) in order to improve transparency of the process. The explicit and measurable aspect of these exercises is sought by defining explicit criteria on which to evaluate the diseases being prioritized. For example, the measure of whether a disease is increasing or decreasing in the population can be captured as a criterion on the current incidence of human cases in the country; however, other concerns must also be evaluated such as what is the severity of the disease, do effective treatments already exist to limit the disease, is the general population already aware of and adopting effective protective behaviour to prevent the disease? If a disease is increasing in the population but not a severe disease or perhaps effective treatment already exists, should we allocate additional resources to this disease? For instance, Lyme disease is currently increasing in Canada, but relatively effective antibiotic treatment exists, how should this disease be prioritized relative to a new disease not yet endemic to Canada for which no treatment exists such as Dengue? Does the priority of a disease differ whether we are discussing allocation of resources in a research context versus a surveillance or prevention and control context? Criteria and how they are used to evaluate diseases are at the crux of the disease prioritization process. Criteria should represent core considerations or values relating to the prioritization objectives and help explicitly track relative differences between the items being prioritized [32]. Additionally, since health decisions in publicly funded health care systems use tax-paying citizen's dollars to operate, in the interest of transparency and accountability, it is important to understand what concerns are held by society both to verify acceptability of potential decisions made following prioritization exercises and understand where differences in values or priorities may be present in order to help bridge existing gaps. For example, does society feel that resources should be prioritized for diseases that have more severe effects on vulnerable groups such as the elderly, very young, or pregnant women versus less severe diseases that affect a greater number of individuals? Examining the impact of different prioritization methodologies on the resulting prioritized list of diseases is also an important issue to explore.

In the current study, we identify criteria for the prioritization of vector-borne diseases applicable in a context of climate change in order to construct a general model for disease prioritization. We then examine differences in concerns expressed by stakeholders with regards to prioritization of public health interventions relating to research, surveillance and prevention and control objectives. Stakeholders working in fields both directly and not directly connected to public health were included in the process to broaden the diversity of perspectives and voices considered in the disease prioritization process, as well as assess similarities and differences in values held. While the inclusion of stakeholders from outside of public health is not yet routine in public health decision making, exercises have been conducted to assess the acceptability and ground feasibility of proposed interventions [33], as well as level of public awareness on a subject, such as the level of Lyme disease awareness in Canada [34], which may be useful in guiding future allocation of resources towards educational outreach activities. The effect of combining current scientific knowledge with stakeholder values on disease prioritization is examined here by means of a pilot prioritization exercise performed with a multicriteria decision aid (MCDA) process using the PROMETHEE (Preference Ranking Organization Method for Enrichment Evaluations) outranking method [35].

MATERIALS AND METHODS

In disease prioritization exercises, criteria are used to systematically take into account concerns relevant to the decision-making context. The use of a participatory multi-stakeholder processes contributes to a broader and transparent selection of decision criteria. Here, a comprehensive review of the disease prioritization literature was conducted to create a synthesis list of the most commonly used prioritization criteria relevant in a context of climate change. These were discussed and validated with a group of stakeholders. The resulting criteria were used in a pilot prioritization exercise using PROMETHEE to examine differences in stakeholder assigned weights and their effect on prioritization under different intervention domains (research, surveillance, prevention and control). These steps are discussed in further details in the following sections and illustrated in a flow chart in Fig 1.

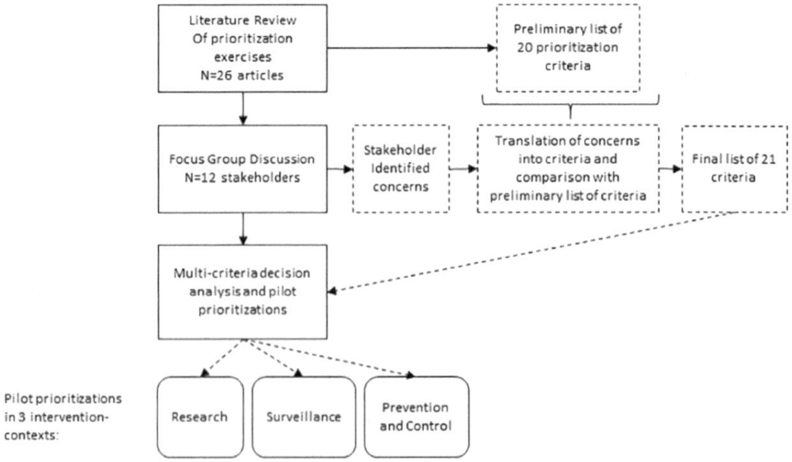

Figure 1: Flow chart of steps conducted to create disease prioritization models.

https://doi.org/10.1371/journal.pone.0190049.g001

Preliminary Criteria Identification

A comprehensive review of infectious disease prioritization studies published between 1990 and 2014 was undertaken in Spring 2014 to identify key criteria that should be considered for inclusion in a generic model (Table 1). A keyword search of the literature was carried out using a scientific database (Pubmed) with combinations of the following keywords: "emerging", "infectious", "communicable", "zoonoses", "disease" and "prioritization". Titles and abstracts were used to identify potentially relevant articles for further data extraction. Articles published in English or French pertaining to a prioritization exercise of infectious disease related items were retained for review. Additionally, relevant peer reviewed and grey literature referenced by articles retained for data extraction were also included in the review if they met the original inclusion criteria (snowball sampling) [36]. Relevant criteria and their related categories as described by citing articles were extracted from reviewed studies. As climate change may alter the season of occurrence and geographical distribution of climate sensitive vector-borne diseases [4,6], criteria pertaining to elements of the disease transmission process that may be affected by climate change were included. Whether or not conditions for transmission are already present for a disease is an important consideration in prioritization of these diseases. Conditions for disease emergence and transmission in a context of climate change can relate to environmental conditions, vector or reservoir conditions. For example, *Anopheles* species of mosquito capable of transmitting malaria are present in Canada and thus represent a vector component necessary for endemic transmission of this disease in Canada. Climate change signals are inherently difficult to separate out from the multitude of other driving forces (such as land use patterns, globalization and associated transport of goods, travel, etc.) which may affect disease transmission patterns. As such, many criteria that pertain to general infectious disease prioritization, such as the underlying vulnerability and susceptibility of the population to a disease or current levels of scientific knowledge and treatment availability, are also relevant in assessing climate sensitive vector-borne diseases and therefore were included in the preliminary list. Criteria were organized into categories following the literature review to facilitate criteria discussion with stakeholders and is presented along with the criteria in the results section.

Table 1: Article selection process for review

https://doi.org/10.1371/journal.pone.0190049.t001

	Steps	Total articles
1	Initial keyword search in Pubmed of studies containing combinations of the following keywords: "emerging", "infectious", "communicable", "zoonotic", "disease" and "prioritization".	N = 1196
2	Title and abstract scan of articles from Step 1 scanned for relevance resulting in 37 studies describing describing disease prioritization studies.	N = 37
3	Related peer reviewed and grey literature articles referenced by articles retained in step 2 were also scanned for relevance (snowball search).	N = 42
4	Final article selection of studies in which prioritization criteria were explicitly listed or described	N = 26*

*Note: In some cases, multiple articles referred to different aspects of the same study

https://doi.org/10.1371/journal.pone.0190049.t001

Focus Group Discussion

Following the literature review, a focus group discussion was held with a small group of stakeholders in Quebec (Canada) in September 2014 to discuss concerns with regards to vector-borne diseases in a context of climate change. Prioritization of diseases was further examined in the context of interventions for research, surveillance as well as prevention and control. Stakeholder invitations were initially sent out to 24 individuals and organizations having previously participated in vector-borne disease consultations by the province. These organizations represent a range of civil, municipal and regional level organizations (including non-governmental environmental rights groups, Sustainable Development, Environment, Wildlife and Parks of Quebec, the Ministry of Agriculture, Fisheries and Food (of Quebec), the Council for the Protection of Sick Patients (of Quebec), municipal representatives from municipalities in the province, representatives from the Health and Welfare Commissioner (of Quebec), academic researchers, representatives from a Quebec seniors rights representative group, regional public health departments and provincial blood donation services) representing the interests of various civil, municipal or regional subsets of the Quebec population. These organizations have all been in existence for over a decade. Stakeholders were selected based on their availability for a concurrent participation in a separate ongoing study on West Nile virus management in Quebec. Implications of this selection are discussed in further detail in the discussion section of this paper. All participating stakeholders gave informed written consent prior to participation in the study. Prior to the stakeholder meeting, participants were invited to reflect on their concerns with regards to managing infectious diseases now and over the course of the next 5 to 10 years in the context of climate change. For example, are diseases with high severity of greater concern than diseases with high incidence or vice versa? What other concerns are of interest, economic concerns, logistic concerns such as treatment availability or knowledge levels of the disease, animal and environmental health impact related concerns? During the focus group discussion, stakeholders began by writing down their prepared concerns. These were then compiled using a modified nominal group technique [37] and discussed with all participants. Following this, the literature identified criteria and their categories were presented in writing and discussed with stakeholders. Stakeholder compiled concerns and literature identified criteria were compared and discussed. Stakeholders were given the opportunity to add additional criteria and clarify or reword the literature identified set in person during the discussion. Measurement scales were also discussed at this time. Following the meeting, stakeholders were given an additional month to reflect on and validate the final list of criteria by means of two rounds of online Delphi review [38]. The online review was conducted using an electronic survey that presented all retained criteria from the in-person discussion and allowed stakeholders to "vote" for the inclusion of individual criteria. Comments on the relevance of the criteria to the prioritization models could also be made at this time. Stakeholders agreed to retain any criteria which received at least one vote from a participating stakeholder. Results from the first online review were compiled and presented to stakeholders to allow further modification before the final validated set was defined. This final set was included in prioritization models pertaining to research, surveillance and prevention and control of vector-borne diseases. This project was reviewed and approved by the Ethical Committee for Health Research of the University of Montreal (Comité d'éthique de la recherche en santé, CERES) (certificate number 14-025-CERES-D).

Criteria Weighting

Following validation of the final list of criteria, stakeholders were asked to weight criteria according to their relative importance with regards to research, surveillance and prevention and control objectives. The purpose of this weighting exercise was to translate stakeholder value systems into numerical weights. In order to do this, stakeholders were given a Microsoft Excel spreadsheet tool and asked to distribute 100 points across the list of decision criteria included in the model. The Excel tool included the finalized list of stakeholder validated criteria, desired effect direction of criteria and measurement scales listed by category with replicated sections for each of the three intervention domains (research, surveillance, prevention and control). The desired effect direction of a criterion refers to what the desired state is with respect to that criterion. For instance, with respect to costs, the desired effect direction is generally to prioritize diseases with the highest cost burden, i.e. maximize this criterion; whereas with respect to the general level of knowledge of the public on a disease, the desired effect direction may be to prioritize diseases that the public has little knowledge of, i.e. minimize this criterion. Stakeholders were asked to weight criteria in accordance with perceived importance taking into account their relative importance overall. Weights of zero were permitted for criteria to allow stakeholders to indicate the absence of importance of criteria if applicable. The difference in relative weights assigned to different categories were compared between the three intervention domains (research, surveillance and prevention and control) and Welch's *t*-test (unequal variances *t*-test) were performed in R (version 3.2.2) (R Core Team (2016), Vienna, Austria, http://www.R-project.org) to test for significant differences in category weights.

Pilot Prioritization of Five Diseases

An exploratory prioritization of five potentially climate-sensitive vector-borne diseases, Lyme, West Nile virus, chikungunya, dengue, and malaria was carried out to examine the effects of criteria weightings described in the previous section on disease rankings for each intervention domain. Only Lyme and West Nile virus have shown a local transmission cycle in Quebec in the last 10 years [39,40]; the other three diseases currently manifest themselves as imported cases only, but local cycles may occur in the coming decades due to climate change [41–43]. Lyme expansion in North America has been linked to climate change [39] and while West Nile virus expansion into North America was not directly linked to climate change, its epidemiology has been shown to be directly sensitive to climatic factors [44]. A literature search was conducted in late fall 2014 in Pubmed using each of the five diseases separately and in combination with the 21 identified criteria in order to assess and score disease performance over the criteria. Only articles published in English or French pertaining to the disease and criteria were reviewed. In other words, given the latest literature, what evaluation does each disease obtain on each of the identified criteria using the specified measurement scale. The same disease assessment scores were used for each intervention domain, however, stakeholders weighted each intervention domain separately and weights were shown to vary depending on the domain. Analysis of disease performance and criteria weights was performed with the PROMETHEE method (Preference Ranking

Organization Method for Enrichment Evaluations) in visual PROMETHEE software (version 1.4.0.0) (VP Solutions software, Brussels, Belgium, http://www.promethee-gaia.net). This MCDA method requires a set of criteria with corresponding weights and diseases performance scores over each of the criteria in order to produce a ranking of the relative performance of all diseases over the combined criteria. The PROMETHEE II method was used as it provides a complete ranking of results without incomparability [45,46]. PROMETHEE methods are pair-wise comparison methods that are part of the outranking class of decision aid methods enabling comparison of multiple items over multiple criteria [45,46]. In traditional multi-criteria type problems, incomparability can occur when one option performs better on certain criteria while another option performs better on very different criteria making comparison challenging [45,46]. PROMETHEE II was designed to overcome this incomparability without scale effects by calculating a net performance score over all criteria [45,46]. Additionally, a GAIA (Geometrical Analysis for Interactive Aid) visual analysis was also used to graphically explore decision maps. A sensitivity analysis was performed to examine the robustness of rankings for the 1st order given weights expressed by stakeholders.

RESULTS

Literature and Stakeholder Identified Criteria

Following an initial keyword search, the titles and abstracts of 1196 articles were scanned for relevance (Table 1). This resulted in 37 studies which were retained for full text review. Five additional articles referenced within the previous set were also reviewed. From this, 26 studies explicitly reporting prioritization criteria were retained for data extraction. A summary of these studies is shown in Table 2. Studies were primarily from high income countries in North America, Europe and Asia. While prioritization exercises have taken place in developing contexts [47], none were found pertaining explicitly to vector-borne diseases in the reviewed time period. An initial list of 122 criteria was extracted from these studies. The number of criteria used per prioritization exercise ranged from as few as 5 to as many as 57. Reported sources included experts, and lists from previous exercises and the literature.

Table 2: Summary of reviewed disease prioritization studies

https://doi.org/10.1371/journal.pone.0190049.t002

Author & year	Country^	Objective	**	Criteria	Weights	#	Item type	Method overview
Carter 1991	Canada	Set priorities for national surveillance (notifiable list)	B	12	No	60	Communicable diseases	Committee (n = 6) scored and discussed. Un-weighted criteria. Cut-off set for inclusion of diseases on notifiable list.
Rushdy et al 1998	UK	Rank diseases to manage resources	C	6	No	41	33 communicable diseases and 8 generic diseases	Expert opinion, questionnaire—assessed by experts in communicable diseases (n = 194)
Doherty 2000	Canada	To inform resource allocation national level	B, D	10	No	43	Communicable diseases	Expert opinion and consensus of subcommittee (n = 6)
Horby et al 2001	UK	Rank diseases to manage resources	B, C	5	No	69	58 pathogens and 11 generic diseases	Expert opinion (n = 518)
Valenciano 2002 (InVS)	France	Determine priorities to improve knowledge, prevention and control of diseases	A, B, C	6	No	37	Non-food borne zoonoses	Expert opinion (n = 10)
WHO 2002 (Dubrovnik pledge) *	WHO—7 eastern European countries	Strengthen infectious disease surveillance systems in 7 countries of South-East Europe	B	8	No	53	Communicable diseases	Expert opinion (n = 24)
Doherty 2006	Canada	Strengthen national surveillance capacities	B	10	No	48	Communicable diseases	Expert opinion and consensus of subcommittee (n = 6)
McKenzie et al 2007	New Zealand	Prioritize wildlife pathogens for surveillance	B	3	No	82	Wildlife pathogens	OIE based risk assessment approach
Krause et al 2008a&b	Germany	Guide research and surveillance strategies of department	A, B	12	Yes	85	Pathogens	Expert opinion (n = 11) and weighted sum aggregation
Cardoen et al 2009	Belgium	Rank food and water-borne pathogens to prioritize resource allocation for management	C	5	Yes	51	Food and water-borne zoonotic pathogens	Expert opinion (n = 35) and weighted sum aggregation
Capek 2010 (InVS)*	France	Rank non-foodborne zoonoses and anticipate emerging threats linked to climate, etc.	A, B, C	6	No	37	Non-food borne zoonoses	Expert opinion (n = 16)
Havelaar et al 2010	The Netherlands	Prioritized emerging zoonoses to support an early warning and surveillance network	B	7	Yes	86	Emerging zoonotic pathogens	MCDA technique. Existing list and expert opinion determined list of pathogens, weighting of criteria based on panel consultation (n = 29)
Pavlin et al 2010	Pacific Island nations	Update list of pathogens to include on urgent NNDL list	B, D	12	Yes	27	Conditions/diseases assessed	Additive model—Sum of scores
Ruzante et al 2010	Canada	Framework to prioritize foodborne risks	D	4	Yes	6	Pathogen-food combinations	MCDA technique—PROMETHEE
Balabanova et al 2011	Germany	Rank infectious diseases for research and surveillance	B	10	Yes	127	Pathogens	Expert opinion (n = 83) and weighted sum aggregation
Humblet et al 2012	Europe	European collaboration and agreement on priority zoonoses for surveillance and eradication	B, C	57	Yes	100	Zoonoses	MCDA technique with Expert scoring (n = 40) with weighted sum aggregation and Monte Carlo simulation
Ng & Sargeant 2012a,b, 2013	Canada	Compare zoonoses priorities between Canada and the US from public and expert perspective	A	21 (59)†	Yes	62	Zoonotic diseases	Criteria elicitation—via conjoint analysis technique conducted with public (n = 1500) and expert (n = 1471) focus groups and surveys, summed using part-worth utility values approach
Cediel et al 2013	Colombia	Prioritize zoonoses for surveillance	B	12	Yes	32	Zoonoses	Delphi (n = 12) and additive model
Del Rio Vilas et al 2013	UK	To inform management of emerging animal health related threats in UK	C	10‡	Yes	111	111 threats, 74 unique	MCDA technique—Developed threat assessment tool
Cox et al 2013	Canada	Test standardised method to prioritise infectious diseases of humans and animals that may emerge in response to CC	A	40	Yes	9	Trialed on 9 test pathogens	MCDA technique—MACBETH and additive model (n = 64)
Kadohira et al 2015	Japan	Surveillance and management of zoonoses	B, C	7	Yes	98	Zoonoses	Author determined criteria, risk profiles generated and reviewed by experts (n = 76) with AHP attributed weights by stakeholder groups (n = 334)
Brookes et al 2014 a&b	Australia	Prioritize exotic pig diseases for management	C	9	Yes	30	Diseases	MCDA technique with stakeholder (n = 81) elicited weight preference via online survey

^ Country targeted by prioritization exercise

* Not peer reviewed

** A = research; B = surveillance; C = prevention & control; D = policy

†59 identified, but only 21 used in prioritization exercises

‡3 models (perception (3 criteria), impacts (4 criteria) and capabilities (3 criteria))

https://doi.org/10.1371/journal.pone.0190049.t002

A number of studies had shared approaches and criteria (e.g.: [7,23,24]; [22,48]; [8,10,27,49]) while other studies used similar concepts, with variations in wording. The most common categories of criteria included: public health impacts, economic or market impacts, animal health impacts (generally pertaining to market impacts but also for animal-welfare), public perception and public health capacity to deal with a disease. These categories were retained in the models and named "Public Health", "Social Impact", "Economic", "Animal and Environmental Health", "Strategic and Operational" (i.e. logistics). Additionally, a "Risk and Epidemiology" category was added to capture epidemic potential, recent disease trends and proportion of susceptible population. Climate sensitive risk and epidemiology were also included in this category. As climate change will likely alter temperature and precipitation patterns with consequences for animal and vector distribution [4], criteria pertaining to existing conditions for disease transmission were included here. Commonly used prioritization criteria and their frequency were tracked across reviewed studies. Recurring relevant criteria were identified and where wording was different but pertaining to the same concept, criteria were combined and synthesized where appropriate into a shorter preliminary list of 20 criteria covering as broad a range of relevant concepts as possible for discussion with stakeholders. This number was chosen in order to present a manageable set for discussion with stakeholders. Retained criteria were then used in a pilot prioritization exercise to examine differences in stakeholder assigned weights under different intervention domains.

Focus Group Discussion

Twelve stakeholders consented to participate in a discussion held on September 29th, 2014 in Montreal, Quebec, Canada on the topic of perspectives and concerns relevant to disease prioritization in a context of climate change. One third of participants were female. All participants were between the ages of 30 and 65. Stakeholders had backgrounds in microbiology, entomology, biology, medicine, veterinary medicine and patient advocacy and hailed from a mix of both provincial and municipal organizations. Stakeholder discussions revealed coherence between stakeholder identified concerns and the literature constructed list of criteria. Further online validation by stakeholders following the initial meeting, resulted in a finalized list of twenty-one criteria (Table 3). Discussions of appropriate measurement scales and direction of desired effect to assess diseases were also held and are included in Table 3.

Table 3: Stakeholder validated list of criteria for the prioritization of climate sensitive vector-borne diseases

https://doi.org/10.1371/journal.pone.0190049.t003

Category	Criteria	Effect direction	Measurement units
Public Health Criteria (PHC)	PHC-01 –Reported yearly incidence of human cases in country	Maximize	0: Nil; 1: Very Low (<5); 2: Low (6–30); 3: Moderate (31-; 100): High (101–500); 5: very high (>500); 6: Unknown
	PHC-02 –Severity of the disease (both physically and mentally)	Maximize	0: Nil; 1: Low severity; 2: Moderate severity; 3: High severity; 4: Very high severity (risk of mortality)
	PHC-03 –Vulnerable groups	Maximize	0: All are vulnerable; 1: Existence of higher risk groups (e.g. 0-5yrs)
	PHC-04 –Potential to increase social inequality *	Maximize	0: No effect on social inequality; 1: Likely to exacerbate social inequality
Social Impact Criteria (SIC)	SIC-01 –Risk perception of the public	Maximize	1: Low perceived importance; 2: Moderate importance; 3: High importance
	SIC-02 –General level of knowledge, attitude and behaviour of the public	Minimize (Diseases for which the public has little knowledge of greater concern)	1: Little or no knowledge; 2: Moderate knowledge (general idea of symptoms); 3: High knowledge (can recognize symptoms and aware of transmission and treatment)
Risk and Epidemiology Criteria (REC)	REC-01 –Existence of favourable conditions for disease transmission	Maximize (diseases for which transmission conditions already favourable of greater concern)	1: Low risk (climate not suitable, no vector and no reservoir hosts); 2: Moderate risk (one of components present, either suitable climate, vector or reservoir host); 3: High risk (all components present–suitable climate, vector and reservoir host—or current or historic transmission)
	REC-02 –Epidemic potential	Maximize	1: Low risk; 2: high risk
	REC-03 –Current global trend of disease over last 5 years	Maximize	1: Stable–little to no recent local or global change in transmission; 2: unstable–recent global changes in transmission; 3: very unstable–recent local changes in transmission
	REC-04 –Proportion of susceptible population	Maximize	1: very low 0–5%; 2: low 5–10%; 3: moderate 10–25%; 4: high 25–50%; 5: very high 50+
Animal and Environmental Health Criteria (AEC)	AEC-01 –Estimated prevalence of yearly animal cases	Maximize (diseases with more cases of greater concern)	0: not transmissible to animals; 1: very low (<5%); 2: low (5–10%); 3: moderate (10–25%); 4: high (25–50%); 5: very high (50+); 6: unknown prevalence
	AEC-02 –Severity of disease	Maximize	0: Not applicable; 1: Low severity; 2: Moderate severity; 3: High severity; 4: Very high severity (risk of mortality)
	AEC-03 –Environmental or animal reservoir stage	Maximize (diseases with environmental stages of greater concern; harder to control)	1: Low risk–no independent stages that can survive in environment, water or reservoir hosts; 2: higher risk–existence of independent stages that can survive in environment, water or reservoir hosts.
Economic Criteria (ECC)	ECC-01 –Cost to provincial government	Maximize	1: low costs; (a few thousand); 2: moderate costs (hundreds of thousands); 3: high costs (millions)
	ECC-02 –Cost to private sector	Maximize	1: low costs (<100$); 2: moderate costs (<1000$); 3: high costs (>1000$)
	ECC-03 –Cost to individuals	Maximize	1: low costs (<100$); 2: moderate costs (<1000$); 3: high costs (>1000$)
Strategic and Operational Criteria (SOC)	SOC-01 –Capacity to detect and diagnose	Minimize	0: no tests, symptoms difficult to recognize; 1: distinct symptoms or existence of tests
	SOC-02 –Existence and effectiveness of current treatments	Minimize	0: no existing treatment; 1: partially effective treatment; 2: highly effective treatment available
	SOC-03 –Level of scientific knowledge of the disease	Minimize (diseases for which little is known of greater concern)	1: low–very little knowledge; 2: moderate–partial yet incomplete knowledge of disease symptoms, transmission, risk factors and treatment; 3: high–symptoms, transmission, risk factors and treatment well known
	SOC-04 –Optimization opportunities	Maximize	0: no opportunities; 1: potential opportunities
	SOC-05– Reportable disease	Maximize	0: not reportable; 1: nationally or internationally reportable

* Criteria added by stakeholders

https://doi.org/10.1371/journal.pone.0190049.t003

Criteria Weighting

Weighting of criteria was done individually by stakeholders and returned to the researchers by email in the weeks following the in-person focus group discussion. Ten of the original twelve stakeholders completed the criteria weighting exercise for each of the three intervention domains (research, surveillance or prevention and control). The relative importance of categories was generally similar for stakeholders across domains with no significant differences found between categories (Fig 2). Individual criteria weights were however found to differ within categories and across intervention domains (Fig 3). For all three intervention domains, the top three weighted categories were consistently "Public Health", "Risk and Epidemiology" and "Strategic and Operational" criteria while the bottom three categories were consistently "Animal and Environmental Health", "Economic" and "Social Impact" criteria. In the top 3, "Public Health" was generally the top weighted category in the subset with "Strategic and Operational" criteria consistently in 3rd place whereas in the bottom 3, the "Animal and Environmental" criteria category was generally top rated while "Social Impact" was either last or tied for last. Despite similarities in the relative importance of categories, differences in individual weights were observed between stakeholders and are reflected in the GAIA visual analysis of projected stakeholder weights (Fig 4). The top weighted criterion varied considerably by individual and by intervention domain though was generally from one of the top 3 weighted categories (i.e. "Public Health", "Strategic and Operational" or "Risk and Epidemiology" categories). The least weighted criterion also varied considerably by individual and intervention domain however, given the large number of criteria, multiple criteria often shared the lowest value but were not limited in origin to only the least weighted categories ("Animal and Environmental Health", "Economic" and "Social Impact"). The weights expressed by stakeholders not directly associated with public health organisations ($n = 3$) were found to vary significantly (though not necessarily together) compared with the rest of the group. This was the case for the weights given to the "Public Health" category for research ($p = 0.011$) and surveillance ($p = 0.016$) interventions as well as the weights given to the "Risk and Epidemiology" category for prevention and control ($p = 0.035$) and research ($p = 0.035$) interventions. In the previously mentioned cases, the stakeholders not directly associated with public health generally attributed less weight to these categories compared to stakeholders directly working in public health. Conversely, the distribution of weights given to the "Social Impact" category for prevention and control interventions ($p = 0.044$) and for the "Strategic and Operational" category for research ($p = 0.028$) interventions were generally found to be higher for stakeholders not directly associated with public health.

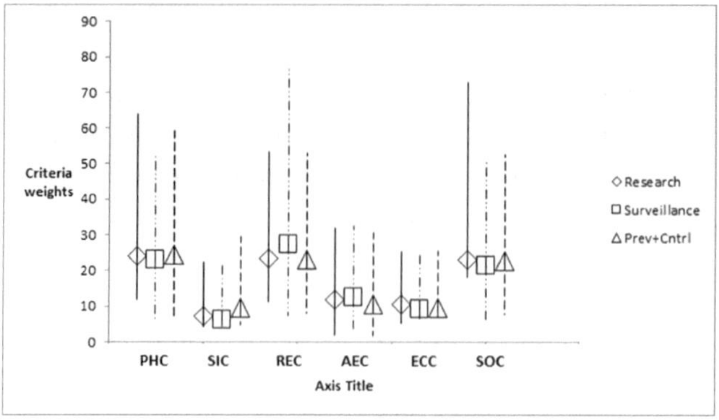

Figure 2: Criteria category weight average comparison by intervention domain.

The span of stakeholder weights is indicated by the vertical lines with shaped makers indicating the intervention specific group means. Criteria categories are shown along the X axis with average weights by category shown along the Y axis. The differences between the weights given to each intervention domain (research, surveillance and prevention & control) were not found to be significantly different for any of the categories. Criteria category Legend (X axis): PHC: Public Health Criteria; SIC: Social Impact Criteria; REC: Risk and Epidemiology Criteria; AEC: Animal and Environmental Health Criteria; ECC: Economic Criteria; SOC: Strategic and Operational Criteria.

https://doi.org/10.1371/journal.pone.0190049.g002

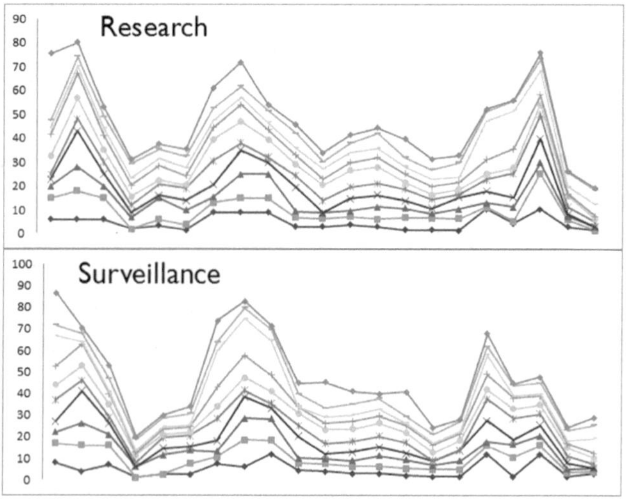

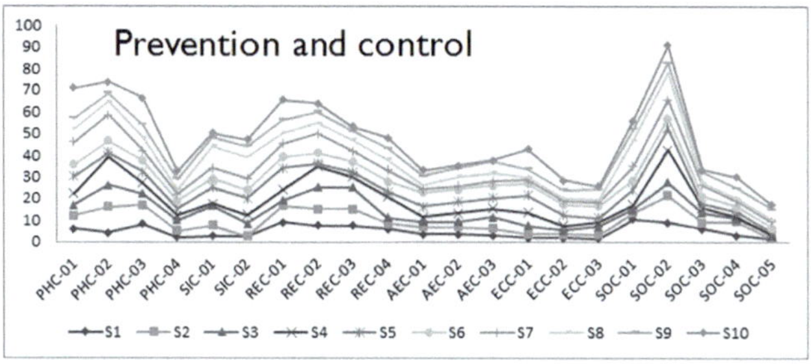

Figure 3: Individual weights by criteria and intervention domain.

Each line in the graph represents each of the 10 Individual stakeholder's weight assignments on all 21 criteria. The relative importance of criteria within each category is seen to vary depending on the intervention domain. For example: the "SOC-03-level of knowledge" criterion received the most weight in the research domain, while the "SOC-01-capacity to detect disease" criterion received the most weight in the surveillance domain. The "SOC-02-Existence of treatment" received the most weight in the prevention and control domain.

https://doi.org/10.1371/journal.pone.0190049.g003

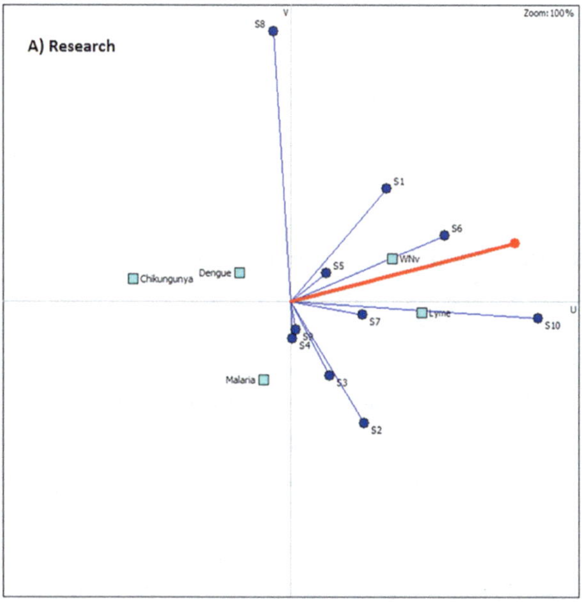

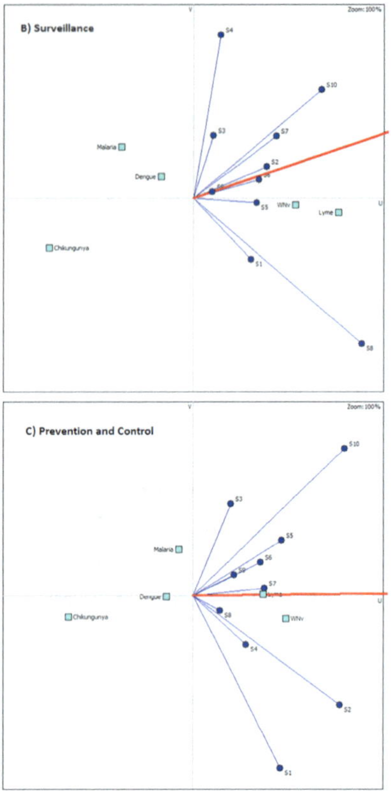

Figure 4: GAIA decision map for all intervention domains.

Panel A) shows the GAIA map for the research domain, B) shows the surveillance domain and C) the prevention and control domain. In each map, the bold red line represents the group decision axis (i.e. consensus ranking) with the filled circle pointing in the direction of the group ranking. Square markers represent the ranking of the different diseases in *k*-dimensional space (where *k* represents the number of criteria) projected onto a 2-dimensional plane. Diseases closest to the group decision axis are prioritized over diseases further away from the decision axis. Stakeholders 1 through 10 are represented by the blue circular markers labelled S1-S10. Stakeholders pointing in the same direction as the group decision axis are most aligned with the group ranking. Stakeholders further away in space from each other and from the group decision axis have more disparate weighting tendencies and hence perspectives. For example, in panel A) S8 shows distinct weight position as compared with the rest of the stakeholders and therefore indicates a different set of values in this context as compared with the rest of the group. Clusters of stakeholders can also be observed occurring in each of the panels and indicate stakeholders with more similar weightings (i.e. perspectives).

https://doi.org/10.1371/journal.pone.0190049.g004

The criterion "existence of favorable conditions for disease transmission" was weighted as the highest or second highest criterion within the "Risk and Epidemiology" category for the majority of stakeholders. However, across prioritization contexts, its relative importance varied depending on the stakeholder. This suggests that additional concerns (such as disease severity, level of scientific knowledge, current incidence and public risk perception) are given priority depending on the context and stakeholder. Criteria from the "Social Impact", "Animal and Environmental Health" as well as "Economic" categories were rarely among the top 5 weighted individual criterion for stakeholders, though these do appear among the top 5 for some stakeholders across each of the intervention domains. The small number of stakeholders and difference in perspective between groups of stakeholders is likely driving this observation.

Pilot Prioritization of Diseases

From the pilot prioritization exercise, *West Nile virus and Lyme* were ranked 1st and 2nd respectively in the research and prevention and control domains, with this order reversed in the surveillance domain. *Dengue, Malaria, and Chikungunya virus* were ranked 3rd, 4th and 5th in the research and surveillance domains with *Malaria* in 3rd and *Dengue* in 4th in prevention and control domain. Diseases were ranked with and without stakeholder assigned weights across intervention domains to assess dominance. No single disease was found to be dominant. Varying the disease evaluations showed sensitivity of disease rankings to their evaluations. Furthermore, different scales would produce different evaluations with potentially different rankings. As the goal of the current project was not to formally assess local disease priorities (for the creation of an official list of priority diseases), but rather to assess differences in stakeholder perspectives and examine the effect of these on potential rankings, a formal systematic assessment of sensitivity to disease scales was not performed. The retained scales allowed us to distinguish between the relative importance of diseases per criterion. Unweighted uni-criterion analysis revealed criteria for which each of the assessed diseases ranked 1st at least once. For instance, *Dengue* ranked 1st on the "capacity to detect and diagnose" criterion, *Chikungunya* ranked 1st on the "level of scientific knowledge of the disease" criterion, *Lyme* ranked 1st on the "current global trend" criterion, *malaria* ranked 1st on the "risk perception of the public" criterion and *West Nile virus* ranked 1st on the "epidemic potential" criterion. Group and individual stakeholder ranking results are shown in Table 4 with corresponding assessment values used (based on context specific data obtained from the literature) shown in the evaluation matrix in Table 5.

Table 4: Pilot prioritization of diseases for the group and by stakeholder for each intervention domain

https://doi.org/10.1371/journal.pone.0190049.t004

	GRP		S1		S2		S3		S4		S5		S6		S7		S8		S9		S10	
Diseases	**Rnk**	**Phi**	**Rnk**	**Phi**	**Rnk**	**Phi**	**Rnk**	**Phi**	**Rnk**	**Phi**	**Rnk**	**Phi**	**Rnk**	**Phi**	**Rnk**	**Phi**	**Rnk**	**Phi**	**Rnk**	**Phi**	**Rnk**	**Phi**
Research																						
West Nile virus (WNV)	1	0.08	2	0.09	3	-0.01	3	0.02	3	-0.00	1	0.10	2	0.17	1	0.10	1	0.09	4	-0.01	2	0.31
Lyme (LYM)	2	0.07	1	0.13	1	0.14	2	0.04	4	-0.03	3	-0.01	1	0.18	2	0.03	4	-0.06	3	0.00	1	0.23
Dengue (DEN)	3	-0.01	3	-0.02	4	-0.04	4	-0.03	2	0.04	4	-0.03	3	-0.07	4	0.01	2	0.08	2	0.01	4	-0.15
Malaria (MAL)	4	-0.02	4	-0.12	2	0.05	1	0.05	1	0.05	2	0.00	4	-0.08	3	0.02	5	-0.18	1	0.02	3	-0.05
Chikungunya (CHIKV)	5	-0.11	5	-0.13	5	-0.13	5	-0.08	5	-0.06	5	-0.06	5	-0.20	5	-0.16	3	0.07	5	-0.03	5	-0.35
Surveillance																						
West Nile virus (WNV)	2	0.10	2	0.02	1	0.15	3	0.02	4	0.00	1	0.15	2	0.10	2	0.15	2	0.26	1	0.04	2	0.16
Lyme (LYM)	1	0.14	1	0.18	2	0.10	2	0.04	3	0.03	2	0.08	1	0.13	1	0.13	1	0.38	2	0.03	1	0.27
Dengue (DEN)	3	-0.02	3	0.00	3	-0.01	4	-0.03	1	0.08	3	-0.01	3	-0.02	3	-0.00	3	-0.13	3	-0.01	4	-0.07
Malaria (MAL)	4	-0.06	5	-0.12	4	-0.07	1	0.05	2	0.06	4	-0.09	4	-0.06	4	-0.05	5	-0.27	4	-0.01	3	-0.03
Chikungunya (CHIKV)	5	-0.16	4	-0.08	5	-0.18	5	-0.08	5	-0.18	5	-0.13	5	-0.15	5	-0.22	4	-0.25	5	-0.04	5	-0.33
Prevention & control																						
West Nile virus (WNV)	1	0.10	1	0.14	1	0.21	3	0.02	1	0.12	1	0.10	2	0.05	1	0.10	2	0.05	1	0.05	2	0.15
Lyme (LYM)	2	0.06	2	0.11	2	0.15	2	0.04	3	-0.02	2	0.07	1	0.09	2	0.03	4	-0.03	2	0.04	1	0.14
Dengue (DEN)	4	-0.02	3	-0.03	4	-0.09	4	-0.03	2	0.02	4	-0.02	3	-0.00	3	0.02	1	0.06	4	-0.02	4	-0.07
Malaria (MAL)	3	-0.01	5	-0.12	3	-0.08	1	0.05	4	-0.03	3	0.02	4	-0.01	4	-0.01	3	-0.02	3	0.01	3	0.06
Chikungunya (CHIKV)	5	-0.13	4	-0.10	5	-0.19	5	-0.08	5	-0.09	5	-0.16	5	-0.13	5	-0.14	5	-0.07	5	-0.07	5	-0.28

GRP–overall group ranking; Rnk–rank; S1-S10 –denotes stakeholders 1 through 10; Phi–net outranking flows (combined positive and negative flows) indicating performance of each disease

https://doi.org/10.1371/journal.pone.0190049.t004

Table 5: Disease evaluation matrix

https://doi.org/10.1371/journal.pone.0190049.t005

Diseases	**Criteria**																				
	PHC1	**PHC2**	**PHC3**	**PHC4**	**SIC1**	**SIC2**	**REC1**	**REC2**	**REC3**	**REC4**	**AEC1**	**AEC2**	**AEC3**	**ECC1**	**ECC2**	**ECC3**	**SOC1**	**SOC2**	**SOC3**	**SOC4**	**SOC5**
West Nile virus (WNv)	2	2	1	1	1	2	3	2	1	5	6	4	2	1	1	1	1	0	3	1	1
Lyme (LYM)	3	2	1	1	1	2	3	1	3	5	6	2	2	2	1	2	1	1	3	1	1
Dengue (DENV)	0	4	0	1	1	1	1	1	2	2	0	1	2	3	2	1	0	1	3	1	1
Malaria (MAL)	0	4	1	2	2	1	2	1	1	2	0	0	2	3	3	1	1	2	3	1	1
Chikungunya (CHIKV)	0	1	0	1	1	1	1	1	2	2	0	1	2	1	1	2	1	0	2	1	1

Disease evaluation matrix showing evaluation scores for each of the five pilot diseases based on context specific data reviewed pertaining to each disease over all criteria. Note: Criteria AEC3, SOC4 and SOC5 are non-discriminating with the above data set due to lack of variation between disease evaluation values but could be discriminating with different diseases or more refined data set. Criteria were retained in the model due to expressed interest of stakeholders. PHC–Public health criteria; SIC–Social impact criteria; REC–Risk and epidemiology criteria; AEC–Animal and environmental health criteria; ECC–Economic criteria; SOC–Strategic and operational criteria

https://doi.org/10.1371/journal.pone.0190049.t005

Sensitivity analysis results with weight stability intervals for all criteria by all stakeholders for the research domain are shown in Table 6 in descending order of stability (from least stable to most stable). Five of the twenty-one criteria were found to be very stable as per the size of their stability intervals spanning almost the entire range of possible values from 0–100 for all stakeholders for the 1st order ranking. This indicates that the rank ordering of diseases would not change for any weight value given to these criteria between 0 and 100. These five criteria were "the existence of a vulnerable group", "potential to increase social inequality", "ability to infect the environment", "optimization opportunities" and "reportable disease". The remaining criteria were found to have relatively small stability intervals (<10 points) for at least one stakeholder indicating high sensitivity to assigned weights by stakeholders. The "current trend", "cost to individuals" and "general knowledge" criteria were found to be the highly sensitive for 7, 6 and 5 stakeholders respectively. This was closely followed by "existence of favourable conditions", "disease severity for animals", "cost to private sector" and "public risk perception" criteria for at least 4 stakeholders..

Table 6: Weight stability intervals in descending order from sensitivity analysis of all stakeholders for the research domain

https://doi.org/10.1371/journal.pone.0190049.t006

Criteria	S1	S2	S3	S4	S5	S6	S7	S8	S9	S10
REC-03	9 (0–10)	6(2–100)	10 (0–11)	5 (0–5.5)	2 (0–8)	4 (0–10)	5 (2–6)	3 (0–4)	5 (0–6)	3 (0–6)
ECC-03	1 (0–2.5)	5 (0–100)	4 (0–6)	5 (0–11)	1 (0–9)	3 (0–11)	2 (0–100)	2 (0–4)	6 (0–8)	3 (0–8)
SOC-02	4 (3–100)	1 (0–10)	6 (0–7)	4 (0–5)	10 (4–100)	8 (2.5–100)	9 (6–11)	16 (15–100)	4 (0–5)	1 (0–100)
REC-01	9 (0–100)	4 (0–100)	3 (0–4.5)	5 (4–10)	10 (1.5–100)	5 (0–100)	6 (4.5–11)	3 (2.5–100)	5 (3.5–8)	9 (0–100)
AEC-02	4 (1.5–100)	3(0–20)	3 (0–3)	5 (0–5.5)	5 (0–100)	4 (0–100)	2 (0–4)	4 (3–100)	4 (0–5)	4 (0–100)
ECC-02	2 (0–14)	5(0–9)	2 (0.5–100)	2 (1–100)	4 (0–10)	4 (0–9)	2 (0–100)	3 (0–4)	4 (2.5–100)	4 (0–23)
SOC-01	10 (0–20)	1 (0–12)	1 (0–8)	5 (0–5.5)	5 (0–14)	6 (0–13)	6 (0–7)	16 (0–17)	4 (0–5)	1 (0–27)
PHC-01	6 (0–100)	9 (4–100)	5 (0–6)	3 (0–6)	2 (0–100)	9 (1–100)	11 (0–100)	3 (2–100)	3 (0–5)	28 (0–100)
ECC-01	2 (0–3)	5(0–13)	4 (2–100)	3 (0–100)	5 (0–11)	4 (0–9)	2 (0–100)	3 (0–3.5)	4 (2.5–100)	7 (0–15)
SIC-02	1 (0–12)	3 (0–7)	6 (4.5–100)	4 (0.5–100)	5 (0–12)	4 (0–11)	5 (0–16)	3 (0–4)	5 (3–100)	3 (0–25)
REC-04	3 (0–13)	4 (0–9)	3 (1–100)	10 (6–100)	5 (0–12)	4 (0–11)	5 (0–100)	3 (0–4)	6 (4–100)	4 (0–60)
AEC-01	3 (0–100)	3(0–100)	3 (0–4)	0 (0–100)	5 (0–100)	3 (0–100)	2 (0–100)	4 (3–100)	3 (0–5)	4 (0–100)
PHC-02	6 (0–18)	12 (0–17)	10 (8.5–100)	15 (11–100)	5 (0–14)	10 (0–18)	12 (3.5–16)	3 (0–4)	4 (1.5–100)	6 (0–31)
SIC-01	3 (0–19)	3 (0–7)	9 (7–100)	1 (0.5–100)	5 (0–12)	6 (0–13)	5 (3–16)	3 (0–20.5)	4 (3–100)	2 (0–24)
REC-02	9 (8–100)	6 (0–13)	10 (0–11)	10 (0–13)	3 (0–100)	7 (0–100)	9 (0–12)	3 (2.5–100)	5 (0–7)	10 (5–100)
SOC-03	10 (0–25)	15 (0–29)	4 (0–13)	10 (0–17)	10 (0–20)	4 (0–21)	5 (1–13.5)	10 (0–12)	5 (0–9)	3 (0–36)
PHC-03	6 (0–100)	9 (0–100)	5 (0–100)	5 (4–100)	5 (0–100)	6 (0–100)	5 (3–100)	3 (2–100)	5 (4–100)	4 (0–100)
PHC-04	2 (1–100)	0 (0–100)	5 (3–100)	2 (0–100)	3 (0–100)	4 (0–100)	3 (0–100)	3 (0–100)	6 (3.5–100)	2 (0–100)
AEC-03	3 (0–100)	3(0–100)	5 (0–100)	5 (4.5–100)	5 (0–100)	4 (0–100)	2 (0–100)	4 (0–100)	6 (0–100)	3 (0–100)
SOC-04	3 (0–100)	3 (0–100)	1 (0–100)	1 (0–100)	3 (0–100)	4 (0–100)	1 (0–100)	3 (0–100)	6 (0–100)	0 (0–100)
SOC-05	1 (0–100)	0 (0–100)	1 (0–100)	0 (0–100)	2 (0–100)	1 (0–100)	1 (0–100)	5 (0–100)	6 (0–100)	1 (0–100)

S1-S10 –denotes stakeholders 1 through 10; Stakeholder assigned weights are given for all criteria followed by the stability interval in parentheses over which the ranking order for the 1st position items are maintained. PHC–Public Health criteria; SIC–Social impact criteria; REC–Risk and epidemiology criteria; AEC–Animal and environmental health criteria; ECC—Economic criteria; SOC–Strategic and operational criteria

https://doi.org/10.1371/journal.pone.0190049.t006

DISCUSSION

The current study solicited Quebec stakeholder perspectives and concerns for vector-borne diseases management under climate change. These concerns were translated into criteria and used to construct prioritization models to rank vector-borne diseases in the context of public health interventions pertaining to research, surveillance and, prevention and control. The prioritization models consisted of 21 criteria and were accompanied by corresponding stakeholder weights that varied depending on the intervention context (research, surveillance or prevention and control). Criteria were distributed across six categories: "public health", "risk and epidemiology", "strategic and operational", "animal and environmental health", "economic" and "social impact" with the relative importance of criteria within these categories varying depending on the intervention context. A pilot prioritization exercise showed how resulting disease priorities also change depending on the intervention context.

What have we learned from years of prioritization exercises? The use of an initial literature review followed by a stakeholder consultation in our study allowed for the inclusion of a broad range of considerations which were factored into our prioritization models. Our literature review suggests that common criteria and categories recur across studies [7–21]. This may be due to shared learning from previously published studies, but may also be representative of core concerns that are shared across scales and regions which translate into a set of common decision criteria that remain applicable across public health contexts and that can be used as a basis for prioritization models as we have done here. Furthermore, these models were found to be equally applicable in three intervention contexts with associated stakeholder specific weights. While our study included similar categories of criteria to previous prioritization exercises, detailed direct comparisons cannot be made between studies since the prioritization objectives and approaches differed.

What is the value of including stakeholders with a variety of backgrounds in the process? Few studies involve the public in their consultation process, and generally involve only a narrow range of participants in the process. Involvement of a diversity of participants (researchers, government, public health personnel, and non-technical citizens) in the criteria and preference elicitation process can help ensure that a broad set of value perspectives are considered. An attempt at including a broader range of voices in the vector-borne disease prioritization process has been done in the current study. While previous studies have contrasted public and expert rankings [28,29], the current study demonstrated a potential method for how to include these voices in the same consultation exercise in order to provide an opportunity for shared knowledge exchange and discussion of concerns between groups. Including multiple stakeholders with diverse backgrounds helps broaden the perspectives considered and contributes to the inclusion of a representative set of societal concerns and perspectives in the decision analysis process. The weighting process also allows a second measure of where stakeholders stand with regards to these concerns. Weights by non public health related organisations were found to be significantly different from the rest of the group and may in part be explained by differences in perceived responsibility or accountability between these groups [50]. Where differences occur between groups may indicate points of entry for future research into how differences in perspective impact intervention acceptance or uptake down the line. While public health officials are a part of society, their knowledge may push them to dismiss a disease which causes great alarm to the general public (e.g. one with high severity but low risk of transmission in the current context). Understanding what is being factored into each group's decision making equation is therefore useful in order to inform future interventions and public health actions.

Remaining Challenges

From a public policy point of view, ensuring that resulting decisions are well aligned with publicly held values should be of interest and is something that is possible to explore with the current approach. The use of formal prioritization approaches is evolving. Who to include in the process is a consideration that should be regularly revisited by decision makers. The challenge becomes how to process the range of viewpoints consulted during analysis. Should different viewpoints be weighted equally? Should all groups be consulted primarily to elicit the most comprehensive range of viewpoints or are related weightings by different stakeholder groups useful and necessary? These questions should be addressed early on in the prioritization exercise and responses are likely to vary depending on the specific prioritization context. Key participant profiles representing the viewpoints of various societal groups can be constructed and has been used before in environmental decision making processes [51,52]. The consistency of weighting tendencies within stakeholder groups was not assessed in the current study given the small number of participants; however, given a larger sampling, the consistency of weighting trends within groups could be further examined.

This work is still in the early stages of understanding the impact of including broader consultation on these issues and requires further exploration to help guide the minimal set of consultations that should be considered. For certain intervention domains, such as surveillance, public health perspectives may be one of the most important to take into account; however, with regards to publicly funded research interventions, it may be of interest to ensure that resulting decision priorities are coherent with socially held values. While public health officials themselves do represent socially held values, their unique perspective may differ from that of society at large and effort should be made to examine where potential differences may exist. Stakeholders outside of public health, especially non-government stakeholders, offer more than just another view and opinion in this process. They help take into account socially held values, but also ensure that interventions are feasible in the field. Their inclusion is a first step in the communication and knowledge transfer process with these organizations, such as municipalities. With regards to the disease prevention and control domain, understanding and attempting to integrate different perspectives may help pinpoint areas where additional information needs exist. For example, diseases which may pose significant health and economic impacts to livestock populations but few to human populations may fall under the radar of human public health officials. Including stakeholders from other domains such as animal health and environmental health may help to ensure that attention is given to such cases. In this exercise, significant differences were found for the "social impact" and "risk and epidemiology" categories suggesting that important differences may exist between public health experts and the general public with regards to risk perception, knowledge and risk assessment of disease threats. Differences in risk perception between the general public and experts have been observed in previous studies [53–55].

Limitations

This study did not aim to predict which diseases might be sensitive to climate change, but rather set out to identify the primary concerns (i.e. criteria) of a cross-section of society with regards to vector-borne disease management in these contexts. Here "cross section of society" refers not to the number of stakeholders engaged, but rather the diversity of organizations consulted. Furthermore, the pilot prioritization presented was for illustration purposes only and should not be interpreted as a formal assessment of local Quebec priorities. Additional data as well as further discussion with experts and stakeholders is required to validate these findings.

With regards to stakeholders included in this study, these groups were invited based on their representation of a cross-section of society at large and meant to represent perspectives from various interest groups beyond that of public health alone (e.g. patient rights group, senior citizens). Of the 24 invited, only 12 were able to participate due to availability. Of those 12, only 10 further completed the weighting exercise, once again due to time constraints. A larger number of participants with different backgrounds or organizational membership would likely have increased the diversity of perspectives included both in terms of criteria considered and weightings assigned to those criteria. Furthermore, while this same group of stakeholders also participated in a concurrent study on West Nile virus, additional evaluations with different stakeholder groups will be needed to assess whether weightings were strongly influenced by this factor. It should be noted that disease rankings are not driven by stakeholder weightings alone, but rather the combined evaluation of literature evidence with stakeholder weights. The relative stability of the rankings was explored in sensitivity analyses presented the results section (Table 6) and shown to be relatively stable for many of the criteria in the final models.

Future Prioritization Exercises

The current study offers preliminary models that can be used as a starting point for future prioritization exercises to build on. Thorough reflection should be held on the purpose or objective of the prioritization exercise and who the most representative stakeholders are to include in that context. The approach used in the current study offers an opportunity to identify concerns held by participating stakeholders and provides a method to examine the effect of these differences on disease rankings. The inclusion of diverse perspectives provides opportunities to identify knowledge gaps and data needs that go beyond an assessment of available scientific evidence alone and thus contributes to a more comprehensive process informing decision. In allocating health resources, a formal process for evaluating diverse evidence and interests notably contributes to enhancing and articulating the rigour, transparency, and acceptability of the decision process, in a manner which is compatible with currently held social values. In other words, what, in practice, decision making will gain from this approach includes: 1) a more comprehensive view of the issue–issues that decision makers did not think existed or did not think were important; 2) a sense of perceived risk (above and beyond what the usual experts / specialists are saying); 3) a level of acceptability by using standard criteria to prioritize health risks; 4) a sense of transparency as to how the prioritization results were obtained; 5) a level of rigour to the results given that the prioritization follows a set of transparent and accepted rules. The approach described here is coherent with the vulnerability and adaptation framework described in the 4th IPCC assessment report [56] and more specifically, the impact assessment stage in terms of providing a mechanism with which to narrow a list of diseases to focus on in the context of climate change.

ACKNOWLEDGMENTS

The authors would like to sincerely thank all stakeholders and experts in Quebec consulted during this project for their participation and contribution as well as the Institut national de santé publique du Québec for the use of their facilities for meeting. The Public Health Agency of Canada provided funding to cover meeting facility and travel cost of participants for focus group discussions in the context of a separate, concurrent study on West Nile virus. VH is funded by the Canadian Institutes of Health Research (CIHR), Ouranos Consortium for research in climatology and adaptation to climate change and the Strategic Training Program in Global Health Research, a partnership of CIHR and the Quebec Population Health Research Network (RRSPQ). This project is also nested within the IRIACC-FACE program funded by IDRC, CIHR, NSERC and SSHRC from Canada. The funding sources had no involvement in the actual research or preparation of the paper.

REFERENCES

1. Ebi KL, Kovats RS, Menne B. An approach for assessing human health vulnerability and public health interventions to adapt to climate change. Environ Health Perspect. 2006;114: 1930–1934. pmid:17185287
2. Baltussen R, Niessen L. Priority setting of health interventions: the need for multi-criteria decision analysis. Cost Eff Resour Alloc. 2006;4: 14. pmid:16923181
3. Gubler DJ, Reiter P, Ebi KL, Yap W, Nasci R, Patz JA. Climate variability and change in the United States: potential impacts on vector—and rodent-borne diseases. Environ Health Perspect. 2001;109.
4. Githeko AK, Lindsay SW, Confalonieri UE, Patz JA. Climate change and vector-borne diseases: a regional analysis. Bull World Health Organ. 2000;78: 1136–1147. pmid:11019462
5. Brookes VJ, Del Rio Vilas VJ, Ward MP. Disease prioritization: what is the state of the art? Epidemiol Infect. 2015;FirstView: 1–12. pmid:25876939
6. Smith KR, Woodward A, Campbell-Lendrum D, Chadee DD, Honda Y, Liu Q, et al. Human Health: Impacts, Adaptation, and Co-Benefits. In: Field CB, Barros VR, Dokken DJ, Mach KJ, Mastrandrea MD, Bilir TE, et al. , editors. Climate Change 2014: Impacts, Adaptation, and Vulnerability Part A: Global and Sectoral Aspects Contributions of Working Group II to the Firth Assessment Report of the Intergovernmental Panel on Climate Change. Cambridge, United Kingdom and New York, NY, USA: Cambridge University Press; 2014. pp. 709–754.
7. Doherty J-A. Final report and recommendations from the national notifiable diseases working group. Can Commun Dis Rep. 2006;32.
8. Krause G. Prioritisation of infectious diseases in public health. Eurosurveillance. 2008;13.
9. Krause G. How can infectious diseases be prioritized in public health? EMBO Rep. 2008;9: S22–S27. pmid:18578019
10. Balabanova Y, Gilsdorf A, Buda S, Burger R, Eckmanns T, Gärtner B, et al. Communicable diseases prioritized for surveillance and epidemiological research: results of a standardized prioritization procedure in Germany, 2011. PLoS ONE. 2011;6: e25691. pmid:21991334

11. Institut de Veille Sanitaire. Définition des priorités dans le domaine des zoonoses non alimentaires 2000–2001. Paris, France: Institut de Veille Sanitaire; 2002 p. 40 p.
12. Cardoen S, Van Huffer X, Berkvens D, Quoilin S, Ducoffre G, Saegerman C, et al. Evidence-based semiquantitative methodology for prioritization of foodborne zoonoses. Foodborne Pathog Dis. 2009;6: 1083–1096. pmid:19715429
13. Havelaar AH, van Rosse F, Bucura C, Toetenel MA, Haagsma JA, Kurowicka D, et al. Prioritizing Emerging Zoonoses in The Netherlands. PLoS ONE. 2010;5: e13965. pmid:21085625
14. Institut de Veille Sanitaire. Définition des priorités dans le domaine des zoonoses non alimentaires 2008–2009. Paris, France: Institut de Veille Sanitaire (InVS); 2010 p. 31 p.
15. Gilsdorf A, Krause G. Prioritisation of infectious diseases in public health: feedback on the prioritisation methodology, 15 July 2008 to 15 January 2009. Euro Surveill. 2011;16.
16. Ng V, Sargeant JM. A quantitative and novel approach to the prioritization of zoonotic diseases in North America: a public perspective. PLoS ONE. 2012;7: e48519. pmid:23133639
17. Cox R, Revie CW, Sanchez J. The Use of Expert Opinion to Assess the Risk of Emergence or Re-Emergence of Infectious Diseases in Canada Associated with Climate Change. PLoS ONE. 2012;7: e41590. pmid:22848536
18. Cox R, Sanchez J, Revie CW. Multi-Criteria Decision Analysis Tools for Prioritising Emerging or Re-Emerging Infectious Diseases Associated with Climate Change in Canada. PLoS ONE. 2013;8.
19. Capek I. Définition des priorités dans le domaine des zoonoses non alimentaires 2008–2009. Institut de Veille Sanitaire; 2010.
20. Brookes VJ, Hernandez-Jover M, Cowled B, Holyoake PK, Ward MP. Building a picture: Prioritisation of exotic diseases for the pig industry in Australia using multi-criteria decision analysis. Prev Vet Med. 2014;113: 103–117. pmid:24211032
21. Brookes VJ, Hernandez-Jover M, Neslo R, Cowled B, Holyoake P, Ward MP. Identifying and measuring stakeholder preferences for disease prioritisation: a case study of the pig industry in Australia. Prev Vet Med. 2014;113: 118–131. pmid:24211106

22. Rushdy A O'Mahony M. PHLS overview of communicable diseases 1997: results of a priority setting exercise. Commun Dis Rep CDR Wkly. 1998;8: S1–12.
23. Carter A. Establishing goals, techniques and priorities for national communicable disease surveillance. Can J Infect Dis. 1991;2: 37–40. pmid:22451750
24. Doherty J-A. Establishing priorities for national communicable disease surveillance. Can J Infect Dis. 2000;11: 21–24. pmid:18159260
25. Kadohira M, Hill G, Yoshizaki R, Ota S, Yoshikawa Y. Stakeholder prioritization of zoonoses in Japan with analytic hierarchy process method. Epidemiol Infect. 2015;143: 1477–1485. pmid:25195643
26. Del Rio Vilas VJ, Voller F, Montibeller G, Franco LA, Sribhashyam S, Watson E, et al. An integrated process and management tools for ranking multiple emerging threats to animal health. Prev Vet Med. 2013;108: 94–102. pmid:22954461
27. Cediel N, Villamil LC, Romero J, Rentaria L, De Meneghi D. setting priorities for surveillance, prevention, and control of zoonoses in Bogota, Colombia. Rev Panam Salud Publica. 2013;33: 316–24. pmid:23764662
28. Ng V, Sargeant JM. A stakeholder-informed approach to the identification of criteria for the prioritization of zoonoses in Canada. PLoS ONE. 2012;7: e29752. pmid:22238648
29. Ng V, Sargeant JM. A quantitative approach to the prioritization of zoonotic diseases in North America: a health professionals' perspective. PLoS ONE. 2013;8: e72172. pmid:23991057
30. Humblet MF, Vandeputte S, Albert A, Gosset C, Kirschvink N, Haubruge E, et al. Multidisciplinary and evidence-based method for prioritizing diseases of food-producing animals and zoonoses. Emerg Infect Dis. 2012;18. pmid:22469519
31. WHO. The Dubrovnik Pledge on Surveillance and Prioritization of Infectious Diseases. Report on a WHO meeting in Bucharest, Romania 21–23 November, 2003. Copenhagen: WHO Regional Office for Europe; 2003 p. 27 p.
32. Belton V, Stewart T. Multiple criteria decision analysis: an integrated approach. 2002;
33. Aenishaenslin C, Michel P, Ravel A, Gern L, Waaub J-P, Milord F, et al. Acceptability of tick control interventions to prevent Lyme disease in

Switzerland and Canada: a mixed-method study. BMC Public Health. 2016;16: 1–10.

34. Aenishaenslin C, Bouchard C, Koffi JK, Pelcat Y, Ogden NH. Evidence of rapid changes in Lyme disease awareness in Canada. Ticks Tick-Borne Dis. 2016;7: 1067–1074. pmid:27665265
35. Brans JP, Vincke P. A Preference Ranking Organisation Method: (The PROMETHEE Method for Multiple Criteria Decision-Making). Manag Sci. 1985;31: 647–656.
36. Greenhalgh T, Peacock R. Effectiveness and efficienty of search methods in systematic review of complex evidence: audit of primary sources. Br Med J. 2005;33: 1064–1065.
37. Jones J, Hunter D. Consensus methods for medical and health services research. BMJ. 1995;311: 376–380. pmid:7640549
38. Sawford K, Dhand N, Toribio J-A, Taylor M. The use of a modified Delphi approach to engage stakeholders in zoonotic disease research priority setting. BMC Public Health. 2014;14: 182. pmid:24552445
39. Ogden NH, Bouchard C, Kurtenbach K, Margos G, Lindsay LR, Trudel L, et al. Active and passive surveillance and phylogenetic analysis of Borrelia burgdorferi elucidate the process of Lyme disease risk emergence in Canada. Environ Health Perspect. 2010;118: 909–914. pmid:20421192
40. Drebot MA, Lindsay R, Barker IK, Buck PA, Fearon M, Hunter F, et al. West Nile virus surveillance and diagnostic: A Canadian perspective. Can J Infect Dis. 2003;14: 105–114. pmid:18159433
41. Berrang-Ford L, McLean JD, Gyorkos TW, Ford JD, Ogden NH. Climate change and Malaria in Canada: a systems approach. Interdiscip Perspect Infect Dis. 2009;2009: 13.
42. Schwartz KL, Giga A, Boggild AK. Chikungunya fever in Canada: fever and polyarthritis in a returned traveller. Can Med Assoc J. 2014;2014.
43. Ogden N, Milka R, Caminade C, Gachon P. Recent and projected future climatic suitability of North America for the Asian tiger mosquito Aedes albopictus. Parasit Vectors. 2014;7: 532. pmid:25441177
44. Paz S. Climate change impacts on West Nile virus transmission in a global context. Philos Trans R Soc Lond B Biol Sci. 2015;370. pmid:25688020

45. Brans JP, Mareschal B. PROMETHEE-GAIA. Une méthodologie d'aide à la décision en présence de critères multiples. l'Université de Bruxelles. Paris, France: Ellipses; 2002.
46. Brans J-P, Mareschal B. Promethee Methods. Multiple Criteria Decision Analysis: State of the Art Surveys. Springer New York; 2005. pp. 163–186.
47. McGregor S, Henderson KJ, Kaldor JM. How Are Health Research Priorities Set in Low and Middle Income Countries? A Systematic Review of Published Reports. PLoS ONE. 2014;9. pmid:25275315
48. Horby P, Hoa N, Pfeiffer D, Wertheim HL. Drivers of Emerging Zoonotic Infectious Diseases. In: Yamada A, Kahn LH, Kaplan B, Monath TP, Woodall J, Conti L, editors. Confronting Emerging Zoonoses. Springer Japan; 2014. pp. 13–26.
49. Pavlin BI, Kool JL, Samo MH, Gerstel L, Working Group on National Notifiable Diseases Prioritization of the Federated States of Micronesia Department of Health and Social Affairs. A standardized process for developing a national notifiable diseases list in a pacific island setting. Asia Pac J Public Health. 2010;22: 279–288. pmid:19666949
50. Abelson J, Giacomini M, Lehoux P, Gauvin F-P. Bringing "the public" into health technology assessment and coverage policy decisions: From principles to practice. Health Policy. 2007;82: 37–50. pmid:16996637
51. Linkov I, Satterstrom FK, Kiker G, Seager TP, Bridges T, Gardner KH, et al. Multicriteria Decision Analysis: A Comprehensive Decision Approach for Management of Contaminated Sediments. Risk Anal Int J. 2006;26: 61–78. pmid:16492181
52. Kiker GA, Bridges TS, Varghese A, Seager TP, Linkov I. Application of multicriteria decision analysis in environmental decision making. Integr Environ Assess Manag. 2005;1: 95–108. pmid:16639891
53. Slovic P. Perception of risk. Science. 1987;236: 280–285. pmid:3563507
54. Krewski D, Lemyre L, Turner MC, Lee JEC, Dallaire C, Bouchard L, et al. Public Perception of Population Health Risks in Canada: Health Hazards and Sources of Information. Hum Ecol Risk Assess Int J. 2006;12: 626–644.
55. Wildavsky A, Dake K. Theories of Risk Perception: Who Fears What and Why? Daedalus. 1990;119: 41–60.

56. Confalonieri UE, Menne B, Akhtar R, Ebi KL, Hauengue M, Kovats RS, et al. Climate Change 2007: Impacts, Adaptation and Vulnerability. Contribution of Working Group II to the Fourth Assessment Report of the Intergovernmental Panel on Climate Change. In: Parry ML, Canziani OF, Palutikof JP, van der Linden PJ, Hanson CE, editors. 2007: Human health. Cambridge, UK: Cambridge University Press; 2007. pp. 391–431.

CHAPTER 5

Data Distribution in Public Veterinary Service: Health and Safety Challenges Push for Context-aware Systems

Laura Contalbrigo[1], Stefano Borgo[2], Giandomenico Pozza[1] and Stefano Marangon[1]

[1] Istituto Zooprofilattico Sperimentale delle Venezie, Viale Dell'Università 10, 35020 Legnaro, (PD), Italy

[2] Laboratory for Applied Ontology, ISTC-CNR, Via alla Cascata, 56/C, 38123 Trento, Italy

Citation: Contalbrigo, L., Borgo, S., Pozza, G. et al. Data distribution in public veterinary service: health and safety challenges push for context-aware systems. BMC Vet Res 13, 397 (2017). https://doi.org/10.1186/s12917-017-1320-0.

ABSTRACT

Background

Today's globalised and interconnected world is characterized by intertwined and quickly evolving relationships between animals, humans and their environment and by an escalating number of accessible data for public health. The public veterinary services must exploit new modeling and decision strategies to face these changes. The organization and control of data flows have become crucial to effectively evaluate the evolution and safety concerns of a given situation in the territory. This paper discusses what is needed to develop modern strategies to optimize data distribution to the stakeholders.

Main text

If traditionally the system manager and knowledge engineer have been concerned with the increase of speed of data flow and the improvement of data quality, nowadays they need to worry about data overflow as well. To avoid this risk an information system should be capable of selecting the data which need to be shown to the human operator. In this perspective, two aspects need to be distinguished: data classification vs data distribution. Data classification is the problem of organizing data depending on what they refer to and on the way they are obtained; data distribution is the problem of selecting which data is accessible to which stakeholder. Data classification can be established and implemented via ontological analysis and formal logic but we claim that a context-based selection of data should be integrated in the data distribution application. Data distribution should provide these new features: (a) the organization of situation types distinguishing at least ordinary vs extraordinary scenarios (*contextualization of scenarios*); (b) the possibility to focus on the data that are really important in a given scenario (*data contextualization by scenarios*); and (c) the classification of which data is relevant to which stakeholder (*data contextualization by users*).

Short conclusion

Public veterinary services, to efficaciously and efficiently manage the information needed for today's health and safety challenges, should contextualize and filter the continuous and growing flow of data by setting suitable frameworks to classify data, users' roles and possible situations.

BACKGROUND

The increasing interdependence between humans, animals and their products as well as the close association with companion animals have encouraged a change in the public health system thinking [1]. The study of public veterinary systems has rapidly grown as a domain in itself supported by the adoption since 1984 of the "One Health" paradigm as an effective strategy for the prevention and control of zoonoses [2, 3]. In an increasingly globalised world, this new approach encompasses zoonotic infections, food safety, the environment and the health delivery systems. The integration of the epidemiological and economic frameworks within the new technological turn called the *Internet of Things* (IoT), led to an escalating number of distributed sensors, which, together with the laboratory data, make available large amounts of information [4]. Indeed, health management systems are fully part of the so-called *big data environment* [5]. Nowadays international organizations like FAO and OIE are developing web based applications to collect, store and analyze data from different sources with the aim to combine epidemiological, spatial and genetic data as well as data about farming systems in order to provide appropriate information to public authorities and scientists on the emergence and spread of animal diseases [6]. This huge amount of data requires a methodological approach to effectively assess and manage the health and safety of the ecosystems [7].

This is a significant challenge, which calls for efficient public veterinary data management systems, for quick and focused data distribution as to better help the stakeholders in their understanding of the situation and decision-making processes [8,9,10].

Every day, public veterinary managers, as well as other stakeholders like farmers, food producers, veterinarians and distributors deal with huge quantities of data on a variety of topics and different organization quality, often with several types and forms. These actors need to use them to understand scenarios in the territory and to make decisions to manage both ordinary and emergency situations. Here, we use the term 'situation' to indicate the state of an area or environment of interest as it is at some point in time. Situations can be classified as positive, negative or neutral by each stakeholder, depending on its role and interests, distinguishing, in particular, between ordinary ones (positive or neutral) and emergency ones (negative).

In ordinary situations, surveillance, monitoring and eradication plans are the basis of veterinary public health. The implementation of these entails the need to elaborate the data collected during previous programs and to integrate them with the new epidemiological information. This is crucial to make an effective risk assessment and to decide on activity plans finalized to guarantee the achievement of sanitary goals with minimum cost and risk for the consumers, the producers and the public veterinary system itself [11,12,13].

Analogously, the management of emergency situations (*mitigation and prevention* - e.g., risk mapping and hazard identification; *preparedness and planning* - e.g., contingency planning, personnel training, warning system; *response and recovery* - e.g., contingency plan activation, epidemiologic surveillance) [14] requires to look at different information types integrating several data sources to correctly identify the kind of incident, the location, the magnitude and to monitor the evolution of the situation (e.g., geographical coordinates, logistic and environmental data, animal, means of transport and people movements, laboratory data). For instance, in veterinary epidemic emergencies the identification of outbreaks and their eradication can be managed effectively only when the competent authorities have quick access to reliable, well processed and pre-analysed information on health-related data [15, 16]. This information should be coherently and rapidly organized and presented in a very intuitive way to be quickly assimilated and analysed by public health managers for decision-making. Otherwise, doubts about the real situation can arise with delay in critical decisions. In the worst-case scenario, data can be distributed in a scattered way, leading to missing information and even misinterpretation [17]. The successful management of a veterinary control system depends on the quality of the data as well as on their correct interpretation. The organization of the data distribution with respect to the targeted data user is clearly a key factor [18,19,20].

Therefore, to be efficient, a veterinary data management system should comply with the needs of the different stakeholders that operate in the same territory although with different perspectives: monitoring, surveillance, screening and modeling of infectious diseases, prevalence estimation, risk-factor study, quantitative risk assessment, product evaluation, commercialization and so on. Each stakeholder is interested in the goals related to his/her duties and these can be quite apart going from health hazard and commercialization approval to animal health monitoring. The stakeholder's purposes indicate his/her interest in the situation and the subset of data relevant to him/her. Furthermore, when two stakeholders need the very same data, they might require them in different formats or granularity.

Moreover, the very same data can acquire different connotations: they can be considered positive by one actor and negative by another [21].

Finally, the meaning of data is often determined in comparison with other data [22]. For example, during an Avian Influenza outbreak in a poultry farm, the poultry producer focuses on the pass/fail result of the laboratory test because this may lead to the stamping out of the entire flock. Instead, the public authority evaluates the same data against a larger period of time, a broader spatial region, considering virus genetic features, comparing this information with data originating from other farms or connected production sites and considering a variety of possible epidemiological scenarios [23,24,25].

These observations show that data interpretation can be challenging and that the traditional approach based on comprehensive reports collecting large numbers of heterogeneous data cannot be a solution anymore, especially if the data relevant for the user are spread across the whole document and reported using granularities not promptly comparable [26, 27]. Of course, beside data interpretation there are other elements that influence the manager in the decision-making process, e.g., scientific background, regulations, time constraints, experience on the field, economic as well as socio-political considerations [28, 29]. While recognizing the importance of the other elements within the whole veterinary system, we claim that there is the need to improve the impact and to correct the interpretation of data taking into account the context in which they are collected and the role of the actors that read them [30]. For these reasons, it is time for the public veterinary data management system to exploit new data modeling and data distribution strategies that can match today's information needs.

Data misleading due to wrong contextualization, distribution of information or representation is a daily problem within the big data environment in public health [31, 32]. Many examples are reported in both human and veterinary medicine. Especially computational epidemiology needs large and flexible datasets for epidemiological models based on records of variable forms and origins (e.g., types of individuals at specific locations and points in time, records of movement of individuals between locations, social media data, digital traces and results of diagnostic test and genomic analysis) [33]. For example, EMPRES-i (FAO's global animal disease information system) offers mapping and graphic functions combining data coming from different databases that do not use the same references to describe host taxonomy and geographical locations, therefore, at the beginning, it was not possible to establish an accurate correspondence between the related references [34]. Moreover, in the veterinary domain, *geographical information systems* (GIS) required spatial data of animal holdings. Often these records lack relevant information, leading to data incompatibility and even inconsistency [35]. Furthermore, various techniques need to be used to enhance decision-making process in the face of data uncertainty as happens when developing information systems based on sparse and incomplete datasets of varying quality as for example the case of GLiPHA (Global Livestock Production and Health Atlas) [36]. Likewise in human medicine, epidemiologists face similar concerns when mapping raw data in epidemiological studies like cancer mortality rate: the generation of spurious spatial features and statistical artifacts may lead to spurious spatial pattern [37, 38]. Regarding misleading interpretation of statistical analysis results, in oncologic clinical trials, it is necessary to identify accurate outcomes and endpoints as well as censoring in the study, otherwise statistical analyses can easily fail [39]. In biology the rapid increase in data volumes due to new molecular technologies like the *next-generation sequencing* (NGS), and the rise of omics data generate new data types and it is still unclear how to process, store and integrate them to produce usable knowledge in accessible formats for healthcare professionals [40,41,42].

It is evident that the selection of the data to distribute, the granularity of the information and the outline of the report should start from the goals of the targeted stakeholder, including its needs and education [43]. Moreover, following the trend of the IoT [44] and the large use of *Information Technology* (IT) devices like "*precision livestock farming*" (PLF), we cannot ignore the problem of integrating and understanding data generated for specific purposes using different sensors and methodologies by farmers, producers and dealers (e.g., cow pedometer, environment sensors, data-logger, video-taping/recording, sensors installed in moving devices, accelerometer) [45,46,47,48]. In this framework it is essential to merge these data using reliable methodologies based on typology and provenance as well as device capacities, location and trustability. While typology and provenance are successfully analyzed via the application of ontological frameworks, see e.g. [49,50,51], contextualization techniques for the distinct stakeholders need to be better understood. As an example, Livestock Geo-wiki, a platform that aims to develop a global livestock information system, provides innovative visualization and analysis tools addressing specific requirements of different groups of users but it does not consider contextualization techniques focused on users' need classification [52].

By introducing a systematic approach for data classification and for the delineation of the stakeholders' perspective, we can deepen the accuracy of data distributed to the stakeholders, like sensitivity and specificity [53], increase their data understanding and trust, and at the same time reduce the risk of data overflow.

The outcomes of the study and contextualization of data have also a positive impact on the integration of information from various datasets, which can be harvested for new purposes beyond early detection of animal diseases or food safety issues, like epidemiological studies, pharmacovigilance or identification of emerging risks [54, 55]. Finally, data contextualization increases awareness of the potential risks and ethical challenges related to data sensitivity, legitimacy requirements, and the relationship between ethics and methodology [56].

MAIN TEXT

The Need of Data Selection Strategies in the Public Veterinary Management System

The public veterinary data management system is an important actor in the territorial network of stakeholders since it functions as an information hub from which the different users receive updated data. This flow of information from the public veterinary data management system and the stakeholders ensures that each actor has a proper understanding of the situation in the field. The core activities of this management system, for what concerns us here, are the acquisition, integration, elaboration, classification and distribution of veterinary data in the field of food safety, animal health and welfare. Information systems are essential to run public veterinary services since veterinary monitoring and surveillance are largely data-oriented to the point that the term *big data environment* has become paradigmatic even in this domain [57, 58]. Public veterinary data management systems have been working with a traditional approach: they collect information from different sources (like sample collectors, public officers, or their own laboratories) and release it to the stakeholders via the use of reports. Reports are data sets organized according to formats fixed once for all. Although there are a few exceptions, mainly due to increased sensibility to privacy concerns, the principle is pretty clear: any data request type is associated with a report type. The same report is released to all the stakeholders no matter the context, their roles or the situation in the field. In short, the management system has no discretional option: it must distribute to every interested party all data relative to a request using some pre-established format.

Today the IT evolution and the new challenges due to the IoT and the so-called big data issue, require to re-assess the public veterinary system organization and strategies. With the exponential grow of data, it is easy to be overwhelmed by the distributed amount of information that has the inherent risk of overpassing the cognitive capacities to correctly read, understand, compare and evaluate them [59]. One could hope that the increasing number of data remains a problem at the IT level since ad hoc services for data filtering (e.g., customization at the client's side or online search facilities) could help to select the needed information for each evaluation and decision step. If that were true, each stakeholder would still want to access all the available information on a given scenario and use its own services to select what to focus on. Unfortunately, this view is naïve. Data are complex entities that have different meanings when looked at in isolation or in collections. Furthermore, the way data are organized, impacts the narrative they give of the territory. For example OpenFluDB, a large database of influenza virus sequences, is isolate-centric rather than sequence-centric like other influenza virus resources. This choice makes easier the association with a comprehensive amount of clinical and epidemiological data [60].

If we rely on services at the stakeholders' side, we cannot guarantee a coherent reading of the situation across the stakeholders in the territory. The risk is that incompatible interpretations may be adopted leading to implement mutually contradictory action plans. The organization of a coordination level across stakeholders could solve this last issue, but it is not practical for at least two reasons. 1) stakeholders (e.g. producers and public officers) have different goals and each would try to enforce its own reading of the data; 2) a coordination level adds an extra lapse of time which is a luxury we cannot afford in emergency situations.

As said, reports with large amounts of data are difficult to read since the user has:

- to look at different sections;
- to identify the subsets of data relevant to its role and knowledge;
- to rewrite them in the granularity of interest;
- to compare these data;
- to find a narrative that explains the data coherently with the territory type and history.

This is especially hard when the user is under pressure, like in emergency scenarios and there is no time to translate from the different standards for data management or even to make confident comparisons and integration of data coming from different data sets with changing measure units, density, frequency, etc. More subtlety, the information in the reports is presented from the general perspective of the veterinary service and considers neither the role and goals of the stakeholders nor the context and motivations for which data have been collected. For example, data on samples collected in a monitoring program for dairy farms in a certain area cannot be reliably used to define the health status of the bovine population in the same area including beef livestock. The correct integration of these two types of data requires knowledge that the stakeholder may not have. To reduce the risk of misjudgment and of missed information, we need to investigate which data should be presented to a given user, in which order and in which format. To answer this, one has to develop methodologies for data selection. Data selection can be done in different ways, e.g., task or function oriented. A selection is task-oriented when only data relevant for a determined task (e.g. decision making) are presented to the operator. The selection is functional when the information is shown depending on the general role the operator has in the system (e.g. surveillance). However, data selection is not an isolated step. Two further steps need to be considered: data classification and data distribution. Data classification is the problem of organizing data according to what they refer to and to the way they are obtained. Data distribution is the problem of choosing which data to distribute to which stakeholder and in which context. Data distribution can be seen as the specific use of a data selection methodology.

Focus on Stakeholders' Classification

In the context of the public veterinary data management system, the classification of data and metadata, the data selection methodologies and the data distribution should be guided by the stakeholder's technical knowledge, role and goals. Note that the role of the user constrains its goals, yet the goals must be explicitly included in the data management since they can change depending on contextual information like available resources. Technical background, role and goal are thus the factors that can be used to balance the tension between full vs limited disclosure in order to optimize data understanding and reduce data overflow. This leads to the issue of user's classification. If the classification of data is based on what data are about and how they are collected, the classification of users is based on their role in the system. These classifications are not affected by the context and are independent from the particular scenario. This means that one can apply the methodologies developed in the applied ontology area to provide guidelines and criteria [49,50,51].

Regarding data selection and distribution, we now look at the management of veterinary laboratory data. Laboratory data furnish a fairly simple case without exogenous factors that could hinder the understanding of the problem. To justify and ground the selection and distribution of data, let us assume that a suitable framework for the classification of the stakeholders is available. This framework should be based on the profiling of the user's types, use this profiling to collect the user's requirements and to characterize their perspectives as well as aims [61, 62]. Given this, the veterinary management system can prepare a data selection methodology for laboratory data that, depending on the interests, action capacities, technical knowledge and goals of the stakeholder, reasons on the usefulness and completeness of some data set for that user. Note that the goal is not to hide information to the user. On the contrary, the goal is to organize and highlight the data relevant to that user so to put him/her in the best position for data comparison and understanding as it happens using the support of data warehouse in the decision making process [63,64,65].

The laboratory data example shows also one point we introduced earlier: the need for context classification. Several research communities have been discussing the use of contexts [66, 67] and yet no general approach has been developed since proposals like [68,69,70] are limited to some application domain or to some types of scenario only. In the case of veterinary services, the context should tell the characteristics of the stakeholders in terms of rights to know, official goals and duties, possibly even information on the procedures it applies and actions it can take. This information should be integrated with the situation in which the stakeholder operates, that is, its environment.

The Notion of Context and Its Impact on the Assessment of Data Knowledge

Being the points cited in the previous paragraph the core elements that a notion of context should manage, a recent proposal by Mizoguchi et al. [71], developed in the area of function context, could help. Mizoguchi and collaborators aim to introduce a general notion of function that applies to both the natural and the engineering domains. They distinguish between three kinds of contexts: systemic, use and design context. The systemic context looks at functionalities of a component from the viewpoint of the system where it is embedded; the use context looks at functionalities of a device in the scenario where the device is used; the design context looks at functionalities of a device from the perspective of the device's designer. The notion of context in the framework of veterinary public health systems can build upon Mizoguchi's notion of use context. The use context is associated with a situation and fixed by an intentional agent, which is a participant in the situation aiming to reach some goal(s). The context is understood as an event where a device, a goal, an agent and the environment are at least partially known. The use context is thus a complex entity analyzed from the user's perspective to identify what function a device (a natural or artifactual part) performs in a certain environment [72].

A veterinary situation presents important differences with respect to Mizoguchi assumptions. Veterinary public health systems deal with information in reports, not physical objects. Furthermore, the process from data collection (like diagnostic activities) to data distribution (like a report release) is not aimed to change the physical world: it starts with the completion of a laboratory test and ends with the users' knowledge update. In this view, we can think of laboratory data and metadata like *information devices* in veterinary public health systems that are used to change not the physical reality but the users' knowledge of physical reality. The analogy is between acting on the world to achieve a physical change and acting on the mind to achieve a knowledge change. This match between user contexts in functional studies and user contexts in veterinary health systems suggests that the data provided by the laboratory (the epidemiologic unit, etc.) have a precise functional role: they make possible to realize the needed changes in the user's mind in terms of situation's knowledge and awareness. The distribution of a set of data, with a suitable format and organization for a user, impacts the user's knowledge about the actual situation and, subsequently, on how the user can act to fulfill its role and duties. In other words, a veterinary management system looks as an agent whose goal is to provide data users with an understanding of the actual situation suitable and reliable for their goals and perspectives, and whose tools are data and metadata which must be carefully tuned (that is, selected and distributed) to achieve the desired result. Following the terminology in [71], we call this special user context the *institutional veterinary context.*

Assuming a data user classification and characterization system is available, covering institutional user(s) and consumers, producers and all legal and economic stakeholders, for each agent the data provider has to select the data and metadata to distribute, to decide the granularity of the data, and to organize the report format. This choice of data selection and distribution is driven by the classification of the data and of the users and by an evaluation of the data and metadata that they expect for their interests and activities including their optimal organization [73,74,75].

Given this setting, data set can be qualified as follows. A data set is *knowledge complete* for a user type in a situation if from this data set and from standard background knowledge for that user type, one can derive how much the area is departing from its standard and acceptable states, the possible risks as well as short and long term consequences. A data set is *action complete* for a user type in a situation if from this data set and from standard background knowledge for that user type, one has all the knowledge needed to develop an optimal plan of actions to achieve its goals [76, 77]. Note that knowledge and action completeness are distinct: the first focuses on understanding the situation and how it may evolve, the latter takes into consideration the user's aims and its action capacities.

Furthermore, we can say that a data set is knowledge (or action) *minimal* for a data user if none of its proper subsets is knowledge (action, respectively) complete for that user in the sense defined above. Then, institutional veterinary contexts help to assess data knowledge and action completeness and minimality for a user by setting a framework to reliably verify effects of data awareness on the user's decisions/actions in different scenarios.

Let us call *controlled* a situation in which the data set provided by the veterinary data management system is knowledge and action complete. Recall that this means that the user can understand the situation in the area of interest and has all the information to act optimally for its goals. Note that a controlled situation is just a situation for which we know how to optimize data flow. The situation itself can be ordinary (positive or neutral) or an emergence (negative). Situations that are *uncontrolled* are the most dangerous: they are either wrongly classified or wrongly estimated to the point that the stakeholder makes wrong decisions and the consumers, animals and environment could be exposed to excessive risks, erratic behaviors of the market operators, and unnecessary economic losses. A flexible data integration and distribution, i.e., the selection of data to distribute contextualized to user types, data interrelation and previous knowledge of the situation, can drastically reduce the class of uncontrolled situations [78, 79]. The flexible integration of data and their contextualization are new conceptual tools that can turn *uncontrolled* situations into *controlled* ones while increasing awareness and confidence in the data users. We call *contextualized* the data of a report filtered to match the interests of a specific user, presented depending on the situation at stake and provided with the needed granularity and organization (Fig. 1). When data are contextualized, the data received by a user are even less prone to be misinterpreted, they can effectively help to properly evaluate the risks and to take the correct decisions.

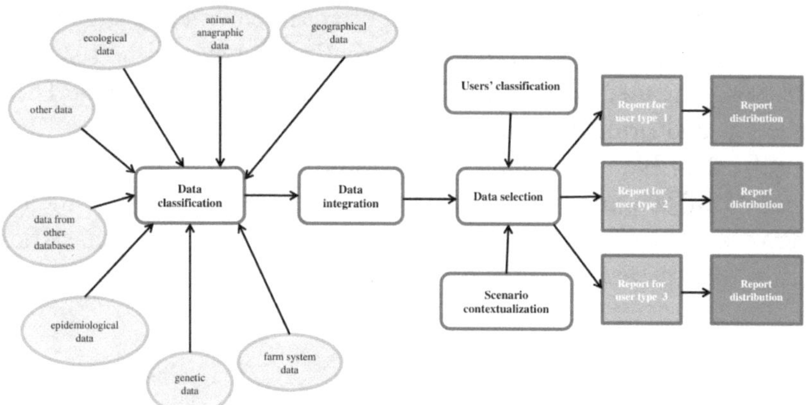

Figure 1: Example of a data flow diagram with a flexible data distribution in accordance with users' classification and scenario contextualization.

In conclusion, modern veterinary data management systems should include services like data classification, selection and distribution that include these further contextualization features: (a) the organization of situation types starting from the distinction between ordinary vs emergence scenarios (*contextualization of scenarios*); (b) the possibility to focus on the data that are important in a given scenario (*data contextualization by scenarios*); and (c) the classification of which data is relevant to which stakeholder (*data contextualization by users*). With this approach, veterinary data providers can become active information hubs enabling the generation of sustainable and optimized data flows suitable for today's information age based on IoT and big data. While the advantages of this approach are clear, further discussions are needed to analyze how the veterinary management system would change in relation to, e.g., ethical challenges like privacy concerns and the balance between public health and the individual rights [80, 81].

SHORT CONCLUSION

In the last decades, the study of health management systems has rapidly increased embracing the public veterinary services. We claimed that these services, to efficaciously and efficiently manage the information needed today, must include classification and contextualization techniques to filter the increasing and continuous flow of data. Such a step can improve the overall quality of the decision processes since these are highly sensitive to the quantity, quality and organization of the received information. This new approach may become particularly important in emergency situations since today the stakeholders need to quickly realize and assess the risk situation while being engaged in assessing large amounts of incoming data. Therefore a coherent and principled notion of context is essential to help filter and promptly distribute the right information to each user. We introduced the notion of institutional veterinary context, among others, which sets a framework to assess the different techniques for flexible data classification, selection and distribution taking into account contextual perspectives.

REFERENCES

1. Van Olmen J, Marchal B, Van Damme W, Kegels G, Hill PS. Health systems frameworks in their political context: framing divergent agendas. BMC Public Health. 2012;12:774.
2. Bidaisee S, Macpherson NLC. Zoonoses and one health: a review of the literature. J Parasitol Res. 2014; https://doi.org/10.1155/2014/874345.
3. Dahal R, Kahn L. Zoonotic diseases and one health approach. Epidemiology. 2014; https://doi.org/10.4172/2161-1165.1000e115.
4. Simonsen L, Gog JL, Olson D, Viboud C. Infectious disease surveillance in the big data era: towards faster and locally relevant systems. JID 2016; doi:10.1093/infdis/jiw376.
5. Lee EC, Aher JM, Goldlust S, Kraemer JD, Lawson AB, Bansal S. Mind the scales: harnessing spatial big data for infectious disease surveillance and inference. JID. 2016; doi:10.1093/infdis/jiw344.
6. Farnsworth ML, Hamilton-West C, Fitchett S, Newman SH, de La Rocque S, De Simone L, et al. Comparing national and global data collection systems for reporting. Outbreaks of H5N1 HPAI. Prev Vet Med. 2010;95:175–85.
7. Cocker R, Rushton J, Mounier-Jack S, Karimuriba E, Lutumba P, Kambarage D, et al. Towards a conceptual framework to support one-health research for policy on emerging zoonoses. Lancet Infect Dis. 2011;11:326–31.
8. Marabelli R. The role of official veterinary service in dealing with new social challenges: animal health and protection, food safety, and the environment. Rev Sci Tech OIE. 2003;22(2):363–71.
9. Capua I, Alexander DJ. The challenge of avian influenza to the veterinary community. Avian Pathol. 2006;35(3):189–205.
10. Wendt A, Kreienbrock L, Campe A. Zoonotic disease surveillance – inventory of systems integrating human and animal disease information. Zoonoses Public Health. 2015;62:61–74.
11. Jameson LJ, Medlock JM. Tick surveillance in great Britain. Vector Borne Zoonotic Dis. 2011;11(4):403–12.
12. Vrbova L, Patrick DM, Stephen C, Robertson C, Koehoorn M, Parmley EJ, Galanis E. Utility of algorithms for the analysis of integrated salmonella surveillance data. Epidemiol Infect. 2016;144(10):2165–75.

13. Umhang G, Comte S, Hormaz V, Boucher JM, Raton V, Favier S, Boué F. Retrospective analyses of fox feces by real-time PCR to identify new endemic areas of Echinococcus multilocularis in France. Parasitol Res. 2016:1–5.
14. Cova TJ. GIS in emergency management. In: Longley PA, Goodchild MF, Maguire DJ, Rhind DW, editors. Geographical information systems: principles, techniques, applications and management. New York: Wiley; 1999. p. 845–58.
15. Iannitti S, Savini L, Palma D, Calistri P, Natale F, Di Lorenzo A, et al. An integrated web system to support veterinary activities in Italy for the management of information in epidemic emergencies. Prev Vet Med. 2014;113:407–16.
16. Drewe AJ, Hoinville LJ, Cook AJC, Floyd T, Gunn G, Stärk KDCSERVAL. A new framework for the evaluation of animal health surveillance. Transbound Emerg Dis. 2015;62:33–45.
17. Dusse F, Jùnior PS, Alves AT, Novais R, Vieira V, Mendoça M. Information visualization for emergency management: a systematic mapping study. Expert Syst Appl. 2016;45:424–37.
18. Dorea CF, Sanchez J, Revie CW. Veterinary syndromic surveillance: current initiatives and potential for development. Prev Vet Med. 2011;101:1–17.
19. Dupuy C, Bronner A, Watson E, Wuyckhuise-Sjouke L, Reist M, Fouillet A, et al. Inventory of veterinary syndromic surveillance initiatives in Europe (triple-S project): current situation and perspectives. Prev Vet Med. 2013;111:220–9.
20. Daniels P, Poermadjaja B, Morrissy C, Ngo TL, Selleck P, Kalpravidh W, et al. Development of veterinary laboratory networks for avian influenza and other emerging infectious disease control: the southeast Asian experience. EcoHealth. 2014;11:44–9.
21. Fürber C. Data Quality. In: Fürber C, editor. Data quality management with semantic technologies. Wiesbaden: springer Fachmedien; 2016. p. 20–55.
22. Clarke R. Big data, big risks. Inf Syst J. 2016;26(1):77–90.
23. Martell-Moran NK, Mauer WA, Kaneene JB. Assessment of avian influenza surveillance and reporting needs of stakeholders in Michigan, 2007. J Am Vet Med Assoc. 2011;238(12):1570–83.

24. Comin A, Klinkenberg D, Marangon S, Toffan A, Stegeman A. Transmission dynamics of low pathogenicity avian influenza infections in Turkey flocks. PLoS One. 2011;6(10):e26935.
25. Dalla Pozza M, Ceolin C, Marangon S. Emergency response following suspicion of an avian influenza outbreak. Zoonoses Public Health. 2008;55(1):50–3.
26. Ferraro S, Braga F, Panteghini M. Laboratory medicine in the new healthcare environment. Clin Chem Lab Med. 2016;54(4):523–33.
27. Huang L, Fernandes H, Zia H, Tavassoli P, Rennert H, Pisapia D, Elemento O. The Precision Medicine Knowledge Base: an online application for collaborative editing, maintenance and sharing of structured clinical-grade cancer mutations interpretations. BioRxiv. 2016:e059824. doi:10.1101/059824.
28. Giangaspero M, Sekiguchi S. Risk assessment of animal infectious diseases and decision making process. Clin Microbiol. 2016;5:242.
29. Del Giudice M, Caputo F, Evangelista F. How are decision systems changing? The contribution of social media to the management of decisional liquefaction. J Decis Syst. 2016:1–13.
30. Dey AK. Understanding and using context. Pers Ubiquit Comput. 2001;5:4–7.
31. Metcalf CJ, Edmunds WJ, Lessler J. Six challenges in modeling for public health policy. Epidemics. 2015;10:93–6.
32. Hansen HK, Porter T. What do big data do in global governance? Glob Gov. 2017;23(1):31–42.
33. Marathe M, Vullikanti AKS. Computational epidemiology. The challenge of developing and using computer models to understand and control the diffusion of disease through populations. Commun ACM. 2013;56(7):88–96.
34. Claes F, Kuznetsov D, Liechti R, Von Dobschuetz S, Truong BD, Gleizes A, et al. The EMPRES-i genetic module: a novel tool linking epidemiological outbreak information and genetic characteristics of influenza viruses. Database. 2014; https://doi.org/10.1093/database/bau008.
35. Ferrè N, Kuhn W, Rumor M, Marangon SA. Conceptual holding model for veterinary applications. Geospat Health. 2014;8(2):583–96.
36. Clements ACA, Pfeiffer DU, Otte MJ, Morteo K, Chen LA. Global

livestock production and health atlas (GLiPHA) for interactive presentation, integration and analysis of livestock data. Prev Vet Med. 2002;56:19–32.

37. Gelman A, Price NP. All maps of parameter estimates are misleading. Stat Med. 1999;18:3221–34.
38. Huck JJ, Whyatt JD, Coulton P. Visualising patterns in spatially-ambiguous point data. JOSIS. 2015;10:47–66.
39. Heller J. Epidemiological and statistical considerations for interpreting and communicating oncology clinical trials. Vet J. 2015;205(2):233–7.
40. Cook CE, Bergman MT, Finn RD, Cochrane G, Birney E, Apweiler R. The European bioinformatics institute in 2016: data growth and integration. Nucleic Acids Res. 2016;44:D20–6.
41. Zhulin IB. Databases and microbiologists. J Bacteriol. 2015;197(15):2458–67.
42. Manzoni C, Kia DA, Vandrovcova J, Hardy J, Wood NW, Lewis PA, Ferrari R. Genome, transcriptome and proteome: the rise of omics data and their integration in biomedical research. Brief Bioinform. 2016:1–17. https://doi.org/10.1093/bib/bbw114.
43. Jelineck HF, Yatsko A, Stranieri A, Venkatraman S, Bagirov A. Diagnostic with incomplete nominal/discrete data. Airman. 2015;4(1):22–35.
44. Atzori L, Iera A, Morabito G. The internet of things: a survey. Comput Netw. 2010;54(15):2787–805.
45. Ruiz-Garcia L, Lunadei L, Barreiro P, Robla JIA. Review of wireless sensor technologies and applications in agriculture and food industry: state of the art and current trends. Sensors. 2009;9(6):4728–50.
46. Halachmi I, Guarino M. Editorial: precision livestock farming: a "per animal" approach using advanced monitoring technologies. Animal. 2016;10(9):1482–3.
47. Terrasson G, Llaria A, Marra A, Voaden S. Accelerometer based solution for precision livestock farming: geolocation enhancement and animal activity identification. IOP Conf Ser Mater Sci Eng. 2016;138(1):012004.
48. Junkan A, Masip-Bruin X, Amla N. Smart Computing and Sensing Technologies for Animal Welfare: A Syst Rev. arXiv. 2016;1609.00627.
49. Borgo S, Masolo C. Ontological foundations of DOLCE. In: Poli R, Healy M, Kameas A, editors. Theory and applications of ontology:

computer applications. Dordrecht: Springer; 2010. p. 279–95. doi:10.1007/978-90-481-8847-5_13.

50. Borgo S, Pozza G. Knowledge objects: a formal construct for material, information and role dependences. KMRP. 2012;10(3):227–36.
51. Pozza G, Borgo S, Oltramari A, Contalbrigo L, Marangon S. Information and organization in public health institutes: an ontology-based modeling of the entities in the reception-analysis-report phases. J Biomed Semant. 2016;7(1):1–15.
52. Robinson TP, Wint GRW, Conchedda G, Van Boeckel TP, Ercoli V, Palamara E, et al. Mapping the global distribution of livestock. PLoS One. 2014;9(5):e96084.
53. Greiner M, Gardner IA. Application of diagnostic tests in veterinary epidemiologic studies. Prev Vet Med. 2000;45:43–59.
54. White RW, Tatonetti NP, Shah NH, Altman RB, Horvitz E. Web-scale pharmacovigilance: listening to signals from the crowd. J Am Med Inform Assoc. 2013;20:404–8.
55. Altman RB. Web-scale pharmacovigilance in the discovery of pharmacological effects and toxicity. San Jose, CA: AAAS Annual Meeting; 2015. https://aaas.confex.com/aaas/2015/webprogram/Paper13785.html. Accessed 10 Nov 2016.
56. Vayena E, Salathé M, Madoff LC, Brownstein JS. Ethical challenges of big data in public health. PLoS Comput Biol. 2015;11(2):e1003904.
57. Pfeiffer DU, Stevens KB. Spatial and temporal epidemiological analysis in the big data era. Prev Vet Med. 2015;122:213–20.
58. Hill AA, Crotta M, Wall B, Good L, O'Brien SJ, Guitian J. Towards an integrated food safety surveillance system: a simulation study to explore the potential of combining genomic and epidemiological metadata. R Soc open sci. 2017;4:160721. https://doi.org/10.1098/rsos.160721.
59. Sivarajah U, Kamal MM, Irani Z, Weerakkody V. Critical analysis of big data challenges and analytical methods. J Bus Res. 2017;70:263–86.
60. Liechti R, Gleizes A, Kuznetsov D, Bougueleret L, Le Mercier P, Bairoch A, Xenarios I. OpenFluDB, a database for human and animal influenza virus. Database 2010. Doi:10-1093/database/baq004.
61. Quintarelli E, Rabosio E, Tanca LA. Principled approach to context schema evolution in a data management perspective. Inform. Syst.

2015;49:65–101.

62. Zapater JJS, Escriva DML, Garcià FRS, Durà JJM. Semantic web service discovery system for road traffic information services. Exp Syst Appl. 2015;42:3833–42.
63. Crauwels APP, de Konig R, Nielen M, Elbers ARW, Dijkhuinzen AA, Tielen MJMA. Concept for a decision support system based on practical experiences from a national disease emergency: the Dutch experience. Acta Vet Scand. 2001;94:61–9.
64. Nielsen AC. Data warehouse for assessing animal health, welfare, risk management and communication. Acta Vet Scand. 2011;53(1):S3.
65. Xu Z, Lee J, Park D, Chung Y. Multidimensional analysis model for highly pathogenic avian influenza using data cube and data mining techniques. Biosyst Eng. 2017;157:109–21.
66. Koç H, Henning E, Jastram S, Starke C. State of the art in context modeling – a systematic literature review. LNBIP. 2014;178:53–64.
67. Mnatsakanyan ZR, Burkom HS, Hashemian MR, Coletta MA. Distributed information fusion models for regional public health surveillance. Inf Fusion. 2012;13:419–32.
68. Dourish P. What we talk about when we talk about context. Personal Ubiquitous Comput. 2004;8:19–30.
69. Wang H-H, Boukiamp F, Elghamrawy T. Ontology-based approach to context representation and reasoning for managing context-sensitive construction information. J Comput Civ Eng. 2011;25(5):331–46.
70. Grill T, Tscheligi M. Towards a multi-perspectival approach of describing context. LNCS. 2011;6967:115–8.
71. Mizoghuci R, Kitamura Y, Borgo S. Towards a unified definition of function. Fr Art Int. 2012;239:103–16.
72. Mizoguchi R, Kitamura Y, Borgo SA. Unifying definition for artifact and biological functions. Appl Ontol. 2016;11(2):129–54.
73. Chen Z-Y, Fan Z-P, Sun M. Behavior-aware user response modeling in social media: learning from diverse heterogeneous data. Eur J Oper Res. 2015;241:422–34.
74. De Andrés J, Pariente B, Gonzalez-Rodriguez M, Lanvin DF. Towards an automatic user profiling system for online information sites: identifying demographic determining factors. Online Inform Rev. 2015;39(1):61–80.

75. Haesen R, Snoeck M, Lemahieu W, Poelmans S. On the definition of service granularity and its architectural impact. LNCS. 2008;5074:375–89.

76. Tempelman-Kluit N, Pearce A. Invoking the user from data to design. Coll Res Libr. 2014;75(5):616–40.

77. Pandey S, Srivastava S. Data driven enterprise UX: a case study of enterprise management systems. LNCS. 2014;8522:205–16.

78. Kumar RL, Stylianou ACA. Process model for analyzing and managing flexibility in information systems. Eur. J Inf Syst. 2014;23(2):151–84.

79. Jablonski S, Ruf T, Wedekind H. Optimization of distributed processing by using a flexible data distribution mechanism. In: Rishe N, Navathe S, Tal D, editors. Proceedings PARBASE-90. International Conference on Databases, parallel architectures and their applications. Miami Beach, Florida, USA: IEEE Computer Society Press; 1990. p. 460–462.

80. Allan PD. Decision granularity: preserving uncertainty information in data consolidations. SPE Annual Technical Conference and Exhibition Society of Petroleum Engineers. 2004; https://doi.org/10.2118/90147-MS.

81. Birkhead GS, Klompas M, Shah NR. Uses of electronic health records for public health surveillance to advance public health. Annu Rev Public Health. 2015;18(36):345–59.

CHAPTER 6

Current Status of Veterinary Public Health Activities in Bangladesh and Its Future Plans

M. Minara Khatun, M. Ariful Islam and M. Mufizur Rahman

Department of Microbiology and Hygiene, Faculty of Veterinary Science, Bangladesh Agricultural University, Mymensingh 2202, Bangladesh

ABSTRACT

Background

Veterinary Public Health (VPH) is a major part of public health in which human health and well-being are the central tasks. In recent years the VPH

Citation: Khatun, M.M., Islam, M.A. & Rahman, M.M. Current status of veterinary public health activities in Bangladesh and its future plans. BMC Vet Res 15, 164 (2019). https://doi.org/10.1186/s12917-019-1879-8.

is gaining increasing importance because immense changes have occurred in animal production processes and agricultural structures. The aim of this paper is to describe the current VPH activities in Bangladesh, its major constraints and future activities plan to ensure safe food production as well as protect the environment and public health.

Main Text

VPH concerns all areas of food production and safety, zoonosis control, environmental protection and animal welfare. In Bangladesh, the VPH unit was established in 1984 by the Directorate of Livestock Services (DLS) for zoonosis control and production of wholesome food of animal origin. Zoonoses are the core domain of VPH. Bangladesh is facing the emergence of zoonotic diseases including anthrax, tuberculosis, brucellosis, salmonellosis, campylobacteriosis, *E. coli* infections, avian influenza, rabies, nipah and dengue virus infections.. Multi-drug resistance bacteria are emerging due to indiscriminate uses of antibiotics in livestock and poultry industries. Lack of proper slaughter houses, antemortem and postmortem inspections of carcasses by qualified veterinarians contributes greatly to unwholesome meat production. The VPH unit has a significant role to play to ensure better public health. However, there are many constraints that affect the VPH services. Absence of VPH services at all administrative level, inadequate budget, lack of qualified personnel, poor lab facilities, absence of legal framework, and the lack of coordination with health department are the major constraints.

Conclusions

The spectrum of VPH issues in Bangladesh is very large. Therefore it is important to carefully set priorities in order to ensure effective and efficient VPH services. Establishment of VPH units at all levels, effective surveillance for zoonotic diseases, institution of legal framework to define role of VPH services, creation of public health awareness, collaborative works with health departments, improving laboratory facilities and training programs for the veterinarian are keys to ensure better VPH services in Bangladesh.

BACKGROUND

The term "Veterinary public health," according to Schwabe [1] was introduced after World War II by public health administrators in the US Public Health Service to designate those areas of public health in which veterinary medicine shares particular interests. In the 1999 joint FAO/WHO expert committee agreed on the following definition of VPH: "VPH is the sum of all contributions to the physical, mental and social well-being of humans through understanding and application of veterinary science". This definition involves attention to the risks at the level of both production and consumption in food of animal origin. The WHO [2] describes VPH as "a component of public health activities devoted to the application of veterinary skills, knowledge and resources to the protection and improvement of human health."

VPH is a major part of public health where human health and human well-being are the central tasks [3]. The new consensus definition of VPH deals not only the physical well-being of humans, but also the role of animals for the mental and social well-being of humans. The importance of VPH in recent years is getting increasing attention because immense changes have occurred in animal production processes and agricultural structures. The trade in animals and products of animal origin is increasing due to the loss of border controls within countries. This trade demands new and elaborate surveillance strategies. The active surveillance of animal diseases, their distribution routes and control of emerging and re-emerging diseases is one of the major tasks of VPH [4]. It includes the analysis of the risk for humans and taking effective interventional steps for the protection of human health.

The scope of VPH is clearly multi-disciplinary, involving veterinarians, health professionals and scientists as well as paraprofessionals who treat, control or prevent diseases of animal origin [5]. This discipline of veterinary medicine is global in its potential impact and contribution as is public health. The needs and opportunities for veterinarians are expanding in a number of organizations including public agencies working on human and animal health [6].

The horizon of VPH is expanding rapidly and based upon the present veterinary skills, knowledge and resources; it has been recognized now as an essential field in public health activities all over the world. The chief aim is to protect and improve human health and welfare. In the past this discipline was evolved only to deal with three different issues: combating animal diseases, performing meat inspection and control of zoonoses [7]. At the present time VPH encompasses a wide variety of professional areas linking the three elements such as control of zoonoses, control of food borne pathogens and chemical residues and conserving the environment [7, 8]. Additionally animal welfare is also managed by the VPH service in the United States [9].

The main function of VPH is to control and prevent zoonotic diseases that can be transferred from animals to humans [10]. Transmission of zoonotic diseases occurs either through consuming contaminated foodstuffs or through contact with infected animals. Zoonoses consist of a wide range of diseases including anthrax, brucellosis, bovine tuberculosis, hydatid disease, echinococcosis, trichinellosis, rabies, highly pathogenic avian influenza, Nipah/Hendra disease and bovine spongiform encephalopathy [11]. A risk based program should be carried out in preventing or controlling the transmission of zoonoses. In order to implement an effective zoonoses control program attention should be paid to ecological, cultural, social and ethical aspects [11].

The protection of foodstuffs of animal origin for human consumption, to guarantee their safety and nutritional quality and to prevent disease transmitted in this way is the main responsibilities of the VPH [10, 12]. Safe food plays an important role on the quality of life, whether domestically produced and consumed, imported or exported. Increased presence of additives, pesticides, antibiotics and hormones in foods of animal origin are other emerging VPH problems [13]. Intentional contamination of food has become a serious threat to consumers. Over the last decades, the food chain approach has been recognized as an important step forward to ensure food safety from production up to consumption [11]. This approach requires the commitment of all players in the food chain, involving producers, traders, processors, distributors, competent authorities as well as consumers. VPH service has the responsibility to conduct ante and postmortem inspection of carcasses to ensure wholesome meat for human consumption [14].

The quality of animal feed is important for the production of safe food for human consumption [11].

A core function of the VPH service is to protect the environment [15]. Air pollution, water pollution and soil pollution can cause a direct risk to human *health*. Environmental pollutants arise from animal waste products including chemicals that may be used during production. In addition practicing vets will also produce potential environmental contaminants in the form of used needles, syringes, animal tissue and other clinical waste [16]. All of these materials have to be dealt with in a safe and controlled way to conserve the environment. Monitoring animal welfare is one of the important functions of the public health veterinarian. Public health veterinarians have the scientific and medical training to ensure animal welfare that enables them to judge the welfare of their patients and clients [17].

MAIN TEXT

VPH Activities in Bangladesh and Its Major Constraints

The VPH unit was established in 1985 at Mohakhali, Dhaka under the Research, Training and Evaluation (RTE) section of DLS. Its mandate is to control of zoonotic diseases and ensure safe and wholesome food of animal origin through sanitary inspection of meat and milk and monitoring antibiotic, hormone and chemical residues at all stages of production and up to final consumption [11].

Control of Zoonotic Diseases

The prevention and control of zoonotic diseases is the core domain of VPH in Bangladesh. The main bacterial zoonotic diseases in Bangladesh are anthrax, tuberculosis, brucellosis, salmonellosis, campylobacteriosis and leptospirosis. The common viral zoonotic diseases are avian influenza, rabies, nipah virus infection, dengue fever, Japanese encephalitis and rota virus infection. The parasitic zoonotic diseases included cysticercosis and echinococcosis. Toxoplamosis, amebiasis, giardiasis, leishmaniasis, cryptosporidiosis are protozoal zoonotic diseases in Bangladesh. Among the zoonotic diseases rabies, nipah virus infection degue fever and anthrax should get the top priority since these disease causes significant human morbidity and mortality every year in Bangladesh [18]. Poverty is recognized as being a major risk factor for zoonoses and food borne illnesses in both rural and urban consumers [19]. The VPH unit is tasked to maintain relations with the Health Department and the Public Health Institute for implementing surveillance and control program for zoonoses and food safety activities. However, the VPH unit of the RTE section of the DLS virtually lacking effective collaborative program together with the Health Department and the National Institute of Preventive and Social Medicine (NIPSOM). Moreover this unit is under budgeted and does not have sufficient skilled manpower. Lack enforcement of policies, laws, regulations and standards are important constraints in prevention and control of zoonotic diseases.

Sanitary Inspection of Meat and Milk

Meat inspection and abattoir management are almost absent in rural areas. People in the rural population do not have the opportunity to buy safe and hygienically produced meats. In major cities, slaughter of small ruminants takes place at the markets, while cattle are usually processed in slaughterhouses or slabs that are managed by city corporations. Municipalities and city corporations are entrusted to carry out meat inspection in their jurisdiction usually with government veterinarians. The structures of the most of slaughter houses are poor and slaughter takes place on the ground in quite unhygienic circumstances. Effluent handing is often appalling and is a public health hazard in itself. In most instances there is no ante-mortem and post-mortem inspection by a veterinarian. With respect to foods of animal origin like milk, sweet meats etc., the sanitary inspectors of health department are the authority of inspection and examination [19]. The sanitary inspectors are not properly trained and not supervised by competent veterinarians. Milk quality control is also performed by the private sector, e.g. corporations marketing milk. The VPH unit has not established links with them.

Monitoring Antibiotic, Hormone and Chemical Residues in Food of Animal Origin

Indiscriminate use of antibiotics, especially in the poultry industry is another public health issue. Antibiotics, used as growth promoters, may contribute to the buildup of bacterial resistance to antibiotics [20]. Consumers should be assured that residual levels of antibiotics, pesticides and hormones in food of animal origin are below maximum permitted levels. The VPH unit has no facility to determine these residues in foods. Legislation and enforcement systems to safeguard food for human consumption are lacking.

Measures need to be Adopted to Address the Public Health Issues in Bangladesh

VPH, food hygiene and zoonoses control in Bangladesh are identified as weak areas that need to be improved. It is important to carefully set priorities in the current VPH service for effectively and efficiently address the public health issues that are listed below:

Surveillance of Zoonotic Disease

Effective surveillance of zoonotic disease needs to be established. It is important to develop surveillance methodology and laboratory facilities. Develop systems to assure that the information collected should be incorporated in the animal health information system. It will be essential to establish risk-based surveillance programs that involve community participation. These are important to perform risk analysis, for which capacity will have to be established at the national level. Single zoonotic/food-borne disease risk analysis and burden assessment can be used as a catalyst for strengthening VPH structures and functions. Joint training of veterinarians and physicians is recommended to better identify local priorities as well as community knowledge, attitudes, and practices. Upazilla Veterinary officer should be responsible for teaching awareness among farmers and consumers, and be responsible for monitoring zoonotic diseases and food safety (slaughterhouses, meat inspection, drug residue, hormone, feed additives testing etc.) and maintaining a liaison with Upazilla health officer and at national level with the VPH department.

Development of Slaughter Facilities and Ensuring Routine Meat Inspection

To prevent environmental pollution, slaughter of animals in any place other than approved for the purpose should be stopped. Internationally acceptable National Standards for Sanitation and Meat Hygiene should be introduced and enforced in all abattoirs, slaughter houses and plants [21,22,23]. These facilities should be inspected yearly jointly by the VPH unit of DLS and Bangladesh food safety authority. Establish meat inspection services as a delegated responsibility of the central veterinary authority. Provision of effective training course for meat inspectors and veterinarians are essential for effective meat inspection [24].

Establishment of Institutional Legal Framework

To establish an efficient, effective and sustainable VPH services, it will require an institutional and legal framework that defines the role of VPH service providers and establishes the functions, responsibilities and control of public, private veterinarians (including the ones working for Non-Government Organizations (NGOs) and private companies and veterinary auxiliaries in the provision of VPH services. This framework should also provide guidelines on the roles, interrelationships and regulations required to link paraprofessionals, private operators and NGOs to the official VPH services [25].

Creation of Public Awareness

Raising awareness about various aspect of public health problems related to food of animal origin at the community level is key for successful implementation of VPH program. Public awareness includes: creating awareness on the risks associated with bad eating habits; risks of drug residues in food and importance of meat inspection.

Monitoring of Antibiotic and Chemical Residues

Through adoption of good agricultural practices, including good animal husbandry and good veterinary practices, it is possible to reduce the necessity for antimicrobials in animal production. Implementation of these activities should be conducted as per the principles laid down in WHO and Office International Epizooties (OIE) guidelines [26]. This will require capacity building, networking and coordination to facilitate implementation. The VPH unit will have to establish the capacity to perform residue testing.

Building Implementation Capacity

Extend and incorporate VPH activities into existing primary human and animal health care services. The VPH section should be strengthened with adequate laboratory facilities and trained veterinarians. This section should establish collaborative training and research programs with NIPSOM. At Upazilla level facilitate the organization of a VPH stakeholder association that represent veterinarians, public health physicians, paraprofessionals, livestock producers, consumers and other relevant groups that may play a role in VPH service delivery. Provide continuing education for technical staff including human and animal health workers and post-graduate training for veterinarians as well as health professionals.

CONCLUSIONS

The scope of VPH issues in Bangladesh is very large and clearly multi-disciplinary. Therefore it will be important to carefully set priorities in order to be effective and efficiently use the scarce available resources. A body of legal framework is required to establish an effective, efficient and sustainable VPH services that defines the role of VPH service providers and establishes the functions, responsibilities and control of public, private veterinarians (including the ones working for NGOs and private companies) and veterinary auxiliaries in the provision of VPH services. Veterinarians are in an ideal position to address the concerns of public health. The involvement of veterinary expertise at the policy level is necessary to ensure adequate planning, design, implementation and supervision of VPH programs and to ensure appropriate integration and collaboration with human health programs. The veterinary professionals contribute to improvement of human health by advancing biomedical and comparative medical research, preventing and addressing zoonotic diseases and enhancing environmental and ecosystem health. More research should be adopted related to public health as well as epidemiology of zoonotic diseases so that upcoming Veterinarians might take the challenges of public health concern. Therefore, it is highly recommended that the VPH section should be strengthened with adequate laboratory facilities and trained veterinarians.

ACKNOWLEDGEMENTS

The authors would like to acknowledge Dr. Stephen M. Boyle (Virginia Tech,USA) for editorial comments of this manuscript.

REFERENCES

1. Schwabe CW. Veterinary medicine and human health, 3rd edn. Williams and Wilkins Press, Baltimore/London. 1984.
2. Future trends in veterinary public health. In: WHO technical report series 907. WHO. Geneva: WHO; 2002.
3. Rahman MM. Innovating veterinary public health challenges of Bangladesh to integrate the concept of 'one world, one health.' Proceedings of the 8th annual scientific conference; 2012. p. 78–98.
4. Mantovani A. Zoonoses control and veterinary public health. Rev Sci Tech. 1992;11:205–18.
5. Ortega C, de Meneghi D, de Balogh K, de Rosa M, Estol L, Leguia G, Fonseca A, Torres M, Caballero-Castillo M. The current importance of veterinary public health: the SAPUVET project. Rev Sci Tech. 2004;23:841–9.
6. MPH Concentration. Veterinary public health. Department of Public Health. https://publichealth.utk.edu/degree-programs/mph/vph.
7. Pal SK, Siddiky NA. Veterinary public health and zoonotic disease control in SAARC countries. www.sac.org.bd/archives/publications/vph&zdc.pdf
8. Gyorffy A, Jozwiak A. A hidden corner of the "one health" concept: one health, the military veterinarian, and education. AARMS. 2016;15(2):107–20.
9. NASPHV - About state public health veterinarians. http://nasphv.org/aboutPHVs.html
10. Arámbulo PV. Veterinary public health: perspectives at the threshold of the 21st century. 3rdEdn. Rev Sci Tech. 1992;11:255–62.
11. FAO's animal production and health division. http://www.fao.org/ag/againfo/programmes/en/a6.html
12. WHO. The veterinary contribution to public health practice. Report of a joint FAO / W HO expert committee on veterinary public health. Tech. Rep. Ser., no. 573. Geneva: WHO; 1975. p. 79.
13. Polydorou K. Organization of veterinary public health activities in countries of the eastern mediterranean region. Rev Sci Tech1992. 11:53–75.
14. Buntain BJ. Emerging challenges in public health protection, food safety and security: veterinary needs in the USDA's food safety and

inspection service. J Vet Med Educ. 2004;31:334–40.

15. Cifuentes EE. Protection of the environment and veterinary public health activities. Rev Sci Tech 1992. 11:191–203.
16. Klooster GV. Formulation of a national livestock policy and action plan of Bangladesh. Draft report of the international consultant on veterinary services and animal health component; 2005. p. 1–33.
17. Ladewig J. The role of the veterinarian in animal welfare. Acta Vet Scand. 2008;50(1):S5.
18. Samad MA. Public health treat caused by zoonotic diseases in Bangladesh. Bangl J Vet Med. 2011;9(2):95–120.
19. Islam MA. Zoonoses in Bangladesh: the role of veterinarian in public health. Bangl J Vet Med. 2014;12(2):93–8.
20. Centre for Disease Control (2012). Plant construction, equipment and operation guidelines to qualify for licensing under the British Columbia meat inspection regulation.https://www2.gov.bc.ca/assets/gov/farming-natural-resources-and-industry/agriculture-and-seafood/food-safety/meat-inspection/abattoirsplantconstructequipoperation.pdf.
21. Standards for Hygiene and Construction of Slaughterhouses (Hansard, 3 August 1972*).* https://api.parliament.uk/historic-hansard/commons/1972/aug/03/standards-for-hygiene-and-construction
22. Skaarup T. Slaughterhouse cleaning and sanitation. Animal production and health paper 53, food and agriculture organization (FAO); 1985.
23. Food and Agriculture Organization. Abattoir development options and designs for hygienic basic and medium-sized abattoirs/recommendations. www.fao.org/docrep/010/ai410e/AI410E08.htm. 2008.
24. Islam KBMS, Mahmuda SSU, Kabir MHB. Antibiotic usage patterns in selected broiler farms of Bangladesh and their public health implications. Journal of Public Health in Developing Countries. 2016;2(3):276–84.
25. Food and Agriculture OrganizationExpert consultation on community based veterinary public health (VPH) systems. Final report of. Rome, Italy; 2003.
26. World Health Organization. WHO global principles for the containment of antimicrobial resistance in animals intended for food. Report of a WHO consultation with FAO and OIE. 2000:1–26.

SECTION 2:
"ONE HEALTH" CONCEPT

CHAPTER 7

Linking Human Health and Livestock Health: A "One-Health" Platform for Integrated Analysis of Human Health, Livestock Health, and Economic Welfare in Livestock Dependent Communities

S. M. Thumbi[1,2], M. Kariuki Njenga[3] , Thomas L. Marsh[1,4], Susan Noh[1,5], Elkanah Otiang[2] , Peninah Munyua[3] , Linus Ochieng[2] , Eric Ogola[2] , Jonathan Yoder[1,4], Allan Audi[2] , Joel M. Montgomery[3] , Godfrey Bigogo[2] , Robert F. Breiman[6], Guy H. Palmer[1] , Terry F. McElwain[1]

[1] Paul G. Allen School for Global Animal Health, Washington State University, Pullman, Washington, United States of America

Citation: Thumbi SM, Njenga MK, Marsh TL, Noh S, Otiang E, Munyua P, et al. (2015) Linking Human Health and Livestock Health: A "One-Health" Platform for Integrated Analysis of Human Health, Livestock Health, and Economic Welfare in Livestock Dependent Communities. PLoS ONE 10(3): e0120761. https://doi. org/10. 1371/ journal. pone. 0120761.

[2] Center for Global Health Research, Kenya Medical Research Institute, Kisumu, Kenya

[3] Centers for Disease Control and Prevention, Nairobi, Kenya

[4] School of Economic Sciences, Washington State University, Pullman, Washington, United States of America

[5] Animal Diseases Research Unit, Agricultural Research Service, United States Department of Agriculture, Pullman, Washington, United States of America

[6] Emory University, Atlanta, Georgia, United States of America

ABSTRACT

Background

For most rural households in sub-Saharan Africa, healthy livestock play a key role in averting the burden associated with zoonotic diseases, and in meeting household nutritional and socio-economic needs. However, there is limited understanding of the complex nutritional, socio-economic, and zoonotic pathways that link livestock health to human health and welfare. Here we describe a platform for integrated human health, animal health and economic welfare analysis designed to address this challenge. We provide baseline epidemiological data on disease syndromes in humans and the animals they keep, and provide examples of relationships between human health, animal health and household socio-economic status.

Method

We designed a study to obtain syndromic disease data in animals along with economic and behavioral information for 1500 rural households in Western Kenya already participating in a human syndromic disease surveillance study. Data collection started in February 2013, and each household is visited bi-weekly and data on four human syndromes (fever, jaundice, diarrhea and respiratory illness) and nine animal syndromes (death, respiratory, reproductive, musculoskeletal, nervous, urogenital, digestive, udder disorders, and skin disorders in cattle, sheep, goats and chickens) are collected. Additionally, data from a comprehensive socio-economic survey is collected every 3 months in each of the study households.

Findings

Data from the first year of study showed 93% of the households owned at least one form of livestock (55%, 19%, 41% and 88% own cattle, sheep, goats and chickens respectively). Digestive disorders, mainly diarrhea episodes, were the most common syndromes observed in cattle, goats and sheep, accounting for 56% of all livestock syndromes, followed by respiratory illnesses (18%). In humans, respiratory illnesses accounted for 54% of all illnesses reported, followed by acute febrile illnesses (40%) and diarrhea illnesses (5%). While controlling for household size, the incidence of human illness increased 1. 31-fold for every 10 cases of animal illness or death observed (95% CI 1. 16–1. 49). Access and utilization of animal source foods such as milk and eggs were positively associated with the number of cattle and chickens owned by the household. Additionally, health care seeking was correlated with household incomes and wealth, which were in turn correlated with livestock herd size.

Conclusion

This study platform provides a unique longitudinal dataset that allows for the determination and quantification of linkages between human and animal health, including the impact of healthy animals on human disease averted, malnutrition, household educational attainment, and income levels.

INTRODUCTION

In livestock-dependent households and communities, the health of humans, livestock, and household economic welfare are closely linked. It is estimated nearly one billion people living on less than 2 dollars a day, 300 million of whom reside in sub-Saharan Africa, depend on livestock as a source of livelihood and nutrition [1]. These figures represent two thirds of the rural poor and one third of the urban poor whose food, income, social status, and store of wealth depend on livestock [2,3]. However, the relationship between livestock health and productivity, and human health and welfare is complex, and understanding it in quantitative terms remains a critical requirement for developing sustainable poverty relief and public health interventions through human and livestock health maintenance and improvement.

The linkages between livestock keeping and human nutrition and health outcomes have been conceptualized[2,4]. Broadly, they include positive effects of keeping livestock that generally improve a household's health and welfare status; and negative effects of livestock ownership that may worsen human health and nutritional status. The positive effects include increased access to nutritious animal source foods (ASFs), such as milk, meat, and eggs, in households owning animals, and higher household cash incomes that increase purchase power for ASF's, food crops, healthcare and education. The consumption of ASF's provides high-quality protein, essential structural fats, and highly bioavailable essential micronutrients (e. g. zinc, iron, calcium, vitamin A, vitamin B-12) that are strongly associated with improved growth, health and cognitive ability of children[5–8], and increased resistance to and recovery from infectious diseases[9,10]. All of these have multiplier effects at the community level through better nourished children becoming more intelligent, healthier and more productive adults[2].

The negative effects associated with livestock keeping may include risk of transmission of zoonotic pathogens from animals to humans e. g. anthrax, leptospirosis, trypanosomiasis, rabies,—many of which are neglected[11], food-borne diseases e. g. cysticercosis, taeniosis, cryptosporidiosis and brucellosis, development of antimicrobial resistance, and chronic diseases such as cardiovascular disease, cancers, and diabetes associated with excessive consumption of the energy-dense high level saturated ASF's. For instance, the rural communities in Western Kenya where this study is based suffer concurrent high levels of poverty[12,13] and high burden of infectious diseases[14], including zoonotic ones such as Q-fever, cysticercosis, cryptosporidium and trypanosomiasis[15–19].

The goal in developing policies aimed at reducing poverty is to maximize the positive linkages of livestock keeping while minimizing the negative effects, especially for people living in poverty and vulnerable groups including pregnant women and children. Quantifying the impact of livestock diseases on human health and welfare, identifying priority diseases including new zoonoses to target for greatest gain amidst limited resources, identifying the factors that enhance or reduce nutritional outcomes, and the use of this information to guide program and policy formulation for improved human health outcomes remain a major challenge[20,21].

Herein we describe a platform for integrated health and economic welfare analysis designed to address this challenge by simultaneously collecting data from a population-based animal syndromic surveillance (PBASS) system and an ongoing human population-based infectious disease surveillance (PBIDS) system in the same rural households in Western Kenya[14], and linking these data to a quantitative assessment of socio-economic dynamics in the same households over time. We provide baseline epidemiological data on the animal and human disease syndromes under investigation, and descriptive data on household demographics, wealth and assets, livestock ownership, household food consumption and expenditures, and investments in health care. Using data from the first year of study, we examine potential linkages between animal and human health and economic welfare and present these, as examples of how the data can be utilized to better understand the complex relationships that impact the rural poor.

MATERIALS AND METHODS

Ethical Review Statement

Ethical clearance for the population based animal syndrome surveillance (PBASS) study was obtained from the Kenya Medical Research Institute Ethical Review Committee and Animal Care and Use Committee (reference number SSC Protocol no. 2250). The population based infectious diseases study (PBIDS) received approval from the Kenya Medical Research Institute Ethical Review Committee (reference SSC Protocol no. 1899) and the Centers for Disease Control and Prevention. Participants over 18 years of age provided a written consent to participate in these studies, and parental consent was provided for children involved in the study.

Study Population

The PBASS study is conducted in Western Kenya within a Health and Demographic Surveillance System (HDSS), which started in September 2001, and is run by the Kenya Medical Research Institute (KEMRI) in collaboration with the Centers for Disease Control and Prevention, Kenya (CDC). The HDSS provides general demographic and health information including population age-structure, in and out migrations, fertility rates, birth and death rates, verbal autopsy, access and utilization of health care for a population of approximately 220,000 individuals, in 54,869 households[22,23]. The HDSS area covers 385 villages that lie to the North-East of Lake Victoria covering Rarieda, Siaya and Gem Districts in Siaya County, an area predominantly inhabited by the Luo ethnic group practicing subsistence farming and fishing. The overall population density for the area is estimated at 325 persons per square kilometre.

Within the HDSS is the Population-Based Infectious Disease Surveillance (PBIDS) platform, which monitors the health of approximately 24,000 persons living in 6,000 households in 33 villages. The PBIDS study was initiated in 2005 to define the burden of the major infectious disease syndromes, describe epidemiologic patterns of disease, identify etiology of disease, and implement and evaluate the health impact of interventions in the study population[14,24]. Data on four disease syndromes (fever, jaundice, diarrhea and respiratory illness) for each household resident are collected through bi-weekly home visits conducted by community interviewers (CIs) in the local language. The CIs have secondary school certificate education, with additional training on the protocol, protection of human subjects, conducting a limited set of physical examinations and administering standardized questions about recent symptoms among household members.

The PBASS study is conducted in 1500 households in 10 villages randomly selected from the 33 villages participating in the PBIDS study (Fig. 1). To be included in the PBASS sample frame, households must be participating in the PBIDS study and provide written informed consent for participation in the PBASS study. All households within the 10 villages and participating in the PBIDS study are eligible for enrolment into PBASS.

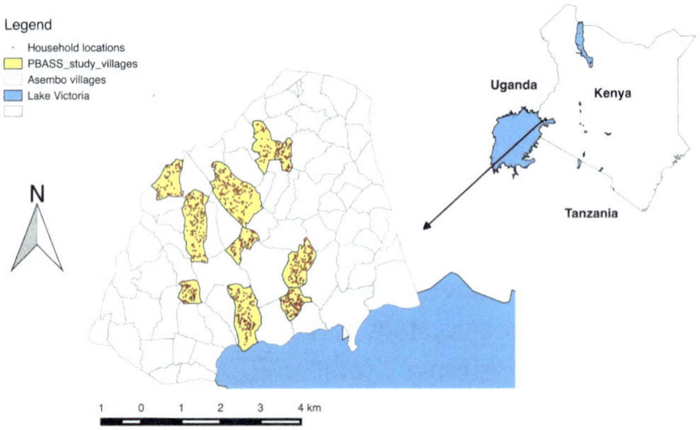

Figure 1: Map of the study area showing the study villages (in yellow) and the location of households (black dots) enrolled in the study.

https://doi. org/10. 1371/journal. pone. 0120761. g001

Once enrolled into the PBASS study, each household is followed longitudinally, and three groups of data collected: animal disease syndrome data, human disease syndrome data, and socio-economic data for each study household.

Animal disease syndrome data.

The PBASS study monitors the health of cattle, goats, sheep and chickens and collects bi-weekly data on 9 syndromes: respiratory system disorders, gastro-intestinal system disorders, nervous system disorders, reproductive system disorders, digestive system disorders, urogenital system disorders, skin disorders, udder disorders, and death. A summary of syndromes and the key signs associated with each syndrome are provided in Table 1.

Syndromes	Clinical signs
Reproductive disorders	abortions, stillbirths, neonatal deaths
Respiratory disorders	cough, nasal discharge, difficulty breathing
Digestive disorders	diarrhea, bloating, inappetence
Urogenital disorders	vaginal discharges, preputial discharges, scrotal swelling
Musculo-skeletal disorders	lameness, recumbency
Skin disorders	hair loss, lumps, itching
Nervous disorders	circling, incoordination
Udder disorders	mastitis (unusual colour or consistency of milk), drop in milk yield
Death	

doi:10.1371/journal.pone.0120761.t001

Table 1: A summary of the syndromes in cattle, goats, sheep and chickens investigated in the PBASS study.

https://doi. org/10. 1371/journal. pone. 0120761. t001

Animal disease syndrome data are collected in two ways: 1) through community interviewers that visit each study household bi-weekly and administer a questionnaire designed to identify if any of the cattle, sheep, goats or chickens in the household have been sick with any of the listed syndromes or have died in the 2 week period preceding the interview, and 2) through a toll-free telephone number provided to all households in the study (>70% of households have access to a mobile phone), allowing the farmers to report cases of animal illness or death directly to the PBASS study team without having to wait for the next bi-weekly visit.

Following alerts of animal illness or death coming through either the CI or the toll-free number, a PBASS animal health team consisting of veterinarians and animal health assistants responds by visiting the affected household within 24 hours. The animal health team conducts a complete clinical examination of each sick animal and collects samples including blood smears, serum samples, milk samples, faecal samples, skin scrapings and any other clinically relevant samples. In cases of death, a full post-mortem examination is conducted. In addition, the team administers a detailed questionnaire that captures data on herd health including herd demographics, breeding history, and specific syndromes/clinical signs observed in the study animals. In order to aid clinical diagnosis for veterinary interventions, a limited set of diagnostics mainly involving microscopy on blood and lymph node smears obtained from sick animals are conducted. Samples collected during these visits are shipped to the KEMRI laboratory for storage for future analysis. A comprehensive diagnostic workup to establish an etiologic diagnosis is not currently conducted but will be carried out in the future to establish aetiologies for the syndromes with the greatest socio-economic impact on households, and to identify zoonotic pathogens that may cause illness in humans.

Human disease syndrome data

In the same households monitored for illnesses and death in animals, human morbidity data is collected bi-weekly, simultaneously with the animal health data. These data are collected as part of the PBIDS study. Briefly, a questionnaire investigating whether any member of household has had any of the 4 syndromes (fever, jaundice, pneumonia/cough, or diarrhoea) in the 2-weeks preceding the current visit is administered. Fever is defined as axillary temperature ≥ 38. 0°C, and diarrhoea as ≥3 looser than normal stools in a 24-hour period. In addition, data on whether any household member who fell ill sought health care, type of care sought, and where health care was sought from is collected. Study participants reporting any of the listed syndromes are referred (or can self-refer) to the St. Elizabeth Mission Hospital in Lwak (located within a 5km radius of study households), where free medical care is provided.

The simultaneous collection of animal health and human health data from the same households enables the linking of the two datasets and evaluation of the relationships between human and animal health.

Socio-economic survey data

A detailed socio-economic survey is carried out every 3 months in each PBASS study household. The three-month return interval represents a balance between survey cost and information loss; however household socioeconomic characteristics either do not tend to change at the same temporal scale as health indicators, and because respondent recall can be (and are) relied on to capture household activity and socio-economic changes between surveys. A team of community interviewers specially trained to conduct household socio-economic surveys collects these data. Table 2 provides a summary of the type of socio-economic data collected.

Table 2: Description of the household socio-economic data collected every 3 months

https://doi. org/10. 1371/journal. pone. 0120761. t002

Group of variables	List of variables
Household demographics	Household identifier; household size; ages of household members, gender, education level, occupation
Household assets	Number and value of farm implements, electronics, bicycles, bikes and motor-vehicles owned; type of housing, absence/presence of a latrine, electricity, type of water source, amount of time taken to fetch water and number of trips per day, source of fuel, person and household member involved in firewood collection, income for each member of the household, savings accounts and balance, loans borrowed, amount and purpose of loan
Household consumption	For each food; milk (from cows, sheep and goats), meat, eggs, fish, maize, cassava, sorghum, banana, pulses, green vegetables, potatoes—amount produced at home, purchased, consumed, and the costs for each food. List of foods fed to children less than 5 years. Home expenses on fuel, clothing, health, education, and any other expense. Records on vaccinations, medication, consultation, transport, and other health costs.
Land and crop inventory	Size of land owned and/or rented, land rent costs, crop losses and value, cost of inputs (seeds, fertilizers, manure, labor) for maize, potato, sorghum, cassava, groundnuts, beans; acreage covered by each crop, size of harvest, quantity sold and sale value for each of the crops, purchases of any of the crops or gifts received
Livestock ownership	Number of calves, heifers, bullocks, cows, bulls, goats, sheep, poultry and donkeys owned on or off farm
Livestock Inventory	Number of calves, heifers, bullocks, cows, bulls, goats, sheep, poultry, donkeys born, dead, sold, consumed, received as gift, given as gifts, purchased or lost since previous visit, livestock value lost
Livestock income	Sale value for livestock sold (cattle, sheep, goats, sheep, donkeys, poultry), amount and sale value of cow, goat and sheep produced, number of eggs produced, number and sale value of eggs sold, any other income from livestock
Livestock expenditure	Purchase costs for cattle, goats, sheep, donkeys, poultry; labor costs, costs for vector (e.g. ticks and tsetse flies) control, helminth control, treatment costs, veterinarian costs, and any other costs for cattle, sheep, goats, donkeys and poultry
Animal health care	Method of tick control, distance covered by cattle for dipping/spraying, frequency of tick control per month, total cost of tick control, vaccinations in animals
Human illness	Sickness among any household members, sum of number of work days missed due to illness, number of health clinic/hospital visits made in the last 3 months, travel time to clinic/hospital in hours, records of vaccination of children under 5

The PBASS study design is summarised in a schematic diagram in Fig. 2.

doi:10.1371/journal.pone.0120761.t002

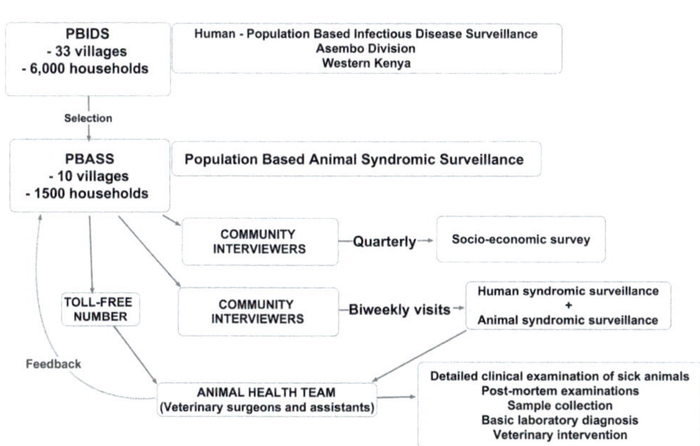

Figure 2: Schematic diagram showing the design of the population based animal syndromic surveillance (PBASS) Study.

All PBASS study households participate in the population based infectious disease surveillance (PBIDS) study, thereby generating human health, animal health, and socio-economic data linkable at the household level.

https://doi. org/10. 1371/journal. pone. 0120761. g002

Data Analysis

Data collection for the PBASS study began in February 2013, and a total of 1500 households provided informed consent to participate in the study. The data on human health, animal health and socio-economic status used in this analysis was collected between February 2013 and February 2014. Pocket digital assistants were used to record the data in the field, which was later downloaded and stored in a Microsoft Access database. Descriptive analyses were conducted to provide baseline numbers for variables on household demographics, animal ownership, and indicators of socio-economic status.

A household illness index for humans and for animals was computed by determining the total number of cases of illnesses in humans, and illnesses and deaths in animals per household observed in first one year of the study. The human illness index was treated as the main outcome variable. To account for the over-dispersion in the household illness index data, negative binomial regression methods were used to estimate the relationship between household illness index and household demographics, socio-economic indicators and illnesses and deaths observed in animals. All data analyses were done using the R platform for statistical computing[25].

RESULTS

Household Characteristics

The total population of people in the PBASS study households was 6,400, with an average family of 4 persons per household. The mean age of the household heads was 53 years, and the distribution of the ages of household heads is shown in Fig. 3. Thirty-three percent of the households had at least one child < 3 years, 42% had at least one child < 5 years, and 56% had one or more children < 10 years old. Up to 85% of the household heads had received some formal education. Most households (80%) were dependent on agriculture for their livelihood, and mainly practiced mixed crop-livestock production system.

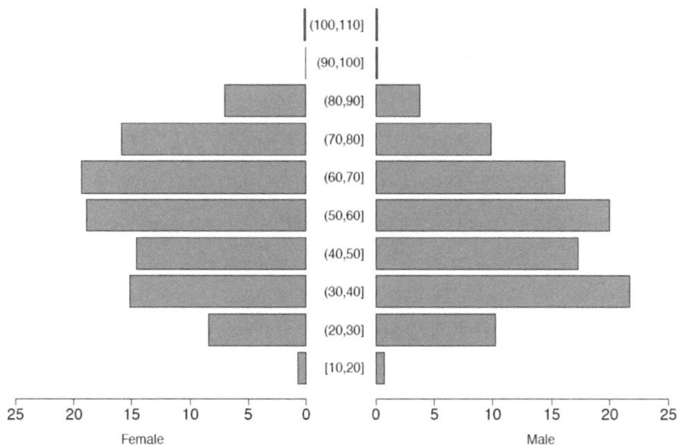

Figure 3: Population pyramid showing the age structure of the household heads by gender in years.

The x-axis represents a percentage of each age group by gender.

https://doi. org/10. 1371/journal. pone. 0120761. g003

Data collected on the ownership of household assets shows the mean household asset value (consisting of assets such as farm implements, electronics, bicycles) to be USD 100 (Kshs. 8,820) with a median household asset value of USD 50 (Kshs. 4,200). Farming was the main source of household income, with 32% of the households reporting to receive some income away from their farms. For households that reported an off-farm income, the monthly average was USD 280 (Kshs. 23,840) and a median of USD 124 (Kshs. 10,500).

Firewood is the main source of cooking energy for 94% of the households, 5% used charcoal, and 1% used natural gas or paraffin. Among the households using firewood, only 3% reported regularly buying firewood, with 97% reporting to collect it from the surrounding environment. Electricity connection in the study area was minimal with only 3% of households being connected to the electricity grid. Most households (82%) have at least one outdoor latrine, while the rest (18%) reported having no latrine available within their homestead.

Forty-three percent of the households obtained water for domestic use from wells, 32% harvested rainwater or accessed it from seasonal streams, while the rest fetched their water from communal reservoirs or from Lake Victoria. Sixty percent of the households spent 1–2 hours each day collecting water, 20% spent under one hour a day, and 20% over 2 hours each day. Only 7% of the enrolled households do not keep livestock. Of the 93% that keep at least one species of livestock, chickens were the most common species (88%), followed by cattle (55%), goats (41%) and sheep (19%), see Table 3. Most households practice mixed small-scale farming with a median of 9 ruminants (cattle, goats and/or sheep) and 11 adult chickens.

Table 3: Proportion of households keeping each species of livestock, and the summary statistics for number of animals per household

https://doi. org/10. 1371/journal. pone. 0120761. t003

Livestock ownership	Frequency (%)	Mean	Median	Maximum
Own least 1 species	1348 (93.4)			
Cattle	800 (55.4)	5	3	98
Goats	587 (40.7)	4	3	30
Sheep	268 (18.6)	5	3	47
Chicken	1272 (88.1)	14	11	100

doi:10.1371/journal.pone.0120761.t003

Ninety-one percent of the households reported growing maize in the planting season preceding the recruitment visit. Details of the other crops grown are provided in Table 4.

Table 4: Showing proportion of households planting each of the common crops in the most recent planting season, and foods consumed in the last 7 days preceding the interview

https://doi. org/10. 1371/journal. pone. 0120761. t004

Crops planted	Frequency (%)
Maize	1312 (91)
Potato	116 (8)
Sorghum	68 (5)
Cassava	61 (4)
Beans	501 (35)
Household consumption	
Cow milk	1085 (75)
Goat milk	3 (0.2)
Eggs	416 (29)
Beef	476 (33)
Goat meat	7 (0.5)
Fish	1275 (88)
Cassava	514 (36)
Sorghum	458 (32)
Banana	210 (15)
Pulses	661 (46)
Onions, carrots	1316 (91)
Green vegetables	1365 (95)
Sweet potato	366 (25)
Irish potato	105 (7)

doi:10.1371/journal.pone.0120761.t004

The sources of animal proteins for the households include fish (88%), cow milk (75%), beef (33%), and eggs (29%), Table 4. Household food consumption for the 7 days preceding the recruitment visit, the type of food consumed and the proportion of families consuming different food types are also presented in Table 4.

To determine if a household's likelihood of consuming animal source foods is associated with livestock ownership, we tested the relationship between milk and egg consumption with the total number of cows and chickens owned, respectively. The likelihood of milk consumption in a household increased by 62% for every increase by one in the number of cows owned, and the likelihood of egg consumption by 28% for every increase in the number of chickens owned (in 10 chicken units) (Table 5).

Table 5: Results of regression between household human illness cases and animal illness and death cases (while accounting for household size), relationship between seeking health care and livestock ownership, household income and assets and livestock ownership, and household animal source food consumption and livestock ownership

https://doi. org/10. 1371/journal. pone. 0120761. t005

Variable	Estimate	Lower CI	Upper CI	*p*-value
Household illness index				
(Animal illness and death/10)	1.31	1.16	1.49	< 0.001
Household size	1.08	1.07	1.1	< 0.001
By syndromes				
Gastro-intestinal tract illness (humans)				
Animal—Gastro-intestinal illness	1.06	1.02	1.12	0.009
Respiratory syndromes				
Animal—respiratory syndromes	1.1	1.04	1.17	0.002
By socio-economic status				
Household illness index				
Household head education level- formal	1.02	0.85	1.21	0.248
Number of cattle owned	1	0.99	1.01	0.241
Household consumption				
Cow milk consumption (yes/no)				
Number of cows owned	1.62	1.44	1.86	< 0.001
Egg consumption (yes/no)				
Number of Chicken owned (10)	1.28	1.17	1.4	< 0.001
Expenditure on health (yes/no)				
Number of cattle owned	1	0.97	1.02	0.906
Number of Chicken owned (10)	1.18	1.09	1.3	< 0.001
Household asset value				
Number of cattle owned	1.1	1.07	1.13	< 0.001

doi:10.1371/journal.pone.0120761.t005

Over half of the households (54%) did not regularly control for ticks in their livestock and only 2% of the households reported vaccinating their cattle against any diseases. For the period of 3 months preceding the recruitment visit, 52% of the respondents reported expenditures on health care for household members (median expenditure of 2 US dollars per month per household). We tested whether the probability of expenditure on human health care was related with livestock ownership. Using the recruitment data, we found no correlation relationship between probabilities of a household spending on health during the last 3 months with the total number of cows, goats or sheep, but there was a correlation with the number of chickens owned [incidence rate ratio 1. 18, 95%CI (1. 09, 1. 3) for every 10 chickens owned].

Livestock Disease Syndromes

Whereas data for all 9 syndromes were collected in cattle, sheep, and goats, household visits for cases in chickens were made only when the mortality was >30% of the flock. Between February 2013 and February 2014, a total of 2193 cases of livestock illness and deaths were recorded. Of these, 75% were illnesses in cattle, sheep and goats and 25% were mortality cases in cattle, sheep, goats, and chicken. Cattle contributed 58% of all cases of illness, with goats and sheep contributing 28% and 14% of the cases respectively. Most of the recorded deaths (54%) occurred in chickens and chicks (only considered if >30% of the flock died), with cattle, goats and sheep contributing 14%, 22% and 10% respectively. Interestingly, the two methods for reporting illnesses or deaths in animals contributed nearly equally. The use of a toll-free number with farmers contacting the PBASS team directly to report cases of illness or deaths accounted for 46% of all the reports, while the bi-weekly visits conducted by the community interviewers accounted for 54% of all reports. Of the illness cases reported among each livestock species, gastro-intestinal tract syndrome cases were the most common, comprising of 52%, 53%, and 34% of total cases in cattle, goats, and sheep respectively. For these three species, respiratory syndromes were the second most common observed, see Fig. 4.

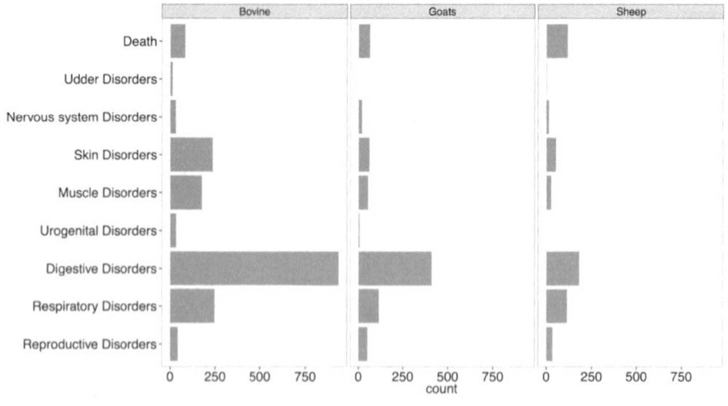

Figure 4: Distribution of the disease syndromes by species for data collected over the first 12 months of the study (February 2013—February 2014).

https://doi. org/10. 1371/journal. pone. 0120761. g004

Human Disease Syndromes

Between February 2013 and February 2014 a total of 38,208 human illness cases were reported, with respiratory illnesses constituting 54. 4%, see Fig. 5.

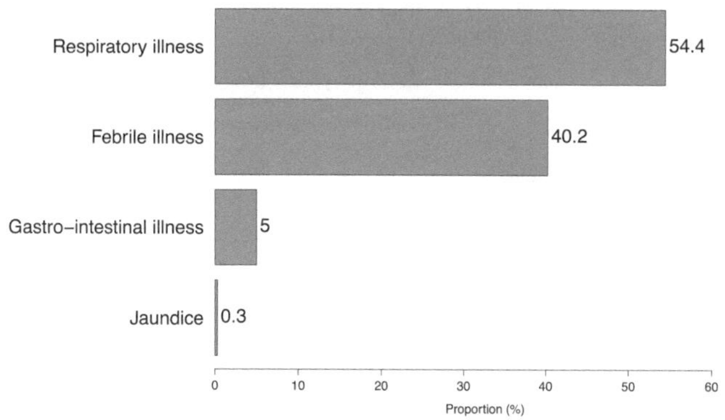

Figure 5: Showing a summary of the distribution of the 4 human syndromes under investigation.

https://doi. org/10. 1371/journal. pone. 0120761. g005

Using the human and animal health data collected over the 12-month period, we tested whether there was a correlation between households reporting illnesses among household members, and those reporting cases of illnesses or death in animals. To do this, we computed a summary statistic for each household, which provided the total number of illnesses reported per household in humans, and in animals over the one-year observation time. The mean number of cases of human illness per household was 26 (median 13, minimum 0, maximum 136), with a mean of 1. 2 cases among livestock (median 0, minimum 0, maximum 24). An individual (human or animal) can provide as many cases to the household aggregate as were recorded through the routine surveillance visits. The distribution of these human and animal cases per household is provided in Fig. 6.

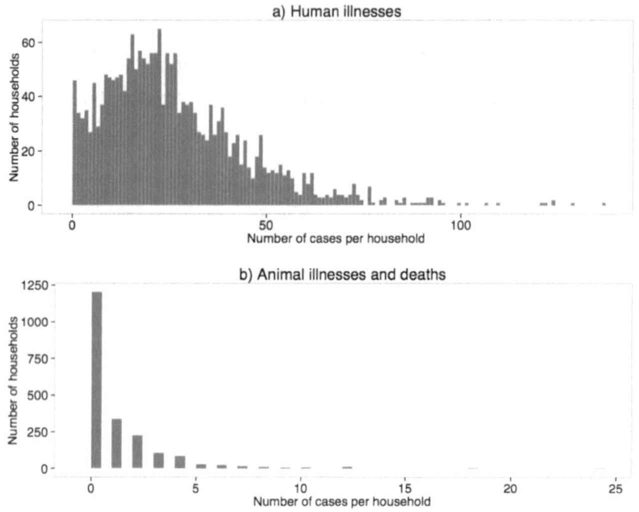

Figure 6: Distribution of the number of human and animal illnesses and deaths per household during the first 12 months of the study.

https://doi. org/10. 1371/journal. pone. 0120761. g006

Using a negative binomial generalised linear model to account for the over-dispersed nature of the data, we tested the hypothesis that there is no correlation between human illness and animal illness cases at the household level, based on the one-year surveillance data collected. The outcome variable for this analysis was the total number of human illnesses per household, and this was regressed against the number of animal illness and deaths cases in the same households. While accounting for the household size, the results suggested that for every 10 cases of animal illness or death observed in a household, the odds of observing a human illness in the same household was 31% greater (95% CI, 16–49%).

When this analysis was extended to determine if there was an association between gastro-intestinal syndromes and respiratory syndromes in humans and animals, we found that there was a 6% (95% CI, 2–12) increase in the probability of gastro-intestinal illness in humans associated with each additional case of gastro-intestinal illness observed in animals in the same households. A 10% (95% CI, 4–17) increase in the probability of respiratory illness case in humans was associated with each additional case of respiratory illness observed in animals (Table 5). There was no evidence of an association between the household illness index and socio-economic factors such as number of cattle owned and education level of the household head. However, the likelihood of health care seeking among household members falling ill increased with the number of cattle owned by 1. 15 times (95% CI 1. 03, 1. 29) for every 10 cattle owned, and with the net off-farm income by 1. 03 times (95% CI 1. 01, 1. 06) for an increase in income of 100 US dollars. Additionally, we found the asset value of a household was positively correlated to the number of cattle owned (see Table 5).

DISCUSSION

The PBASS study platform is designed to investigate and understand the links between human health, animal health, and specific socio-economic variables. The key strength of this study is the simultaneous collection of longitudinal data on human health, animal health, and socio-economics from the same households over time. This "One-Health" approach that allows for an integrated human-animal health and economic framework for assessing causes of poor human health have been previously proposed[26–28].

The study presumes that human health and animal health may be linked through 3 main pathways: i) a socio-economic pathway where improved livestock production is associated with healthy livestock leading to improved household incomes and wealth, access to education and health care, ii) a nutritional pathway where owning healthy livestock increases access to animal source foods that in turn reduces the risk of malnutrition and disease, and iii) a zoonotic pathway where healthier livestock are less likely to transmit zoonotic and food-borne infections. Our data is linkable at the household level, making it possible to identify the pathways by which human and animal health are related, determine the magnitude and direction of effect of each pathway, and test interventions in the animal sector that would increase human opportunity, including human health, wealth and welfare.

Although the results we present are based on analysis of only the first 12 months of data, we find strong associations between the cumulative number of human illnesses in a household (human illness index) and the cumulative number of animal illnesses and death (animal illness index) in the same households. Additionally, for the common respiratory and gastro-intestinal illnesses, households with high numbers of animal cases of each of these syndromes have high numbers of the similar syndromes in humans. We hypothesize that the observed relationships between households with high numbers of human illnesses and those with high numbers of animal illnesses may be driven by either the zoonotic pathway, with pathogens shared between humans and the animals they keep, or driven by factors that increase a household's illness index. In the latter case, illnesses may be related to the environment in which the humans and animals co-reside. Disease clustering is known to be associated with environments that are suitable for survival of parasites and the vectors that transmit them, and where disparity conditions promote disease[29]. Although the role of environment and environmental factors on human and animal illness is not investigated in this current study, this will be a subject of analysis moving forward. Human and animal illnesses may be mediated through human and animal health management, nutrition, and healthcare-seeking activities, which in turn are mediated by wealth, income, as well as credit and liquidity constraints. This concept of social and economic determinants of health is well recognized, with greater likelihood of poor health associated with people of lower socioeconomic status[30].

We find a correlation between a household's asset accumulation and health seeking behaviour, and the number of animals owned. Families owning large herds of cattle had assets of a greater value compared to those owning fewer animals. It is possible that asset accumulation is a result of increased income from livestock, or asset accumulation and livestock accumulation are independent but influenced by a third source of income. Such dynamics will be investigated using the multi-year data available from this longitudinal study. Asset accumulation has been related to increased income from livestock[31]. In relation to health, the likelihood of health seeking following human illness increased with number of cattle owned. These examples demonstrate the potential pathways that may relate livestock ownership to human income and wealth, and access to health care.

The hypotheses that human illnesses are driven by zoonotic infections will be tested by determining the etiologies of the human syndromes and animal syndromes observed, and the risk factors associated with specific infectious diseases identified. Currently, only limited diagnostic work is being performed to determine the etiology of the observed animal disease syndromes, until the syndromes associated with the highest economic impact at the household can be determined.

The collection of human health and animal health data at the household level is done at least every 2 weeks. Although the 2-week inter-visit interval may be associated with some recall decay[24], the interval is optimal in balancing the risk of excessive recall bias associated with longer intervals between consecutive household visits, the risk of participant fatigue with more frequent visits, and the demand on resources required for surveillance through home visits. The use of a toll-free number as an extra tool for surveillance in livestock allows the farmer to reach the veterinary team whenever an animal falls sick. This surveillance tool allows for the capture of active clinical cases, and reduces the risk of under-estimating the incidence of illnesses associated with recall bias. From this study, approximately half of all recorded cases of illness and death in animals were reported through the toll-free number, a relatively inexpensive, and efficient alternative surveillance tool. The use of mobile phone-based tools for surveillance of diseases has expanded in the last decade[32,33]. The uptake and success of these tools is dependent on a number of factors including costs of data transmission, ease of use by the data collectors, and availability of infrastructure such as a good mobile network. In our study, successful use of mobile phones to report disease events can be attributed to a high proportion (> 70%) of mobile phone ownership among study households, a strong mobile network signal (most households receive mobile network signals), and the use of the toll-free option, which takes away the cost of the phone-call from the farmer.

The use of toll-free reporting system alone would possibly miss disease events that farmers view as less serious to warrant reporting. Such cases missed by the toll-free system are picked during the bi-weekly home visits. However, the timely response by the PBASS veterinary team to disease events reported by the farmer, and the offer of free veterinary advice (clinical diagnosis and limited care) for syndromes relevant to the study have been an effective motivator for farmers to report disease events. The combination of a toll-free number and biweekly visits allows analysis of the quality of disease incidence data collected by each pathway, with each providing a quality control check. Over time, we will better assess whether one method is preferable over the other. However, we postulate that with proper feedback offered to the farmers, the use of a toll-free reporting system may be an attractive and cost-effective tool to increase disease surveillance, especially in resource-limited areas.

The platform allows for the study of shocks on household's health and wealth that may be associated with morbidity and mortality among livestock attributable to infectious diseases. Data on health of livestock through the surveillance system will likely be related to household expenditures on health, health-care seeking following illness in a member of the household, and income and asset changes over the same period of time. The longitudinal nature of data collection provides a chance to relate changes in household assets and consumption with illness and death events in animals. Data will be analysed by syndromes to determine which of the livestock syndromes under study have the greatest economic impacts on households, and the results used to prioritize which interventions will not only reduce livestock disease burden but maximize protection of household assets and livelihoods.

Analysis of data on consumption of animal source foods indicates households that keep cattle and those that keep chickens have an increased likelihood of consuming milk and eggs respectively. While such a correlation might seem intuitive, it is not uncommon that milk and eggs are used for household income to purchase less expensive sources of food with less nutritive value. The consumption of animal source foods is important as it has been found to be positively associated with increased cognitive development and a reduced risk of malnutrition and stunting in children[34–36]. In order to better understand the relationships between livestock ownership and health, and human nutrition, this study is collecting monthly anthropometric data from children under 5 years of age, accompanied by records of nutritional intakes of these same children for a period of at least one year. Linked to the human health data, these outcome variables will be used to investigate the nutritional pathways and determine the impact of accessing animal source foods on disease susceptibility and overall child health, nutritional status, and growth and development.

Our preliminary analysis shows a positive association between the likelihood of spending money on health and the number of chickens owned, and no association with the number of small or large stock owned. Whereas it is plausible to relate health spending to livestock ownership, it is curious why the relationship is significant with poultry alone and not with the other livestock species. To illustrate some of the complexity of household economic decisions that can be examined using our platform, note that the average amount of health spending is relatively small (approximately 5 US dollars a month) and it may be that different livestock species serve different livelihood roles. These spending patterns and associations with specific livestock species may be related to gender roles and livestock ownership. For instance, women are known to have more rights over chicken and sometimes small stock than they have over larger livestock, and are likely to spend significantly higher amounts of their income on families compared to men[37]. There is need for additional studies and analysis to provide information on the intra-household dynamics of ownership, management and marketing of livestock products to draw stronger inferences on relationship between livestock ownership and productivity and human health and nutrition.

Livestock are economically substantial assets. The sale of larger (more valuable) stock may be reserved to cushion against larger household financial needs such as education fees for children, whereas the sale of a chicken may represent a smaller "withdrawal" for the purpose of paying these smaller expenditures. These complexities that drive household spending may be responsible for whether a family gets into or out of poverty and disease[38]. Continuing longitudinal data from longer periods of follow-up will allow for more robust testing of hypotheses on the role different livestock species in meeting specific household needs such as nutrition, cash, education, and socio-cultural needs of livestock keepers.

It is important to note that while animal illness is included in these regressions as an explanatory variable for human illness, these results should not be taken to imply directional causality of livestock health on human health, but instead should be interpreted carefully as qualitative evidence of a (potentially complex) relationship between human and animal health.

In conclusion, the PBASS-PBIDS platform produces a unique opportunity to link human health, animal health and socio-economic data, collected frequently and simultaneously in the same households over the same period of time. These data allow for the investigation, understanding and quantification of the pathways by which human health and welfare are linked to animal health, and provide a platform for testing hypotheses related to "one-health" including scientific enquiries focusing on specific diseases, co-infections and their interactions. While it is tempting to suggest that intervention in animals to control gastrointestinal or respiratory illness may have its most important direct impact on human illness, data on household welfare may indicate that animal health interventions, not necessarily related to a specific disease syndrome, increases household wealth, which, in turn may have the greatest impact on human health. Such data and analyses may have novel contributions to guide health policy and investment in resource constrained areas.

ACKNOWLEDGMENTS

We thank the PBIDS and PBASS community interviewers, animal health technicians and staff for the enthusiasm and tireless effort in the field collection of human health, animal health and socio-economic data. Special thanks to the study coordinators George Aol and Ratwar Okoth for their excellent supervision and support in the PBIDS project. We thank the KEMRI HDSS team, especially Dr. Frank Odhiambo, Dr. Nyaguara Amek, and David Obor for all their help with identification of study compounds and households. We thank the study participants for their continued support and willingness to continue participation in the study. We thank Prof. Jonathan Rushton and Dr. Meslin François whose excellent comments greatly improved the work. This manuscript is published with the approval of the Director of KEMRI. The findings and conclusions are those of the authors and do not necessarily represent the views of the Centers for Disease Control and Prevention.

AUTHOR CONTRIBUTIONS

Conceived and designed the experiments: GHP TFM MKN TLM SN EO PM RFB SMT LO JY. Performed the experiments: SMT EO LO JMM GB. Analyzed the data: SMT LO AA MKN TFM. Contributed reagents/materials/analysis tools: SMT AA LO. Wrote the paper: SMT MKN GHP TLM SN EO PM RFB LO JMM GB AA JY TFM.

REFERENCES

1. Staal S, Poole J, Baltenweck I, Mwacharo J. Targeting strategic investment in livestock development as a vehicle for rural livelihoods, Report to Bill and Melinda Gates Foundation. ILRI, Nairobi; 2009.
2. Randolph TF, Schelling E, Grace D, Nicholson CF, Leroy JL, Cole DC, et al. Invited review: Role of livestock in human nutrition and health for poverty reduction in developing countries. J Anim Sci. 2007;85(11):2788–800. pmid:17911229
3. Perry B, Grace D. The impacts of livestock diseases and their control on growth and development processes that are pro-poor. Philos Trans R Soc Lond B Biol Sci. 2009;364(1530):2643–55. pmid:19687035
4. Bradford G. Contributions of animal agriculture to meeting global human food demand. Livest Prod Sci. 1999 Jun;59(2–3):95–112.
5. Sigman M, Neumann C, Baksh M, Bwibo N, McDonald MA. Relationship between nutrition and development in Kenyan toddlers. J Pediatr. 1989 Sep;115(3):357–64. pmid:2769494
6. Nicholson CF, Mwangi L, Staal SJ, Thornton PK. Dairy Cow Ownership and Child Nutritional Status in Kenya. 2003 AAEA Annual Meetings. Montréal, Québec, Canada; 2003.
7. Whaley SE, Sigman M, Neumann C, Bwibo N, Guthrie D, Weiss RE, et al. The impact of dietary intervention on the cognitive development of Kenyan school children. J Nutr. 2003 Nov;133(11 Suppl 2):3965S–3971S. pmid:14672297
8. Lannotti L, Lesorogol C. Animal milk sustains micronutrient nutrition and child anthropometry among pastoralists in Samburu, Kenya. Am J Phys Anthropol. 2014 Jun 18;00(January).
9. Bhaskaram P. Micronutrient malnutrition, infection, and immunity: an overview. Nutr Rev. 2002 May;60(5 Pt 2):S40–5. pmid:12035857
10. Hughes S, Kelly P. Interactions of malnutrition and immune impairment, with specific reference to immunity against parasites. Parasite Immunol. 2006 Dec;28(11):577–88. pmid:17042929
11. Mableson HE, Okello A, Picozzi K, Welburn SC. Neglected zoonotic diseases-the long and winding road to advocacy. PLoS Negl Trop Dis. 2014 Jun;8(6):e2800. pmid:24901769
12. Thornton PK, Kruska RL, Henninger N, Kristjanson PM, Reid RS, Atieno F, et al. Mapping Poverty and Livestock in the Developing

World. Nairobi, Kenya: International Livestock Research Institute; 2002.

13. Okwi PO, Ndeng'e G, Kristjanson P, Arunga M, Notenbaert A, Omolo A, et al. Spatial determinants of poverty in rural Kenya. Proc Natl Acad Sci U S A. 2007 Oct 23;104(43):16769–74. pmid:17942704
14. Feikin DR, Olack B, Bigogo GM, Audi A, Cosmas L, Aura B, et al. The Burden of Common Infectious Disease Syndromes at the Clinic and Household Level from Population-Based Surveillance in Rural and Urban Kenya. Beeson JG, editor. PLoS One. 2011 Jan 18;6(1):e16085. pmid:21267459
15. Knobel DL, Maina AN, Cutler SJ, Ogola E, Feikin DR, Junghae M, et al. Coxiella burnetii in humans, domestic ruminants, and ticks in rural western Kenya. Am J Trop Med Hyg. 2013 Mar;88(3):513–8. pmid:23382156
16. Eshitera EE, Githigia SM, Kitala P, Thomas LF, Fèvre EM, Harrison LJS, et al. Prevalence of porcine cysticercosis and associated risk factors in Homa Bay District, Kenya. BMC Vet Res. 2012 Jan;8:234. pmid:23217158
17. Githigia SM, Willingham AL, Njeruh FM. Palpable lingual cysts, a possible indicator of porcine cysticercosis, in Teso District, Western Kenya. 2007;15(August):206–12.
18. Kotloff KL, Nataro JP, Blackwelder WC, Nasrin D, Farag TH, Panchalingam S, et al. Burden and aetiology of diarrhoeal disease in infants and young children in developing countries (the Global Enteric Multicenter Study, GEMS): a prospective, case-control study. Lancet. 2013 Jul 20;382(9888):209–22. pmid:23680352
19. Von Wissmann B, Machila N, Picozzi K, Fèvre EM, deC Bronsvoort BM, Handel IG, et al. Factors associated with acquisition of human infective and animal infective trypanosome infections in domestic livestock in Western Kenya. PLoS Negl Trop Dis. 2011 Jan;5(1):e941. pmid:21311575
20. Zinsstag J, Schelling E, Wyss K, Mahamat MB. Potential of cooperation between human and animal health to strengthen health systems. Lancet. 2005 Dec 17;366(9503):2142–5. pmid:16360795
21. Morse SS, Mazet JA, Woolhouse M, Parrish CR, Carroll D, Karesh WB, et al. Prediction and prevention of the next pandemic zoonosis. Lancet. Elsevier Ltd; 2012 Dec;380(9857):1956–65. pmid:23200504

22. Adazu K, Lindblade K a, Rosen DH, Odhiambo F, Ofware P, Kwach J, et al. Health and demographic surveillance in rural western Kenya: a platform for evaluating interventions to reduce morbidity and mortality from infectious diseases. Am J Trop Med Hyg. 2005 Dec;73(6):1151–8. pmid:16354829
23. Odhiambo FO, Laserson KF, Sewe M, Hamel MJ, Feikin DR, Adazu K, et al. Profile: the KEMRI/CDC Health and Demographic Surveillance System—Western Kenya. Int J Epidemiol. 2012 Aug;41(4):977–87. pmid:22933646
24. Feikin DR, Audi A, Olack B, Bigogo GM, Polyak C, Burke H, et al. Evaluation of the optimal recall period for disease symptoms in home-based morbidity surveillance in rural and urban Kenya. Int J Epidemiol. 2010 Apr;39(2):450–8. pmid:20089695
25. R Core Team. R: A Language and Environment for Statistical Computing. Vienna, Austria; {ISBN} 3-900051-07-0, 2014.
26. Mazet JAK, Clifford DL, Coppolillo PB, Deolalikar AB, Erickson JD, Kazwala RR. A "one health" approach to address emerging zoonoses: the HALI project in Tanzania. PLoS Med. 2009 Dec;6(12):e1000190. pmid:20016689
27. Narrod C, Zinsstag J, Tiongco M. A one health framework for estimating the economic costs of zoonotic diseases on society. Ecohealth. 2012 Jun 7;9(2):150–62. pmid:22395956
28. Grace D. The business case for One Health. Onderstepoort J Vet Res. 2014 Jan;81(2):E1–6. pmid:25134174
29. Singer M, Bulled N. Interlocked infections: the health burdens of syndemics of neglected tropical diseases. Ann Anthropol Pract. 2012 Nov 12;36(2):328–45.
30. Marmot M, Friel S, Bell R, Houweling T a J, Taylor S. Closing the gap in a generation: health equity through action on the social determinants of health. Lancet. 2008 Nov 8;372(9650):1661–9. pmid:18994664
31. Argent J, Augsburg B, Rasul I. Livestock asset transfers with and without training: Evidence from Rwanda. J Econ Behav Organ. Elsevier B. V. ; 2014 Aug;
32. Aanensen DM, Huntley DM, Feil EJ, Al-Own F, Spratt BG. EpiCollect: linking smartphones to web applications for epidemiology, ecology and community data collection. PLoS One. 2009 Jan;4(9):e6968. pmid:19756138

33. Madder M, Walker JG, VAN Rooyen J, Knobel D, Vandamme E, Berkvens D, et al. e-Surveillance in Animal Health: use and evaluation of mobile tools. Parasitology. 2012 Apr 13;1–12.
34. Murphy S, Allen L. Nutritional Importance of Animal Source Foods. J Nutr. 2003;3932–5.
35. Darapheak C, Takano T, Kizuki M, Nakamura K, Seino K. Consumption of animal source foods and dietary diversity reduce stunting in children in Cambodia. Int Arch Med. International Archives of Medicine; 2013 Jul 17;6(1):29.
36. Jin M, Iannotti LL. Livestock production, animal source food intake, and young child growth: The role of gender for ensuring nutrition impacts. Soc Sci Med. Elsevier Ltd; 2014 Mar;105:16–21. pmid:24606793
37. Njuki J, Sanginga P. Gender and livestock: Issues, challenges and opportunities. Bridging the Gender Gap: Women, Livestock Ownership and Markets in Eastern and Southern Africa. Nairobi, Kenya: International Livestock Research Institute; 2013.
38. Kristjanson P, Krishna A, Radeny M, Nindo W. Pathways out of Poverty in Western Kenya and the Role of Livestock. Policy. Rome (Italy): FAO; 2004 p. 23. Report No. : 14.

CHAPTER 8

Bartonella spp. - A Chance to Establish One Health Concepts in Veterinary and Human Medicine

Yvonne Regier, Fiona O'Rourke and Volkhard A. J. Kempf

Institute for Medical Microbiology and Infection Control, University Hospital, Goethe-University, Frankfurt am Main, Germany

ABSTRACT

Infectious diseases remain a remarkable health threat for humans and animals. In the past, the epidemiology, etiology and pathology of infectious agents affecting humans and animals have mostly been investigated in

---

Citation: Regier, Y., O'Rourke, F. & Kempf, V.A.J. Bartonella spp. - a chance to establish One Health concepts in veterinary and human medicine. Parasites Vectors 9, 261 (2016). https://doi.org/10.1186/s13071-016-1546-x.

separate studies. However, it is evident, that combined approaches are needed to understand geographical distribution, transmission and infection biology of "zoonotic agents". The genus *Bartonella* represents a congenial example of the synergistic benefits that can arise from such combined approaches: *Bartonella* spp. infect a broad variety of animals, are linked with a constantly increasing number of human diseases and are transmitted via arthropod vectors. As a result, the genus *Bartonella* is predestined to play a pivotal role in establishing a One Health concept combining veterinary and human medicine.

BACKGROUND

The threat of infectious diseases to mankind has never been greater than today. For the first time, political leaders of the 41st "G7 summit" in Schloss Elmau/Germany on June 7–8, 2015, set the topic "global health" (including infectious diseases) as one of the key issues on their agenda. In the past, health issues played only a minor part in such international economic summits. However, governments have now realized that public health is an essential prerequisite for education, working capacity and therefore the economic prosperity of societies.

In this regard, it is important to recognize that human health and animal health are closely linked. An estimated 75 % of emerging infectious diseases are zoonotic and 28 % vector-borne [1]. Global warming represents an additional factor promoting the spread of these diseases as the geographic range of some vectors and reservoir hosts expands in response to a changing climate [2].

To respond to these challenges, the One Health concept aims to establish interdisciplinary collaborations between medical, veterinary and environmental researchers as well as public health officials for the early detection of health hazards affecting both humans and animals and to fight them on multiple levels. The genus *Bartonella* represents a prototypical example for zoonotic pathogens as *Bartonella* species are infectious agents for humans and animals. High pathogen prevalence and severe courses of infection raise the importance to investigate possible routes of transmission and to combat infections.

THE GENUS *BARTONELLA*: A DIVERSE AND EXPANDING GROUP OF BACTERIA

The bacterial genus of *Bartonella* is comprised of Gram-negative, slow growing and facultative intracellular pathogens that infect mainly mammalian hosts and are often transferred via blood sucking arthropod vectors. *Bartonella* infections of humans and animals are often characterized by an intraerythrocytic bacteremia. At least 20 species are known to cause host-specific intraerythrocytic infections in their specific mammalian reservoir hosts, including the human-specific pathogens *Bartonella quintana* and *Bartonella bacilliformis,* the agents of trench fever and Oroya fever, respectively. A secondary tissue phase can be associated with development of vasculoproliferative lesions, e.g. bacillary angiomatosis (*Bartonella henselae, B. quintana*) or verruga peruana (*B. bacilliformis*) and may play a role in various other dermal conditions [3–7].

Molecular epidemiology techniques have revealed a remarkable diversity within the genus *Bartonella*. A wide variety of *Bartonella* spp. specialized to various mammalian hosts and transferred by specific arthropod vectors have been identified over the years and the prevalence of infection appears to be widespread across species and geographic regions. At least 13 species of *Bartonella* have been identified as pathogenic to humans with three species responsible for most of the clinically relevant infections in humans: *B. bacilliformis*, *B. quintana* and *B. henselae* [6].

Bartonella spp. infections are often chronic or asymptomatic in their reservoir hosts. Bacteria have been shown to infect erythrocytes, endothelial cells, macrophages and even human stem cells [8–17]. The infection of erythroctyes is host-specific and mediated by the so-called "Trw"-type 4 secretion system which facilitates host-restricted adhesion to erythrocytes [18]. Localized tissue manifestations may occur in reservoir and incidental hosts and the growth of bacteria in vascular tissue can lead to angioproliferative tumors and inflammation [5, 6, 12, 19]. The ability of *Bartonella* spp. to persist within immunoprivileged intracellular habitats is probably a key factor contributing to the establishment of chronic infections; however, the cyclic release of bacteria to the blood stream or the hemolytic activity of some species can also result in dramatic illnesses such as trench fever or Oroya fever, respectively [20]. The presence of *Bartonella* spp. in the blood stream of infected hosts or within the erythrocytes also facilitates their transfer via ingestion along with the blood meal of arthropod vectors [5, 21].

BARTONELLA SPP. INFECTIONS IN ANIMALS

Cat Infections

Cats are the main reservoir host for the species *B. henselae, B. clarridgeiae* (both of which can cause cat scratch disease) and *B. koehlerae* (a causative agent of endocarditis in humans) [22–24]. Infected cats are often clinically asymptomatic although they suffer from relapsing bacteremia over long periods of time [25]. Co-infections with more than one *Bartonella* species are not uncommon [26–29].

Transmission of *Bartonella* spp. among cats occurs via arthropod vectors, predominantly fleas. Uninfected cats kept together with infected cats in a specific ectoparasite-free environment do not become seropositive emphasizing the importance of arthropod vectors in the transmission of disease. Furthermore, the transmission via arthropod vectors appears to be essential as no direct transmission of *B. henselae* from cat to cat has been documented experimentally and flea-prevention measures have been shown to be effective in preventing pathogen transmission [30–35]. Prevalence of infection is highest in warm, moist areas with a higher ectoparasite burden (0 % in Norway *versus* 68 % in the Phillipines) [8, 22, 30, 36–38]. Up to 50 % of all cats (stray and pet) living in regions where fleas are endemic, harbor bacteremic *Bartonella* infections [26, 28]. Usually, cats are bacteremic for weeks or months but even longer infection intervals are possible. Young cats are more likely to be bacteremic than old cats and stray cats more than pet cats [8, 22, 23]. Cats were tested in several regions of Spain, for *B. henselae* seroreactivity and 50 % were shown to be positive. It is known, however, that serum antibodies have a limited value for the detection of active infections. In the same study, DNA of *Bartonella* spp. was detected in 4.4 % of the examined cat fleas [39]. *Bartonella* spp. have also been isolated from cat blood in various other locations throughout the world (e.g. from San Francisco/USA, North Carolina/USA, Hawaii/USA, Japan, Sydney, New Zealand, the Netherlands, France, Indonesia and Germany) [26, 28, 40–49].

Those *Bartonella* strains which have been isolated from healthy cats were usually not of the same genetic background as those strains detected from infected humans. Some of the feline strains were never found on patients and therefore might be less pathogenic for humans [50, 51].

Although healthy cats can be infected with *B. henselae* and *B. clarridgeiae* for months or even years, there is evidence that cats may also suffer from the persistent infection [26]. Especially infections with *Bartonella* spp. that are not believed to be specifically adapted to the cat as reservoir host (e.g. *Bartonella vinsonii* subsp. *berkhoffii*) can result in more serious clinical symptoms, e.g. osteomyelitis [30, 31, 52]. Several seroepidemiological studies indicated a correlation between seroreactivity and stomatitis, kidney- and urinary tract-diseases and uveitis [26, 53–55]. Another survey found that stomatitis is associated with the detection of *Bartonella* spp. but not with seroreactivity and revealed no association with uveitis, neurological symptoms and chronic kidney diseases; however, a weak association between seroreactivity and idiopathic feline lower urinary tract disease was detected [56]. Cats experimentally infected with *B. henselae* or *B. clarridgeiae* were found to suffer from fever, eosinophilia, lymphadenomegaly and anemia. Perinatal transmission was not described but reproductive disorders were observed. Furthermore, some cats suffered from transient neurological disorders, endocarditis and focal myocarditis [26, 57–62].

Isolation of *Bartonella* spp. has been possible from cats whose owners suffered from cat scratch disease and bacillary angiomatosis and *B. clarridgeiae* was isolated from a kitten that had caused cat scratch disease in a veterinarian [25, 26, 40, 63]. *Bartonella quintana* was found in the mouth of a domestic cat and there are reported cases of humans suffering from *B. quintana* infections where no louse infestation was verifiable but contact with cats was reported [30, 64]. In one case *B. qintana* was detected in a woman and two cats, one of which had bitten the woman previously providing further evidence for the incidental zoonotic transmission of *Bartonella* between animals and humans [65]. Antimicrobial treatment for pathogen eradication in cats is not broadly recommended; therefore, ectoparasite-control (e.g. collars containing acaricides) is crucial as the main instrument to lower the *Bartonella* prevalence in cats and therefore reduces the risk of pathogen-transmission to humans [30, 35].

DOG INFECTIONS

Dogs represent an incidental host for *Bartonella* and two species are known to cause clinically apparent infections: *B. vinsonii* subsp. *berkhoffii*, causing endocarditis, arrhythmias, myocarditis, granulomatous lymphadenitis and granulomatous rhinitis, and *B. henselae* causing peliosis hepatis [66–71]. In a study from the United States surveying *Bartonella* bacteremia in dogs, *B. henselae* was found in 30 of 61 infected dogs [72]; however, there are also rare cases in which other *Bartonella* spp. have caused disease in dogs: *B. clarridgeiae*, *B. washoensis* and *B. quintana* were isolated from dogs suffering from endocarditis [8, 22, 23]. To date, all *Bartonella* spp. identified in sick dogs are also known as pathogenic or potentially pathogenic infectious agents for humans and this observation led to the suggestion that dogs might act as useful sentinel species and important comparative models for human infections [22, 73].

Domestic dogs are generally incidental hosts for *B. henselae* with a reported seroprevalence of ~10 % in healthy dogs in the United States and ~27 % in sick dogs [30, 74]. Similar to cat epidemiology, seroprevalence increases in warmer regions [30]. *Bartonella henselae*, *B. quintana*, *B. vinsonii* subsp. *berkhoffii* and *B. bovis* have been detected in mouth swabs from dogs and there is some evidence that dogs may be able to transmit *B. henselae* to humans via bites [22, 30, 37, 75]. Because of the prolonged bacteremia of *B. vinsonii* subsp. *berkoffii* in dogs, they are suspected to represent the reservoir host of these bacteria and seroreactivity of dogs against *B. vinsonii* subsp. *berkoffii* is found worldwide [22, 26, 32]. In Gabon, *B. clarridgeiae* was also isolated from ~2 % of the examined dogs indicating that these animals may represent a potential reservoir host for *Bartonella* spp. in Africa [76].

Serological surveys suggest that *B. vinsonii* subsp. *berkhoffii* may also cause immuno-mediated hemolytic anemia, neutrophil or granulomatous meningoencephalitis, neutrophil polyarthritis and uveitis in dogs [8,22]. *Bartonella vinsonii* subsp. *berkhoffii* can cause endocarditis, especially in large breed dogs with a predisposition for aortic valve involvement. Intermittent lameness and fever of unknown origin can occur several months before endocarditis. Myocarditis without an associated endocarditis is also possible and may result in arrhythmias, syncope or sudden death [26, 67]. To detect *Bartonella* spp. as the causative agent of infectious endocarditis in dogs, diagnostic PCRs should be performed from blood or heart valve specimen as blood cultures often remain negative [77, 78]. High antibody-titers and characteristic lesions in echocardiography are also suspicious for *Bartonella* endocarditis. In most cases, *Bartonella* infect the aortic valve causing aortic insufficiency leading to severe chronic heart failure and arrhythmias [66–68, 77, 79, 80].

INFECTIONS OF OTHER MAMMALS

There are many publications describing *Bartonella* infections of numerous mammals and even reptiles. For instance, *Bartonella* spp. have been detected in a wide variety of wild and domestic animals throughout the world including, e.g. mountain lions, bobcats, coyotes, grey foxes, elks, mule deer, cougars, rabbits, several rodent species, cattle, belugas, bats and porpoises. However, it is unclear which diseases if any are associated with such infections and if these animals play a role as potential reservoir hosts. [26, 30, 81–92].

BARTONELLA SPP. INFECTIONS OF HUMANS

The first human pathogenic *Bartonella* species to be identified in the early 1900s was *B. bacilliformis*. This human-specific bacterium causes a biphasic disease characterised by a primary hemolytic fever ("Oroya fever") with high mortality (up to 90 %), followed by a chronic vasculoproliferative tissue phase ("verruga peruana"). Pathogens are transmitted by the sand fly (*Lutzomyia verrucarum*). The human body louse (*Pediculus humanus humanus*) transmits *B. quintana*, a second human pathogenic *Bartonella* species which emerged as a major agent of disease causing debilitating cyclic fever ("trench fever") during World War I. Today, trench fever occurs mainly in the homeless population or among drug addicts. Endocarditis, generalized lymphadenopathy and bacillary angiomatosis are symptoms of *B. quintana* infections in immunocompromised people [93–100].

Of the three most significant human pathogenic *Bartonella* species *B. henselae* is the most common symptomatic infection causing agent identified in the modern clinical setting. *Bartonella henselae* infection is the cause of multiple clinical entities in humans and infections result in differential disease outcomes often depending on the immune status of the patient. Humans become infected via the scratches or bites of infected cats contaminated with infected flea feces or are directly contaminated with infected blood. Dogs are also suspected to be an additional reservoir for *B. henselae* transmission to humans [41]. In immunocompetent patients, infections normally cause cat scratch disease which is often self-limiting with no need for antibiotic treatment. Typically, two to three weeks after infection, a unilateral lymphadenitis in the draining region of the lymph node near the site of inoculation can be observed. In ~10 % of the cases, the lymph node forms a fistula where pus is draining. Other symptoms include chronic lymph node swelling, fever, headache, skin and mucosal lesions near the inoculation site and splenomegaly. "Blood-culture negative"-endocarditis, oculoglandular involvement ("Parinaud's syndrome"), encephalopathy, neuroretinitis and osteomyelitis are described as complications of infection [101]. Recurring or systemic infections can be treated with macrolides. In immunocompromised hosts, chronic infections can occur, leading to angioproliferative diseases like bacillary angiomatosis and peliosis hepatis which can be fatal if not treated [6, 19].

Several *Bartonella* spp. have been reported as cause of fever of unknown origin and culture-negative endocarditis in humans and animals [102–105]. In humans, endocarditis caused by *B. henselae*, *B. quintana*, *B. elizabethae*, *B. vinsonii* subsp. *berkhoffii*, *B. vinsonii* subsp. *arupensis*, *B. koehlerae*, *B. alsatica*, *B. washoensis* and Candidatus *B. mayotimonensis* have been reported [24, 106–114]; however, human endocarditis cases are most often associated with *B. henselae* and *B. quintana* [79, 80, 115]. In most cases, high anti-*Bartonella*-IgG antibody-titers can be detected [102, 116].

Co-infections with more than one *Bartonella* spp. (even in immunocompetent patients) [117–119] and with other zoonotic bacterial species have been reported. Co-infections with *Borrelia burgdorferi* (*sensu lato*) and *B. henselae* were described in patients with atypical neuroborreliosis [120–122]. Furthermore, surveys showed the occurrence of co-infections with *B. henselae* in people suffering from persistent symptoms after borreliosis treatment where ticks might have been the source of infection [121]. The transmission of multiple pathogens via co-infected vectors might contribute to atypical disease progression and should be considered for the diagnosis of tick-borne diseases [121, 123, 124]. However, it must be stated that the occurrence of chronic, atypical tick-borne co-infections in patients with chronic, nonspecific illnesses is highly controversially discussed. As reviewed by Lantos & Wormser, in most reported cases of *Bartonella* and *Borrelia* co-infections, laboratory diagnostics were not properly performed [125].

Different population groups are exposed to animals and arthropod vectors in variable dimensions. In particular, veterinarians, veterinary technicians or zookeepers might be at increased risk of infection with *Bartonella* spp. [119, 126]. For example, one case of *B. vinsonii* subsp. *berkhoffii*, transmission to a veterinarian was likely caused by a needle puncture injury [127]. *Bartonella* infections have even been suspected to have been a contributing factor in the death of two veterinarians in 2013 [128]. In an epidemiological study, *Bartonella* DNA was also detected in the blood of

28 % of veterinary workers whereas no *Bartonella* was detected in control subjects [126]. The prevalence of *Bartonella* infections has also been found to be elevated in other risk groups. In a recent study in Germany *B. henselae* IgG antibodies were found in ~45 % of forestry workers which may be due to a higher contact with arthropods which is inevitable during forest work [129]. From a One Health perspective, the identification of possible vectors and means of *Bartonella* transmission is crucial to reduce occupational hazards in certain risk groups and prevent such cases of *Bartonella* transmission in future.

Blood transfusion has also been identified as a risk factor for the transmission of *Bartonella* infections. The transmission of infection via blood transfusion was first shown 20 years ago in cats [58] and a very recent study from Brazil has also indicated a ~3 % prevalence of *Bartonella* spp. in asymptomatic human blood donors. Remarkably, the results of this study found that professionals with animal contact were seven times more likely to harbor *Bartonella* than other blood donors and individuals with cat contact or a history of tick bite were three to four times more likely to be infected with *Bartonella* spp. [130]. Considering that patients receiving blood transfusions are already in a state of weakened health, screening of blood donors for *Bartonella* infections, especially in certain risk groups should be considered to prevent transmission of infection.

VECTOR TRANSMISSION OF *BARTONELLA* SPP.

The transmission cycle of bartonellosis is typical for vector-borne diseases. Typically, infections are characterized by persistent intraerythrocytic bacteremia within the reservoir host. Infected blood is ingested by the blood sucking arthropod vector and subsequently transmitted to a further reservoir or incidental host. To date, vector-competence of several arthropods has been proven for *Bartonella* spp. transmission and additional vectors competencies are suspected in many more.

FLEA TRANSMISSION OF *BARTONELLA* SPP.

The cat flea (*Ctenocephalides felis*) represents the main vector for *B. henselae* infection among cats. Its vector-competence for *B. henselae* transmission is experimentally proven and its presence is essential for the maintenance of *B. henselae* infection within the cat population. The contamination of the flea-feeding wound, or other wounds such as scratches or bites with contaminated flea feces has been identified as an important transmission route among hosts including cats and humans [30, 33, 96, 131–133]. Bacteria are replicating in the flea's intestine and are secreted with the feces over the life-span of the flea (~12 days). The excreted flea feces contain *B. henselae* within 24 h of a blood meal [134].

Further supporting the importance of the flea as a vector of *B. henselae* transmission, epidemiological studies have shown an increased risk of *B. henselae* infection in cats suffering from flea infestation and the use of flea-prevention collars has also been shown to be effective in preventing the transmission of *B. henselae* infection from cat to cat [35]. Once infected, *B. henselae* bacteremia in cats can last for weeks, months or even longer than one year supporting further vector-transmission [22, 23, 29, 39, 132, 135, 136]. In addition to *B. henselae*, cats are susceptible to infections with *B. quintana*, *B. koehlerae*, *B. clarridgeiae*, *B. vinsonii* subsp. *berkhoffii* and *B. bovis* which have also been detected in cat fleas. With the exception of *B. bovis*, these species can also be pathogenic to humans [6, 30, 52, 137]. Flea control is highly recommended in endemic areas to reduce the pathogen exposition of cats and humans [39]. In addition to the cat flea various other flea species may also play important roles in *Bartonella* transmission. *Bartonella* spp. were detected in several flea species collected from bats and different rodents [30, 96, 138–144]. Vector-competence, however, has not been confirmed experimentally for these species.

LOUSE TRANSMISSION OF *BARTONELLA* SPP.

The human body louse (*Pediculus humanus humanus*) represents the vector of human to human-transmission of *B. quintana*. Environmental factors supporting louse infestation such as unhygienic living conditions lead to an increased risk of infection. In the past, infections with *B. quintana* were a severe medical problem in the trenches and in prisoners of war camps of World War I from which the name "trench fever" arises. Today, mostly homeless people or drug addicts are affected resulting in the term "urban trench fever" [99, 145]. The vector becomes infected when adult lice feed on bacteremic hosts. *Bartonella quintana* reaches the louse gut and can infect humans when bite sites or other wounds are contaminated with infected louse feces [30, 64, 65, 96, 97, 146]. *Bartonella* spp. have also been detected from several other louse species (e.g. *Neohaematopinus sciuri*, *Hoplopleura sciuricola, Pediculus humanus capitis* and others) which may also serve as vectors [96, 139, 147, 148].

SAND FLY TRANSMISSION OF *BARTONELLA* SPP.

The sand fly (*Lutzomyia verrucarum*) transmits *B. bacilliformis* from humans to humans and its vector-competence has been proven experimentally [96, 98, 149–151]. The occurrence of the disease is strictly limited to the Peruvian Andes where the vector is endemic. However, it should be considered that climate change may extend the distribution area of this vector and thereby increase the spread of *B. bacilliformis*.

TICK TRANSMISSION OF *BARTONELLA* SPP.

Ticks are known to act as vectors for many different bacterial, protozoan and viral pathogens. Hard ticks (e.g. *Ixodes* spp., *Dermacentor* spp.) usually feed three times during their life-cycle and can possibly be infected with different pathogens during every blood meal. Hosts can be bitten by ticks several times during their lifetime presenting multiple opportunities for pathogen transmission [152–154]. Several studies have detected the presence of *Bartonella* spp. in various tick species from around the world [26, 84, 93, 120, 121, 123, 138, 152, 155–168]. The prevalence of *Bartonella* DNA in hard ticks in Europe has been shown to be as high as 40 % [158]. In a recent study conducted in Finland, ticks were found to contain no detectable *Bartonella* DNA whereas DNA of *Borrelia* spp. was found frequently at ~19 % [169]. On the other hand, *Bartonella* DNA was detected in ~2 % of ticks collected in a recent study from Austria [170]. Figure 1 displays the percentage of ticks found to harbor *Bartonella* in different studies. Overall, in ~15 % of ticks studied *Bartonella* DNA was detectable.

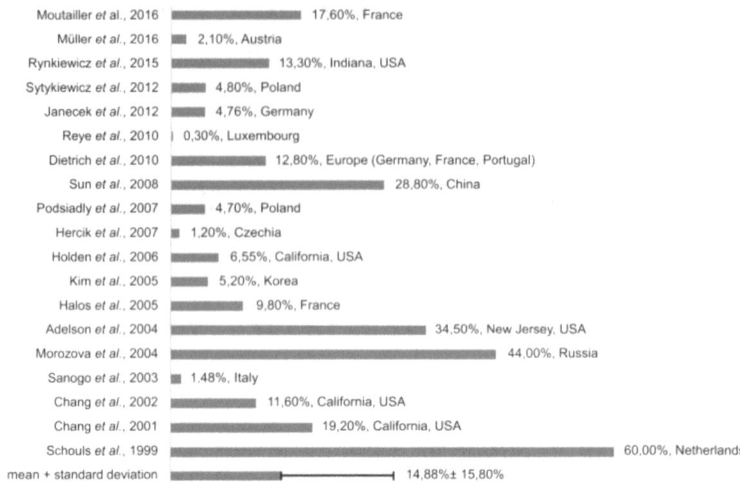

Fig. 1: Percentage of ticks found to harbor *Bartonella* spp. DNA (literature review). Presence of *Bartonella* spp. was determined via molecular techniques. Overall, *Bartonella* spp. were found in ~14.88 ± 15.80 % of ticks. Countries in which ticks were collected are depicted within the diagram.

Vector competence has been demonstrated experimentally by the use of artificial tick-feeding procedures for *B. henselae* [171] and a murine *B. birtlesii* infection model [172]; however, vector competence of naturally infected ticks has still not been confirmed.

Bartonella DNA has been detected in hard ticks removed from dogs. However, as DNA was detectable in only some but not all ticks removed from one particular dog, the infection of the tick may have been acquired from another source previously [173]. Furthermore, several studies indicate co-transmission of *Bartonella* with other tick-borne pathogens (e.g. *Ehrlichia, Babesia*) in dogs [66, 174–178]. In a study surveying dogs with endocarditis from California, all dogs that were infected with *Bartonella* were also seroreactive to *Anaplasma phagocytophilum,* another tick-borne pathogen [77].

In two cases, *B. henselae* DNA was detected in ticks collected from the home of patients which suffered from Lyme disease and which did not respond to a *Borrelia*-specific antibiotic therapy. In another study, *Bartonella* DNA was detectable in human blood after tick bites and recently, *B. henselae* and three other animal-associated *Bartonella* species (*B. doshiae, B. schoenbuchensis* and *B. tribocorum*) were isolated from patients suffering from undifferentiated chronic illness and who had reported tick bites [121, 179, 180]. Several case reports of *B. henselae* infections of humans have been published where no or very limited cat contact was reported, limiting the possibility of transmission via cats or cat fleas. Authors concluded that transmission via arthropod vectors (e.g. ticks) may provide an alternative explanation [96, 181].

The most important reservoir hosts for tick-borne pathogens are small rodents as they are the preferred hosts of tick larvae and nymphs. Several *Bartonella* spp. have been detected in these small mammals further supporting the possibility that ticks may represent a vector for *Bartonella* transmission [84, 87, 152–154, 162, 182, 183]. *Bartonella* spp. have also been isolated from cattle and mule deer in North America. As ruminants are rarely infested with fleas, ticks seem to be more likely to transmit these pathogens to these animals [81].

Nevertheless, it must be mentioned that the transmission of *Bartonella* spp. via ticks to humans and animals is still controversially discussed. Clearly, *Bartonella* DNA which was found in several tick species in multiple studies does not prove the presence of viable bacteria. Therefore, some researchers doubt heavily that *Bartonella* spp. are transmitted by ticks [125, 184]. Furthermore, concerns about the relevance of experimental tick transmission studies performed with an artificial feeding system [171] have been raised [184]: the amount of colony forming units in the blood was criticized to be much higher than it would be in natural infected bacteremic cats and the *B. henselae* strain that was used is not representative for *Bartonella* strains found in nature [184]. Building on these points, the authors conclude that neither of these studies demonstrated transmission of *Bartonella* spp. from ticks to mammalian hosts [125]. At least for *B. birtlesii*, tick transmission was proven in a murine infection model [172] whereas a *bona fide* tick transmission of *B. henselae* has not been demonstrated so far.

THE ROLE OF OTHER ARTHROPODS IN THE TRANSMISSION OF *BARTONELLA* SPP.

Bartonella species have been found in biting flies collected from cattle in California: *B. bovis* was detected in a horn fly (*Haematobia* spp.) and *B. henselae* was found in a stable fly (*Stomoxy* spp.) [185]. Several studies found *Bartonella* spp. in mites collected from rodents and bats from Korea, Egypt and Costa Rica [96, 162, 186]. Deer keds (*Lipoptena mazamae* and *Lipoptena cervi*) were shown to be infected with *B. henselae* and *B. schoenbuchensis* [30, 187–189]. *Lipoptena* species usually feed on deer but were also found on horses, cattle and humans. *Bartonella schoenbuchensis* was detected in *Lipoptena cervi* from a deer (*Capreolus capreolus*) in Germany and is suspected to be the causative agent of deer ked dermatitis in humans [190]. *Bartonella* was also found in several other species of the family *Hippoboscidae* indicating that they may play a role in the transmission of these bacteria [96, 191]. However, no experimental transmission studies with these species have been performed nor do any data exist on the transmission of *B. schoenbuchensis* by *Lipoptena* spp. to humans.

THE NEED FOR SCIENTIFIC ONE HEALTH APPROACHES IN *BARTONELLA* RESEARCH

When discussing transmission of *Bartonella* spp. from animals to humans, e.g. via arthropod vectors, a more integrative approach elucidating *Bartonella* prevalence in vectors as well as the infection status of animals and humans would clearly help to increase understanding of *Bartonella* infection dynamics, infection risk and prevent speculative and non-evidence-based conclusions.

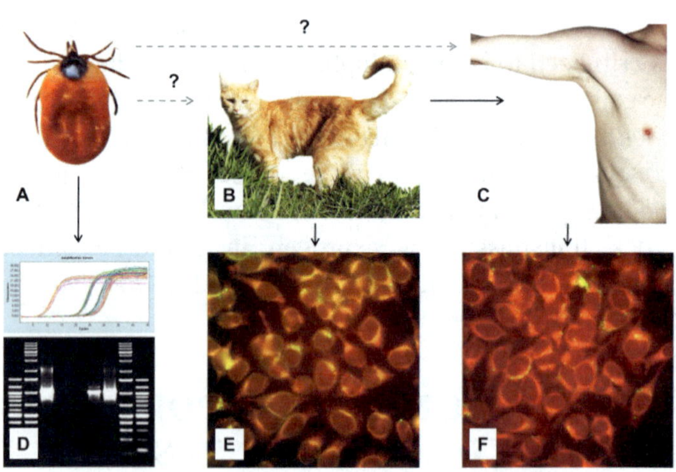

Figure 2: One Health concept for detecting of *Bartonella* infections in humans and domestic animals. Fleas transmit *B. henselae* to cats. Transmission of *B. henselae* by ticks, e.g. *Ixodes ricinus* a to cats b or humans c is assumed but controversially discussed. *Bartonella* DNA can be detected in ticks via real-

time PCR or conventional PCR d. Finally, *B. henselae* infections of cats and humans can be diagnosed by indirect immunofluorescence analysis (anti-B. henselae-IgG: green, e, f).

Such an approach might, for instance, include the investigation of the prevalence of *Bartonella* DNA or (even more reliable) of viable *Bartonella* species in feeding ticks, combined with parallel detection of these pathogens via direct detection or seroprevalence in animals (e.g. pets) and humans (e.g. pet owners). Figure 2 shows this concept of such a One Health approach.

For instance, a coordinated set of data might include (i) the pathogen-DNA status of questing ticks (with bacterial, e.g. *Bartonella* 16S-rDNA sequences; analyzed by conventional PCRs or metagenomics analysis), (ii) the direct detection of these pathogen(s) or detection of pathogen-specific (e.g. *Bartonella*) antibodies in animals (arguing for infection of the pet), and (iii) the direct detection of these pathogen(s) if feasible or the determination of the respective pathogen-specific antibodies in humans (e.g. pet owner) in parallel (arguing for a previous or present infection).

First incidental results from a clinical case revealed interesting findings: In a female, adult, half-engorged *I. ricinus* tick (black forest, Germany) which was feeding for ~1–2 days on a cat, *Bartonella*-DNA was detected via nested-PCR. Sequence analysis revealed most likely the presence of *B. henselae* 16S-rDNA (99 % sequence homology). For medical reasons, a serum sample from the cat was taken (because of unspecific illness). Immunofluorescence testing revealed cat antibody titers of 1:640 whereas no specific *B. henselae*-antibodies were detected in the pet owner. This setting might be interpreted as follows: The questing *I. ricinus* tick was probably ingesting *B. henselae*-containing cat blood as anti-*B. henselae* IgG antibodies were detectable. The pet owner, however, had no serological evidence for having been exposed to *B. henselae*. As a further option, a chronic *B. henselae* infection of the cat might additionally be confirmed (e.g. via PCR-analysis of peripheral blood). The application of such One Health approaches in prospective scientific studies would be useful to assess the real risk of transmission of *Bartonella* spp. from pets to pet owners and to clarify the role of ticks in this process.

CONCLUSIONS

The cumulative data collected in many studies and conducted in several countries throughout the world indicate that infections with *Bartonella* spp.

might represent an underappreciated danger for human and animal health. A great deal more research is needed to specify arthropod vectors and characteristics of diseases caused by *Bartonella* species. To date, strict ectoparasite-control is highly recommended to lower the risk of *Bartonella* infection from arthropod vectors to domestic animals and pets, thereby preventing pathogen transmission from animals to human owners. Overall, these results demonstrate that reliable data about vector transmission of *Bartonella* spp. from animals to humans can only be generated through the application of scientific One Health approaches which take into account the epidemiological factors and interactions of humans, animals and their environments as an integrated system.

ACKNOWLEDGEMENTS

The authors thank Wibke Ballhorn, Corinna Illich, Carmen Jung, Rebecca Kaufmann, Heike Podlich, Claudia Stolz and Yael Wiegand for excellent technical assistance. This paper has been sponsored by Bayer Animal Health in the framework of the 11th CVBD World Forum Symposium.

REFERENCES

1. Taylor LH, Latham SM, Woolhouse ME. Risk factors for human disease emergence. Philos Trans R Soc Lond B Biol Sci. 2001;356:983–9.
2. Mills JN, Gage KL, Khan AS. Potential influence of climate change on vector-borne and zoonotic diseases: a review and proposed research plan. Environ Health Perspect. 2010;118:1507–14.
3. Anderson BE, Neuman MA. *Bartonella* spp. as emerging human pathogens. Clin Microbiol Rev. 1997;10:203–19.
4. Maurin M, Birtles R, Raoult D. Current knowledge of *Bartonella* species. Eur J Clin Microbiol Infect Dis. 1997;16:487–506.
5. Dehio C. *Bartonella*-host-cell interactions and vascular tumour formation. Nat Rev Microbiol. 2005;3:621–31.
6. Kaiser PO, Riess T, O'Rourke F, Linke D, Kempf VAJ. *Bartonella* spp.: throwing light on uncommon human infections. Int J Med Microbiol. 2011;301:7–15.
7. Rossi MA, Balakrishnan N, Linder KE, Messa JB, Breitschwerdt EB. Concurrent *Bartonella henselae* infection in a dog with panniculitis and owner with ulcerated nodular skin lesions. Vet Dermatol. 2015;26:60–3.
8. Boulouis HJ, Chang CC, Henn JB, Kasten RW, Chomel BB. Factors associated with the rapid emergence of zoonotic *Bartonella* infections. Vet Res. 2005;36:383–410.
9. Kempf VA, Schaller M, Behrendt S, Volkmann B, Aepfelbacher M, Cakman I, et al. Interaction of *Bartonella henselae* with endothelial cells results in rapid bacterial rRNA synthesis and replication. Cell Microbiol. 2000;2:431–41.
10. Kempf VA, Volkmann B, Schaller M, Sander CA, Alitalo K, Riess T, et al. Evidence of a leading role for VEGF in *Bartonella henselae*-induced endothelial cell proliferations. Cell Microbiol. 2001;3:623–32.
11. Mändle T, Einsele H, Schaller M, Neumann D, Vogel W, Autenrieth IB, et al. Infection of human CD34+ progenitor cells with *Bartonella henselae* results in intraerythrocytic presence of *B. henselae*. Blood. 2005;106:1215–22.
12. O'Rourke F, Mändle T, Urbich C, Dimmeler S, Michaelis UR, Brandes RP, et al. Reprogramming of myeloid angiogenic cells by *Bartonella henselae* leads to microenvironmental regulation of pathological angiogenesis. Cell Microbiol. 2015;17:1447–63.

13. Maeno N, Oda H, Yoshiie K, Wahid Fujimura T, Matayoshi S. Live *Bartonella henselae* enhances endothelial cell proliferation without direct contact. Microb Pathog. 1999;27:419–27.
14. Resto-Ruiz SI, Schmiederer M, Sweger D, Newton C, Klein TW, Friedman H, et al. Induction of a potential paracrine angiogenic loop between human THP-1 macrophages and human microvascular endothelial cells during *Bartonella henselae* infection. Infect Immun. 2002;70:4564–70.
15. Kyme PA, Haas A, Schaller M, Peschel A, Iredell J, Kempf VAJ. Unusual trafficking pattern of *Bartonella henselae*-containing vacuoles in macrophages and endothelial cells. Cell Microbiol. 2005;7:1019–34.
16. Schülein R, Guye P, Rhomberg TA, Schmid MC, Schröder G, Vergunst AC, et al. A bipartite signal mediates the transfer of type IV secretion substrates of *Bartonella henselae* into human cells. Proc Natl Acad Sci U S A. 2005;102:856–61.
17. Lu YY, Franz B, Truttmann MC, Riess T, Gay-Fraret J, Faustmann M, et al. *Bartonella henselae* trimeric autotransporter adhesin BadA expression interferes with effector translocation by the VirB/D4 type IV secretion system. Cell Microbiol. 2013;15:759–78.
18. Vayssier-Taussat M, Le Rhun D, Deng HK, Biville F, Cescau S, Danchin A, et al. The Trw type IV secretion system of *Bartonella* mediates host-specific adhesion to erythrocytes. PLoS Pathog. 2010;6:e1000946.
19. Koehler JE, Quinn FD, Berger TG, LeBoit PE, Tappero JW. Isolation of *Rochalimaea* species from cutaneous and osseous lesions of bacillary angiomatosis. N Engl J Med. 1992;327:1625–31.
20. Foucault C, Rolain JM, Raoult D, Brouqui P. Detection of *Bartonella quintana* by direct immunofluorescence examination of blood smears of a patient with acute trench fever. J Clin Microbiol. 2004;42:4904–6.
21. Pulliainen AT, Dehio C. Persistence of *Bartonella* spp. stealth pathogens: from subclinical infections to vasoproliferative tumor formation. FEMS Microbiol Rev. 2012;36:563–99.
22. Chomel BB, Boulouis HJ, Maruyama S, Breitschwerdt EB. *Bartonella* spp. in pets and effect on human health. Emerg Infect Dis. 2006;12:389–94.
23. Chomel BB, Boulouis HJ, Breitschwerdt EB. Cat scratch disease and other zoonotic *Bartonella* infections. J Am Vet Med Assoc. 2004;224:1270–9.

24. Avidor B, Graidy M, Efrat G, Leibowitz C, Shapira G, Schattner A, et al. *Bartonella koehlerae*, a new cat-associated agent of culture-negative human endocarditis. J Clin Microbiol. 2004;42:3462–8.
25. Kordick DL, Wilson KH, Sexton DJ, Hadfield TL, Berkhoff HA, Breitschwerdt EB. Prolonged *Bartonella* bacteremia in cats associated with cat-scratch disease patients. J Clin Microbiol. 1995;33:3245–51.
26. Breitschwerdt EB, Kordick DL. *Bartonella* infection in animals: carriership, reservoir potential, pathogenicity, and zoonotic potential for human infection. Clin Microbiol Rev. 2000;13:428–38.
27. Gurfield AN, Boulouis HJ, Chomel BB, Heller R, Kasten RW, Yamamoto K, et al. Coinfection with *Bartonella clarridgeiae* and *Bartonella henselae* and with different *Bartonella henselae* strains in domestic cats. J Clin Microbiol. 1997;35:2120–3.
28. Heller R, Artois M, Xemar V, de Briel D, Gehin H, Jaulhac B, et al. Prevalence of *Bartonella henselae* and *Bartonella clarridgeiae* in stray cats. J Clin Microbiol. 1997;35:1327–31.
29. Kordick DL, Brown TT, Shin K, Breitschwerdt EB. Clinical and pathologic evaluation of chronic *Bartonella henselae* or *Bartonella clarridgeiae* infection in cats. J Clin Microbiol. 1999;37:1536–47.
30. Mosbacher ME, Klotz S, Klotz J, Pinnas JL. *Bartonella henselae* and the potential for arthropod vector-borne transmission. Vector Borne Zoonotic Dis. 2011;11:471–7.
31. Chomel BB, Kasten RW, Floyd-Hawkins K, Chi B, Yamamoto K, Roberts-Wilson J, et al. Experimental transmission of *Bartonella henselae* by the cat flea. J Clin Microbiol. 1996;34:1952–6.
32. Kordick DL, Breitschwerdt EB. Persistent infection of pets within a household with three *Bartonella* species. Emerg Infect Dis. 1998;4:325–8.
33. Guptill L. Bartonellosis. Vet Clin North Am Small Anim Pract. 2003;33:809–25.
34. Bradbury CA, Lappin MR. Evaluation of topical application of 10 % imidacloprid-1 % moxidectin to prevent *Bartonella henselae* transmission from cat fleas. J Am Vet Med Assoc. 2010;236:869–73.
35. Lappin MR, Davis WL, Hawley JR, Brewer M, Morris A, Stanneck D. A flea and tick collar containing 10 % imidacloprid and 4.5 %

flumethrin prevents flea transmission of *Bartonella henselae* in cats. Parasit Vectors. 2013;6:26.

36. Jameson P, Greene C, Regnery R, Dryden M, Marks A, Brown J, et al. Prevalence of *Bartonella henselae* antibodies in pet cats throughout regions of North America. J Infect Dis. 1995;172:1145–9.
37. Skerget M, Wenisch C, Daxboeck F, Krause R, Haberl R, Stuenzner D. Cat or dog ownership and seroprevalence of ehrlichiosis, Q fever, and cat-scratch disease. Emerg Infect Dis. 2003;9:1337–40.
38. Solano-Gallego L, Hegarty B, Espada Y, Llull J, Breitschwerdt E. Serological and molecular evidence of exposure to arthropod-borne organisms in cats from northeastern Spain. Vet Microbiol. 2006;118:274–7.
39. Gracia MJ, Marcén JM, Pinal R, Calvete C, Rodes D. Prevalence of *Rickettsia* and *Bartonella* species in Spanish cats and their fleas. J Vector Ecol. 2015;40:233–9.
40. Koehler JE, Glaser CA, Tappero JW. *Rochalimaea henselae* infection. A new zoonosis with the domestic cat as reservoir. JAMA. 1994;271:531–5.
41. Keret D, Giladi M, Kletter Y, Wientroub S. Cat-scratch disease osteomyelitis from a dog scratch. J Bone Joint Surg. 1998;80:766–7.
42. Demers DM, Bass JW, Vincent JM, Person DA, Noyes DK, Staege CM, et al. Cat-scratch disease in Hawaii: etiology and seroepidemiology. J Pediatr. 1995;127:23–6.
43. Maruyama S, Nogami S, Inoue I, Namba S, Asanome K, Katsube Y. Isolation of *Bartonella henselae* from domestic cats in Japan. J Vet Med Sci. 1996;58:81–3.
44. Branley J, Wolfson C, Waters P, Gottlieb T, Bradbury R. Prevalence of *Bartonella henselae* bacteremia, the causative agent of cat scratch disease, in an Australian cat population. Pathol. 1996;28:262–5.
45. Flexman JP, Lavis NJ, Kay ID, Watson M, Metcalf C, Pearman JW. *Bartonella henselae* is a causative agent of cat scratch disease in Australia. J Infect. 1995;31:241–5.
46. Joseph AK, Wood CW, Robson JM, Paul SL, Morris AJ. *Bartonella henselae* bacteraemia in domestic cats from Auckland. N Z Vet J. 1997;45:185–7.

47. Bergmans AM, de Jong CM, van Amerongen G, Schot CS, Schouls LM. Prevalence of *Bartonella* species in domestic cats in The Netherlands. J Clin Microbiol. 1997;35:2256–61.

48. Marston EL, Finkel B, Regnery RL, Winoto IL, Graham RR, Wignal S, et al. Prevalence of *Bartonella henselae* and *Bartonella clarridgeiae* in an urban Indonesian cat population. Clin Diagn Lab Immunol. 1999;6:41–4.

49. Arvand M, Klose AJ, Schwartz-Porsche D, Hahn H, Wendt C. Genetic variability and prevalence of *Bartonella henselae* in cats in Berlin, Germany, and analysis of its genetic relatedness to a strain from Berlin that is pathogenic for humans. J Clin Microbiol. 2001;39:743–6.

50. Breitschwerdt EB, Broadhurst JJ, Cherry NA. *Bartonella henselae* as a cause of acute-onset febrile illness in cats. J Feline Med Surg. 2015. doi:10.1177/2055116915600454.

51. Gil H, Escudero R, Pons I, Rodríguez-Vargas M, García-Esteban C, Rodríguez-Moreno I, et al. Distribution of *Bartonella henselae* variants in patients, reservoir hosts and vectors in Spain. PLoS One. 2013;8:e68248.

52. Varanat M, Travis A, Lee W, Maggi RG, Bissett SA, Linder KE, et al. Recurrent osteomyelitis in a cat due to infection with *Bartonella vinsonii* subsp. *berkhoffii* genotype II. J Vet Intern Med. 2009;23:1273–7.

53. Ueno H, Hohdatsu T, Muramatsu Y, Koyama H, Morita C. Does coinfection of *Bartonella henselae* and FIV induce clinical disorders in cats? Microbiol Immunol. 1996;40:617–20.

54. Glaus T, Hofmann-Lehmann R, Greene C, Glaus B, Wolfensberger C, Lutz H. Seroprevalence of *Bartonella henselae* infection and correlation with disease status in cats in Switzerland. J Clin Microbiol. 1997;35:2883–5.

55. Lappin MR, Black JC. *Bartonella* spp infection as a possible cause of uveitis in a cat. J Am Vet Med Assoc. 1999;214:1205–7.

56. Sykes JE, Westropp JL, Kasten RW, Chomel BB. Association between *Bartonella* species infection and disease in pet cats as determined using serology and culture. J Feline Med Surg. 2010;12:631–6.

57. Guptill L, Slater L, Wu CC, Lin TL, Glickman LT, Welch DF, et al. Experimental infection of young specific pathogen-free cats with Bartonella henselae. J Infect Dis. 1997;176:206–16.

58. Kordick DL, Breitschwerdt EB. Relapsing bacteremia after blood transmission of *Bartonella henselae* to cats. Am J Vet Res. 1997;58:492–7.
59. O'Reilly KL, Bauer RW, Freeland RL, Foil LD, Hughes KJ, Rohde KR, et al. Acute clinical disease in cats following infection with a pathogenic strain of *Bartonella henselae* (LSU16). Infect Immun. 1999;67:3066–72.
60. Guptill L, Slater LN, Wu CC, Lin TL, Glickman LT, Welch DF, et al. Evidence of reproductive failure and lack of perinatal transmission of *Bartonella henselae* in experimentally infected cats. Vet Immunol Immunopathol. 1998;65:177–89.
61. Perez C, Hummel JB, Keene BW, Maggi RG, Diniz PP, Breitschwerdt EB. Successful treatment of *Bartonella henselae* endocarditis in a cat. J Feline Med Surg. 2010;12:483–6.
62. Chomel BB, Wey AC, Kasten RW, Stacy BA, Labelle P. Fatal case of endocarditis associated with *Bartonella henselae* type I infection in a domestic cat. J Clin Microbiol. 2003;41:5337–9.
63. Kordick DL, Hilyard EJ, Hadfield TL, Wilson KH, Steigerwalt AG, Brenner DJ, et al. *Bartonella clarridgeiae*, a newly recognized zoonotic pathogen causing inoculation papules, fever, and lymphadenopathy (cat scratch disease). J Clin Microbiol. 1997;35:1813–8.
64. La VD, Tran-Hung L, Aboudharam G, Raoult D, Drancourt M. *Bartonella quintana* in domestic cat. Emerg Infect Dis. 2005;11:1287–9.
65. Breitschwerdt EB, Maggi RG, Sigmon B, Nicholson WL. Isolation of *Bartonella quintana* from a woman and a cat following putative bite transmission. J Clin Microbiol. 2007;45:270–2.
66. Chomel BB, Mac Donald KA, Kasten RW, Chang CC, Wey AC, Foley JE, et al. Aortic valve endocarditis in a dog due to *Bartonella clarridgeiae*. J Clin Microbiol. 2001;39:3548–54.
67. Breitschwerdt EB, Atkins CE, Brown TT, Kordick DL, Snyder PS. *Bartonella vinsonii* subsp. *berkhoffii* and related members of the alpha subdivision of the Proteobacteria in dogs with cardiac arrhythmias, endocarditis, or myocarditis. J Clin Microbiol. 1999;37:3618–26.
68. Breitschwerdt EB, Kordick DL, Malarkey DE, Keene B, Hadfield TL, Wilson K. Endocarditis in a dog due to infection with a novel *Bartonella* subspecies. J Clin Microbiol. 1995;33:154–60.

69. Kordick DL, Swaminathan B, Greene CE, Wilson KH, Whitney AM, O'Connor S, et al. *Bartonella vinsonii* subsp. *berkhoffii* subsp. nov., isolated from dogs; *Bartonella vinsonii* subsp. *vinsonii* and emended description of *Bartonella vinsonii*. Int J Syst Bacteriol. 1996;46:704–9.
70. Pappalardo BL, Brown T, Gookin JL, Morrill CL, Breitschwerdt EB. Granulomatous disease associated with *Bartonella* infection in 2 dogs. J Vet Int Med. 2000;14:37–42.
71. Kitchell BE, Fan TM, Kordick D, Breitschwerdt EB, Wollenberg G, Lichtensteiger CA. Peliosis hepatis in a dog infected with *Bartonella henselae*. J Am Vet Med Assoc. 2000;216:519–23.
72. Perez C, Maggi RG, Diniz PP, Breitschwerdt EB. Molecular and serological diagnosis of *Bartonella* infection in 61 dogs from the United States. J Vet Int Med. 2011;25:805–10.
73. Breitschwerdt EB, Linder KL, Day MJ, Maggi RG, Chomel BB, Kempf V. Koch's postulates and the pathogenesis of comparative infectious disease causation associated with *Bartonella* species. J Comp Pathol. 2013;148:115–25.
74. Solano-Gallego L, Bradley J, Hegarty B, Sigmon B, Breitschwerdt E. *Bartonella henselae* IgG antibodies are prevalent in dogs from southeastern USA. Vet Res. 2004;35:585–95.
75. Duncan AW, Maggi RG, Breitschwerdt EB. *Bartonella* DNA in dog saliva. Emerg Infect Dis. 2007;13:1948–50.
76. Gundi VA, Bourry O, Davous B, Raoult D, La Scola B. *Bartonella clarridgeiae* and *B. henselae* in dogs, Gabon. Emerg Infect Dis. 2004;10:2261–2.
77. MacDonald KA, Chomel BB, Kittleson MD, Kasten RW, Thomas WP, Pesavento P. A prospective study of canine infective endocarditis in northern California (1999-2001): emergence of *Bartonella* as a prevalent etiologic agent. J Vet Intern Med. 2004;18:56–64.
78. Millar B, Moore J, Mallon P, Xu J, Crowe M, Mcclurg R, et al. Molecular diagnosis of infective endocarditis - a new Duke's criterion. Scand J Infect Dis. 2001;33:673–80.
79. Raoult D, Fournier PE, Drancourt M, Marrie TJ, Etienne J, Cosserat J, et al. Diagnosis of 22 new cases of *Bartonella* endocarditis. Ann Intern Med. 1996;125:646–52.
80. Fournier PE, Lelievre H, Eykyn SJ, Mainardi JL, Marrie TJ, Bruneel F, et al. Epidemiologic and clinical characteristics of *Bartonella*

quintana and *Bartonella henselae* endocarditis: a study of 48 patients. Med Chem Commun. 2001;80:245–51.

81. Chang CC, Chomel BB, Kasten RW, Heller RM, Kocan KM, Ueno H, et al. *Bartonella* spp. isolated from wild and domestic ruminants in North America. Emerg Infect Dis. 2000;6:306–11.
82. Chang C, Yamamoto K, Chomel BB, Kasten RW, Simpson DC, Smith CR, et al. Seroepidemiology of *Bartonella vinsonii* subsp. *berkhoffii* infection in California coyotes, 1994-1998. Emerg Infect Dis. 1999;5:711–5.
83. Heller R, Kubina M, Mariet P, Riegel P, Delacour G, Dehio C, et al. *Bartonella alsatica* sp. nov., a new *Bartonella* species isolated from the blood of wild rabbits. Int J Syst Bacteriol. 1999;49:283–8.
84. Chang CC, Hayashidani H, Pusterla N, Kasten RW, Madigan JE, Chomel BB. Investigation of *Bartonella* infection in ixodid ticks from California. Comp Immunol Microbiol Infect Dis. 2002;25:229–36.
85. Birtles RJ, Harrison TG, Molyneux DH. *Grahamella* in small woodland mammals in the U.K.: isolation, prevalence and host specificity. Ann Trop Med Parasitol. 1994;88:317–27.
86. Birtles RJ, Canales J, Ventosilla P, Alvarez E, Guerra H, Llanos-Cuentas A, et al. Survey of *Bartonella* species infecting intradomicillary animals in the Huayllacallan Valley, Ancash, Peru, a region endemic for human bartonellosis. Am J Trop Med Hyg. 1999;60:799–805.
87. Ellis BA, Regnery RL, Beati L, Bacellar F, Rood M, Glass GG, et al. Rats of the genus *Rattus* are reservoir hosts for pathogenic *Bartonella* species: an Old World origin for a New World disease? J Infect Dis. 1999;180:220–4.
88. Fichet-Calvet E, Jomaa I, Ben IR, Ashford RW. Patterns of infection of haemoparasites in the fat sand rat, *Psammomys obesus*, in Tunisia, and effect on the host. Ann Trop Med Parasitol. 2000;94:55–68.
89. Heller R, Riegel P, Hansmann Y, Delacour G, Bermond D, Dehio C, et al. *Bartonella tribocorum* sp. nov., a new *Bartonella* species isolated from the blood of wild rats. Int J Syst Bacteriol. 1998;48:1333–9.
90. Hofmeister EK, Kolbert CP, Abdulkarim AS, Magera JM, Hopkins MK, Uhl JR, et al. Cosegregation of a novel *Bartonella* species with *Borrelia burgdorferi* and *Babesia microti* in *Peromyscus leucopus*. J Infect Dis. 1998;177:409–16.

91. Kosoy MY, Regnery RL, Tzianabos T, Marston EL, Jones DC, Green D, et al. Distribution, diversity, and host specificity of *Bartonella* in rodents from the Southeastern United States. Am J Trop Med Hyg. 1997;57:578–88.

92. Veikkolainen V, Vesterinen EJ, Lilley TM, Pulliainen AT. Bats as reservoir hosts of human bacterial pathogen, *Bartonella mayotimonensis* Emerg Infect Dis. 2014;20:960–7.

93. Hercik K, Hasova V, Janecek J, Branny P. Molecular evidence of *Bartonella* DNA in ixodid ticks in Czechia. Folia Microbiol. 2007;52:503–9.

94. Ihler GM. *Bartonella bacilliformis*: dangerous pathogen slowly emerging from deep background. FEMS Microbiol Lett. 1996;144:1–11.

95. Raoult D, Roux V. The body louse as a vector of reemerging human diseases. Clin Infect Dis. 1999;29:888–911.

96. Billeter SA, Levy MG, Chomel BB, Breitschwerdt EB. Vector transmission of *Bartonella* species with emphasis on the potential for tick transmission. Med Vet Entomol. 2008;22:1–15.

97. Brouqui P, Raoult D. Arthropod-borne diseases in homeless. Ann N Y Acad Sci. 2006;1078:223–35.

98. Townsend CHT. Resumen de la labores en el Peru sobre el *Phlebotomus verrucarum* y su agencia en la transmission de la verruga. Annales Zoologici Aplicada. 1914;44.

99. Brouqui P, Lascola B, Roux V, Raoult D. Chronic *Bartonella quintana* bacteremia in homeless patients. N Engl J Med. 1999;340:184–9.

100. Spach DH, Kanter AS, Dougherty MJ, Larson AM, Coyle MB, Brenner DJ, et al. *Bartonella* (*Rochalimaea*) *quintana* bacteremia in inner-city patients with chronic alcoholism. N Engl J Med. 1995;332:424–8.

101. Maggi RG, Kempf VAJ, Chomel BB, Breitschwerdt EB. *Bartonella.* In: Versalovic J, Jorgensen JH, Funke G, Warnock DW, Landry ML, Carroll KC. Manual of Clinical Microbiology, 10th Edition. Washington DC: American Society of Microbiology; 2011. p. 786

102. Chomel BB, Kasten RW, Williams C, Wey AC, Henn JB, Maggi R, et al. *Bartonella* endocarditis: a pathology shared by animal reservoirsand patients. Ann N Y Acad Sci. 2009;1166:120–6.

103. Lamas CDC, Ramos RG, Lopes GQ, Santos MS, Golebiovski WF, Weksler C, et al. *Bartonella* and *Coxiella* infective endocarditis in Brazil: Molecular evidence from excised valves from a cardiac surgery referral center in Rio de Janeiro, Brazil, 1998 to 2009. Int J Infect Dis. 2013;17:e65–6.

104. Tsukahara M, Tsuneoka H, Iino H, Murano I, Takahashi H, Uchida M. *Bartonella henselae* infection as a cause of fever of unknown origin. J Clin Microbiol. 2000;38:1990–1.

105. Guiyedi V, Haddad H, Okome-Nkoumou M, Gire F, Ongali B, Lore P, et al. Cat-scratch disease in adult hospitalized for prolonged-Fever associated with multiple lymphadenopathies and weight loss. Open Microbiol J. 2013;7:152–5.

106. Hadfield TL, Warren R, Kass M, Brun E, Levy C. Endocarditis caused by Rochalimaea henselae. Hum Pathol. 1993;24:1140–1.

107. Spach DH, Callis KP, Paauw DS, Houze YB, Schoenknecht FD, Welch DF, et al. Endocarditis caused by *Rochalimaea quintana* in a patient infected with human immunodeficiency virus. J Clin Microbiol. 1993;31:692–4.

108. Daly JS, Worthington MG, Brenner DJ, Moss CW, Hollis DG, Weyant RS, et al. *Rochalimaea elizabethae* sp. nov. isolated from a patient with endocarditis. J Clin Microbiol. 1993;31:872–81.

109. Roux V, Eykyn SJ, Wyllie S, Raoult D. *Bartonella vinsonii* subsp. *berkhoffii* as an agent of afebrile blood culture-negative endocarditis in a human. J Clin Microbiol. 2000;38:1698–700.

110. Fenollar F, Sire S, Raoult D. *Bartonella vinsonii* subsp. *arupensis* as an agent of blood culture-negative endocarditis in a human. J Clin Microbiol. 2005;43:945–7.

111. Raoult D, Roblot F, Rolain JM, Besnier JM, Loulergue J, Bastides F, et al. First isolation of *Bartonella alsatica* from a valve of a patient with endocarditis. J Clin Microbiol. 2006;44:278–9.

112. Kosoy M, Murray M, Gilmore RD, Bai Y, Gage KL. *Bartonella* strains from ground squirrels are identical to *Bartonella washoensis* isolated from a human patient. J Clin Microbiol. 2003;41:645–50.

113. Jeanclaude D, Godmer P, Leveiller D, Pouedras P, Fournier PE, Raoult D, et al. *Bartonella alsatica* endocarditis in a French patient in close contact with rabbits. Clin Microbiol Infect. 2009;15:110–1.

114. Lin EY, Tsigrelis C, Baddour LM, Lepidi H, Rolain JM, Patel R, et al. *Candidatus Bartonella mayotimonensis* and Endocarditis. Emerg Infect Dis. 2010;16:500–3.

115. La Scola B, Raoult D. Culture of *Bartonella quintana* and *Bartonella henselae* from human samples: a 5-year experience (1993 to 1998). J Clin Microbiol. 1999;37:1899–905.

116. Fournier PE, Mainardi JL, Raoult D. Value of microimmunofluorescence for diagnosis and follow-up of *Bartonella* endocarditis. Clin Diagn Lab Immun. 2002;9:795–801.

117. Maggi RG, Mozayeni BR, Pultorak EL, Hegarty BC, Bradley JM, Correa M, et al. *Bartonella* spp. bacteremia and rheumatic symptoms in patients from Lyme disease–endemic Region. Emerg Infect Dis. 2012;18:783–91.

118. Breitschwerdt EB, Maggi RG, Lantos PM, Woods CW, Hegarty BC, Bradley JM. *Bartonella vinsonii* subsp. *berkhoffii* and *Bartonella henselae* bacteremia in a father and daughter with neurological disease. Parasit Vectors. 2010;3:29.

119. Maggi RG, Mascarelli PE, Pultorak EL, Hegarty BC, Bradley JM, Mozayeni BR, et al. *Bartonella* spp. bacteremia in high-risk immunocompetent patients. Diagn Microbiol Infect Dis. 2011;71:430–7.

120. Halos L, Jamal T, Maillard R, Beugnet F, Le Menach A, Boulouis HJ, et al. Evidence of *Bartonella* sp. in questing adult and nymphal *Ixodes ricinus* ticks from France and co-infection with *Borrelia burgdorferi sensu lato* and *Babesia* sp. Vet Res. 2005;36:79–87.

121. Eskow E, Rao RVS, Mordechai E. Concurrent Infection of the Central Nervous System by *Borrelia burgdorferi* and *Bartonella henselae.* Arch Neurol. 2001;58:1357.

122. Podsiadly E, Chmielewski T, Tylewska-Wierzbanowska S. *Bartonella henselae* and *Borrelia burgdorferi* infections of the central nervous system. Ann N Y Acad Sci. 2003;990:404–6.

123. Holden K, Boothby JT, Kasten RW, Chomel BB. Co-detection of *Bartonella henselae*, *Borrelia burgdorferi*, and *Anaplasma phagocytophilum* in *Ixodes pacificus* ticks from California, USA. Vector Borne Zoonotic Dis. 2006;6:99–102.

124. Belongia EA. Epidemiology and impact of coinfections acquired from *Ixodes* ticks. Vector Borne Zoonotic Dis. 2002;2:265–73.

125. Lantos PM, Wormser GP. Chronic coinfections in patients diagnosed with chronic Lyme disease: a systematic review. Am J Med. 2014;127:1105–10.

126. Lantos PM, Maggi RG, Ferguson B, Varkey J, Park LP, Breitschwerdt EB, et al. Detection of *Bartonella* species in the blood of veterinarians and veterinary technicians: a newly recognized occupational hazard? Vector Borne Zoonotic Dis. 2014;14:563–70.

127. Oliveira AM, Maggi RG, Woods CW, Breitschwerdt EB. Suspected needle stick transmission of *Bartonella vinsonii* subspecies *berkhoffii* to a veterinarian. J Vet Intern Med. 2010;24:1229–32.

128. Breitschwerdt EB. Did *Bartonella henselae* contribute to the deaths of two veterinarians? Parasit Vectors. 2015;8:317.

129. Jurke A, Bannert N, Brehm K, Fingerle V, Kempf VAJ, Kömpf D, et al. Serological survey of *Bartonella* spp., *Borrelia burgdorferi*, *Brucella* spp., *Coxiella burnetii*, *Francisella tularensis*, *Leptospira* spp., *Echinococcus*, Hanta-, TBE- and XMR-virus infection in employees of two forestry enterprises in North Rhine-Westphalia, Germany, 2011-2013. Int J Med Microbiol. 2015;305:652–62.

130. Diniz PPV, Velho PENF, Pitassi LHU, Drummond MR, Lania BG, Barjas-Castro ML, et al. Risk Factors for *Bartonella* species Infection in Blood Donors from Southeast Brazil. PLoS Negl Trop Dis. 2016;10:e0004509.

131. Foil L, Andress E, Freeland RL, Roy AF, Rutledge R, Triche PC, et al. Experimental infection of domestic cats with *Bartonella henselae* by inoculation of *Ctenocephalides felis* (*Siphonaptera: Pulicidae*) feces. J Med Entomol. 1998;35:625–8.

132. Chomel BB, Kasten RW. Bartonellosis, an increasingly recognized zoonosis. J Appl Microbiol. 2010;109:743–50.

133. Finkelstein JL, Brown TP, O'Reilly KL, Wedincamp J, Foil LD. Studies on the growth of *Bartonella henselae* in the cat flea (Siphonaptera: Pulicidae). J Med Entomol. 2002;39:915–9.

134. Bouhsira E, Franc M, Boulouis HJ, Jacquiet P, Raymond-Letron I, Liénard E. Assessment of Persistence of *Bartonella henselae* in *Ctenocephalides felis*. Appl Environ Microbiol. 2013;79:7439–44.

135. Guptill L. Bartonellosis. Vet Microbiol. 2010;140:347–59.

136. Abbott RC, Chomel BB, Kasten RW, Floyd-Hawkins KA, Kikuchi Y, Koehler JE, et al. Experimental and natural infection with *Bartonella henselae* in domestic cats. Comp Immun Microbiol Infect Dis. 1997;20:41–51.

137. Rolain JM, Franc M, Davoust B, Raoult D. Molecular detection of *Bartonella quintana*, *B. koehlerae*, *B. henselae*, *B. clarridgeiae*, *Rickettsia felis*, and *Wolbachia pipientis* in cat fleas, France. Emerg Infect Dis. 2003;9:338–42.

138. Sanogo YO, Zeaiter Z, Caruso G, Merola F, Shpynov S, Brouqui P, et al. *Bartonella henselae* in *Ixodes ricinus* ticks (*Acari: Ixodida*) removed from humans, Belluno province, Italy. Emerg Infect Dis. 2003;9:329–32.

139. Durden LA, Ellis BA, Banks CW, Crowe JD, Oliver JH. Ectoparasites of gray squirrels in two different habitats and screening of selected ectoparasites for bartonellae. J Parasitol. 2004;90:485–9.

140. Bown KJ, Bennet M, Begon M. Flea-borne *Bartonella grahamii* and *Bartonella taylorii* in bank voles. Emerg Infect Dis. 2004;10:684–7.

141. Bermond D, Heller R, Barrat F, Delacour G, Dehio C, Alliot A, et al. Bartonella birtlesii sp. nov., isolated from small mammals (*Apodemus* spp.). Int J Syst Evol Microbiol. 2000;50:1973–9.

142. Reeves WK, Rogers TE, Dasch GA. *Bartonella* and *Rickettsia* from fleas (*Siphonaptera: Ceratophyllidae*) of prairie dogs (*Cynomys* spp.) from the western United States. J Parasitol. 2007;93:953–5.

143. Reeves WK, Rogers TE, Durden LA, Dasch GA. Association of *Bartonella* with the fleas (*Siphonaptera*) of rodents and bats using molecular techniques. J Vector Ecol. 2007;32:118–22.

144. Abbot P, Aviles AE, Eller L, Durden LA. Mixed infections, cryptic diversity, and vector-borne pathogens: evidence from *Polygenis* fleas and *Bartonella* species. Appl Environ Microbiol. 2007;73:6045–52.

145. Jackson LA, Spach DH. Emergence of *Bartonella quintana* infection among homeless persons. Emerg Infect Dis. 1996;2:141–4.

146. Regnery R, Tappero J. Unraveling mysteries associated with cat-scratch disease, bacillary angiomatosis, and related syndromes. Emerg Infect Dis. 1995;1:16–21.

147. Sasaki T, Poudel SK, Isawa H, Hayashi T, Seki N, Tomita T, et al. First molecular evidence of *Bartonella quintana* in *Pediculus humanus*

capitis (*Phthiraptera: Pediculidae*), collected from Nepalese children. J Med Entomol. 2006;43:110–2.

148. Reeves WK, Szumlas DE, Moriarity JR, Loftis AD, Abbassy MM, Helmy IM, et al. Louse-borne bacterial pathogens in lice (Phthiraptera) of rodents and cattle from Egypt. J Parasitol. 2006;92:313–8.
149. Battistini TS. Estudios sobre la verruga peruana. La Acción Médica. 1929
150. Battistini TS. La verrue péruvienne: sa transmission par le Phlébotome. Revue Sud-Americaine de Médicine et de Chirurgie. 1931;2:719–24.
151. Hertig M. Phlebotomus and Carrion's disease. Am J Trop Med Hyg. 1942;22:1–80.
152. Rynkiewicz EC, Hemmerich C, Rusch DB, Fuqua C, Clay K. Concordance of bacterial communities of two tick species and blood of their shared rodent host. Mol Ecol. 2015;24:2566–79.
153. Benson MJ, Gawronski JD, Eveleigh DE, Benson DR. Intracellular symbionts and other bacteria associated with deer ticks (*Ixodes scapularis*) from Nantucket and Wellfleet, Cape Cod, Massachusetts. Appl Environ Microbiol. 2004;70:616–20.
154. Steiner FE, Pinger RR, Vann CN, Grindle N, Civitello D, Clay K, et al. Infection and co-infection rates of *Anaplasma phagocytophilum* variants, *Babesia* spp., *Borrelia burgdorferi*, and the rickettsial endosymbiont in *Ixodes scapularis* (Acari: Ixodidae) from sites in Indiana, Maine, Pennsylvania, and Wisconsin. J Med Entomol. 2008;45:289–97.
155. Sun J, Liu Q, Lu L, Ding G, Guo J, Fu G, et al. Coinfection with four genera of bacteria (*Borrelia*, *Bartonella*, *Anaplasma*, and *Ehrlichia*) in *Haemaphysalis longicornis* and *Ixodes sinensis* ticks from China. Vector Borne Zoonotic Dis. 2008;8:791–5.
156. Vayssier-Taussat M, Moutailler S, Michelet L, Devillers E, Bonnet S, Cheval J, et al. Next generation sequencing uncovers unexpected bacterial pathogens in ticks in western Europe. PLoS One. 2013;8:e81439.
157. Angelakis E, Billeter SA, Breitschwerdt EB, Chomel BB, Raoult D. Potential for tick-borne bartonelloses. Emerg Infect Dis. 2010;16:385–91.
158. Dietrich F, Schmidgen T, Maggi RG, Richter D, Matuschka FR, Vonthein R, et al. Prevalence of *Bartonella henselae* and *Borrelia*

burgdorferi sensu lato DNA in *Ixodes ricinus* ticks in Europe. Appl Environ Microbiol. 2010;76:1395–8.

159. Schouls LM, van de Pol I, Rijpkema SG, Schot CS. Detection and identification of *Ehrlichia*, *Borrelia burgdorferi sensu lato*, and *Bartonella* species in Dutch *Ixodes ricinus* ticks. J Clin Microbiol. 1999;37:2215–22.
160. Chang CC, Chomel BB, Kasten RW, Romano V, Tietze N. Molecular evidence of *Bartonella* spp. in questing adult *Ixodes pacificus* ticks in California. J Clin Microbiol. 2001;39:1221–6.
161. Morozova OV, Cabello FC, Dobrotvorsky AK. Semi-nested PCR detection of *Bartonella henselae* in *Ixodes persulcatus* ticks from Western Siberia, Russia. Vector Borne Zoonotic Dis. 2004;4:306–9.
162. Kim CM, Kim JY, Yi YH, Lee MJ, Cho MR, Shah DH, et al. Detection of *Bartonella* species from ticks, mites and small mammals in Korea. J Vet Sci. 2005;6:327–34.
163. Podsiadly E, Chmielewski T, Sochon E, Tylewska-Wierzbanowska S. *Bartonella henselae* in *Ixodes ricinus* ticks removed from dogs. Vector Borne Zoonotic Dis. 2007;7:189–92.
164. Adelson ME, Rao RVS, Tilton RC, Cabets K, Eskow E, Fein L, et al. Prevalence of *Borrelia burgdorferi*, *Bartonella* spp., *Babesia microti*, and *Anaplasma phagocytophila* in *Ixodes scapularis* ticks collected in Northern New Jersey. J Clin Microbiol. 2004;42:2799–801.
165. Janecek E, Mietze A, Goethe R, Schnieder T, Strube C. *Bartonella* spp. infection rate and *B. grahamii* in ticks. Emerg Infect Dis. 2012;18:1689–90.
166. Reye AL, Hubschen JM, Sausy A, Muller CP. Prevalence and seasonality of tick-borne pathogens in questing *Ixodes ricinus* ticks from Luxembourg. Appl Environ Microbiol. 2010;76:2923–31.
167. Sytykiewicz H, Karbowiak G, Werszko J, Czerniewicz P, Sprawka I, Mitrus J. Molecular screening for *Bartonella henselae* and *Borrelia burgdorferi sensu lato* co-existence within *Ixodes ricinus* populations in central and eastern parts of Poland. Ann Agric Environ Med. 2012;19:451–6.
168. Moutailler S, Valiente Moro C, Vaumourin E, Michelet L, Tran FH, Devillers E, et al. Co-infection of ticks: The rule rather than the exception. PLoS Negl Trop Dis. 2016;10:e0004539.

169. Sormunen JJ, Penttinen R, Klemola T, Hanninen J, Vuorinen I, Laaksonen M, et al. Tick-borne bacterial pathogens in southwestern Finland. Parasit Vectors. 2016;9:168.
170. Müller A, Reiter M, Schötta AM, Stockinger H, Stanek G. Detection of *Bartonella* spp. in *Ixodes ricinus* ticks and *Bartonella* seroprevalence in human populations. Ticks Tick Borne Dis. 2016. doi:10.1016/j.ttbdis.2016.03.009.
171. Cotté V, Bonnet S, Le Rhun D, Le Naour E, Chauvin A, Boulouis HJ, et al. Transmission of *Bartonella henselae* by *Ixodes ricinus*. Emerg Infect Dis. 2008;14:1074–80.
172. Reis C, Cote M, Le Rhun D, Lecuelle B, Levin ML, Vayssier-Taussat M, et al. Vector competence of the tick *Ixodes ricinus* for transmission of *Bartonella birtlesii*. PLoS Negl Trop Dis. 2011;5:e1186.
173. Podsiadly E, Karbowiak G, Tylewska-Wierzbanowska S. Presence of *Bartonella* spp. in *Ixodidae* ticks. Clin Microbiol Infect. 2009;15:120–1.
174. Tuttle AD, Birkenheuer AJ, Juopperi T, Levy MG, Breitschwerdt EB. Concurrent bartonellosis and babesiosis in a dog with persistent thrombocytopenia. J Am Vet Med Assoc. 2003;223(1306-10):1280–1.
175. Pappalardo BL, Correa MT, York CC, Peat CY, Breitschwerdt EB. Epidemiologic evaluation of the risk factors associated with exposure and seroreactivity to *Bartonella vinsonii* in dogs. Am J Vet Res. 1997;58:467–71.
176. Breitschwerdt EB, Hegarty BC, Hancock SI. Sequential evaluation of dogs naturally infected with *Ehrlichia canis*, *Ehrlichia chaffeensis*, *Ehrlichia equi*, *Ehrlichia ewingii*, or *Bartonella vinsonii*. J Clin Microbiol. 1998;36:2645–51.
177. Honadel TE, Chomel BB, Yamamoto K, Chang C, Farver TB. Seroepidemiology of *Bartonella vinsonii* subsp *berkhoffii* exposure among healthy dogs. J Am Vet Med Assoc. 2001;219:480–4.
178. Kordick SK, Breitschwerdt EB, Hegarty BC, Southwick KL, Colitz CM, Hancock SI, et al. Coinfection with multiple tick-borne pathogens in a Walker Hound kennel in North Carolina. J Clin Microbiol. 1999;37:2631–8.
179. Morozova OV, Chernousova N, Morozov IV. Detection of the *Bartonella* DNA by the method of nested PCR in patients after tick bites in Novosibirsk region. Mol Gen Mikrobiol Virusol. 2005;4:14–7.

180. Vayssier-Taussat M, Moutailler S, Féménia F, Raymond P, Croce O, La Scola B, et al. Identification of Novel Zoonotic Activity of *Bartonella* spp., France. Emerg Infect Dis. 2016;22:457–62.
181. Lucey D, Dolan MJ, Moss CW, Garcia M, Hollis DG, Wegner S, et al. Relapsing illness due to *Rochalimaea henselae* in immunocompetent hosts: implication for therapy and new epidemiological associations. Clin Infect Dis. 1992;14:683–8.
182. van Buskirk J and Ostfeld RS. Controlling Lyme-disease by modifying the density and species composition of tick hosts. Ecol Appl. 1995;5:1133–40.
183. Keesing F, Holt RD, Ostfeld RS. Effects of species diversity on disease risk. Ecol Lett. 2006;9:485–98.
184. Telford SR, Wormser GP. *Bartonella* spp. transmission by ticks not established. Emerg Infect Dis. 2010;16:379–84.
185. Chung CY, Kasten RW, Paff SM, van Horn BA, Vayssier-Taussat M, Boulouis HJ, et al. *Bartonella* spp. DNA associated with biting flies from California. Emerg Infect Dis. 2004;10:1311–3.
186. Reeves WK, Dowling AP, Dasch GA. Rickettsial agents from parasitic *dermanyssoidea* (Acari: Mesostigmata). Exp Appl Acarol. 2006;38:181–8.
187. Dehio C. Molecular and cellular basis of bartonella pathogenesis. Annu Rev Microbiol. 2004;58:365–90.
188. Reeves WK, Nelder MP, Cobb KD, Dasch GA. *Bartonella* spp. in deer keds, *Lipoptena mazamae* (Diptera: Hippoboscidae), from Georgia and South Carolina, USA. J Wildl Dis. 2006;42:391–6.
189. Matsumoto K, Berrada ZL, Klinger E, Goethert HK, Telford 3rd SR. Molecular detection of *Bartonella schoenbuchensis* from ectoparasites of deer in Massachusetts. Vector Borne Zoonotic Dis. 2008;8:549–54.
190. Dehio C, Sauder U, Hiestand R. Isolation of *Bartonella schoenbuchensis* from *Lipoptena cervi*, a blood-sucking arthropod causing deer ked dermatitis. J Clin Microbiol. 2004;42:5320–3.
191. Halos L, Jamal T, Maillard R, Girard B, Guillot J, Chomel B, et al. Role of *Hippoboscidae* flies as potential vectors of *Bartonella* spp. infecting wild and domestic ruminants. Appl Environ Microbiol. 2004;70:6302–5.

CHAPTER 9

Erratum to: *Bartonella* spp. - A Chance to Establish One Health Concepts in Veterinary and Human Medicine

Yvonne Regier, Fiona O’Rourke and Volkhard A. J. Kempf

Institute for Medical Microbiology and Infection Control, University Hospital, Goethe-University, Frankfurt am Main, Germany

Erratum

Unfortunately, the original version of this article [1] contained an error. The name of the author Fiona O’Rourke was incorrectly spelt as Fiona O´Rourke. The correct spelling is included here, and has been updated in the original article.

Citation: Regier, Y., O’Rourke, F. & Kempf, V.A.J. Erratum to: Bartonella spp. - a chance to establish One Health concepts in veterinary and human medicine. Parasites Vectors 9, 330 (2016). https://doi.org/10.1186/s13071-016-1604-4.

REFERENCES

1. Regier Y, O'Rourke F, Kempf VAJ. *Bartonella* spp. - a chance to establish One Health concepts in veterinary and human medicine. Parasit Vectors. 2016;9:261. doi:10.1186/s13071-016-1546-x.

CHAPTER 10

A System Dynamics Approach to Understanding the One Health Concept

Tai Xie[1,2], Wenbao Liu[2] , Benjamin D. Anderson[3] , Xiaorong Liu[1], Gregory C. Gray[3]

[1] Faculty of Health Service, Second Military Medical University, Shanghai, China

[2] Faculty of Naval Medicine, Second Military Medical University, Shanghai, China

[3] Division of Infectious Diseases and Duke Global Health Institute, Duke University, Durham, North Carolina, United States of America

ABSTRACT

There have been many terms used to describe the One Health concept, including movement, strategy, framework, agenda, approach, among others. However, the inter-relationships of the disciplines engaged in the One Health concept have not been well described. To identify and better elucidate

Citation: Xie T, Liu W, Anderson BD, Liu X, Gray GC (2017) A system dynamics approach to understanding the One Health concept. PLoS ONE 12(9): e0184430. https://doi.org/10.1371/journal.pone.0184430.

the internal feedback mechanisms of One Health, we employed a system dynamics approach. First, a systematic literature review was conducted via searches in PubMed, Web of Knowledge, and ProQuest with the search terms: 'One Health' and (concept* or approach*). In addition, we used the HistCite® tool to add significant articles on One Health to the library. Then, of the 2368 articles identified, 19 were selected for evaluating the inter-relationships of disciplines engaged in One Health. Herein, we report a visually rich, theoretical model regarding interactions of various disciplines and complex problem descriptors engaged in One Health problem solving. This report provides a conceptual framework for future descriptions of the interdisciplinary engagements involved in One Health.

INTRODUCTION

The One Health concept is not new [1, 2] perhaps dating back to the late nineteenth century when medical observations were made that human and animal health were closely linked [3, 4]. Today the One Health concept has often been cited an effective approach to complex public health problems that involve multiple disciplines. Often human health, animal health and environmental health are examined as closely linked. Examples of such complex problems include emerging infectious diseases, food safety, and selection of antimicrobial resistant pathogens. In particular, emerging zoonotic disease problems are now remarkable common [5] and very complex. Controlling these complex problems often requires interdisciplinary approaches such as *One Health* [6, 7]. Despite the One Health concept's growing popularity and acceptance by the professional community, the definition of the term remains imprecise. For example, One Health has been variously defined as an initiative [8], a movement [9], a strategy [10], a framework [11], an agenda [12, 13], an approach, a collaborative effort, among others. As a milestone in the history of One Health, the symbolic 'umbrella', developed by One Health Sweden in cooperation with the One Health Initiative Autonomous Pro Bono Team, illustrates the broad One Health scope. According to the 'umbrella', the goal of One Heath is to promote a worldwide strategy for the well-being in all aspects of health of people, animals and the environment [2, 14]. As this definition is broad, many disciplines might contribute One Health efforts [15]. Yet, the internal relationships of these disciplines remain a bit amorphous. To better understanding the One Health concept, we sought to review the literature and study key One Health publications with a *system dynamics approach*, providing a systematic representation of the numerous components and how they interact.

MATERIALS AND METHODS

System Dynamics

System dynamics (SD) is a methodology first developed by J.W. Forrester [16] that uses visual and mathematical modeling to simulate and understand complex system problems. Systems thinking is the notion that the system must have structure and the structure determines the system's patterns of behavior. As such, the patterns of behavior will lead to the system's output. Therefore, an understanding of system structure helps to better deliver the desired results. Or alternatively stated, thoroughly understanding a complex problem involves discovering the system's structure, which often involves feedback connections between interdependent components. *Qualitative system dynamics* is used to explain the system's internal feedback loops to make its relationships easier to understand. This approach has also been successfully used to enhance the development of health policies and programs [17, 18], enable modeling of health care systems [19], and guide chronic disease prevention and interventions [20–22].

A system dynamics theoretical model can be constructed using the following series of steps:

- Step 1. **Problem Identification.** Make clear the problem you want to study.
- Step 2. **Behavior Analysis.** Carefully identify the elements or components of the system. Analyze the influence of the most crucial behavioral factors involved in the system, either positive or negative, and attempt to explain the cause of the identified problem.
- Step 3. **Model Hypothesis.** Propose a hypothesis explaining the results of the second step, and illustrate the model hypothesis in the form of a causal loop, showing the clear mathematical feedback relationships.
- Step 4. **Model Evaluation.** Evaluate the model and make sure that it reflects the behaviors and relationships presenting in the system. If it does not, revisit the steps above, else proceed to step 5.

- Step 5. **Conclusion and Discussion.** Draw conclusions from the system dynamics model.

Systematic Literature Search for One Health Articles

To identify scientific articles pertaining to the One Health concept, we performed searches in multiple abstract databases, including PubMed, Web of Science, and ProQuest, in February and March 2017. In PubMed the query string used was '("one health") AND (concept* OR approach*)'. In Web of Science, it was 'TS = ("one health") AND TS = ((concept*) OR approach*)' with the language restriction—English-language only. In ProQuest, we used 'all("one health") AND all(concept* OR approach*)' with the language restriction—English-language only and exclusion source types from newspapers and social media reports. All references were extracted into a reference manager—EndNote (version X7.4, Bld 8818) and screened independently by two researchers (TX and WL). This systematic review conforms to the PRISMA statement. Articles were included if they were peer-reviewed and met four additional criteria: 1) addressed the concept of 'One Health', not one health care plan, one health center, or other unrelated topics described by the words "one health"; 2) focused on the concept of One Health overall, not just briefly mentioned the term or only covered a particular practice of One Health; 3) were in the form of an article or review; 4) were written in English. In full test screening, the exclusion criteria were: 1) One Health practice in a particular disease or district; 2) One Health concept or approach mentioned in limited sentences.

To ensure that we did not omit important literature using the aforementioned search strategy, we also used the HistCite® tool to identify significant articles in One Health. HistCite® is a science tool that visualizes the citation relationships between papers to identify key publications in the results returned by a Web of Science search. We used the query string 'One Health'. Articles were ranked by each publication›s Local Citation Score (LCS), which describes how many times it has been cited by the other publications returned by the search. The results were analyzed with HistCite® software version 12.3.17 (Thomson Reuters, New York City, NY, USA). The default setting for searched articles in the HistCite® software is 30 but we increased this default number to 100 articles.

RESULTS

The initial search identified 2368 articles: 750 from PubMed, 976 from Web of Science, 542 form ProQuest, and 100 from the HistCite® analysis (Fig 1). One thousand and one hundred thirty-five duplicate reports were manually excluded using reference management software. The abstracts of the remaining 1233 articles were reviewed. Of these, 550 articles were excluded for not addressing the concept of One Health, for originating from newspapers or social media platforms, or for being written in a language other than English. An addition, 106 articles were excluded for being comments or letters or from books. The full texts of the remaining 577 articles were reviewed, after which 561 reports were excluded for lacking a primary focus on One Health or the One Health concept. Three additional report were added during the full text review of articles. In total, 19 remaining publications were included in the subsequent analyses (Table 1).

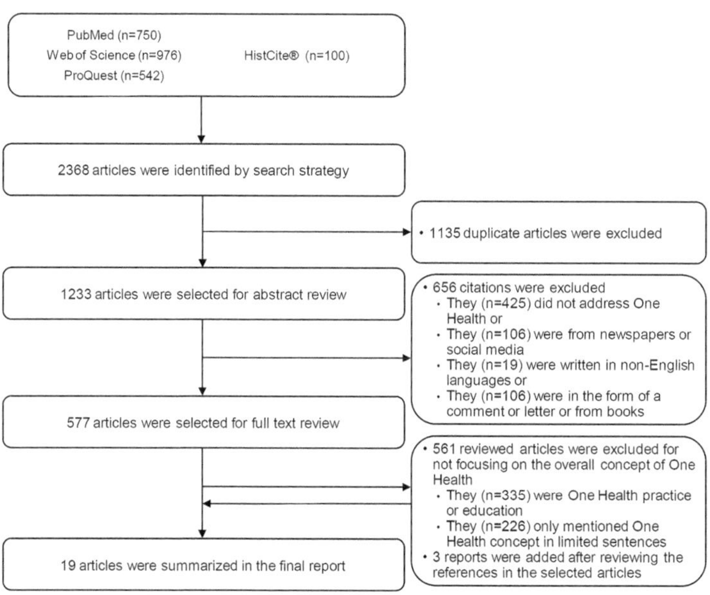

Figure 1: Flow diagram of the literature search process.

Using the search strategy, a total of 2368 articles were identified. Duplicate articles were removed. Nineteen publications were used in the final analyses. See text for more details.

https://doi.org/10.1371/journal.pone.0184430.g001

Table 1: Nineteen key One Health articles selected for component analyses

https://doi.org/10.1371/journal.pone.0184430.t001

Publications	Components
Cardiff et al, 2008 [24]	pathology, comparative pathology, genomic biology, food safety, emerging infectious disease
Frank, 2008[25]	veterinary medicine, food security, food safety, emerging infectious diseases, ecosystem protection, comparative medical, human physical/mental health, public health, human medicine, bio-engineering, animal science, environmental science, wildlife
Kahn et al, 2008[4]	veterinary medicine, food supply, intensive agricultural practices, exotic animals global trade, consumption of "bush meat", human population pressures, comparative medicine
Gibbs et al, 2009[9]	emerging disease, medicine and veterinary medicine, wildlife, public health, education, cross-species disease transmission, media publications
Monath et al, 2010 [10]	zoonotic agents, veterinary medical, laboratory animal research, public health, environmental health, biomedical,research, clinical medicine and surgery, biomedical research, ecosystem management, public health, food and agricultural systems safety, biosecurity
Coker et al, 2011[11]	public health, economic security, social stability, health service organisation, patterns and provision and access, fiscal systems, regulation and governance, information systems
Zinsstag et al, 2011 [26]	veterinary medicine, public health, animals and wildlife
Rabozzi et al, 2012 [27]	public health, clinicians, public health workers, veterinary medicine and veterinary public health officials
Zinsstag et al, 2012 [28]	zoonotic infections, livestock and wildlife, global public health, livestock production and wildlife conservation
Calistri et al, 2013 [29]	infectious diseases, globalization, animal products trade, wild animals, bush meat, zoonoses agents, public health policy, food-agricultura, veterinary medicine
Atlas, 2013[3]	infectious diseases, vaccination, epidemiology, medical education and clinical care, global travel and commerce, medical education and clinical care
Gibbs, 2013[7]	public health, biomedical research, global food safety and security, ecosystem health, caring for animals, public health, food safety, epidemiology, population medicine, foreign animal diseases
Bidaisee et al, 2014 [30]	zoonoses, food safety, agriculture, infectious diseases, livestock into wildlife, scientific and policy challenges, the social, cultural, political norms, veterinary medicine, public health, clinical practice
Evans et al, 2014[6]	epidemiological globalisation, pathogen adaptation, changing human demographics, evolving animal production systems, climate change, water, pollution and environmental contaminants, food safety, food sufficiency and insecurity, the universal global condition of rapid environmental change, new drugs, biodiversity, epidemics, pests, food security, economic prosperity
Gibbs, 2014[2]	veterinary profession, medical profession, wildlife specialists, environmentalists, health policy analysts, social scientists, humanities scholars
Stephen et al, 2014 [31]	infectious disease, veterinary, medical and environment sectors, wildlife, animal welfare, food safety and security, drugs and vaccines, public health, training and research
Lerner et al, 2015 [15]	infection biology, contagious diseases, zoonotic infections, evolutionary medicine, comparative medicine, translational medicine, biology, human medicine, veterinary medicine, public health, environmental chemistry, health economy, zoonotic diseases, epidemics and toxicants, food-producing animals, pet ownership, education
Stadtlander, 2015 [32]	medicine, veterinary medicine, microbiology, public health, biogeography, ecology, environmental, biology
Kingsley, 2016[33]	parasitologists, biosecurity, infection control risks, zoonotic infections, human and veterinary medicine, food security, healthy diets, climate change

https://doi.org/10.1371/journal.pone.0184430.t001

Problem Identification (Step 1)

From the articles identified by the literature review, it was difficult to determine a generally accepted definition for the One Health concept, including the scope and breadth of its practice. This lack of consensus supported the value of our system dynamics theoretical model in developing a conceptual framework to better understand One Health. This concept of One Health served as the root or first level of the theoretical model.

Behavior Analysis (Step 2)

All 19 analyzed reports agreed that the One Health concept contained at least three primary domains—human, animal, and environment/ecosystem. These three domains were added as the second level of the theoretical model. However, the articles markedly differed regarding their description of the scope of One Health. This discrepancy is likely due to the ongoing rapid development of One Health and the addition of new disciplines that are embracing it as a way forward [23]. Table 1 summarizes the various and diverse disciplines or scopes mentioned in each of the 19 publications, which highlights the broad interpretation that could result in defining the One Health concept. These disciplines and scopes were included as the third level of the theoretical model.

Model Hypothesis (Step 3)

In the theoretical model, the direction of influence is indicated by the polarity of a causal loop, represented by connecting arrows, which may be positive ("+") or negative ("-"). A positive causal link indicates that both the causative and the resultant factors increase or decrease in the same direction. A negative causal link indicates that the two linked factors change in opposite directions. Additionally, should two or more elements in the model have a reinforcing or non-reinforcing relationship, the model can be annotated such that these associations can be visualized. This was done by placing a "R" within a clockwise cycling arrow between elements that did have a reinforcing relationship, and a "B" for those elements that did not.

In the first level of our model, we demonstrate the connection between disease with health using an arrow that shows the direction of this relationship. We then added the human, animal, and ecosystem domains (second level) to the model. These terms are represented by the red squares which are used to denote stock elements. Next, the third level elements were added, denoted by squares with grey color, which were also derived from the reviewed articles. Then, variable elements associated with each of the three One Health domains were added, denoted by circular shapes. These variable elements are assigned as descriptive components of the stocks. Last, causal links as previously described, were drawn between the elements. This model (we have learned the "One Health Cosmos", Fig 2) was constructed using AnyLogic® PLE software, version 7.3.4 (AnyLogic North America, LLC, Chicago, IL, USA).

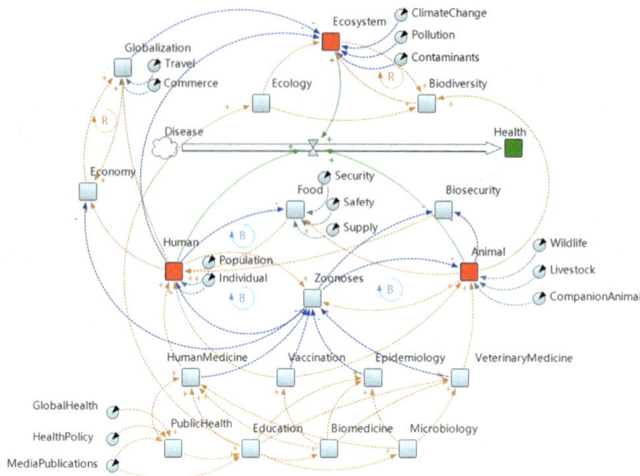

Figure 2: "One Health Cosmos".

One Health Cosmos shows the relationships between the various disciplines and complex problem descriptors that are reported to fall within the One Health concept. Squares and circles represent nodes, and the arrows connecting nodes represent causal links. Brown color is used to show positive causal link which also have a "+" sign besides the arrowhead. Negative causal link is portrayed with a blue color and "-" sign besides the arrowhead. A positive causal link means that both the causative and the resultant factors increase or decrease in the same direction. A negative causal link indicates that the two linked factors change in opposite directions. The positive reinforcing loop has a "R" in the clockwise cycle. A negative reinforcing loop opposite has a "B" in the counterclockwise cycle. A big arrow shows the direction of this relationship between disease and health through One Health.

https://doi.org/10.1371/journal.pone.0184430.g002

Model Evaluation (Step 4)

Our analysis focused upon peer-reviewed articles specifically discussing the One Health concept, which were selected as described by the search criteria. Some articles were excluded because they were not focused on the overall concept of One Health, just briefly mentioned the concept, or only covered a particular practice of One Health (exclusion criteria). While these papers may have been excluded, it is important to note that many of the specific topics they discuss, such as zoonoses and food safety, in the context of One

Health practice, are still accounted for in our model. Thus, the model likely benefits from the removal of those reports so that the conceptual framework is centered on more specific interpretations of the One Health concept.

A term that has been used synonymously to One Health is comparative/translational medicine, which in fact was used before the One Health term was introduced [2, 3, 6, 25]. However, comparative/translational medicine was not used as a search term in this model given that the One Health concept has instead been more readily accepted as the more common term to use. Furthermore, compared to comparative/translational medicine, One Health has a broader application in scope [15].

DISCUSSION

From our analysis of the One Health literature and through the construction of the One Health Cosmos using a system dynamics theoretical model, we were able to derive multiple inferences.

Our results make it clear that One Health represents a wide-ranging synergistic field that is rapidly growing. This is reflected in the number of One Health articles that have been published in the past 25 years, which has seen an average publication increase of 14.6% per year during this time [23]. As a result, the interpretation of the One Health concept remains unclear because its internal relationships between this growing list of various components have not been systematically described.

Our model indicates that humans and animals are closely connected by zoonoses and the food supply chain [34, 35]. Some common zoonoses include avian influenza [36, 37], leishmaniasis [38], Middle East respiratory syndrome coronavirus prevention [39], rabies [40, 41] etc. Additionally, the rising demand for animal protein for human consumption, resulting in a shift towards larger, more complex animal production operations, has spurred the increased use of antimicrobials. In fact, experts estimate that from 2010 and 2030, the global consumption of antimicrobials will increase by 67% [42]. As such, investing more resources into the research and control of zoonotic diseases and complex food production systems could be an effective strategy for improving the overall health of humans and animals.

Our model also shows that zoonoses are closely associated with the economy, and can result in extensive economic losses when large zoonotic diseases outbreaks among humans and animals occur. For example, after the 2014 outbreak of Ebola virus, it was estimated that the affected countries of

Guinea, Liberia, and Sierra Leone alone lost $2.2 billion in gross domestic product (GDP) the following year [43]. Additionally, the 2014 highly pathogenic avian influenza (HPAI) H5N2 outbreak in the United States, resulted in an estimated economic loss of $3.3 billion [44]. Hence, given the strength of association between zoonoses and the economy as demonstrate by the model, investing in prevention measures could result in an overall cost savings as new potential outbreaks are avoided [45].

Also, the health of companion animals is a growing concern. Pet keeping is a traditional custom for Western countries and becoming a new trend for non-Western countries. Pets could also bring disease to pet owners or non-pet owners who lives in close with them [46], such as echinococcosis, toxocarosis [47–49]. The One Health concept is required for a good understanding of the parasites' biology and epidemiology and the education of pet owners.

We also found that education plays a foundational role in the development of the One Health concept. One Health education can be divided into education of those already working in the relevant professional disciplines, and of students seeking professional qualifications to enter one of these disciplines [2]. As One Health continues to be accepted as a way forward, there will be an increasing need to integrate the One Health concept into related disciplines' education [50, 51]. Our model currently links education to epidemiology, human medicine, veterinary medicine, ecology, microbiology, and biomedicine. However, the depth of current education in One Health is unknown. Seems likely that education is not standardized across disciplines.

This report has a number of limitations. Despite our comprehensive approach we likely missed some important One Health publications in the literature review. Likewise, we may have failed to identify some important factors or components engaged in One Health activities.

To date, our analyses is the first to apply a system dynamics approach to understanding the One Health concept. Such an approach is useful in further elucidating the complex connections and causal factors that currently comprise One Health. By showing the inter-relationships between the various components of One Health, offering a visual example of the conceptual framework, the One Health Cosmos can help facilitate a more concise interpretation of the One Health concept for the development of future research and training programs.

ACKNOWLEDGMENTS

The funders of the study had no role in study design, data analysis, data interpretation, or writing of the report. All authors had full access to all data in the study and had final responsibility for the decision to submit for publication.

REFERENCES

1. Gibbs P. Origins of One Health and One Medicine. The Veterinary record. 2014;174(6):152. http://dx. doi. org/10. 1136/vr. g1372. pmid:24509397.
2. Gibbs EPJ. The evolution of One Health: a decade of progress and challenges for the future. Veterinary Record. 2014;174(4):85–91. Epub 2014/01/28. pmid:24464377.
3. Atlas RM. One Health: Its Origins and Future. One Health: The Human-Animal-Environment Interfaces in Emerging Infectious Diseases: The Concept and Examples of a One Health Approach. Current Topics in Microbiology and Immunology. 3652013. p. 1–13.
4. Kahn LHMDMPHMPPF, Kaplan BDVM, Monath TPMD, Steele JHDVMMPH. Teaching "One Medicine, One Health". The American Journal of Medicine. 2008;121(3):169. pmid:18328295
5. Taylor LH, Latham SM, Woolhouse ME. Risk factors for human disease emergence. Philosophical transactions of the Royal Society of London Series B, Biological sciences. 2001;356(1411):983–9. pmid:11516376.
6. Evans BR, Leighton FA. A history of One Health. Revue scientifique et technique (International Office of Epizootics). 2014;33(2):413–20.
7. Gibbs SEJ, Gibbs EPJ. The historical, present, and future role of veterinarians in One Health. Current topics in microbiology and immunology. 2013;365:31–47. http://dx. doi. org/10. 1007/82_2012_259. pmid:22911439
8. King LJ, Anderson LR, Blackmore CG, Blackwell MJ, Lautner EA, Marcus LC, et al. Executive summary of the AVMA One Health Initiative Task Force report. Journal of the American Veterinary Medical Association. 2008;233(2):259–61. pmid:18627228.
9. Gibbs EP, Anderson TC. One World—One Health' and the global challenge of epidemic diseases of viral aetiology. Veterinaria italiana. 2009;45(1):35–44. pmid:20391388.
10. Monath TP, Kahn LH, Kaplan B. Introduction: One Health Perspective. ILAR journal. 2010;51(3):193–8. pmid:21131719
11. Coker R, Rushton J, Mounier-Jack S, Karimuribo E, Lutumba P, Kambarage D, et al. Towards a conceptual framework to support one-health research for policy on emerging zoonoses. The Lancet infectious diseases. 2011;11(4):326–31. Epub 2011/03/08. pmid:21376670.

12. Lee K, Brumme ZL. Operationalizing the One Health approach: the global governance challenges. Health Policy Plan. 2013;28(7):778–85. pmid:23221123.
13. Craddock S, Hinchliffe S. One world, one health? Social science engagements with the one health agenda. Social science & medicine (1982). 2015;129:1–4. pmid:25434985.
14. One-Health-Initiative. One Health Initiative—About One Health http://www. onehealthinitiative. com/about. php: One Health Initiative; 2010 [cited 2016 (Accessed 2016. 08. 20)]. http://www. onehealthinitiative. com/about. php.
15. Lerner H, Berg C. The concept of health in One Health and some practical implications for research and education: what is One Health? Infection ecology & epidemiology. 2015;5:25300. Epub 2015/02/11. pmid:25660757.
16. Forrester JW. Industrial Dynamics—a Major Breakthrough for Decision Makers. Harvard Bus Rev. 1958;36(4):37–66.
17. Royston G, Dost A, Townshend J, Turner H. Using system dynamics to help develop and implement policies and programmes in health care in England. System Dynamics Review. 1999;15(3):293–313.
18. Ghaffarzadegan N, Lyneis J, Richardson GP. How small system dynamics models can help the public policy process. System Dynamics Review. 2011;27(1):22–44.
19. Günal MM, Pidd M. Discrete event simulation for performance modelling in health care: a review of the literature. Journal of Simulation. 2010;4(1):42–51.
20. Homer JB, Hirsch GB. System dynamics modeling for public health: background and opportunities. Am J Public Health. 2006;96(3):452–8. pmid:16449591.
21. Homer J, Hirsch G, Minniti M, Pierson M. Models for collaboration: how system dynamics helped a community organize cost—effective care for chronic illness. System Dynamics Review. 2004;20(3):199–222.
22. Hirsch G, Homer J, Evans E, Zielinski A. A system dynamics model for planning cardiovascular disease interventions. Am J Public Health. 2010;100(4):616–22. pmid:20167899.

23. Manlove KR, Walker JG, Craft ME, Huyvaert KP, Joseph MB, Miller RS, et al. "One Health" or Three? Publication Silos Among the One Health Disciplines. Plos Biol. 2016;14(4). pmid:27100532
24. Cardiff RD, Ward JM, Barthold SW. 'One medicine-one pathology': are veterinary and human pathology prepared? Laboratory Investigation. 2008;88(1):18–26. pmid:18040269
25. Frank D. One world, one health, one medicine. Canadian Veterinary Journal-Revue Veterinaire Canadienne. 2008;49(11):1063–+.
26. Zinsstag J, Schelling E, Waltner-Toews D, Tanner M. From "one medicine" to "one health" and systemic approaches to health and well-being. Preventive veterinary medicine. 2011;101(3–4):148–56. Epub 2010/09/14. pmid:20832879.
27. Rabozzi G, Bonizzi L, Crespi E, Somaruga C, Sokooti M, Tabibi R, et al. Emerging zoonoses: the "one health approach". Safety and health at work. 2012;3(1):77–83. MEDLINE:22953235. pmid:22953235
28. Zinsstag J, Mackenzie JS, Jeggo M, Heymann DL, Patz JA, Daszak P. Mainstreaming One Health. EcoHealth. 2012;9(2):107–10. http://dx. doi. org/10. 1007/s10393-012-0772-8. pmid:22777051
29. Calistri P, Iannetti S, Danzetta ML, Narcisi V, Cito F, Di Sabatino D, et al. The Components of 'One World—One Health' Approach. Transboundary and emerging diseases. 2013;60:4–13. pmid:24589096
30. Bidaisee S, Macpherson CNL. Zoonoses and One Health: A Review of the Literature. Journal of parasitology research. 2014:874345-. ZOOREC:ZOOR15010050282. pmid:24634782
31. Stephen C, Karesh WB. Is One Health delivering results? Introduction. Revue Scientifique Et Technique-Office International Des Epizooties. 2014;33(2):375–9.
32. Stadtlander CTKH. One Health: people, animals, and the environment. Infection ecology & epidemiology. 2015;5. http://dx. doi. org/10. 3402/iee. v5. 30514.
33. Kingsley P, Taylor EM. One Health: competing perspectives in an emerging field. Parasitology. 2016:1–8. Epub 2016/01/29. pmid:26817944.
34. Rock M, Buntain BJ, Hatfield JM, Hallgrimsson B. Animal-human connections, "one health," and the syndemic approach to prevention. Social science & medicine (1982). 2009;68(6):991–5. Epub 2009/01/23. pmid:19157669.

35. Yates-Doerr E. The world in a box? Food security, edible insects, and "One World, One Health" collaboration. Social science & medicine (1982). 2015;129:106–12. pmid:24973999.
36. Capua I. Joining the dots on the emergence of pandemic influenza. Journal of clinical virology: the official publication of the Pan American Society for Clinical Virology. 2013;58(2):342–3. Epub 2013/05/07. pmid:23643193.
37. Sims LD, Peiris M. One Health: The Hong Kong Experience with Avian Influenza. One Health: The Human-Animal-Environment Interfaces in Emerging Infectious Diseases: The Concept and Examples of a One Health Approach. Current Topics in Microbiology and Immunology. 3652013. p. 281–98.
38. Palatnik-de-Sousa CB, Day MJ. One Health: the global challenge of epidemic and endemic leishmaniasis. Parasites & vectors. 2011;4:197. Epub 2011/10/12. pmid:21985335.
39. Zumla A, Dar O, Kock R, Muturi M, Ntoumi F, Kaleebu P, et al. Taking forward a 'One Health' approach for turning the tide against the Middle East respiratory syndrome coronavirus and other zoonotic pathogens with epidemic potential. International journal of infectious diseases: IJID: official publication of the International Society for Infectious Diseases. 2016. Epub 2016/06/21. pmid:27321961.
40. Manoj S, Mukherjee A, Johri S, Kumar KV. Recovery from rabies, a universally fatal disease. Mil Med Res. 2016;3:21. pmid:27429788;
41. Perez de Diego AC, Vigo M, Monsalve J, Escudero A. The One Health approach for the management of an imported case of rabies in mainland Spain in 2013. Euro surveillance: bulletin Europeen sur les maladies transmissibles = European communicable disease bulletin. 2015;20(6). Epub 2015/02/20. pmid:25695478.
42. Van Boeckel TP, Brower C, Gilbert M, Grenfell BT, Levin SA, Robinson TP, et al. Global trends in antimicrobial use in food animals. Proc Natl Acad Sci U S A. 2015;112(18):5649–54. pmid:25792457;
43. CDC. Cost of the Ebola Epidemic Centers for Disease Control and Prevention: Centers for Disease Control and Prevention; 2016 [updated August 8, 2016; cited 2016 (Accessed 2016. 08. 25)]. http://www. cdc. gov/vhf/ebola/outbreaks/2014-west-africa/cost-of-ebola. html.
44. Greene JL. Update on the Highly-Pathogenic Avian Influenza Outbreak of 2014–2015. Congressional Research Service. 2015:1–18.

45. Rushton J, Häsler B, De Haan N, Rushton R. Economic benefits or drivers of a'One Health'approach: Why should anyone invest? Onderstepoort Journal of Veterinary Research. 2012;79(2):75–9.
46. Rock MJ, Adams CL, Degeling C, Massolo A, McCormack GR. Policies on pets for healthy cities: a conceptual framework. Health promotion international. 2015;30(4):976–86. Epub 2014/04/04. pmid:24694682.
47. Deplazes P, van Knapen F, Schweiger A, Overgaauw PAM. Role of pet dogs and cats in the transmission of helminthic zoonoses in Europe, with a focus on echinococcosis and toxocarosis. Veterinary parasitology. 2011;182(1):41–53. pmid:21813243
48. Dantas-Torres F, Otranto D. Dogs, cats, parasites, and humans in Brazil: opening the black box. Parasites & vectors. 2014;7:22. Epub 2014/01/16. pmid:24423244.
49. Maia C, Ramos C, Coimbra M, Bastos F, Martins A, Pinto P, et al. Bacterial and protozoal agents of feline vector-borne diseases in domestic and stray cats from southern Portugal. Parasites & vectors. 2014;7. pmid:24655431
50. Lane SL. Integrated environmental education: Introducing One Health concepts into veterinary technician education [1590047]. Ann Arbor: Montreat College; 2015.
51. Barrett MA, Bouley TA, Stoertz AH, Stoertz RW. Integrating a One Health approach in education to address global health and sustainabilrty challenges. Frontiers in Ecology and the Environment. 2011;9(4):239–45. http://dx. doi. org/10. 1890/090159.

CHAPTER 11

A Veterinary Perspective on One Health in the Arctic

Christian Sonne[1], Robert James Letcher[2] , Bjørn Munro Jenssen[1,3,4], JeanPierre Desforges[1] , Igor Eulaers[1] , Emilie AndersenRanberg[1] , Kim Gustavson[1] , Bjarne Styrishave[5] and Rune Dietz[1]

[1]Department of Bioscience, Arctic Research Centre (ARC), Aarhus Univer sity, Faculty of Science and Technology, Frederiksborgvej 399, 4000 Roskilde, Denmark

[2] Ecotoxicology and Wildlife Health Division, Environment and Climate Change Canada, National Wildlife Research Centre, Carleton University, Ottawa, ON K1A 0H3, Canada

[3] Department of Biology, Norwegian University of Science and Technology, 7491 Trondheim, Norway

[4] Department of Arctic Technology, The University Centre in Svalbard, 9171 Longyearbyen, Norway

Citation: Sonne, C., Letcher, R.J., Jenssen, B.M. et al. A veterinary perspective on One Health in the Arctic. Acta Vet Scand 59, 84 (2017). https://doi.org/10.1186/s13028-017-0353-5.

[5] Toxicology Laboratory, Department of Pharmacy, Faculty of Health and Medical Sciences, University of Copenha gen, Universitetsparken 2, 2100 Copenhagen, Denmark

ABSTRACT

Exposure to long-range transported industrial chemicals, climate change and diseases is posing a risk to the overall health and populations of Arctic wildlife. Since local communities are relying on the same marine food web as marine mammals in the Arctic, it requires a One Health approach to understand the holistic ecosystem health including that of humans. Here we collect and identify gaps in the current knowledge of health in the Arctic and present the veterinary perspective of One Health and ecosystem dynamics. The review shows that exposure to persistent organic pollutants (POPs) is having multiple organ-system effects across taxa, including impacts on neuroendocrine disruption, immune suppression and decreased bone density among others. Furthermore, the warming Arctic climate is suspected to influence abiotic and biotic long-range transport and exposure pathways of contaminants to the Arctic resulting in increases in POP exposure of both wildlife and human populations. Exposure to vector-borne diseases and zoonoses may increase as well through range expansion and introduction of invasive species. It will be important in the future to investigate the effects of these multiple stressors on wildlife and local people to better predict the individual-level health risks. It is within this framework that One Health approaches offer promising opportunities to survey and pinpoint environmental changes that have effects on wildlife and human health.

BACKGROUND

One Health aims to improve health and well-being through the prevention of risks and the mitigation of effects of crises that originate at the interface between humans, animals and their various environments (http://www.onehealthglobal.net). Given the close relationship between wildlife, sled dogs and humans in the North Atlantic, in combination with long-term biomonitoring programs using multiple veterinary tools over the last three decades, an Arctic One Health approach requires focus on few key species and study areas that interlink wildlife, human and ecosystem health. The present review will therefore discuss clinical and veterinary studies on polar bears (*Ursus maritimus*), ringed seals (*Pusa hispida*), narwhals (*Monodon monoceros*), killer whales (*Orcinus orca*), domestic sled dogs (*Canis *familiaris*), and humans (*Homo sapiens*), and how shared information on these provide a base from which One Health can be operated. The focus will be on exposure of Arctic wildlife and Inuit populations to long-range transported contaminants and its interactions with climate change and infectious diseases (zoonoses).

Search Strategy

We conducted this literature review to give a systematic overview of the current knowledge on environmental contaminants and zoonosis in the Arctic and how it links to climate change in a multiple stressors framework. From our own working knowledge in the field, we identified "persistent organic pollutants (POPs)", "mercury (Hg)" and "zoonoses" relevant for the Arctic and the link to "climate change" as overall themes/keywords. We used ScienceDirect, PubMed, Google, Google Scholar, ISI Web of Knowledge/ Web of Science and Springer Link to locate peer-reviewed scientific articles and reports, using the thematic keywords "immune", "endocrine", "neuro", "bone", "diseases" and "multiple stressors" either alone or in combination. We also used dissertations and AMAP reports to obtain information from the grey literature. Although we attempted to be systematic within the scope of the review, we acknowledge that this is not an exhaustive representation of all the material that may exist in the scientific literature.

The Arctic Environment as a Sink for Pollutants

Hunting and fishing have always been an important part of Arctic human existence in this harsh northern environment [1,2,3,4,5,6], and their intensity has increased with a growing Arctic human population bringing along more effective hunting methods, including larger and faster boats and fishing vessels, riffles and gillnets. Adding to these local stressors on Arctic ecosystems, anthropogenic pollution has become a ubiquitous problem that is particularly relevant in the Arctic environment [6,7,8]. In addition to elevated mercury (Hg), an array of anthropogenic chlorinated, brominated, and fluorinated persistent organic pollutants (POPs), all alien substances, have been introduced to the Arctic [1, 7, 9,10,11,12,13,14]. More specifically, the 1850s marked the period of early industrialization and was associated with elevated Hg emissions [10], whereas the onset of the 1940s marked the onset of large-volume worldwide use of industrial chemicals and pesticides. POPs and Hg typically originate from industrial and household emissions at temperate regions and are transported via global atmospheric and oceanic pathways that result in deposition in the Arctic environment [6, 7, 9, 11, 12, 15].

Arctic fish and wildlife rely on energy-rich fatty tissues as their main energy source in the harsh Arctic environment [16, 17]. Fatty tissues typically host various natural lipophilic compounds, such as specific vitamins (A and D) and endogenous steroid hormones transported by portal and periphery blood supply among others [8, 18,19,20,21,22,23]. However, most POPs and methyl-Hg, the bioavailable chemical from of Hg, are highly lipophilic as well and are therefore readily stored in lipid-rich tissues. Additional low excretion of these compounds results in a net intake of POPs and Hg over time, referred to as bioaccumulation, and is moreover transferred from prey to predator along the food chain resulting in biomagnification. As a result, apex predators such as polar bears, Arctic fox (*Vulpes lagopus*), seal spp., whale spp. and seabird spp., are exposed to the highest concentrations occurring the Arctic environment, already a major sink for POPs and Hg as described above. Finally, indigenous northerners and their dogs are ultimate sinks due to their traditional consumption of the aforementioned wildlife [7, 13, 14, 24, 25]. Newer perfluorinated compounds [poly- and perfluoralkyl substances (PFASs)], in particular perfluorooctane sulfonate (PFOS) and other long-chained poly- and perfluorocarboxylic acid (PFCAs) are proteinophilic and also biomagnify due to high resistance to biological degradation [7].

Long-range transported pollutants have been extensively monitored in the Arctic due to the high exposure of Inuit populations, resulting from their consumption of a marine diet consisting especially of apex marine predators high in contaminants [2, 3, 6]. Such biomonitoring activities have shown that, among these POPs, polychlorinated biphenyls (PCBs) continue to dominate and are of the greatest exposure concern, despite their ban decades ago [7]. However, other high-concentration POPs, amongst which organochlorine pesticides (OCPs), brominated flame retardants (BFRs), PFASs, and Hg can also be found at concentrations that raise concerns for the health of top predators and humans [7, 11, 26,27,28].

Arctic Wildlife and Human Health

POPs and Hg pose a health threat to Arctic top predators and humans because the compounds and their biotransformation metabolites have structural similarities to endogenous compounds. These anthropogenic compounds have been classified as endocrine disruptors or cellular toxicants acting via non-endocrine pathways, and thus negatively affect immune and neuro-endocrine functioning, growth and development, reproduction and general fitness [7, 8, 29]. Since the compounds potentially target different organ-tissues, the dietary exposure causes chronic and combined stress manifested through several health effects at the organism level [8, 30, 31].

There is evidence that high exposure poses a great risk to neonatal individuals during critical periods of development. Seasonal cycles of energy requirements for fasting, breeding, lactation, and migration lead to increased intake or catabolism of adipose tissue causing pulsed exposure to bioavailable contaminants circulating in the blood [32, 33]. In polar bears, for example, up to 70% of the total organochlorine body burden is transported from mother to offspring during lactation, resulting in cub adipose tissue concentrations that are approximately three times higher than those in their mothers [32, 34,35,36,37]. A female polar bears' very first cubs are believed to be especially vulnerable since high contaminant exposure can affect normal development and growth [7, 8, 31]. In a meta-study, it has recently been modelled that chlorinated and brominated POPs, singularly or collectively, were far better predictors of declines in population densities in 14 polar bear subpopulations than were human population density, harvest rate and sea ice extention [38]. Indeed, circumarctic polar bear subpopulations are under influence of immunological, reproductive and carcinogenic consequences from POP exposure [27]. From a population conservation point of view, contaminants that reduce pregnancy, fecundity and survival in both males and females are among the most important to monitor in different Arctic subpopulations of polar bears, as well as other top predators and northerners relying on the same food web [39, 40].

East Greenland polar bears, killer whales, narwhals and ringed, harp (*Pagophilus groenlandicus*) and hooded seals (*Cystophora cristata*) carry very high concentrations of POPs and Hg [7, 11, 12, 15, 41]. Since Greenlanders in this region traditionally ingest significant quantities of adipose tissue from these species, they are among the Arctic people carrying the highest POP burdens [5, 6, 43, 44]. Such high exposure is likely to pose a health risk based on available literature on dioxin toxic equivalency factor (TEQ)ÿQXw]iK/ and tolerable daily intake (TDI) guidelines [44, 45]. Human exposure to contaminants in Greenland has been evaluated from chemical analyses of prey species and food intake [5, 46, 47], showing that the TDI was exceeded for chlordane (CHL) by a factor of 3–6, while PCB exposure did not. However, none of these studies reported on polar bear and ringed seal blubber important to people's exposure in East Greenland, where POP loads are known to be four times larger than in west Greenland. According to Nielsen et al. [48] it is recommended that Greenlanders reduce their exposure to PCBs and CHL by reducing their blubber intake. The Arctic Monitoring and Assessment Programme (AMAP) is therefore concerned about Arctic human health within a contaminant exposure context and their studies do support observations that Greenland hunters are particularly exposed to high PCB concentrations due to frequent ingestion of polar bear, killer whale, narwhal and seal tissues [6, 44, 49,50,51]. Studies from the Russian Arctic have shown that dioxin, furans and PCB exposure of neo- and prenatal children exceeded TEQ TDI levels by up to 33 times in the year 2000 [5, 52]. It has been shown that blood concentrations of PFASs including PFOS in male Inuits from East Greenland can be two to three times higher as compared to the Faroese population where local exposure has already been attributed with effects on the immune system [53,54,55].

Although certain PFASs have been associated with developmental and hormonal effects, immunotoxicity, and tumour growth in rodents [56], the impact of these compounds on human health appear to be inconclusive [56, 57]. Of the PCBs found in Greenlanders, the congeners CB-77, CB-126 and CB-169 attain a coplanar configuration similar to the very toxic dioxins and furans, and are in fact commonly found in Arctic wildlife [3, 5, 58]. These coplanar PCB congeners are characteristically highly potent inducers of aryl hydrocarbon hydroxylase activity [3, 5]. Furthermore, for Greenlanders, significant correlations were found between blood contaminant concentrations and calculated daily intake of POPs [51]. Hg exposure of Inuit people is also of great health concern and has been recognised as a neuro-endocrine and immune health problem in the societies of Faroe Islands, West Greenland (Avanersuaq, Thule) and Canada [6, 11, 59].

Biomonitoring of Wildlife Health

Several of the environmental contaminants, such as PCBs and Hg are regulated by international agreements over the last 15 years through international treaties and conventions including the Stockholm Convention on POPs (http://chm.pops.int) and Minamata Convention on Hg (http://www.mercuryconvention.org). However, over the last two decades, the concentrations of the highly toxic PCBs, chlordane pesticides and Hg have remained essentially unchanged or even increased in polar bears inhabiting contaminant hot spots, such as Greenland and Hudson Bay [10, 12,13,14, 60, 61]. This is likely due to climate change effects on food web interactions, generational transfer, and continued secondary and unintentional emissions [12, 13, 17, 62]. For Hg, body burdens even appear to be continuously increasing in most top predators in the central Arctic reaching up to 20-fold baseline levels of pre-industrialisation [10, 11, 60, 63].

For three decades the AMAP program and associated subprograms have therefore monitored the health of Arctic wildlife and humans [1, 3, 5,6,7, 41, 60]. For the purpose of studying contaminant concentrations, spatial and temporal trends, and human exposure, these programs have used ringed seals and polar bears as key monitoring species [7, 64]. From the East Greenland region, a large number of organ–tissue samples from polar bears have been obtained since the 1980s as part of the traditional hunt [12, 42]. Likewise, from Svalbard, Canada and Alaska, adipose tissue and blood has been archived, facilitating the spatial and temporal trends which are further supported by similar samples from ringed seals [7]. With respect to East Greenland polar bears, histopathological, gross morphological and bone composition and morphological investigations have been carried out since 1999 [8, 31, 65, 66]. These studies have provided a unique opportunity to investigate the potential organ-specific effects of POP exposure [8]. Similar to polar bears, analyses of bone density and histopathology have been carried out on West Greenland ringed seal populations [67, 68] and on Alaska polar bears, ringed seals and whale spp. [69,70,71,72].

Interpretations and conclusions in wildlife health studies linked to contaminant exposure are typically based on correlational and descriptive interpretations, unavoidably confounded by a plethora of factors affecting the physiological state of a free-ranging animal. To improve the understanding of the effects of exposure to real-world contaminant cocktails experimental exposure studies have been performed on sled dogs and domesticated Arctic foxes, being possible surrogate model species for *Canidae* spp. In captive sled dogs and Arctic fox studies, which both included a cohort fed a naturally POP contaminated diet of minke whale (*Balaenoptera acutorostrata*) blubber, it has been possible to define and compare POP exposed and unexposed reference groups in direct relation to an array of effects such as on reproductive organs and other internal organs, the skeletal system, immune and endocrine systems, and POP dietary accumulation, biotransformation and toxicokinetics [7, 8, 31].

Biological Effects

In the following sections we review the available literature of One Health in the Arctic (Fig. 1). We present and discuss the results from empirical studies of wildlife and compare these with controlled studies of sledge dogs and arctic foxes in the context of One Health and health effects in humans. The following sections are divided into first specific organ-systems and after that the relatively sparse information on multiple effects health effects are presented and discussed.

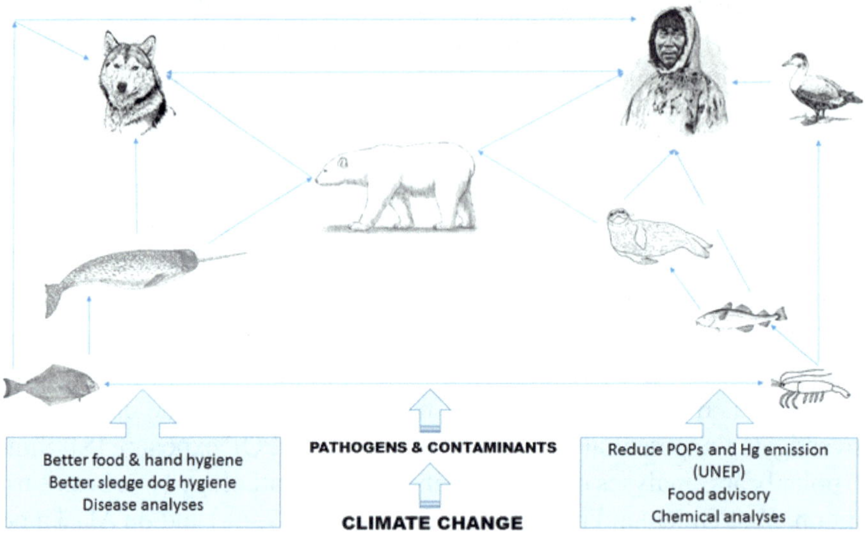

Figure 1: A simplification of the marine food web and One Health in the Arctic. Climate change affect the presence of pathogens and composition of contaminants in Arctic wildlife that is handled and eaten by Inuits and sled dogs transferring contaminants and pathogens into the local societies. Inuits and sled dogs share microbiomes which increases the risk of zoonotic infections. Actions to be taken is given in callouts.

Chronic Diseases

Chronic diseases, including diabetes, upper respiratory and recurrent middle ear infections, cancer, and osteoporosis, are becoming an epidemic in the Arctic and may be partially caused by chronic POP exposure and associated endocrine disruption [3, 5, 73]. It is apparent that subsistence hunters in East Greenland are exposed to mixtures of POPs that include both toxic parent compounds and also their derived metabolites [74]. Some POPs can induce (hepatic) cytochrome (CYP) 450 enzymes in Arctic people that may interfere with normal metabolic processes and homeostasis of various intrinsic hormones and vitamins influencing immunity and reproduction [1, 3, 5, 75, 76]. Likewise, Sandau et al. [77] found that metabolism of PCBs was significantly negatively correlated with thyroid hormones, namely free thyroxine, in northern peoples in the Ungava region in northern Quebec, Canada. These findings of associations between POP exposure, induction of CYP-450 enzymes, and changes in thyroid hormone concentrations, may play a role in the development of chronic diseases [5].

Bone mineral density

Studies of polar bears suggest that osteoporosis may be a problem for especially the male bears that do not have the evolutionary physiological mechanisms that females have, putting them at high risk of POP related bone mineral density declines [8, 66, 78]. The studies of polar bears have shown reductions in bone mineral density due to PCB exposure all supported by studies on seals in the Baltic Sea and alligators in Florida, USA [79,80,81].

A single study has been published on osteoporosis in relation to POP exposure in Greenlanders [82]. The study focused on quantitative ultrasound parameters (QUS) at the right calcaneus of 153 people from Southwest Greenland and found blood plasma CB-153 concentrations strongly and negatively associated with the three QUS parameters. While the relationship was no longer significant when normalizing for age and weight, people in Southwest Greenland belong to some of the lowest POP exposed Greenlanders and effects on bone composition are therefore not likely to occur [5, 6]. Another study of Cree women from Eastern James Bay in Canada showed that bone stiffness index was negatively associated to blood concentrations of CB-105 and CB-118 [83].

Endocrine disruption

Endocrine disruption from POP exposure is likely to be a challenge in Arctic wildlife [30, 84]. Polar bears have been in focus over the last decades and both steroid hormones and thyroid hormones seem to be impacted by POP parent compounds and their metabolites, mimicking hormone activity in both the transport pathways and receptor mechanisms and affecting overall health and survival [8, 30]. These investigations have been backed up by controlled experiments on sled dogs and Arctic foxes [8, 84]. Likewise, in Inuit peoples, disruption by POPs of the thyroid and steroid hormone endocrine axis is influencing physiological endpoints with effects on reproduction, cancer, and immunity [6, 85]. Studies on polar bears from Svalbard indicate that plasma steroid levels may be affected by POP exposure, particularly PCBs and their hydroxylated metabolites (OH-PCBs). In females, pregnenolone and androstenedione in the blood appeared to be significantly negatively correlated to several hydroxylated PCBs [86], indicating that these POPs may affect early or perhaps rate-limiting steps in steroidogenesis. The data also indicated that CYP-17 may be a target for OH-PCBs and consequently the reproductive potential of female polar bears. PCBs also appear to affect male plasma steroids, in particular androgen levels in Svalbard polar bears [87]. Concentrations of the most masculinizing steroids, dihydrotestosterone and testosterone, were negatively correlated to levels of a large number of PCB congeners.

Reproductive organs

Studying reproductive organs in wildlife is often based on necropsy samples as clinical investigations are rather difficult. Necropsy of human reproductive organs from Arctic populations has for ethical reasons not been studied, though the investigation of endocrinology and reproductive organs in wildlife and domesticated species (sled dogs and Arctic fox) can give some important information. For example, the presence of orchitis in polar bear testes [88] may be of importance while studying testicular dysgenesis syndrome in Inuits. Seasonal hormone and semen production in Arctic fox was found to be affected by POPs [89, 90]. Similarly, there are indications that the size of polar bear penis bone, testicles and ovaries may be inversely correlated to PCB concentrations [66, 88]. Such information may be of use in human medicine when autopsies are not an option and when POPs are suspected of having endocrine disruptive effects in humans.

Immune effects

Environmental contaminants have been shown to modulate all measurable aspects of cellular and humoral immunity in marine mammals [89]. Resistance against acute and chronic diseases also depends on the optimal function of the immune system [91, 92]. For years, POPs have been known to modulate immunity, and while the mechanisms of immune suppression are not fully understood, it includes both humoral cell-mediated systems [3, 92,93,94,95]. Several studies have reported immune effects in Arctic wildlife, including polar bears, ringed seals, sled dogs and Arctic foxes [8], demonstrating that contaminants are reaching levels that can cause significant changes in physiology and immune fitness, with important consequences for resistance to disease.

Polar bear immunity was assessed in a comprehensive study in Svalbard and Hudson Bay, and negative associations were found between PCB exposure and serum immunoglobulin G (IgG) levels, antibody titres against influenza virus and reovirus following immunizations, and lymphocyte proliferation [96, 97]. In a controlled study on Greenland sled dogs, exposed pups, but not adults, showed reduced and delayed IgG antibody production with circulating levels of IgG in all pups correlating to blood concentrations of several POPs [8]. Hepatic mRNA expression of interleukin-1β (IL-1β), an important pro-inflammatory cytokine, in ringed seals from Svalbard were positively correlated with hepatic POP levels [98]. Similarly, hepatic IL-1 mRNA expression was positively correlated with blubber PCB burdens in 41 ringed seals in northern Labrador, Canada [99]. These studies illustrate that in vivo real life exposure to contaminants cause measurable changes in immune function in Arctic wildlife, and thus likely humans.

In vitro experiments, where immune cells are exposed to contaminants under laboratory conditions in order to characterise effect levels, have also been carried-out in Arctic wildlife species. East Greenland ringed seal leukocytes were exposed to four PCB congeners (CB-138, -153, -169, and -180) and two PFASs, i.e. PFOS and perfluorooctanoic acid (PFOA), and it was found that PCBs but not PFASs cause significant suppression of lymphocyte proliferation at relevant environmental concentrations for seals [100]. Beluga whale (*Delphinapterus leucas*) leukocytes were also used to show that low levels of Hg exposure can cause significant reductions in lymphocyte proliferation and intracellular thiol production, and significant induction of metallothionein [101]. Altogether, studies of free-ranging animals and in vitro experiments suggest high contaminant loads in the Arctic can induce immune suppression which affect the ability to respond to intruding infectious pathogens as previously suggested for marine mammals [7, 8, 31, 89] and Inuit and other northern peoples [6]. For example, prenatal exposure to PCBs has been suspected to play a role in the relatively high incidence of acute respiratory infections and middle ear inflammation in Inuit children [6, 102,103,104,105,106].

Neurological effects

Multiple environmental contaminants can cross the blood–brain barrier and exert effects on the brain functioning thus causing neurobehavioral effects. Through various mechanisms, contaminants can influence mating and other reproductive behaviours, motivation, communication, aggression, dominance and other social behaviours, as well as learning and other cognitive abilities

[29, 107]. Field studies of behavioural effects of POPs in polar bears are difficult to conduct due to logistical and ethical constraints. However, several POPs reported in brain tissue of polar bears [74, 108,109,110,111,112] are confirmed or suspected developmental neurotoxicants in humans and experimental animals [29, 107, 113, 114]. POPs might affect brain function or development through many mechanisms, for instance by interacting with brain neurotransmitter systems [113]. Also, the ability of some POPs to induce epigenetic changes could present a mechanistic pathway of neurodevelopmental perturbations [115,116,117]. Associations between neurochemical and epigenetic biomarkers and Hg levels in brain tissue have been reported for polar bears [11, 118, 119]. Despite relatively low concentrations of Hg, significant negative correlations were found between both Hg concentrations and *N*-methyl-D-aspartic acid (NMDA) glutamate receptors. NMDA glutamate and genomic methylation is important for animal health, behaviour, reproduction, and survival, and their reduction may have population-level effects for polar bears [120].

Another proposed mechanism of developmental neurotoxicity is through thyroid hormone disruption; thyroid hormones are essential for proper neurodevelopment of the foetus and early neonate [121, 122]. Thus, the high levels of POPs reported in 4-month old polar bear cubs [34] and their associated thyroid disrupting properties in cubs [123, 124] raises concern for neurodevelopmental effects in polar bears. The relatively high concentrations of several POPs reported in polar bear brains may cause adverse effects, with a possible heightened susceptibility during the more sensitive foetal and neonatal stages of brain development [29, 125]. Similarly to humans, this could alter behavioural traits and reduce cognitive abilities related to memory and learning in offspring. For polar bears, this could functionally reduce hunting skills or alter mating behaviour, and thus ultimately affect reproduction and survival.

Multiple Stressors

In the previous section we discussed the important risk posed by environmental contaminants in Arctic wildlife and human populations for a wide range of physiological health endpoints. Arctic ecosystems, however, are being stressed by more than contaminants, resulting in a situation of multiple cumulative stress for wildlife and humans. Two major additional aspects to consider in the study of Arctic health is climate change and infectious diseases. Climate change has a duel impact, acting through alteration of food web pathways for contaminants [61, 126] and spreading and virulence of zoonotic diseases associated to sea ice conditions [127,128,129]. A pollution induced increase in disease rates due to immunotoxic effects of POPs and Hg can increase the likelihood and risk of disease transfer from animals to humans (zoonoses) as the proportion of infected Arctic wildlife increases [30, 89, 130,131,132]. A large volume of marine and terrestrial wildlife is consumed by humans in the Arctic, often raw and inadequately frozen, and this likely increases the risk of zoonotic diseases [8].

Climate change and contaminant exposure

Arctic wildlife have received considerable focus as they, depending on the regional subpopulation, are threatened most dramatically by climate change due to observed and projected loss of sea ice, which has important implications for ice-associated hunters like polar bears [133,134,135]. Modeling has shown that southernmost polar bear subpopulations in the Hudson Bay are at greatest risk, and will struggle to persist throughout this century [136]. In fact, models have also predicted two-thirds of the world's polar bears could disappear if greenhouse gas emissions continue to increase as predicted [134, 137]. This has been linked to the occupation of large home range sizes and the requirement of higher energetic costs and thus higher feeding rates, which can lead to increasing blood PCB concentrations [138, 139]. In some regions, the decline of sea ice extent has resulted in changes in the presence of seal species that polar bears prey upon, and this has been shown to cause increased bioaccumulation of certain POPs as more contaminated prey are being consumed [61, 140]. Dietary shift towards feeding on plants, berries and caribou (*Rangifer tarandus*) and seabird eggs [141,142,143] will most likely decrease and increase, respectively, the exposure to POPs. Furthermore, climate warming induced migration of warm water adapted fish species [144, 145] may act as bio-vectors increasing contaminant levels in marine Arctic ecosystems [146, 147], ultimately causing increased bioaccumulation and biomagnification of these compounds to humans and other high trophic marine wildlife [30, 126]. McDonald et al. [147] conducted a review on ecological impacts of global climate change on POP and Hg pathways and exposures in arctic marine ecosystems, and documented that lower sea ice extent mediated dietary changes were associated with higher contaminant levels in some populations of polar bears, ringed seals, and thick-billed murres (*Uria lomvia*), but the influence of changing trophic interactions on POP levels and trends varied widely in both magnitude and direction.

Climate change and infectious diseases

Climate change not only threatens to alter contaminant dynamics by changing Arctic ecosystems, but these same factors influencing the presence and extent of different species in the Arctic will have implications for the introduction of novel infectious diseases to the region. Climate change has been deemed the most important factor in the emergence of infectious diseases, and nowhere else in the world is climate change occurring as fast as in the polar regions [148, 149]. A warming climate may profoundly affect disease dynamics in the Arctic by changing the species composition and northward invasion of disease vectors and transport of pathogens [149, 150]. In addition, increased survival of infected animals during milder winters may further increase the risk of pathogen reservoir in marine mammals including that of zoonosis [149, 150]. Moreover and as discussed above immunotoxic contaminants may increase disease-related mortality and morbidity of Arctic marine mammals [129].

Perspectives and Recommendations

Further efforts are required to understand the toxicokinetics and toxicodynamics of POPs and Hg in Greenland wildlife and peoples in this changing Arctic in order to better predict the individual-level health risks associated with contaminant exposure. Arctic top predators are sentinels for humans as they consume the same diet and act as potential vectors for zoonotic transfer to humans due to harvesting. Combining correlational studies on wildlife health with experimental work on surrogate species, such as the sled dog, will allow better understanding of the proximate toxic pathways behind exposure to contaminants and infectious diseases, their interactions, and the driving role of a rapidly changing climate. Doing this offers a promising One Health approach to survey and pinpoint environmental change and multiple stressors that may have effects on wildlife and human health [7, 8, 31, 151,152,153]. The warming Arctic climate is suspected to influence abiotic and biotic long-range transport and exposure pathways of contaminants to the Arctic. As a result there will be likely increases in POP exposure of Arctic wildlife and human populations, while exposure to vector-borne diseases and zoonoses may increase as well through range expansion and introduction from invasive species. Broad, and nevertheless in-depth studies on the occurrence and human health risk of Arctic zoonoses, and their interactive effects with climate change and contaminant exposure are pending, as well as an increased effort to educate the relevant groups of the public regarding safe handling of wildlife.

AUTHORS' CONTRIBUTIONS

CS designed the review and produced the 1st draft and further revisions and submission of manuscript. RJL dealt with the precision of environmental contaminants, BMJ with the multiple stressors' framework, JPD with the immune sections, IE with the ecological framework, EAR with zoonosis, KG with PBPK modelling, BS with steroid hormones and RD with modelling and ecosystem effects. All authors read and approved the final manuscript.

REFERENCES

1. AMAP. Arctic Monitoring and Assessment Programme: AMAP assessment report—arctic pollution issues 1998; Oslo 1998. http://www.amap.no.
2. AMAP. Arctic Monitoring and Assessment Programme: AMAP assessment 2002—human health in the Arctic 2003; Oslo 2003. http://www.amap.no.
3. AMAP. Arctic Monitoring and Assessment Programme: AMAP assessment 2002—persistent organic pollutants in the Arctic 2004; Oslo 2004. http://www.amap.no.
4. AMAP. Arctic Monitoring and Assessment Programme: AMAP assessment 2002—heavy metals in the Arctic 2005; Oslo 2005. http://www.amap.no.
5. AMAP. Arctic Monitoring and Assessment Programme: AMAP assessment 2009—human health in the Arctic 2009; Oslo 2009. http://www.amap.no.
6. AMAP. Arctic Monitoring and Assessment Programme: AMAP assessment 2013—human health in the Arctic 2013; Oslo 2015. http://www.amap.no.
7. Letcher RJ, Bustnes JO, Dietz R, Jenssen BM, Jørgensen EH, Sonne C, et al. Effects assessment of persistent organohalogen contaminants in Arctic wildlife and fish. Sci Total Environ. 2010;408:2995–3043.
8. Sonne C. Health effects from long-range transported contaminants in Arctic top predators: an integrated review based on studies of polar bears and relevant model species. Environ Int. 2010;36:461–91.
9. Muir DCG, de Wit CA. Trends of legacy and new persistent organic pollutants in the circumpolar Arctic: overview, conclusions, and recommendations. Sci Total Environ. 2010;408:3044–51.
10. Dietz R, Outridge PM, Hobson KA. Anthropogenic contribution to mercury levels in present-day Arctic animals—a review. Sci Total Environ. 2009;407:6120–31.
11. Dietz R, Basu N, Braune B, O'Hara T, Letcher R, Scheuhammer T, et al. What are the toxicological effects of mercury in Arctic biota? Sci Total Environ. 2013;443:775–90.
12. Dietz R, Rigét FF, Sonne C, Born EW, Bechshøft T, McKinney MA, et al. Part 2: three decades (1984–2010) of flame retardant trends in East

Greenland polar bears (*Ursus maritimus*). Environ Int. 2013;59:494–500.

13. Letcher RJ, Morris AD, Dyck M, Sverko E, Reiner E, Blair DAD, et al. Legacy and (re)emerging halogenated persistent organic pollutants in polar bears from a contamination hotspot in the Arctic, Hudson Bay Canada. Sci Total Environ. 2018;610–611:121–36.
14. Rigét F, Vorkamp K, Bossi R, Sonne C, Letcher RJ, Dietz R. Twenty years of monitoring of persistent organic pollutants in Greenland biota. A review. Environ Pollut. 2016;217:114–23.
15. Pedro S, Boba C, Dietz R, Sonne C, Rosing-Asvid A, Hansen M, et al. Blubber-depth distribution and bioaccumulation of PCBs and organochlorine pesticides in Arctic-invading killer whales. Sci Total Environ. 2017;601–602:237–46.
16. Butt CM, Berger U, Bossi R, Tomy GT. Levels and trends of poly- and perfluorinated compounds in the Arctic environment. Sci Total Environ. 2010;408:2936–65.
17. Houde M, De Silva AO, Muir DCG, Letcher RJ. Monitoring of perfluorinated compounds in aquatic biota: an updated review. Environ Sci Technol. 2011;45:7962–73.
18. Bechshøft TØ, Jakobsen J, Sonne C, Dietz R. Distribution of vitamin A (retinol) and E (α-tocopherol) in polar bear kidney: implications for biomarker studies. Sci Total Environ. 2011;409:3508–11.
19. Bechshøft TØ, Sonne C, Jakobsen J, Rigét FF, Born EW, Letcher RJ, et al. Vitamins A and E in liver, kidney, and whole blood from organohalogen exposed East Greenland polar bears sampled 1994–2008: reference values and temporal trends. Polar Biol. 2016;39:743–54.
20. Kirkegaard M, Sonne C, Dietz R, Jenssen BM, Letcher RJ. Organohalogens in a whale blubber supplemented diet reduce hepatic retinol and renal tocopherol concentrations in Greenland sled dogs. J Toxicol Environ Health Part A. 2010;73:773–86.
21. Kirkegaard M, Sonne C, Dietz R, Jenssen BM, Leifsson PS, Jensen JEB, et al. Testosterone concentrations and male genital organ morphology in Greenland sled dogs (*Canis familiaris*) dietary exposed to organohalogen contaminants. Toxicol Environ Chem. 2010;92:955–67.

22. Sonne C, Kirkegaard M, Jacobsen J, Jenssen BM, Letcher RJ, Dietz R. Altered 25-hydroxyvitamin D3 in liver tissue from Greenland sledge dogs (*Canis familiaris*) dietary exposed to organohalogen polluted minke whale (*Balaenoptera acuterostrata*) blubber. Ecotoxicol Environ Saf. 2014;104:403–8.
23. Sonne C, Dietz R, Rigét FF, Letcher RJ, Munk Pedersen K, Styrishave B. Steroid hormones in blood plasma from Greenland sledge dogs (*Canis familiaris*) dietary exposed to organohalogen polluted minke whale (*Balaenoptera acuterostrata*) blubber. Toxicol Environ Chem. 2014;96:273–86.
24. Sonne C, Andersen-Ranberg E, Rajala EL, Agerholm JS, Bonefeld-Jørgensen E, Desforges JP, et al. Prevalence of antibodies against *Brucella* spp. in West Greenland polar bears (*Ursus maritimus*) and East Greenland muskoxen (*Ovibos moschatus*). Polar Biol **(In review)**.
25. Sonne C, Andersen-Ranberg E, Rajala EL, Agerholm JS, Bonefeld-Jørgensen E, Desforges JP, et al. Seroprevalence for *Brucella* spp. in Baltic ringed seals (*Phoca hispida*) and East Greenland harp (*Pagophilus groenlandicus*) and hooded (*Cystophora cristata*) seals. Vet Immunol ImmunoPathol **(In review)**.
26. Brown TM, Macdonald RW, Muir DCG, Letcher RJ. The distribution and trends of persistent organic pollutants and mercury in marine mammals from Canada's Eastern Arctic. Sci Total Environ. 2017;618:500–17. https://doi.org/10.1016/j.scitotenv.2017.11.052.
27. Dietz R, Gustavson K, Sonne C, Desforges JP, Rigét FF, McKinney MA, et al. Physiologically-based pharmacokinetic modelling of immune, reproductive and carcinogenic effects from contaminant exposure in polar bears (*Ursus maritimus*). Environ Res. 2015;140:45–55.
28. Landrigan PJ, Fuller R, Acosta NJR, Adeyi O, Arnold R, Basu N, et al. The lancet commission on pollution and health. 2017.
29. Grandjean P, Landrigan PJ. Developmental neurotoxicity of industrial chemicals. Lancet. 2006;368(9553):2167–78.
30. Jenssen BM, Dehli Villanger G, Gabrielsen KM, Bytingsvik J, Ciesielski TM, Sonne C, et al. Anthropogenic flank attack on polar bears: interacting consequences of climate warming and pollutant exposure. Front Ecol. 2015;3:1–7.

31. Sonne C, Letcher RJ, Bechshøft TØ, Rigét FF, Muir DCG, Leifsson PS, et al. Two decades of biomonitoring polar bear health in Greenland: a review. Acta Vet Scan. 2012;54:S15.
32. Polischuk SC, Norstrom RJ, Ramsay MA. Body burdens and tissue concentrations of organochlorines in polar bears (*Ursus maritimus*) vary during seasonal fasts. Environ Pollut. 2002;118:29–39.
33. Tartu S, Bourgeon S, Aars J, Andersen M, Polder A, Thiemann GW, Welker JM, Routti H. Sea ice-associated decline in body condition leads to increased concentrations of lipophilic pollutants in polar bears (*Ursus maritimus*) from Svalbard, Norway. Sci Total Environ. 2017;576:409–19.
34. Bytingsvik J, Lie E, Aars J, Derocher AE, Wiig Ø, Jenssen BM. PCBs and OH-PCBs in polar bear mother-cub pairs: a comparative plasma levels in 1998 and 2008. Sci Total Environ. 2012;417:117–28.
35. Dietz R, Rigét FF, Sonne C, Letcher RJ, Born EW, Muir DCG. Seasonal and temporal trends in polychlorinated biphenyls and organochlorine pesticides in East Greenland polar bears (*Ursus maritimus*), 1990–2001. Sci Total Environ. 2004;331:107–24.
36. Dietz R, Rigét FF, Sonne C, Muir DCG, Backus S, Born EW, et al. Age and seasonal variability of polybrominated diphenyl ethers in free-ranging East Greenland polar bears (*Ursus maritimus*). Environ Pollut. 2007;146:177–84.
37. Muir DCG, Backus S, Derocher AE, Dietz R, Evans TJ, Gabrielsen GW, et al. Brominated flame retardants in polar bears (*Ursus maritimus*) from Alaska, the Canadian Arctic, East Greenland, and Svalbard. Environ Sci Technol. 2006;40:449–55.
38. Nuijten RJM, Hendriks AJ, Jenssen BM, Schipper AM. Circumpolar contaminant concentrations in polar bears (*Ursus maritimus*) and potential population-level effects. Environ Res. 2016;151:50–7.
39. Pavlova V, Grimm V, Dietz R, Sonne C, Vorkamp K, Rigét FF, et al. Modelling population level effects of PCB contamination in East Greenland polar bears. Arch Environ Contam Toxicol. 2016;70:143–54.
40. Pavlova V, Nabe-Nielsen J, Dietz R, Sonne C, Grimm V. PCB contamination as a potential driver of mate—finding associated Allee effect in Svalbard polar bears: the implications from an individual based model. Proc Biol Sci. 2016;283:1843.

41. AMAP. Arctic Monitoring and Assessment Programme: AMAP assessment 2018—persistent organic pollutants in the Arctic 2018; Oslo 2018. http://www.amap.no.
42. Dietz R, Rigét FF, Sonne C, Born EW, Bechshøft T, McKinney MA, Letcher RJ. Part 1: three decades (1984–2010) of legacy contaminant trends in East Greenland polar bears (*Ursus maritimus*). Environ Int. 2013;59:485–93.
43. Lindh CH, Rylander L, Toft G, Axmon A, Rignell-Hydbom A, Giwercman A, et al. Blood serum concentrations of perfluorinated compounds in men from Greenlandic Inuit and European populations. Chemosphere. 2012;88:1269–75.
44. Sonne C, Dietz R, Letcher RJ. Chemical cocktail party in East Greenland: a first time evaluation of human organohalogen exposure from consumption of ringed seal and polar bear tissues and possible health implications. Toxicol Environ Chem. 2013;95:853–9.
45. Wielsoe M, Kern P, Bonefeld-Jorgensen EC. Serum levels of environmental pollutants is a risk factor for breast cancer in Inuit: a case control study. Environ Health. 2017;16:16.
46. Deutch B, Dyerberg J, Pedersen HS, Asmund G, Møller P, Hansen JC. Dietary composition and contaminants in north Greenland, in the 1970s and 2004. Sci Total Environ. 2006;370:372–81.
47. Johansen P, Muir DCG, Asmund G, Rigét FF. Human exposure to contaminants in the traditional Greenland diet. Sci Total Environ. 2004;331:189–206.
48. Nielsen E, Larsen JC, Ladefoged O. Risk assessment of contaminant intake from traditional Greenland food items. Danish Veterinary and Food Administration; 2006. p. 180. http://www.dfvf.dk.
49. AMAP. Arctic Monitoring and Assessment Programme: AMAP assessment 2011—mercury in the Arctic 2011; Oslo 2011. http://www.amap.no.
50. Bonefeld-Jørgensen E. Biomonitoring in Greenland: human biomarkers of exposure and effects—a short review. Rural Remote Health. 2010;10:1362.
51. Sonne C, Gustavson K, Rigét FF, Krüger T, Dietz R, Bonefeld-Jørgensen E. Physiologically based pharmacokinetic modeling of contaminants in Greenlanders: a new tool for monitoring levels and possible health effects? Environ Int. 2014;64:91–7.

52. Polder A, Gabrielsen GW, Odland JØ, Savinova TN, Tkachev A, Løken KB, Skaare JU. Spatial and temporal changes of chlorinated pesticides, PCBs, dioxins (PCDDs/PCDFs) and brominated flame retardants in human breast milk from northern Russia. Sci Total Environ. 2008;391:41–54.
53. Bonefeld-Jørgensen E, Long M, Bossi R, Ayotte P, Asmund G, Krüger T, et al. Perfluorinated compounds are related to breast cancer risk in Greenlandic Inuit: a case control study. Environ Health. 2011;10:88.
54. Grandjean P, Andersen EW, Budtz-Jørgensen E, Nielsen F, Mølbak K, Weihe P, et al. Serum vaccine antibody concentrations in children exposed to perfluorinated compounds. JAMA. 2012;307:391–7.
55. Long M, Bossi R, Bonefeld-Jørgensen EC. Levels and temporal trends of perfluoroalkyl acids in human serum from Greenland Inuit. Int J Circumpolar Health. 2012;71:17998.
56. Rand AA, Mabury SA. Is there a human health risk associated with indirect exposure to perfluoroalkyl carboxylates (PFCAs)? Toxicology. 2017;375:28–36.
57. Wilkinson JL, Hooda PS, Barker J, Barton S, Swinden J. Ecotoxic pharmaceuticals, personal care products, and other emerging contaminants: a review of environmental, receptor-mediated, developmental, and epigenetic toxicity with discussion of proposed toxicity to humans. Crit Rev Environ Sci Technol. 2016;46:336–81.
58. Sonne C, Leifsson PS, Dietz R, Born EW, Letcher RJ, Kirkegaard M, et al. Enlarged clitoris in wild polar bears (*Ursus maritimus*) can be misdiagnosed as pseudohermaphroditism. Sci Total Environ. 2005;337:45–58.
59. Grandjean P, Satoh H, Murata K, Eto K. Adverse effects of methylmercury: environmental health research implications. Environ Health Perspect. 2010;118:1137–45.
60. Dietz R, Born EW, Rigét FF, Aubail A, Sonne C, Drimmei RC, et al. Temporal trends and future predictions of mercury concentrations in Northwest Greenland polar bear (*Ursus maritimus*) hair. Environ Sci Technol. 2011;45:1458–65.
61. McKinney MA, Iverson SJ, Fisk AT, Sonne C, Rigét FF, Letcher RJ, et al. Global change effects on the long-term feeding ecology and contaminant exposures of East Greenland polar bears. Glob Change Biol. 2013;19:2360–72.

62. Letcher RJ, Boisvert G, Dyck M, Sonne C, Dietz R. Polar bears from two Arctic pollution hot spots: Comparison of recent levels of established and new perfluorinated sulfonic and carboxylic acids and precursors. Sci Total Environ. 2017 **(In review)**.
63. Rigét FF, Braune B, Bignert A, Wilson S, Aars J, Andersen M, et al. Temporal trends of Hg in Arctic biota, an update. Sci Total Environ. 2011;409:3520–6.
64. Rigét FF, Bossi R, Sonne C, Vorkamp K, Dietz R. Trends of perfluorochemicals in Greenland ringed seals and polar bears: indications of decreasing trends. Chemosphere. 2013;93:1607–14.
65. Pertoldi C, Sonne C, Wiig Ø, Baagøe HJ, Loeschcke V, Bechshøft TØ. East Greenland and Barents Sea polar bears (*Ursus maritimus*): adaptive variation between two populations using skull morphometrics as an indicator of gene flow. Hereditas. 2012;149:99–107.
66. Sonne C, Dyck M, Rigét FF, Bech-Jensen JE, Hyldstrup L, Letcher RJ, et al. Penile density and globally used chemicals in Canadian and Greenland polar bears. Environ Res. 2015;137:287–91.
67. Andersen-Ranberg. Metals and flukes in West Greenland ringed seals (*Pusa hispida*)—a histopathological, toxicological, parasitological and molecular study. M.Sc. thesis 2014, University of Copenhagen; p. 109.
68. Sonne-Hansen C, Dietz R, Leifsson PS, Hyldstrup L, Rigét FF. Cadmium toxicity to ringed seals (*Phoca hispida*)—an epidemiological study of possible cadmium induced nephropathy and osteodystrophy in ringed seals (*Phoca hispida*) from Qaanaaq in Northwest Greenland. Sci Total Environ. 2002;295:167–81.
69. Woshner VM. Concentrations and interactions of selected elements in tissues of four marine mammal species harvested by Inuit hunters in arctic Alaska, with an intensive histologic assessment, emphasizing the beluga whale. University of Illinois at Urbana—Champaign 2000; Urbana, Illinois 2000, Ph.D. Dissertation.
70. Woshner VM, O'Hara TM, Bratton GR, Suydam RS, Beasley VR. Concentrations and interactions of selected essential and non-essential elements in bowhead and beluga whales of Arctic Alaska. J Wildl Dis. 2001;37:693–710.
71. Woshner VM, O'Hara TM, Bratton GR, Beasley VR. Concentrations and interactions of selected essential and non-essential elements in ringed seals and polar bears of Arctic Alaska. J Wildl Dis. 2001;37:711–21.

72. Woshner VM, Knott K, Wells R, Willetto C, Swor R, O'Hara T. Mercury and selenium in blood and epidermis of bottlenose dolphins (*Tursiops truncatus*) from Sarasota Bay, FL: interaction and relevance to life history and hematologic parameters. EcoHealth. 2008;5:1–11.
73. Donaldson SG, Van Oostdam J, Tikhonov C, Feeley M, Armstrong B, Ayotte P, et al. Environmental contaminants and human health in the Canadian Arctic. Sci Total Environ. 2010;408:5165–234.
74. Gebbink WA, Sonne C, Dietz R, Kirkegaard M, Rigét FF, Born EW, et al. Tissue-specific congener composition of organohalogen and metabolite contaminants in East Greenland polar bears (*Ursus maritimus*). Environ Pollut. 2008;152:621–9.
75. Lagueux J, Pereg D, Ayotte P, Dewailly E, Poirier GG. Cytochrome P450 CYP1A1 enzyme activity and DNA adducts in placenta of women environmentally exposed to organochlorines. Environ Res. 1999;80:369–82.
76. Medehouenou TC, Larochelle C, Dumas P, Dewailly E, Ayotte P. Determinants of AhR-mediated transcriptional activity induced by plasma extracts from Nunavik Inuit adults. Chemosphere. 2010;80:75–82.
77. Sandau CD, Ayotte P, Dewailly E, Duffe J, Norstrom RJ. Pentachlorophenol and hydroxylated polychlorinated biphenyl metabolites in umbilical cord plasma of neonates from coastal populations in Québec. Environ Health Perspect. 2002;110:411–7.
78. Sonne C, Dietz R, Born EW, Rigét FF, Kirkegaard M, Hyldstrup L, et al. Is bone mineral composition disrupted by organochlorines in East Greenland polar bears (*Ursus maritimus*)? Environ Health Perspect. 2004;112:1711–6.
79. Herlin M, Kalantari F, Stern N, Sand S, Larsson S, Viluksela M, et al. Quantitative characterization of changes in bone geometry, mineral density and bio-mechanical properties in two rat strains with different Ah-receptor structures after long-term exposure to 2,3,7,8-tetrachlorodibenzo-*p*-dioxin. Toxicology. 2010;29:1–11.
80. Lind PM, Bergman A, Olsson M, Orberg J. Bone mineral density in male Baltic grey seal (*Halichoerus grypus*). Ambio. 2003;32:385–8.
81. Lind PM, Milnes MR, Lundberg R, Bermudez D, Örberg J, Guillette LJ. Abnormal bone composition in female juvenile American alligators from a pesticide-polluted lake (Lake Apopka, Florida). Environ Health Perspect. 2004;112:359–62.

82. Côté S, Ayotte P, Dodin S, Blanchet C, Mulvad G, Petersen HS, et al. Plasma organochlorine concentrations and bone ultrasound measurements: a cross-sectional study in peri-and postmenopausal Inuit women from Greenland. Environ Health. 2006;5:33.
83. Paunescu AC, Dewailly E, Dodin S, Nieboer E, Ayotte P. Dioxin-like compounds and bone quality in Cree women of Eastern James Bay (Canada): a cross-sectional study. Environ Health. 2013;12:54.
84. Sonne C, Letcher RJ, Jenssen BM, Desforges JP, Eulaers I, Andersen-Ranberg E, et al. Use of dogs to monitor remote locations: sledge dogs and the Artic environment. Springer Book **(In review)**.
85. Dallaire R, Dewailly E, Ayotte P, Muckle G, Laliberte C, Bruneau S. Effects of prenatal exposure to organochlorines on thyroid hormone status in newborns from two remote coastal regions in Quebec, Canada. Environ Res. 2008;108:387–92.
86. Gustavson L, Ciesielski TM, Bytingsvik J, Styrishave B, Hansen M, Lie E, et al. Hydroxylated polychlorinated biphenyls decrease circulating steroids in female polar bears (*Ursus maritimus*). Environ Res. 2015;138:191–201.
87. Ciesielski TM, Hansen IT, Bytingsvik J, Hansen M, Lie E, Aars J, et al. Relationships between POPs, biometrics and circulating steroids in male polar bears (*Ursus maritimus*) from Svalbard. Environ Pollut. 2017;230:598–608.
88. Sonne C, Leifsson PS, Dietz R, Born EW, Letcher RJ, Hyldstrup L, et al. Xenoendocrine pollutants may reduce size of sexual organs in East Greenland polar bears (*Ursus maritimus*). Environ Sci Technol. 2006;40:5668–74.
89. Desforges JPW, Sonne C, Levin M, Siebert U, De Guise S, Dietz R. Immunotoxic effects of environmental pollutants in marine mammals. Environ Int. 2016;86:126–39.
90. Sonne C, Torjesen PA, Berg KA, Fuglei E, Muir DCG, Jenssen BM, et al. Exposure to persistent organic pollutants reduces testosterone concentrations and affects sperm viability and morphology during the mating peak-period in a controlled experiment on farmed Arctic foxes (*Vulpes lagopus*). Environ Sci Technol. 2017;51:4673–80.
91. Janeway CA, Travers P, Walport M, Shlomchik M. Immune biology—the immune system in health and disease. 5th ed. New York: Garland Publishing, Taylor and Francis; 2001. p. 732.

92. Vos JG, Luster MI. Immune alterations. In: Kimbrough RD, Jensen AD, editors. Halogenated biphenyls, terphenyls, naphthalenes, dibenzodioxins, and related products. Amsterdam: Elsevier; 1989.
93. Hertz-Picciotto I, Park HY, Dostal M, Kocan A, Trnovec T, Sram R. Prenatal exposures to persistent and non-persistent organic compounds and effects on immune system development. Basic Clin Pharmacol Toxicol. 2008;102:146–54.
94. Klecha AJ, Barreiro Arcos ML, Frick L, Genaro AM, Cremaschi G. Immune-endocrine interactions in autoimmune thyroid diseases. Neuroimmunomodulation. 2008;15:68–75.
95. Tryphonas H. Immunotoxicity of polychlorinated-biphenyls—present status and future considerations. Exp Clin Immunogenet. 1994;11:149–62.
96. Lie E, Larsen HJS, Larsen S, Johansen GM, Derocher AE, Lunn NJ, et al. Does high organochlorine (OC) exposure impair the resistance to infection in polar bears (*Ursus maritimus*)? Part I: effect of OCs on the humoral immunity. J Toxicol Environ Health A. 2004;67:555–82.
97. Lie E, Larsen HJS, Larsen S, Johansen GM, Derocher AE, Lunn NJ, et al. Does high organochlorine (OC) exposure impair the resistance to infection in polar bears (*Ursus maritimus*)? Part II: possible effect of OCs on mitogen- and antigen-induced lymphocyte proliferation. J Toxicol Environ Health A. 2005;68:457–84.
98. Routti H, Arukwe A, Jenssen BM, Letcher RJ, Nyman M, Bäckman C, et al. Comparative endocrine disruptive effects of contaminants in ringed seals (*Phoca hispida*) from Svalbard and the Baltic Sea. Comp Biochem Physiol C Toxicol Pharmacol. 2010;152:306–12.
99. Brown TM, Ross PS, Reimer KJ, Veldhoen N, Danger NJ, Fisk AT, et al. PCB related effects thresholds as derived through gene transcript profiles in locally contaminated ringed seals (*Pusa hispida*). Environ Sci Technol. 2014;48:12952–61.
100. Levin M, Gebhard E, Jasperse L, Desforges JP, Dietz R, Sonne C, et al. Immunomodulatory effects of exposure to polychlorinated biphenyls and perfluoroalkyl acids in East Greenland ringed seals (*Pusa hispida*). Environ Res. 2016;151:244–50.
101. Frouin H, Loseto LL, Stern G, Haulena M, Ross PS. Mercury toxicity in beluga whale lymphocytes: limited effects of selenium protection. Aquat Toxicol. 2012;109:185–93.

102. Parkinson AJ. The Arctic human health initiative: a legacy of the international polar year 2007–2009. Int J Circumpolar Health. 2013;72:10.

103. Dallaire F, Dewailly E, Vézina C, Ayotte P, Muckle G, Bruneau S. Effect of prenatal exposure to polychlorinated biphenyls on incidence of acute respiratory infections in preschool Inuit children. Environ Health Perspect. 2006;114:1301–5.

104. Dewailly É, Ayotte P, Bruneau S, Laliberté C, Gingras S, Belles-Isles M, et al. Susceptibility to infections and immune status in Inuit infants exposed to organochlorines. Environ Health Perspect. 2000;108:205–11.

105. Dallaire R, Dewailly E, Ayotte P, Forget-Dubois N, Jacobson SW, Jacobson JL, et al. Growth in Inuit children exposed to polychlorinated biphenyls and lead during fetal development and childhood. Environ Res. 2014;134:17–23.

106. Weisglas-Kuperus N. Neurodevelopmental, immunological and endocrinological indices of perinatal human exposure to PCBs and dioxins. Chemosphere. 1998;37:1845–53.

107. Grandjean P, Landrigan PJ. Neurobehavioural effects of developmental toxicity. Lancet. 2014;13:330–8.

108. Greaves AK, Letcher RJ, Sonne C, Dietz R, Born EW. Tissue-specific concentrations and patterns of perfluoroalkyl carboxylates and sulfonates in East Greenland polar bears. Environ Sci Technol. 2012;46:11575–83.

109. Greaves AK, Letcher RJ, Sonne C, Dietz R. Brain region distribution and patterns of bioaccumulative perfluoroalkyl carboxylates and sulfonates in East Greenland polar bears (*Ursus maritimus*). Environ Toxicol Chem. 2013;32:713–22.

110. Krey A, Kwan M, Chan HM. Mercury speciation in brain tissue of polar bears (*Ursus maritimus*) from the Canadian Arctic. Environ Res. 2012;114:24–30.

111. Pedersen KE, Basu N, Letcher RJ, Greaves AK, Sonne C, Dietz R, et al. Brain region-specific perfluoroalkylated sulfonate (PFSA) and carboxylic acid (PFCA) accumulation and neurochemical biomarker responses in East Greenland polar bears (*Ursus maritimus*). Environ Res. 2015;138:22–31.

112. Pedersen KE, Basu N, Letcher RJ, Sonne C, Dietz R, Styrishave B. Per- and polyfluoroalkyl substances (PFASs)—new endocrine disruptors in polar bears (*Ursus maritimus*)? Environ Int. 2016;138:22–31.

113. Fonnum F, Mariussen E. Mechanisms involved in the neurotoxic effects of environmental toxicants such as polychlorinated biphenyls and brominated flame retardants. J Neurochem. 2009;111:1327–47.

114. Mariussen E. Neurotoxic effects of perfluoroalkylated compounds: mechanisms of action and environmental relevance. Arch Toxicol. 2012;86:1349–67.

115. Bollati V, Baccarelli A. Environmental epigenetics. Heredity. 2010;105:105–12.

116. Crews D. Epigenetic modifications of brain and behavior: theory and practice. Horm Behav. 2011;59:393–8.

117. Mill J, Petronis A. Pre- and peri-natal environmental risks for attention-deficit hyperactivity disorder (ADHD): the potential role of epigenetic processes in mediating susceptibility. J Child Psychol Psychiatry. 2008;49:1020–30.

118. Basu N, Scheuhammer AM, Sonne C, Dietz R, Letcher RJ. Is mercury in the environment of neurotoxic concern to polar bears? Environ Toxicol Chem. 2009;28:133–40.

119. Pilsner JR, Lazarus AL, Nam D, Letcher RJ, Scheuhammer T, Sonne C, et al. Mercury-associated DNA hypomethylation in polar bear brains via the LUminometric Methylation Assay (LUMA): a sensitive method to study epigenetics in wildlife. Mol Ecol. 2010;19:307–14.

120. Siegel GJ, Albers RW, Brady ST, Price DL. Basic neurochemistry. Molecular, cellular and medical aspects. 7th ed. Burlington: Elsevier Academic; 2006.

121. Ahmed EI, Zehr JL, Schulz KM, Lorenz BH, Don Carlos LL, Sisk CL. Pubertal hormones modulate the addition of new cells to sexually dimorphic brain regions. Nat Neurosci. 2008;11:995–7.

122. Zoeller RT, Crofton KM. Mode of action: developmental thyroid hormone insufficiency—neurological abnormalities resulting from exposure to propylthiouracil. Crit Rev Toxicol. 2005;35:771–81.

123. Bytingsvik J, Simon E, Leonards PEG, Lamoree M, Lie E, Aars J, et al. Transthyretin-binding activity of contaminants in blood from polar bear (*Ursus maritimus*) cubs. Environ Sci Technol. 2013;47:4778–86.

124. Simon E, van Velzen M, Brandsma SH, Lie E, Løken K, de Boer J, et al. Effect-directed analysis to explore the polar bear exposome: identification of thyroid hormone disrupting compounds in plasma. Environ Sci Technol. 2013;47:8902–12.
125. Rodier PM. Developing brain as a target of toxicity. Environ Health Perspect. 1995;103:73–6.
126. McKinney MA, Pedro S, Dietz R, Sonne C, Fisk AT, Roy D, et al. A review of ecological impacts of global climate change on persistent organic pollutant and mercury pathways and exposures in arctic marine ecosystems. Curr Zool. 2015;61:617–28.
127. Burek KA, Gulland FMD, O'Hara TM. Effects of climate change on Arctic marine mammal health. Ecol Appl. 2008;18:S126–34.
128. Harvell CD, Mitchell CE, Ward JR, Altizer S, Dobson AP, Ostfeld RS, et al. Climate warming and disease risks for terrestrial and marine biota. Science. 2002;296:2158–62.
129. Dudley JP, Hoberg EP, Jenkins EJ, Parkinson AJ. Climate change in the North American Arctic: a One Health perspective. EcoHealth. 2015;12:713–25.
130. Bradley M, Kutz SJ, Jenkins E, O'Hara TM. The potential impact of climate change on infectious diseases of Arctic fauna. Int J Circumpolar Health. 2005;64:468–77.
131. Greer A, Ng V, Fisman D. Climate change and infectious diseases in North America: the road ahead. CMAJ. 2008;178:715–22.
132. Jenkins EJ, Castrodale LJ, de Rosemond SJ, Dixon BR, Elmore SA, Gesy KM, et al. Tradition and transition: parasitic zoonoses of people and animals in Alaska, northern Canada, and Greenland. Adv Parasitol. 2013;82:33–204.
133. Derocher AE, Aars J, Amstrup SC, Cutting A, Lunn NJ, Molnár PK, et al. Rapid ecosystem change and polar bear conservation. Conserv Lett. 2013;6:368–75.
134. Durner GM, Douglas DC, Nielson RM, Amstrup SC, McDonald TL, Stirling I, et al. Predicting 21st-century polar bear habitat distribution from global climate models. Ecol Monogr. 2009;79:25–58.
135. Molnár PK, Derocher AE, Klanjscek T, Lewis MA. Predicting climate change impacts on polar bear litter size. Nat Commun. 2011;2:186.

136. Hamilton SG, de la Guardia LC, Derocher AE, Sahanatien V, Tremblay B, Huard D. Projected polar bear sea ice habitat in the Canadian Arctic Archipelago. PLoS ONE. 2014;9:e113746.
137. Amstrup SC, Marcot BG, Douglas DC. A Bayesian network modeling approach to forecasting the 21st century worldwide status of polar bears. In: DeWeaver ET, Bitz CM, Tremblay L-B, editors. Arctic sea ice decline: observations, projections, mechanisms, and implications, vol. 180. Washington, DC: American Geophysical Union; 2018. p. 213–68.
138. Olsen GH, Mauritzen M, Derocher AE, Sørmo EG, Skaare JU, Wiig O, et al. Space-use strategy is an important determinant of PCB concentrations in female polar bears in the Barents Sea. Environ Sci Technol. 2003;37:4919–24.
139. van Beest FM, Aars J, Routti H, Lie E, Andersen M, Pavlova V, Sonne C, et al. Spatiotemporal variation in home range size of female polar bears and correlations with individual contaminant load. Polar Biol. 2016;39:1479–89.
140. McKinney MA, Peacock E, Letcher RJ. Sea ice-associated diet change increases the levels of chlorinated and brominated contaminants in polar bears. Environ Sci Technol. 2009;43:4334–9.
141. Gormezano LJ, Rockwell RF. What to eat now? Shifts in polar bear diet during the ice-free season in western Hudson Bay. Ecol Evol. 2013;3:3509–23.
142. Iles DT, Petersen SL, Gormezano LJ, Koons DN, Rockwell RF. Terrestrial predation by polar bears: not just a wild goose chase. Polar Biol. 2013;36:1373–9.
143. Iverson SA, Gilchrist HG, Smith PA, Gaston AJ, Forbes MR. Longer ice-free seasons increase the risk of nest depredation by polar bears for colonial breeding birds in the Canadian Arctic. Proc R Soc B. 2014;281:20133128.
144. MacKenzie BR, Payne MR, Boje J, Hoyer JL, Siegstad H. A cascade of warming impacts brings bluefin tuna to Greenland waters. Glob Chang Biol. 2014;20:2484–91.
145. Valdimarsson H, Astthorsson OS, Palsson J. Hydrographic variability in Icelandic waters during recent decades and related changes in distribution of some fish species. ICES J Mar Sci. 2012;69:816–25.

146. Carrie J, Wang F, Sanei H, Macdonald RW, Outridge PM, Stern GA. Increasing contaminant burdens in an arctic fish, burbot (*Lota lota*), in a warming climate. Environ Sci Technol. 2010;44:316–22.
147. Macdonald RW, Harner T, Fyfe J. Recent climate change in the Arctic and its impact on contaminant pathways and interpretation of temporal trend data. Sci Total Environ. 2005;342:5–86.
148. IOM. Global climate change and extreme weather events: understanding the contributions to infectious disease emergency. Washington, DC: The National Academic Press, Institute of Medicine; 2008. p. 280.
149. Parkinson AJ, Butler JC. Potential impacts of climate change on infectious diseases in the Arctic. Int J Circumpolar Health. 2005;64:478–86.
150. Tryland M, Nesbakken T, Robertson L, Grahek-Ogden D, Lunestad BT. Human pathogens in marine mammal meat—a northern perspective. Zoonoses Publ Health. 2013;61:377–94.
151. Jenssen BM. Marine pollution: the future challenge is to link human and wildlife studies. Environ Health Perspect. 2003;111:A198–9.
152. Jenssen BM. Endocrine-disrupting chemicals and climate change: a worst-case combination for Arctic marine mammals and seabirds? Environ Health Perspect. 2006;114:76–80.
153. Weihe P, Debes F, Halling J, Petersen MS, Muckle G, Odland JØ, et al. Health effects associated with measured levels of contaminants in the Arctic. Int J Circumpolar Health. 2016;75:33805.

CHAPTER 12

Exploring Semantic Deep Learning for Building Reliable and Reusable One Health Knowledge from PubMed Systematic Reviews and Veterinary Clinical Notes

Mercedes Arguello-Casteleiro[1], Robert Stevens[1] , Julio Des-Diz[2] , Chris Wroe[3] , Maria Jesus Fernandez-Prieto[4] , Nava Maroto[5] , Diego Maseda-Fernandez[6,7], George Demetriou[1] , Simon Peters[8] , Peter-John M. Noble[9] , Phil H. Jones[9] , Jo Dukes-McEwan[10], Alan D. Radford[9] , John Keane[1,11] and Goran Nenadic[1,11,12]

[1]School of Computer Science, University of Manchester, Manchester, UK

[2] Hospital do Salnés, Villagarcía de Arousa, Pontevedra, Spain

Citation: Arguello-Casteleiro, M., Stevens, R., Des-Diz, J. et al. Exploring semantic deep learning for building reliable and reusable one health knowledge from PubMed systematic reviews and veterinary clinical notes. J Biomed Semant 10, 22 (2019). https://doi.org/10.1186/s13326-019-0212-6.

[3] BMJ, Tavistock Square, London, UK

[4] Salford Languages, University of Salford, Salford, UK

[5] Departamento de Lingüística Aplicada a la Ciencia y a la Tecnología, Universidad Politécnica de Madrid, Madrid, Spain

[6] Midcheshire Hospital Foundation Trust, NHS England, Crewe, UK

[7] School of Medical Sciences, University of Manchester, Manchester, UK

[8] School of Social Sciences, University of Manchester, Manchester, UK

[9] Small Animal Veterinary Surveillance Network, University of Liverpool, Liverpool, UK

[10] Small Animal Teaching Hospital, University of Liverpool, Liverpool, UK

[11] Manchester Institute of Biotechnology, University of Manchester, Manchester, UK

[12] Health eResearch Centre, University of Manchester, Manchester, UK

ABSTRACT

Background

Deep Learning opens up opportunities for routinely scanning large bodies of biomedical literature and clinical narratives to represent the meaning of biomedical and clinical terms. However, the validation and integration of this knowledge on a scale requires cross checking with ground truths (i.e. evidence-based resources) that are unavailable in an actionable or computable form. In this paper we explore how to turn information about diagnoses, prognoses, therapies and other clinical concepts into computable knowledge using free-text data about human and animal health. We used a Semantic Deep Learning approach that combines the Semantic Web technologies and Deep Learning to acquire and validate knowledge about 11 well-known medical conditions mined from two sets of unstructured free-text data: 300 K PubMed Systematic Review articles (the PMSB dataset) and 2.5 M veterinary clinical notes (the VetCN dataset). For each target condition we obtained 20 related clinical concepts using two deep learning methods applied separately on the two datasets, resulting in 880 term pairs (target term, candidate term). Each concept, represented by an n-gram, is mapped to UMLS using MetaMap; we also developed a bespoke method for mapping short forms (e.g. abbreviations and acronyms). Existing ontologies were used to formally represent associations. We also create ontological modules and illustrate how the extracted knowledge can be queried. The evaluation was performed using the content within BMJ Best Practice.

Results

MetaMap achieves an F measure of 88% (precision 85%, recall 91%) when applied directly to the total of 613 unique candidate terms for the 880 term pairs. When the processing of short forms is included, MetaMap achieves an F measure of 94% (precision 92%, recall 96%). Validation of the term pairs with BMJ Best Practice yields precision between 98 and 99%.

Conclusions

The Semantic Deep Learning approach can transform neural embeddings built from unstructured free-text data into reliable and reusable One Health knowledge using ontologies and content from BMJ Best Practice.

BACKGROUND

One Health is an approach to achieve better public health outcomes by combining efforts from different disciplines and domains [1]. It entails the recognition that the health of animals and the environment are essential for human health, and – specifically – that human medicine can benefit from veterinary medicine, as animals develop many of the same diseases as humans do [2]. Zoonotic infections and anti-microbial resistance are examples that have received much of the attention recently. A recent study by Stroud et al. [3] compiled several examples of One Health cases where human medicine can benefit from veterinary studies. However, while both biomedical and clinical knowledge about human and animal health are growing, they remain isolated silos. This paper investigates to what extent it is possible to acquire One Health knowledge from the evidence-based biomedical literature and veterinary clinical narratives.

We focus on using Semantic Deep Learning (SemDeep) [4, 5] – an emerging area combining the Semantic Web resources and technologies and Deep Learning for this task. Deep Learning is an area of machine learning that applies neural networks to "*learn representations of data with multiple levels of abstraction*" [6]. The Semantic Web embodies standards and tools for publishing and processing meta-data, with ontologies at its core. This paper proposes the use of ontologies as a way of specifying the meaning of semantically related terms derived from neural language models obtained via Deep Learning. We explore how to turn information about diagnoses, prognoses, therapies, as well as other clinical/healthcare entities into computable knowledge "*integrating individual clinical expertise and best external evidence*" [7]. There are two main challenges to achieving actionable or computable knowledge about human diseases or syndromes:

Multiple evidence-based resources – such as BMJ Best Practice [8], DynaMed Plus [9] or UptoDate [10] – are consulted by clinicians on a daily basis to assist clinical decision making at point-of-care. Healthcare professionals need to provide effective and safe patient care, and this also implies complying with Clinical Practice Guidelines (CPGs) like those provided by the UK National Institute for Health and Care Excellence (NICE) [11]. CPGs tend to be evidence-based and have been defined as "*systematically developed statements to assist practitioners and patient decisions about appropriate health care for specific circumstances*" [12]. Typically, evidence-based resources, such as the NICE CPGs or BMJ Best Practice, consist of human-readable text that is updated periodically and is intended only for expert-to-expert communication. A fundamental obstacle for achieving interoperability between evidence-based resources is lack of "grounding" or "normalisation" [13], i.e. the fact that biomedical/clinical terms appearing in the NICE CPGs or BMJ Best Practice are not mapped to specific terminological entries like the Unified Medical Language System (UMLS) [14] Metathesaurus. Normalisation may help with periodic updates of evidence-based resources from the biomedical literature, as PubMed/MEDLINE articles are indexed with Medical Subject Headings (MeSH) [15], which is included in the UMLS Metathesaurus. Unfortunately, annotating biomedical articles with MeSH terms is difficult and expensive [16] and normalisation of evidence-based resources is not being carried out.

World-leading terminologies, such as the Systematized Nomenclature of Medicine Clinical Terms (SNOMED CT) [17], are included within the UMLS Metathesaurus and provide multiple terms for expressing a biomedical concept. However, SNOMED CT does not contain formal or semi-formal descriptions that help understand how a disease develops and which treatment approaches can be considered. For example, medication statements that state which drugs treat a disease or syndrome are not included in SNOMED CT. Hence, even automatic acquisition of a partial set of SNOMED CT concepts relevant for a disease or syndrome (e.g. medications/drugs and clinical findings) remains an unmet need.

This paper investigates whether a SemDeep approach can help address both challenges. On the one hand, the neural language models can be applied to PubMed Systematic Reviews [17], which is a large collection of evidence-based articles (e.g. clinical trials, systematic reviews, and CPGs) available in PubMed/MEDLINE. Vector representations of the terms ("word embeddings" or "neural embeddings") can be learnt from this unstructured text corpus, where semantically related terms will end up close in the representational space. On the other hand, if the terms that participate in the associations derived from these distributional similarities are normalised (i.e. terms with vector representations are mapped to UMLS Metathesaurus concepts and SNOMED CT concepts), that would allow: a) the expansion of the knowledge that exists in SNOMED CT; and b) automatic derivation of SNOMED CT subsets of semantically related concepts that can be reused and shared.

In this paper we focus on 11 well-known medical conditions that affect both humans and animals: heart failure, asthma, epilepsy, glaucoma, chronic kidney disease, osteoarthritis, anaemia, arthritis, diabetes mellitus, hypertension, and obesity. We adhere to "comparative/translational medicine" [18] and investigate the added value of One Health by combining: a) the Systematic Reviews Subset of PubMed, and b) a large set of veterinary clinical narratives collected by the Small Animal Veterinary Surveillance Network (SAVSNET) [19]. We rely on existing ontologies (lemon (Lexicon Model for Ontologies) [20] and OBAN (Open Biomedical AssociatioNs) [21]) to formally represent associations (semantically related term pairs) derived from neural embeddings. However, as Deep Learning algorithms have a "*black-box representation*" [22], their wider acceptance and adoption in the biomedical and clinical domain requires confidence and trust. To build such confidence (through transparency and interpretability), we use evidence-based resources (namely BMJ Best Practice) to verify if the associations (the semantic relatedness) captured by neural embeddings are reliable for human medicine. We note that clinicians consult multiple evidence-based resources, so BMJ Best Practice cannot be taken as the only gold standard that provides ground-truth. However, in practical terms, when there is a lack of external evidence from systematic research about the meaningful association of two terms (e.g. a medical condition and a treatment), the term pair should be considered as unrelated.

Related Work

Neural embeddings learnt from the biomedical literature or clinical narrative corpora have been widely used from many tasks, but they pose a challenge when measuring their quality. On the one hand, the biomedical/ clinical domain requires background knowledge that makes crowd-worker evaluation (e.g. users of Amazon Mechanical Turk) unsuitable. On the other hand, there are no similarity and relatedness benchmarks developed for well-known medical conditions per se. Currently, there are four main standard benchmarks that are specific to the medical/clinical domain and suitable for a semantic similarity and relatedness task: Caviedes and Cimino [23] with 10 medical term pairs; Pedersen et al. [24] with 30 medical term pairs; Pakhomov et al. [25] with 101 clinical term pairs; and Pakhomov et al. [26] with 724 medical term pairs (the last two available at [27]). In total, these standard benchmarks provide less than 1 K term pairs. It should be noted that similarity and relatedness benchmarks were used to evaluate traditional distributional semantic models [28] – e.g. Latent Semantic Analysis (LSA) [29] or Latent Dirichlet Allocation (LDA) [30]. Faruqui et al. [31] emphasise that the lack of standard evaluation methods for neural embeddings was the trigger to create new benchmark datasets (e.g. Simlex-999 [32] and SimVerb-3500 [33] that are outside of the biomedical/clinical domain) and highlight that "*the use of word similarity tasks for evaluation of word vectors is not sustainable and calls for further research on evaluation methods*".

A common characteristic of biomedical/clinical documents is that "*longer words and phrases are frequently mapped onto a shorter form such as abbreviations or acronyms for efficiency of communication*" [34]. For example, "heart failure" (long form) can appear as "HF" (short form). Another issue is that "*the number of abbreviations and the average number of definitions per abbreviation*" is ever growing [34]. For example, "HF" (short form) can have multiple meanings, and therefore, refer to multiple senses besides "heart failure", such as "Hispanic female" or "high-fat" or "Hartree-Fock" or "hemofiltration". Although short forms (abbreviations and acronyms) are present in UMLS, several studies [35,36,37] have shown UMLS to have shortcomings when mapping short forms to long forms. Sense inventories have been created such as SaRAD [38], ADAM [39], and more recently Allie [40] – in May 2018, Allie contained 840 K of short forms. These sense inventories and their algorithms assume that the short and long form co-occur in the biomedical literature, such as MEDLINE/PubMed abstracts; however, they rarely co-occur in clinical narratives [41, 42]. Sense inventories from clinical documents are fewer in number than sense inventories from the biomedical literature and contain fewer short forms. Wu et al. [43] highlight that "*accurate identification of clinical abbreviations is a challenging task and advanced abbreviation recognition modules are needed for existing clinical NLP systems*". Dealing with short forms is therefore a challenge that requires an approach to deal with terms appearing in both biomedical and clinical documents.

The paper is organised as follows. In the next section, we present a SemDeep approach that builds on our previous work [44,45,46,47]. We first introduce the datasets, and then present the approach that deals with two NLP tasks: a semantic similarity and relatedness task, and a named-entity recognition (NER) task. As part of the SemDeep pipeline, we show how to reuse existing ontologies to formally represent the associations derived from neural embeddings in OWL. We also create ontological modules with the SNOMED CT ontology and illustrate how to query the extracted knowledge using the SPARQL query language [48], which exploit the underlying ontological representations and can be executed using Jena ARQ [49].

Materials and Methods

Data

The PubMed Systematic Reviews dataset (the PMSB dataset): we downloaded the MEDLINE/PubMed baseline files for 2015 and up-to-date files through 8th June 2016. Applying the PubMed Systematic Reviews filter [17], a subset of 301,201 PubMed publications published between 2000 and 2016 was obtained. We extracted titles and available abstracts. We note that this dataset was also used in our previous study on sepsis [46].

The SAVSNET dataset (the VetCN dataset): a collection of 2,465,420 de-identified and non-empty veterinary clinical narratives was obtained from SAVSNET on 20th October 2017. SAVSNET is an initiative by the British Small Animal Veterinary Association and the University of Liverpool that collects free-text consultations notes across around 500 UK veterinary premises in real-time.

We used the UMLS Metathesaurus with close to 3.7M biomedical/clinical concepts with 203 sources contributing to concept names (as of May 2018), including SNOMED CT and the Veterinary Extension for SNOMED CT (VetSCT) [50]. We used MetaMap 2016v2 (released September 2016) and the SNOMED CT version released in January 2017 in OWL to maximise compatibility among versions. When using the UMLS Terminology Services [14] to access the UMLS Metathesaurus online, we select the UMLS2016AB version. As we cannot release a subset of the UMLS Metathesaurus or SNOMED CT, to replicate the results obtained with our method it is necessary to use the UMLS API [51] and the OWL API [52].

SemDeep Pipeline

Below we describe each step of the pipeline intended to acquire and validate knowledge about medical conditions from the unstructured text datasets.

Step 1: computing n-grams from unstructured text

We employed word2phrase within the word2vec software package [53] to compute n-grams from the unstructured text datasets (e.g. PMSB or VetCN). It should be noted that text within consultation notes (e.g. "today pain++") contains a significant number of misspellings, local abbreviations and short forms, and lacks the grammatical correctness found within biomedical literature. After applying word2phrase, the character "_" appears within tokens that co-occur repeatedly together (e.g. "heart_failure" is a bigram with two tokens). Each n-gram has a frequency count. We note that an n-gram captures words or tokens that appear together in a textual corpus with a certain frequency. Therefore, an n-gram can capture words and phrases as well as combinations of tokens that may not correspond exactly to meaningful unit when looked in isolation phrase, e.g. "(COPD)_is_a".

Step 2: creation of neural embeddings

We create neural embeddings with CBOW (Continuous Bag-of-Words) and Skip-gram [54] using the word2vec software. The input for CBOW and Skip-gram is the unstructured text with n-grams, where the character "_" typically denotes the presence of an n-gram of size greater than one (e.g. a bigram or a trigram). The output for CBOW and Skip-gram is typically: 1) a lexicon (i.e. a list of n-grams) that is present in the unstructured text and for which the vector representations have been learnt; and 2) a binary file that contains neural embeddings, i.e. real-number representations for the terms in the lexicon. When producing neural embeddings, there are a small number of hyperparameters that need to be tuned – we used the hyperparameters configuration described in our previous work [55].

Step 3: obtaining term pairs for the semantic similarity and relatedness task

Taking the vector for a specific target term (e.g. a disease or syndrome) and applying the cosine similarity, a list of top ranked terms (highest cosine values) can be obtained from the created neural embeddings. These top ranked terms are candidate terms, i.e. terms that need to be judged as semantically similar or related to the target term. Some authors agree that "*semantic similarity represents a special case of semantic relatedness*" [24]; Hill et al. [32] interpret "relatedness" as "association", where the strongest similarity relation is synonymy; this interpretation is applied in this study.

Target terms selection: to choose the target terms, we first select the n-grams that pass the threshold of 1K frequency count and MetaMap has assigned them UMLS Metathesaurus concepts with the Semantic Type "T047|Disease or Syndrome". The final selection of target terms needs to be done manually as: a) MetaMap may erroneously assign a UMLS Metathesaurus concept to an n-gram, particularly if the n-gram is a short form such as "HF"; and b) the well-known medical conditions selected as target terms should be covered by BMJ Best Practice – the preferred gold standard in this study.

Number of top ranked candidate terms per target term: to the best of our knowledge, no published study justifies the number of terms selected with the highest cosine value. Different studies used different numbers: from three [56], ten [57, 58], 40 [59] to a range of numbers (e.g. 5, 10, 20, 40, and 100) [60]. We limit the list of candidate terms to the 20 n-grams with the highest cosine value.

Step 4: named entity recognition

Named Entity Recognition (NER) consists of "*identifying specific words or phrases ('entities') and categorizing them*" [61]. We use MetaMap to categorise the n-grams into one or more of 133 broad categories (Semantic Types) from the UMLS Metathesaurus. To determine if MetaMap supplies a correct CUI for an n-gram, detailed guidelines were developed that intend to favour compatibility with the SNOMED CT Compositional Grammar [62]:

- Single Map (SM) – MetaMap provides a single CUI that captures the full meaning of the n-gram. This case corresponds to "Simple Expression" [62], i.e. a single concept identifier. For example, the n-gram "septic_shock" is mapped only to UMLS CUI=C0036983.
- Multiple Maps (MM) – MetaMap provides multiple CUIs that capture the meaning of the n-gram, and one or more focus concepts may appear among the CUIs provided. This case may also correspond to "Simple Expression" [62], although it more often corresponds to "Expression with Refinements" or "Multiple

Focus Concepts" [62]. Selection of focus concept(s) is guided by six principles described in the Additional file 2.

- Incorrectly Mapped (IM) – MetaMap provides one or more CUIs, however, none captures the meaning of the n-gram. For example, the n-gram "HF" is not mapped to "C0018801|Failure, Heart (Heart failure)".
- Not Mapped (NM) – MetaMap does not provide any CUI. For example, the n-gram "HFpEF" is not mapped to "C3889077|Heart failure with preserved ejection fraction".

Three domain experts (two biomedical terminologists and a medical consultant) inspected the results of MetaMap considering the above-mentioned guidelines and assigned to each candidate term (n-gram) one of the above values {SM, MM, IM, NM}. Beside the candidate term (n-gram), we also provide the target terms (n-grams) in lower case to provide local context. When an n-gram is incorrectly mapped or not mapped, a CUI for the n-gram is manually assigned. On the one hand, the guidelines presented aim to reduce the number of CUIs assigned to each n-gram. On the other hand, the n-grams are the result of statistical NLP and may contain short forms, and therefore, decomposing the n-grams into lexico-semantic units of meaning has proved demanding in our previous work [45,46,47] (even with years of experience doing clinical coding). Hence, the three domain experts worked together to identify the minimal semantic constituents of the n-grams according to the guidelines, i.e. determining systematically the focus concept(s).

Dealing with Short Forms

We created a short form detector to identify n-grams with or without one or more clinically meaningful short forms. The detector is based on a hybrid approach: it contains if-then-else heuristic rules and utilises two lists of terms, i.e. a list of measurement units compiled from three resources [63,64,65] and the rank frequency list of the British National Corpus of written and spoken English [66] with 7726 words. The underlying assumption is that a measurement unit can have one or more short forms. For example, "mmHg" is a short form for the long form "millimetre of mercury".

Figure 1 depicts a flowchart outlining how the short form detector that assigns one of the following labels to an n-gram:

- "SF-U" when an n-gram contains a unit of measurement. The n-gram is mapped to the UMLS Metathesaurus concept "C1519795|Unit of Measure".

- "SF-NU" when an n-gram contains a number with a unit of measurement. The n-gram is mapped to the UMLS Metathesaurus concept "C0242485|Measurement".
- "SF" when an n-gram contains a short form token that is not a measurement unit or a measurement unit and a number.
- No label when an n-gram does not contain a short form.

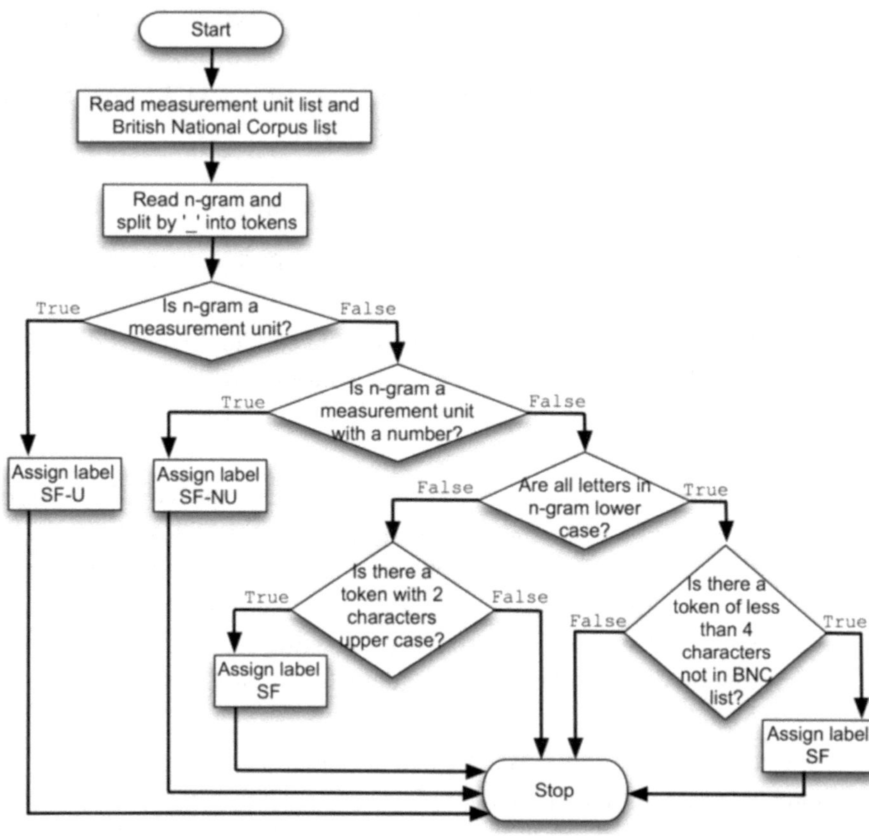

Figure 1: *Flowchart of the short form detector introduced* – a diagrammatic representation outlining how the short form detector assigns the labels {SF-U, SF-NU, SF}. If no label is assigned, this means that the n-gram has no clinically meaningful short form(s).

For those n-grams with a short form that is not a measurement unit or a measurement unit and a number, the domain experts manually utilised Allie as the preferred sense inventory, for expanding short forms into long forms. The reasons for using Allie are: a) it contains a much larger number of short forms than the UMLS SPECIALIST Lexicon; b) it has long forms for a short form ranked based on appearance frequency in PubMed/MEDLINE abstracts; and c) for each long form the research area and co-occurring abbreviations are provided, thus aiding disambiguation.

The short form detector can make two errors, and the domain experts will assign the following labels to an n-gram:

- "SF-I" denotes that a short form identified in an n-gram was assessed as not clinically meaningful, i.e. "incorrect".
- "SF-NF" denotes that a clinically meaningful short form was not identified in an n-gram, i.e. "not found".

Experiment Set-up and Performance Measures

We investigate the impact of the short forms on the performance of MetaMap via two experiments:

- Experiment 1 (EXP1): we expose the candidate terms directly to MetaMap, i.e. the lists with the 20 top-ranked n-grams (i.e. the 20 n-grams with the highest cosine value) are taken as input for MetaMap.
- Experiment 2 (EXP2): we expose firstly the candidate terms (n-grams) to the short form detector described above, and for those n-grams with one or more short forms, we expand each short form into the corresponding long form by utilising Allie. Once the short forms are replaced with the long forms, we take the modified/expanded candidate terms (n-grams) as input for MetaMap.

For both experiments, we assume that the candidate terms are biomedical or clinical terms, and thus, there should be no True Negatives (TNs). We use the conventional evaluation measures of Precision, Recall, and F measure [67] to calculate the MetaMap performance. As Pratt and Yetisgen-Yildiz [68], we have weaker MetaMap precision and recall where exact matches (typically Single Map) and partial matches (typically Multiple Maps) are equally counted, i.e. SM and MM are considered as True Positives (TPs). An Incorrectly Mapped (IM) is interpreted as False Positive (FP). A Not Mapped (NM) is interpreted as a False Negative (FN). Hence, we calculate precision as TP/(TP+FP) [67] and recall as TP/(TP+FN) [67]. To calculate F measure, we use equal weighting of precision and recall, calculating F measure as (2 x Precision x Recall)/(Recall + Precision) [67]. We compute precision, recall, and F measure for each well-known medical condition under study (i.e. target term), and then, average each evaluation measure over all to obtain an overall measure of performance (a.k.a. macro-averaging) [67]. We also report micro-averaging, i.e. making a single contingency table for all data [67] – all lists of the 20 n-grams with highest cosine value that are the input to MetaMap.

Following Smucker et al. [69], we measure statistical significance of the difference in the mean average precision, recall, and F measure to judge if there is a statistically significant improvement in performance for EXP2 (short form detection and expansion into long form before applying MetaMap to the unique candidate terms) when compared with the performance for EXP1 (applying MetaMap to the unique candidate terms). We use the Student's paired t-test [70] as implemented in scikit-learn [71] to compare performance of EXP1 and EXP2.

All unique candidate terms (n-grams) are exposed to the short form detector. We also compute precision, recall, and F measure for the short form detector considering all candidate terms (micro-averaging) and the capability of the short form detector to identify n-grams with or without one or more clinically meaningful short forms.

Step 5: validation of the term pairs mapped to UMLS Metathesaurus concept pairs using BMJ best practice

This step validates the candidate terms obtained in Step 3 for the semantic similarity and relatedness task using BMJ Best Practice, which covers prevention, diagnosis, treatment and prognosis for well-known medical conditions. It contains "top down" knowledge manually extracted by medical experts and is based both on the latest clinical practice guidelines and underlying research evidence. Hence, BMJ Best Practice can be considered an evidence-based gold standard, supporting frontline clinicians.

The validation involves four domain experts. The three experts from the previous step (two terminologists and a medical consultant) together validate the concept pairs considering BMJ Best Practice and external resources (whenever necessary). A health informatician – who works with the content of BMJ Best Practice – contributes to the final stages of validation of the concept pairs and (additionally) provides feedback to the editors of BMJ Best Practice. To avoid bias, the domain experts validate the concept pairs without knowing: a) the dataset from which the concept pairs were derived; and b) the neural language models applied to the dataset. To avoid further hints about the underlying dataset, the target terms (n-grams) are presented in lower case.

The outcome of this step is a set of 3-tuples (target concept, candidate concept, validation label). The target and candidate concepts are UMLS Metathesaurus concepts representing the focus concept(s) of the n-grams for the term pairs (target term, candidate term). The validation label indicates how the matching between a candidate concept and a term from BMJ Best Practice is performed and reflects the amount of domain knowledge required to perform such a match. We distinguish six cases when matching a candidate concept name from the UMLS Metathesaurus to a term appearing in BMJ Best Practice:

- *Itself* – a candidate term may have as its focus concept(s) the same UMLS Metathesaurus concept as the target term (i.e. the well-known medical condition), and therefore, they will not be matched to terms appearing in BMJ Best Practice. This case denotes synonymy (i.e. the strongest similarity relation).
- *Relatedness by exact/approximate match* – candidate concept names as they appear in the UMLS Metathesaurus may match exactly or approximately terms appearing in BMJ Best Practice, where the biomedical/clinical meaning is the same. This typically denotes a similarity relation between the candidate concept and the BMJ term. This case can be interpreted as "normalising" BMJ Best Practice terms. For example, "pyrexic" (adjective) when used to refer to patients with "pyrexia" (noun) can be interpreted as "pyrexia". From a linguistic point of view, "pyrexic" is a morphological variant of "pyrexia", and thus, this is an example of an approximate match where the biomedical/clinical meaning is the same. Some "implicit knowledge" may be needed for this case.
- *Relatedness by inexact match (hypernym/hyponym) or (hyponym/hypernym)* – candidate concept names as they appear in the UMLS Metathesaurus may have is-a relations with the terms appearing in BMJ Best Practice or vice versa. An is-a relation is also similar to a hypernym/hyponym relation or generalisation/specialisation relation and it denotes a similarity relationship. For example, "bacterial sepsis" is-a type of "sepsis", where "bacterial sepsis" is the hyponym (specialisation) and "sepsis" is the hypernym (generalisation). Hypernym/hyponym relations can be used to build semantic taxonomies (a.k.a. hierarchies). Some "implicit knowledge" may be needed for this case.

- *Relatedness by inexact match (background knowledge)* – candidate concept names as they appear in the UMLS Metathesaurus may have similarity (similar meaning or is-a relations) or relatedness (association) relations with the terms appearing in BMJ Best Practice, although the relations are not obvious for someone lacking biomedical/clinical knowledge. To make the "implicit knowledge" explicit, one or more excerpts of external evidence from systematic research or terminological resources are provided. This case may apply transitivity, i.e. "if A is related to B and B is related to C, then A is related to C". Therefore, by making known one or more terms (call them B) it is feasible to make transparent how a UMLS Metathesaurus concept A related to a term C appearing in BMJ Best Practice. In other words, by making B known, and how A and C relate to B, implicit knowledge becomes "explicit".
- *Unrelated: not clinically meaningful* – An n-gram can capture combinations of words or tokens that can be mapped to a focus UMLS Metathesaurus concept(s), although it may not be interpreted per se as clinically meaningful in connection with a given medical condition. For example, "guided" is a unigram for which alternative UMLS Metathesaurus concepts are available to represent multiple meanings or senses: "C0181090|Guide (Professional guide)"; "C0302614|Guide (Guide device)"; and "C1706050|Guide (Guide Device Component)". However, for "sepsis", any sense for the n-gram "guided" per se is not clinically meaningful.
- *Unrelated: excluded* – candidate concept names as they appear in the UMLS Metathesaurus may not have up-to-date clinically meaningful association (relatedness) relations with the terms appearing in BMJ Best Practice, even if there is implicit knowledge that can justify the association. A typical example of this case are treatments or therapies that became known as ineffective or have adverse effects. For example: "*The only novel anti-sepsis agent to successfully complete a phase 3 sepsis trial, human recombinant activated protein C, was recently taken off the market after a follow up placebo-controlled trial (PROWESS SHOCK) failed to replicate the results of the initial registration trial (PROWESS) performed 10 yr earlier.*" [72]

The process of assigning the six validation labels introduced above is iterative. A focus concept for a candidate term and a term from BMJ Best Practice may have been allocated the "Relatedness by Inexact match (background knowledge)" label to indicate that one or more excerpts of external evidence from systematic research or terminological resources are needed to establish "relatedness". However, a closer inspection of the "implicit knowledge" that has become explicit may change the label into: "Relatedness by exact/approximate match", or "Relatedness by Inexact match (hypernym/hyponym) or (hyponym/hypernym)", or "Unrelated: Excluded". The cases under the "Relatedness by Inexact match (background knowledge)" label may be further refined, for example, by considering the relationship between cause and effect.

The six validation labels introduced can be used for calculating precision by considering: a) the last two "unrelated" labels representing False Positives (FPs); and b) the "itself" and the labels starting with "Relatedness by" representing True Positives (TPs).

Step 6: formal representation of the knowledge acquired and validated

We use OWL-DL to formally represent concept names, concept expressions, and terminological axioms. Figures 2 and 3 overview the two core ontologies that will be populated, i.e. the extended lemon core ontology (called here the lemonEXT) [73] and the modified OBAN core ontology (called here the OBANmod) [74]. Both core ontologies reused the USTG (UMLS Semantic Types and Groups) core ontology in OWL-DL that we created programmatically and utilised in [45, 46]. The USTG core ontology represents in OWL the information publically available at [75].

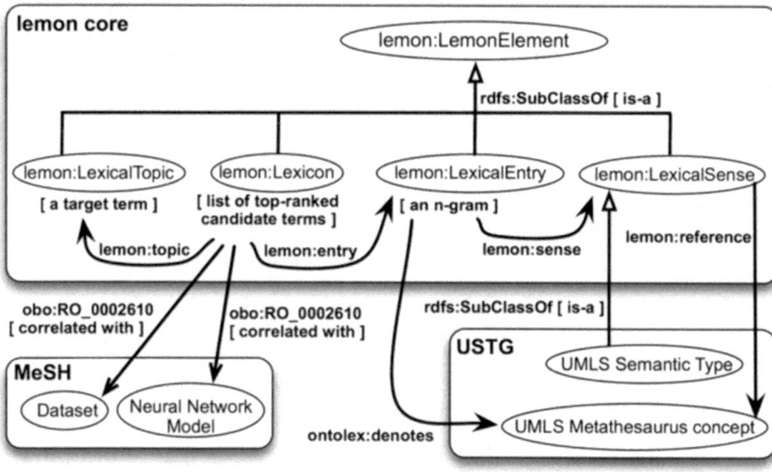

Figure 2: Overview of the extended version of the lemon core ontology (called here the lemonEXT) used for this study.

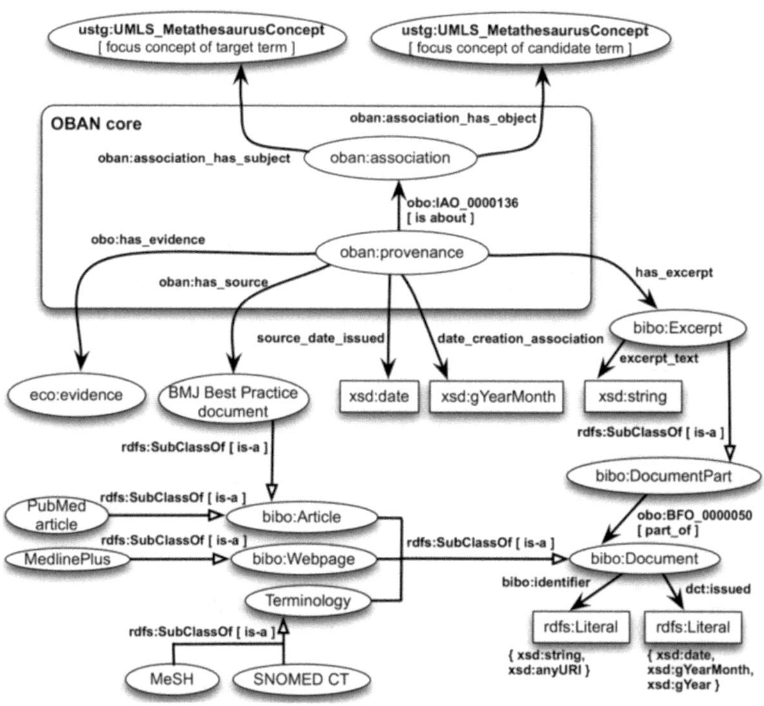

Figure 3: Overview of the modified version of the OBAN core ontology (called here the OBANmod) used for this study.

The USTG core ontology represents formally the UMLS Semantic Types and Groups as well as the part-whole relations among them by reusing the OWL object properties "part_of" (obo:BFO_0000050) and "has_part" (obo:BFO_0000051) from the Basic Formal Ontology (BFO) [76]. The USTG core ontology also contains the UMLS Metathesaurus concept, an OWL class we created that can have as a subclass any Metathesaurus concept from the UMLS. A new addition to the USTG core ontology is the OWL annotation property "hasDbXrefInSCT" to create annotation assertion axioms that act as cross-reference between the UMLS Metathesaurus and SNOMED CT. The annotation property "hasDbXrefInSCT" is a sub-annotation property of the annotation property "database_cross_reference" (oboInOwl:hasDbXref) from the oboInOwl meta-model [77]. The USTG core ontology has a total of 593 axioms (class count: 151; individual count: 0) and its Description Logic (DL) expressivity is ALEI.

Table 1 shows the axiom patterns in Manchester OWL Syntax [78] for populating programmatically the main OWL Classes of lemonEXT and OBANmod core ontologies. In this study, a pattern (a.k.a. axiom pattern) can represent a set of OWL axioms.

Table 1: *The axiom patterns in Manchester OWL Syntax for populating the main classes of the two core ontologies:* **the extended lemon core ontology** (lemonEXT); and the modified OBAN core ontology (OBAMmod). Each axiom pattern in the second column contains variables, which can be easily identified as they start with the character "?". The third column exemplifies OWL individuals that are the result of populating the axiom patterns introduced in the second column

Brief description of axiom pattern	**Axiom pattern in Manchester OWL Syntax**	**Example of populating the Axiom pattern in Manchester OWL Syntax**
Individual of the OWL class lemon:LexicalTopic	Individual:? x Annotations: label? lbx@en Types: 'Lexical topic'	Individual: heart_failure Annotations: label "heart_failure"@en Types: 'Lexical topic'
Individual of the OWL class lemon:LexicalEntry	Individual:? y Annotations: label? lby@en Types: 'Lexical entry' Facts: denotes? CUI	Individual: beta-blockers Annotations: label "beta-blockers"@ en Types: 'Lexical entry' Facts: denotes 'Adrenergic beta-Antagonists'

Individual of the OWL class lemon:Lexicon	Individual:? z Annotations: label? lbz@en Types: Lexicon Facts: 'correlated with'? dataset, 'correlated with'? model, entry? y1, … … … entry? y20	Individual: PMSB_CBOW_heart_failure Annotations: label "PMSB_CBOW_heart_failure"@en Types: Lexicon Facts: 'correlated with' 'PMSB dataset', 'correlated with' CBOW, entry beta-blockers, … … … entry 'cardiac_resynchronization_therapy_(CRT)'
Individual of the OWL class oban:association	Individual:? Cpair Annotations: rdfs:label? lbCpair@en Types: association Facts: oban:association_has_object? CUI1, oban:association_has_subject? CUI2	Individual: '(C0017601,C0020581)' Annotations: label "(C0017601,C0020581)"@en Types: association Facts: 'association has object' Hyphema, 'association has subject' Glaucoma
Individual of the OWL class oban: provenance	Individual:? ECpair Annotations: rdfs:label? lbECpair@en Types: provenance Facts: 'has source'? BMJdoc, 'has excerpt'? BMJterm, 'has evidence'? lbEvidence, 'has excerpt'? EEexcerpt, 'is about'? Cpair, 'date creation association'? d1^^string, 'source date issued'? d2^^string	Individual: '(C0017601,C0020581,Relatedness by inexact match (background knowledge))' Annotations: label "(C0017601,C0020581,Relatedness by inexact match (background knowledge))"@en Types: provenance Facts: 'has source' 'BMJ Best Practice: Open-angle glaucoma', 'has excerpt' trabeculotomy, 'has evidence' 'Relatedness by inexact match (background knowledge)', 'has excerpt' EEexcerpt_70, 'is about' '(C0017601,C0020581)', 'date creation association' "May-2018"^^string, 'source date issued' "09-Dec-2016"^^string

The Extended Lemon Core Ontology (lemonEXT)

In Fig. 2, the concept "Lexical Topic" represents the target term (i.e. an n-gram corresponding to a given medical condition); the concept "Lexicon" represents the lexicon, i.e. a list of the 20 top-ranked terms (n-grams with the highest cosine value for the target term) obtained with CBOW and Skip-gram; and the concept "Lexical Entry" represents an n-gram in the lexicon. We reuse two concepts from MeSH: "D064886|Data Set" and "D016571|Neural Networks (Computer)". The latter is a MeSH heading that has the entry term "Neural Network Models", which is the term that appears in Fig. 2. The two MeSH concepts are represented as OWL classes and they are connected with the lemon concept "Lexicon" by reusing the OWL object property "correlated with" (obo:RO_0002610) from the Relations Ontology (RO) [79] .

As depicted in Fig. 2, the concept "Lexical entry" from lemon is connected to the UMLS Metathesaurus concept from the USTG core ontology with the OWL object property "denotes" from the Ontology Lexicon (Ontolex) ontology [80]. We made the OWL Class "Lexical sense" from lemon a superclass of the OWL Class UMLS Semantic Type from the USTG core ontology. As one or more UMLS Metathesaurus concept(s) is the focus concept(s) for each n-gram, one or more UMLS Metathesaurus concept(s) from the UTSG ontology captures the senses or meanings of a lexical entry. Taking into account the UMLS Semantic Type(s) assigned to each UMLS Metathesaurus concept, it is possible to categorised the n-grams based on the OWL class descriptions within the USTG ontology – e.g. "C0036983|Septic Shock" is a subclass of "T046|Pathologic Function" – and therefore, we follow the "semantics by reference" principle from [81] that says: "*the expressivity and the granularity at which the meaning of words can be expressed depend on the meaning distinctions made in the ontology*".

The lemonEXT core ontology has a total of 643 axioms (class count: 158; individual count: 0) and its DL expressivity is ALEI. Once the lemonEXT is populated, it is possible to create SPARQL SELECT queries retrieving for each target term (i.e. n-gram) the candidate concepts, i.e. the UMLS Metathesaurus concepts that are the focus concepts of candidate terms (n-grams). We built three queries (see the Additional file 4 for details) that retrieve candidate concepts based on few UMLS Semantic Types and a UMLS Semantic Group. The UMLS Semantic Types and Group chosen intend to bring forward candidate concepts related to the diagnoses and management of a given medical condition (i.e. the target term):

- The UMLS Semantic Types "T059|Laboratory Procedure"; "T060|Diagnostic Procedure"; and "T061|Therapeutic or Preventive Procedure". These are the three subtypes of the UMLS Semantic Type "T058|Health Care Activity" [82]. We call the query q1 and it utilises these four UMLS Semantic Types.

- The UMLS Semantic Types "T034|Laboratory or Test Result"; and "T184|Sign or Symptom". These are the two subtypes of the UMLS Semantic Type "T033|Finding" [82]. We call the query q2 and it utilises these three UMLS Semantic Types.
- The UMLS Semantic Group "Chemicals & Drugs" (a.k.a. CHEM) that contains UMLS Semantic Types that are chemicals taking into account their structural and functional perspective [82]. Some of the UMLS Metathesaurus concepts belonging to CHEM are typically drug treatments (medications). We call the query q3 and it utilises the Semantic Group CHEM.

The Modified Version of the OBAN Core Ontology (OBANmod)

An OBAN association relates biomedical entities (e.g. X, Y) "without enforcing directionality on the link" [21] (i.e. "X is associated with Y" or "Y is associated with X") and separating the association between entities from its provenance [21]. Although we can safely state that there is a clinically meaningful association for the term pair (heart failure, pulmonary edema) it does not mean that all patients with "heart failure" will also have "pulmonary edema". Instead, we should interpret the term pair (heart failure, pulmonary edema) as a "sometimes true" association relationship [21] that can be represented in OWL as an OBAN association.

Figure 3 presents a modified version of the OBAN core ontology for this study considering:

- The OBAN association is between two UMLS Metathesaurus concepts represented as OWL Classes as in [21], and we also use punning. It should be noted that OWL 2 DL relaxes the separation between classes and individuals.
- The validations labels introduced in Step 5 are represented as OWL Classes that have as super-classes OWL Classes from the Evidence and Conclusion Ontology (ECO) [83] (release of 2018-04-06). A total of five OWL Classes from ECO have been reused as well as the subclass axioms for them in the ECO.
- The provenance that validates the "sometimes true" association relationship is BMJ Best Practice. To represent a document from BMJ Best Practice, we first reuse the OWL Class "Document" from the Bibliographic Ontology Specification ontology (BIBO) [84], and then, create a new subclass with the name "BMJ Best

Practice document". Therefore, "BMJ Best Practice for chronic congestive heart failure" [85] will be an OWL instance of the OWL Class "BMJ Best Practice document".

- Excerpts from biomedical or clinical resources are needed for this study, such as: 1) the term(s) appearing in BMJ Practice that are key when validating the concept pairs; or 2) additional information that make the "implicit knowledge" explicit, such as excerpts from the scientific literature (e.g. PubMed articles or MedlinePlus [86] Webpages) or from terminologies like SNOMED CT [62]. To represent an excerpt, we reuse the OWL Class "Excerpt" from the BIBO. In the BIBO, an "Excerpt" is part of a "Document". We create the annotation property "excerpt_text" to store terms from BMJ Best Practice or lines of text that are considered pertinent when making the "implicit knowledge" explicit. Therefore, in this study, every OBAN association will have at least one instance of the OWL Class "Excerpt", i.e. BMJ term(s) that are key to validate the focus concept pairs for the term pairs.
- The replacement and modification of OWL data properties from OBAN to better fit the current study: a) the OWL data property "date_creation_association" where we store the month and year when the association was validated; b) the OWL data property "source_date_issued" where we store the last update of BMJ Best Practice document used to validate the association; and c) relax the xsd:dateTime declarations from OBAN as sometimes it proves difficult to trace the publication date for a PubMed paper or the release day for a terminological resource.

The OBANmod core ontology has a total of 779 axioms (class count: 181; individual count: 6) and its DL expressivity is ALEHI (D). Once the OBANmod is populated, it is possible to refine the SPARQL queries q1 to q3 into q1V to q3V (see the Additional file 4 for details), where the candidate concepts (i.e. UMLS Metathesaurus concepts that are the focus concepts for the candidate terms) should have an up-to-date clinically meaningful association to the target concepts (the selected diseases or syndromes) according to BMJ Best Practice (i.e. human medicine), and thus, the UMLS Metathesaurus concept pairs should have the validation labels introduced in Step 5 starting with "Relatedness by" or the validation label "Itself".

Extracting Locality-based Modules

Our aim is the extraction of locality-based modules from the SNOMED CT ontology that are: 1) much smaller in size (i.e. number of axioms) than the SNOMED CT ontology; 2) as specific as possible for a given medical condition while being logically sound according to OWL-DL; and 3) can be reused and shared among organisations.

We extract a locality-based module (a.k.a. upper module) [87] per target term using as signature all the SNOMED CT concept identifiers mapped to UMLS CUI pairs validated with BMJ Best Practice content. We use the reasoner FaCT++ [88] for the method ModuleType BOT in the OWL API [52] as we did in [47]. A DL reasoner, like FaCT++, can calculate inferred information (e.g. inferred subsumption hierarchy) from the asserted information, i.e. the axioms within an ontology. A locality-based module contains at least all the (entailed) super-classes of an OWL class included in the signature [87] as well as all axioms relevant to the meaning of the OWL Classes in the signature. A SNOMED CT concept may have one or more attribute-value pairs [62], where the value of the pair is typically another SNOMED CT concept. Attribute-value pairs are considered relevant to the meaning of a SNOMED CT concept.

A locality-based module keeps the SNOMED CT top-level hierarchies for the OWL Class extracted, which is expected by the clinicians, and is likely to be smaller than the SNOMED CT ontology. The SNOMED CT ontology corresponding to the January 2017 release contains a total of 1.5 M axioms (Class count: 325 K; individual count: 0; Object property count: 80; SubClassOf axioms count: 246 K; and EquivalentClasses axioms count: 79 K) and its DL expressivity is ALER.

Multiple SPARQL queries can be built seeking to gain insights into diseases and syndromes of significance for both human medical and veterinary healthcare, i.e. One Health knowledge. For this study, the SPARQL SELECT queries q1V to q3V presented in Step 6 intend to retrieve reliable knowledge for human medicine (UMLS CUI pairs validated with BMJ Best Practice content) about the diagnosis and management of well-known medical conditions that affect humans and animals. We created the SPARQL SELECT queries q1VU to q3VU (see the Additional file 4 for details) that combine the results of the queries q1V to q3V over each dataset VetCN and PMSB. Hence, we report the number of UMLS Metathesaurus concepts pairs with up-to-date clinically meaningful associations for human medicine, although the source data can be from veterinary medicine (i.e. the

SAVSNET veterinary clinical narratives) or from human medical science (i.e. the PubMed Systematic Reviews).

The results of the queries q1VU to q3VU as well as the results of the queries q1V to q3V are quantitative. Hence, we can quantify to what extent One Health can provide added value when compared with a conventional approach that will keep both datasets VetCN and PMSB separated for being part of either veterinary medicine or medical science.

A UMLS Metathesaurus concept can be mapped to none, one or more than one SNOMED CT concepts. We created the SPARQL SELECT query q1VM to q3VM (see the Additional file 4 for details) to retrieve from the OBANmod those UMLS CUI pairs validated with BMJ Best Practice content, where the candidate concept is mapped to at least one SNOMED CT concept. Using the OBANmod and the asserted information within each locality-based module as the default graph, we created the SPARQL SELECT query q1VS to q1VS (see the Additional file 4 for details) that retrieves those SNOMED CT concept pairs (OWL Classes) mapped to UMLS CUI pairs validated with BMJ Best Practice content. For these SNOMED CT pairs, it is possible to go beyond the knowledge captured in the OBAN "sometimes true" association relationships.

Each locality-based module created for a well-known medical condition contains asserted as well as inferred knowledge that can expand/enrich the results from the queries q1VS to q3VS by exploiting the transitive closure of rdfs:subClassOf for the SNOMED CT concepts in OWL. The SPARQL SELECT queries q1VR to q3VR (see the Additional file 4 for details) use as the default graph the inferred model obtained with the DL reasoner FaCT++ for each locality-based module. The queries q1VR to q3VR retrieve the OWL Classes that are asserted and inferred descendants of the those SNOMED CT concepts that are mapped to candidate concepts of the SNOMED CT pairs retrieved from the SPARQL SELECT queries q1VS to q3VS.

RESULTS

We start by illustrating and reporting the results obtained for each step of the SemDeep pipeline using the PMSB and VetCN datasets. Next, we combine the results from the SemDeep pipeline to investigate to what extent the One Health approach can provide added value.

A SemDeep pipeline

Step 1: computing n-grams from unstructured text

As in our previous work [44,45,46,47], we employ word2phrase to obtain n-grams and we preserve numbers and punctuation marks including parenthesis as they appear in the unstructured text. However, the original text of the PMSB dataset is not converted to lower case as in [44,45,46,47]. The PMSB dataset contains 447 M terms (words and tokens as they appear in the text), and after obtaining the n-grams, this number reduces significantly to 46 M. This means that a high number of tokens/words appear to be repeatedly collocated.

The original text of the VetCN dataset was converted to lower case before computing n-grams as many examples were found within the unstructured text of indiscriminate alternation between lowercase and uppercase. The VetCN dataset contains 149 M terms (words and tokens as they appear in the text), and after obtaining the n-grams, this reduces to 103 M. A plausible reason for the modest numeric reduction is presence of spelling variations and/or errors for the same term. To confirm this hypothesis, we utilise a probabilistic spelling corrector [89] that provides alternative spellings for a word or token appearing within a corpus; the hypothesis was confirmed. For example, the term "vomiting" appears in VetCN with more than 50 spelling variations such as "vomiteting" or "vomittimng" or "vomirtting" or "vomikting" (to mention just a few), which can be considered spelling errors. Furthermore, by close inspection of the unstructured text, it can be observed that short forms like "v" are used instead of the long form "vomiting".

Step 2: creation of neural embeddings

We use the word2vec implementations for CBOW and Skip-gram and apply the same hyperparameter configuration as our previous study [55]. To compute the neural embeddings, we use a Dell PowerEdge R430 with 100GB RAM and 32 virtual CPUs Intel Xeon E5–2690 v4 at 2.6 GHz. With this, creating the neural embeddings with PMSB takes 17 min for Skip-gram and 2 min for CBOW; for VetCN, creating the neural embeddings takes 29 min for Skip-gram and 2 min for CBOW.

For both PMSB and VetCN, we obtained a list of terms (n-grams) with a frequency count greater than 5 and with vector representations of real numbers. For PMSB, 423 K terms have vector representations; for VetCN, 488 K terms have vector representations.

Step 3: obtaining term pairs for the semantic similarity and relatedness task

Table 2 shows the target terms that are the subject for this study, i.e. 11 n-grams with the same UMLS CUI corresponding to the medical conditions – UMLS Semantic Type "T047|Disease or Syndrome" – that appear in both PMSB and VetCN with a frequency count greater than 1 K. For each target term in Table 2, it was feasible to find a document in BMJ Best Practice (last column in Table 2) about the medical condition.

Table 2: The target terms for PMSB and VetCN datasets

Target terms for this study and their concept identifiers in UMLS and SNOMED CT				**BMJ Best Practice document**
UMLS CUI	**SNOMED CT identifier**	**VetCN dataset n-gram (frequency count)**	**PMSB dataset n-gram (frequency count)**	
C0018801	84,114,007	heart_failure (1292)	heart_failure (4615)	Chronic congestive heart failure
C0004096	195,967,001	asthma (1194)	asthma (8891)	Asthma in adults
C0014544	84,757,009	epilepsy (1164)	epilepsy (3521)	Generalised seizure
C0017601	23,986,001	glaucoma (1657)	glaucoma (1635)	Open-angle glaucoma
C1561643	709,044,004	ckd (2698)	CKD (1550)	Chronic kidney disease
C0029408	396,275,006	osteoarthritis (1765)	osteoarthritis (1991)	Osteoarthritis
C0002871	271,737,000	anaemia (1414)	anaemia (1154)	Assessment of anaemia
C0003864	3,723,001	arthritis (8276)	arthritis (1023)	Rheumatoid arthritis
C0011849	73,211,009	diabetes (3660)	diabetes (12846)	Type 2 diabetes in adults
C0020538	38,341,003	hypertension (1132)	hypertension (8365)	Essential hypertension
C0028754	414,916,001	obesity (1763)	obesity (10030)	Obesity in adults

The last column contains the names and references of BMJ Best Practice documents used for validation in Step 5 (see details within the section Materials and methods). The first column contains the UMLS CUI mapped to a target term (n-gram) with the aid of MetaMap. The second column shows the SNOMED CT identifier mapped to the UMLS CUI with the aid of the UMLS API. The third column displays the target terms from the VetCN dataset, i.e. the n-grams with their frequency counts in the corpus appear within brackets. The fourth column shows the target terms from PMSB dataset with the same format of the third column. All target terms (i.e. n-grams) are identical for both datasets except one. The well-known medical condition "chronic kidney disease" with UMLS CUI="C1561643" has the n-gram "CKD" (i.e. a short form with all the characters in upper case) in the PMSB dataset; while in VetCN dataset it has the n-gram "ckd". The difference in these two target terms "CKD" and "ckd" happens as in Step 1, VetCN corpus is transformed to lower case while PMSB corpus is not

We limit the list of the candidate terms (n-grams) to the 20 top-ranked terms. As there are 11 chosen medical conditions (i.e. target terms), the Additional file 1 contains 880 term pairs (target term, candidate term): the worksheet "VetCN" with the 440 term pairs using CBOW and Skip-gram with VetCN; and the worksheet "PMSB" with the 440 term pairs using CBOW and Skip-gram with PMSB.

By visual inspection of the Additional file 1, we observe that the cosine values are systematically higher for the candidate terms obtained from the neural embeddings created with Skip-gram than the cosine values for CBOW. Hence, CBOW learns the vectors quicker than Skip-gram (as reported in Step2), although the vectors obtained for the 20 top-ranked candidate terms are less similar to the target term (lower cosine value) for CBOW than for Skip-gram.

Step 4: named entity recognition (NER) task

The NER task can be interpreted as a grounding or normalisation to unveil the semantic meaning of the n-grams, and thus, provide more detail on the extent of semantic overlap between the candidate terms obtained from both datasets. Considering the guidelines introduced in Step 4 (section Materials and methods), UMLS CUIs are assigned to the n-grams to represent the focus concepts corresponding to all 613 unique candidate terms for the 880 term pairs. Table 3 shows only 20 n-grams that are the only common candidate terms among the 880 term pairs. From Table 3, there is at least one candidate term for all the 11 well-known medical conditions (target terms) that is common to both datasets.

Table 3: The 20 n-grams that are the only common candidate terms among the 880 term pairs from both VetCN and PMSB datasets

Neural language model (unstructured text dataset)	Target term	Candidate term
CBOW(VetCN)|CBOW(PMSB)	asthma	rhinitis
CBOW(VetCN)|Skip-gram(VetCN)|CBOW(PMSB)|Skip-gram(PMSB)	anaemia	anaemic
CBOW(VetCN)|Skip-gram(VetCN)|CBOW(PMSB)|Skip-gram(PMSB)	anaemia	anemia
CBOW(VetCN)|Skip-gram(VetCN)|CBOW(PMSB)|Skip-gram(PMSB)	hypertension	blood_pressure
CBOW(VetCN)|Skip-gram(VetCN)|CBOW(PMSB)|Skip-gram(PMSB)	hypertension	high_blood_pressure
CBOW(VetCN)|Skip-gram(VetCN)|CBOW(PMSB)|Skip-gram(PMSB)	glaucoma	cataract
Skip-gram(VetCN)|CBOW(PMSB)|Skip-gram(PMSB)	glaucoma	intraocular_pressure
CBOW(VetCN)|Skip-gram(VetCN)|CBOW(PMSB)	heart_failure	congestive_heart_failure
CBOW(VetCN)|Skip-gram(VetCN)|CBOW(PMSB)|Skip-gram(PMSB)	diabetes	diabetes_mellitus
CBOW(VetCN)|Skip-gram(VetCN)|CBOW(PMSB)|Skip-gram(PMSB)	arthritis	osteoarthritis
CBOW(PMSB)|CBOW(VetCN)	osteoarthritis|arthritis	glucosamine
Skip-gram(VetCN)|CBOW(PMSB)	osteoarthritis	hip_osteoarthritis
CBOW(VetCN)|Skip-gram(VetCN)|CBOW(PMSB)|Skip-gram(PMSB)	ckd|hypertension	hypertensive
CBOW(VetCN)|Skip-gram(VetCN)|CBOW(PMSB)|Skip-gram(PMSB)	ckd|CKD	kidney_disease
CBOW(VetCN)|CBOW(PMSB)	ckd|CKD	proteinuria
CBOW(VetCN)|Skip-gram(VetCN)|CBOW(PMSB)	ckd|CKD	renal
CBOW(VetCN)|Skip-gram(VetCN)|CBOW(PMSB)|Skip-gram(PMSB)	obesity	obese
Skip-gram(VetCN)|CBOW(PMSB)|Skip-gram(PMSB)	epilepsy	partial_seizures
CBOW(VetCN)|Skip-gram(VetCN)|CBOW(PMSB)|Skip-gram(PMSB)	epilepsy	seizure
CBOW(VetCN)|Skip-gram(VetCN)|CBOW(PMSB)|Skip-gram(PMSB)	epilepsy	seizures

The character '|' that appears in the first column separates the different neural language models. The grey background indicates that the target term and the candidate term has the same focus concept, i.e. same UMLS CUI

The columns I, J, and K from both worksheets "VetCN" and "PMSB" from the Additional file 1 have values if the candidate term (n-gram) contains a short form correctly or incorrectly identified by the short form detector. The column I will have the values {SF, SF-U, SF-NU} if a clinically meaningful short form has been correctly identified and the values {SF-I, SF-NF} if the short form detector has made an error. For example, "US" is a short form for the long form "United States", which per se is not clinically meaningful. Typical examples of the "SF-NF "error are short forms with four characters or more appearing in n-grams from VetCN. For example, "echo" is the abbreviation (i.e. short form) for the long form "echocardiography". Table 4 has the micro-averaging precision, recall, and F measure for the short form detector.

Table 4: Performance of the short form detector with VetCN and PMSB datasets

Data set	**Unique candidate terms (n-grams)**	**SF-U + SF-NU + SF**	**SF-I**	**SF-NF**	**n-grams with no clinically meaningful short forms**	**Detect n-grams with one or more clinically meaningful short forms**			**Detect n-grams with no clinically meaningful short forms**		
						P	**R**	**F**	**P**	**R**	**F**
VetCN	300	57	1	14	228	98.28	80.28	88.37	94.21	99.56	96.82
PMSB	333	75	2	0	256	97.40	100	98.68	100	99.22	99.61
						97.78	90.41	93.95	97.07	99.36	98.20

To assess the capability of the short form detector to identify candidate terms (n-grams) with one or more clinically meaningful short forms: the value of the column "SF-U + SF-NU + SF" is interpreted as TP; the value of the column "SF-I" is interpreted as FP; and the value of the column "SF-NF" is interpreted as FN. To assess the capability of the short form detector to identify candidate terms (n-grams) with no clinically meaningful short forms: the value of the column "n-grams with no clinically meaningful short forms" is interpreted as TP; the value of the column "SF-I" is interpreted as FN; and the value of the column "SF-NF" is interpreted as FP. The last row shows the micro-averaging values taking into account the total of 613 unique candidate terms (n-grams) for the 880 term pairs. Abbreviations: P=precision; R=recall; and F=F measure

Tables 5 and 6 show the performance of MetaMap in experiments 1 and 2 for the candidate terms (n-grams) obtained for each target term from the neural embeddings created with CBOW and Skip-gram. To retrace the MetaMap output, the domain experts' assessment, and the performance calculations, the Additional file 1 contains:

- The worksheet "SF to LF" – it has the 63 long forms for 80 short forms (including variants of the short forms) within the candidate terms (n-grams) for the VetCN and PMSB datasets.
- Within both worksheets "VetCN" and "PMSB" – the column G has the UMLS CUIs for the focus concepts of the candidate terms (n-grams). The labels (i.e. {SM, MM, IM, NM}) assigned in EXP-1 by the domain experts to MetaMap output appear within the column H and the labels assigned in EXP-2 by the domain experts to MetaMap output appear in column L.
- The worksheet "MetaMap performance" – it contains the number of TP, FP, and FN obtained and used to calculate precision, recall, and F measure for MetaMap in EXP-1 and EXP-2.

Table 5: MetaMap performance for the candidate terms from VetCN dataset

Target term (n-gram)	**Candidate terms (20 top-ranked n-grams) from CBOW neural embeddings for a target term**						**Candidate terms (20 top-ranked n-grams) from Skip-gram neural embeddings for a target term**					
	MetaMap Experiment 1			**MetaMap Experiment 2**			**MetaMap Experiment 1**			**MetaMap Experiment 2**		
	P	**R**	**F**	**P**	**R**	**F**	**P**	**R**	**F**	**P**	**R**	**F**
anaemia	84.21	94.12	88.89	95.00	100.00	97.44	94.74	94.74	94.74	95.00	100.00	97.44
arthritis	93.33	73.68	82.35	100.00	75.00	85.71	94.44	89.47	91.89	100.00	90.00	94.74
asthma	100.00	90.00	94.74	100.00	95.00	97.44	89.47	94.44	91.89	100.00	100.00	100.00
ckd	68.75	73.33	70.97	100.00	95.00	97.44	62.50	71.43	66.67	100.00	100.00	100.00
diabetes	76.47	81.25	78.79	100.00	90.00	94.74	88.89	88.89	88.89	94.74	94.74	94.74
epilepsy	100.00	90.00	94.74	100.00	95.00	97.44	100.00	90.00	94.74	100.00	95.00	97.44
glaucoma	87.50	77.78	82.35	94.74	94.74	94.74	93.33	73.68	82.35	94.74	94.74	94.74
heart_failure	73.68	93.33	82.35	95.00	100.00	97.44	84.21	94.12	88.89	100.00	100.00	100.00
hypertension	71.43	62.50	66.67	100.00	95.00	97.44	72.22	86.67	78.79	100.00	100.00	100.00
obesity	75.00	100.00	85.71	85.00	100.00	91.89	84.21	94.12	88.89	89.47	94.44	91.89
osteoarthritis	94.74	94.74	94.74	100.00	95.00	97.44	85.00	100.00	91.89	85.00	100.00	91.89
	84.10	84.61	83.85	97.25	94.07	95.38	86.27	88.87	87.24	96.27	97.17	96.63

Table 6: MetaMap performance for the candidate terms from PMSB dataset

Target term (n-gram)	Candidate terms (20 top-ranked n-grams) from CBOW neural embeddings for a target term						Candidate terms (20 top-ranked n-grams) from Skip-gram neural embeddings for a target term					
	MetaMap Experiment 1			MetaMap Experiment 2			MetaMap Experiment 1			MetaMap Experiment 2		
	P	R	F	P	R	F	P	R	F	P	R	F
anaemia	90.00	100.00	94.74	85.00	100.00	91.89	84.21	94.12	88.89	94.74	94.74	94.74
arthritis	88.89	88.89	88.89	89.47	94.44	91.89	100.00	100.00	100.00	95.00	100.00	97.44
asthma	76.47	81.25	78.79	72.22	86.67	78.79	63.16	92.31	75.00	68.42	92.86	78.79
CKD	100.00	100.00	100.00	100.00	100.00	100.00	90.00	100.00	94.74	90.00	100.00	94.74
diabetes	63.16	92.31	75.00	68.42	92.86	78.79	75.00	100.00	85.71	80.00	100.00	88.89
epilepsy	85.00	100.00	91.89	95.00	100.00	97.44	90.00	100.00	94.74	95.00	100.00	97.44
glaucoma	90.00	100.00	94.74	100.00	100.00	100.00	84.21	94.12	88.89	100.00	100.00	100.00
heart_failure	85.00	100.00	91.89	90.00	100.00	94.74	73.68	93.33	82.35	90.00	100.00	94.74
hypertension	95.00	100.00	97.44	100.00	100.00	100.00	84.21	94.12	88.89	95.00	100.00	97.44
obesity	100.00	95.00	97.44	100.00	100.00	100.00	94.74	94.74	94.74	95.00	100.00	97.44
osteoarthritis	90.00	100.00	94.74	100.00	100.00	100.00	90.00	100.00	94.74	100.00	100.00	100.00
	87.59	96.13	91.41	90.92	97.63	93.96	84.47	96.61	89.88	91.20	98.87	94.70

The table shows the performance of MetaMap in Experiment 1 (applying MetaMap to the candidate terms) and Experiment 2 (short form detection and expansion into long form before applying MetaMap to the candidate terms) for each target term (n-gram for a well-known medical condition). The candidate terms are a list of the 20 top-ranked terms (highest cosine value) obtained from the created neural embeddings with CBOW or Skip-gram taking the vector for a target term. The last row shows the average of each evaluation measure over all 11 medical conditions under study to get an overall measure of performance (a.k.a. macro-averaging). Abbreviations: P=precision; R=recall; and F=F measure

The table shows the performance of MetaMap in Experiment 1 (applying MetaMap to the candidate terms) and Experiment 2 (short form detection and expansion into long form before applying MetaMap to the candidate terms) for each target term (n-gram for a well-known medical condition). The candidate terms are a list of the 20 top-ranked terms (highest cosine value) obtained from the created neural embeddings with CBOW or Skip-gram taking the vector for a target term. The last row shows the average of each evaluation measure over all 11 medical conditions under study to get an overall measure of performance (a.k.a. macro-averaging). Abbreviations: P=precision; R=recall; and F=F measure

The last row of Tables 5 and 6 shows the macro-averaging precision, recall and F measure. Taking the 613 unique candidate terms for the 880 term pairs, the micro-averaging precision of MetaMap is 84.68% for EXP-1 and 92.41% for EXP-2; the micro-averaging recall of MetaMap is 91.44% for EXP-1 and 96.48% for EXP-2; hence, the micro-averaging F measure of MetaMap is 87.93% in EXP-1 and 94.40% in EXP-2.

The macro-averaging and the micro-averaging values for precision, recall, and F measure are consistently higher in EXP-2 than in EXP-1. However, when examining the performance of MetaMap in Tables 5 and 6, two exceptions are noticeable (two rows in Table 6 with a grey background) where a drop in performance for EXP-2 can be observed:

- The target term "anaemia" has precision and F measure higher in EXP-1 than in EXP-2. Both n-grams "g/dL" and "g/dl" can be mapped by MetaMap in EXP-1 to the UMLS Metathesaurus concept "C0439267|Gram per Deciliter", which is more specific than the UMLS Metathesaurus concept "C1519795|Unit of Measure". The UMLS CUI=C1519795 is automatically assigned by the short form detector in EXP-2.

- The target term "arthritis" has precision and F measure higher in EXP-1 than in EXP-2. The n-gram "Cox-2_inhibitors" is mapped by MetaMap in EXP-1 to the UMLS Metathesaurus concept "C1257954|Cyclooxygenase 2 Inhibitors". As "Cox-2" is the short form for the long form "cycloxygenase-2", the n-gram "Cox-2_inhibitors" was expanded into "cycloxygenase-2_inhibitors" for EXP-2. However, MetaMap in EXP-2 did not map the expanded n-gram to the UMLS CUI=C1257954.

Following Smucker et al. [69], we apply the Student's paired t-test [70] taking the values from Tables 5 and 6 for EXP-1 and EXP-2 to measure the statistical significance of the difference in the mean average precision, recall, and F measure. We obtain the following p-values: 0.0001046 for precision, 0.001077 for recall, and 0.00001191 for F measure. As the p-values obtained with the Student's paired t-test are significantly small, we reject that "*the difference in averages could be due to chance*" [67].

The outcome of this step is a set of 342 unique UMLS Metathesaurus concept pairs (target concept, candidate concept) that represent the focus concept pairs for the 880 n-gram pairs (target term, candidate term). The next step will determine if there is a clinically meaningful association for each UMLS Metathesaurus concept pair.

Step 5: validation of the term pairs mapped to UMLS Metathesaurus concept pairs using BMJ best practice

The Additional file 3 contains the 3-tuples (target concept, candidate concept, validation label) for the VetCN dataset (worksheet "VetCN") and the PMSB dataset (worksheet "PMSB"). Among the 35 unique UMLS Metathesaurus concept pairs that are common to both datasets, there are 10 concept pairs with the validation label "`Itself", i.e. one per chosen medical condition (target term) with the only exception being "chronic kidney disease" (i.e. ckd). Table 7 displays the 25 unique UMLS Metathesaurus concept pairs common to both datasets and with validation labels different than "itself".

Table 7: The 25 unique UMLS Metathesaurus concept pairs that are common to both VetCN and PMSB datasets and have validation labels different to "itself"

Target term (n-gram) lower case	UMLS CUI pair (Target, Candidate)	Validation label "Relatedness by..."	BMJ Best Practice term	UMLS Metathesaurus concept name for focus concept of candidate term
heart_failure	(C0018801, C0018802)	Exact/Approximate match	congestive heart failure	Congestive heart failure
asthma	(C0004096, C0035455)	Inexact match (hyponym/hypernym)	allergic rhinitis	Rhinitis
asthma	(C0004096, C0032285)	Inexact match (background knowledge)	bacterial infections (Mycoplasma pneumoniae or Chlamydia pneumoniae)	Pneumonia
epilepsy	(C0014544, C0036572)	Exact/Approximate match	seizures	Seizures
epilepsy	(C0014544, C0751495)	Exact/Approximate match	focal seizures	Seizures, Focal
glaucoma	(C0017601, C0015392)	Exact/Approximate match	eye	Eye
glaucoma	(C0017601, C0021888)	Exact/Approximate match	Intra-ocular pressure	Physiologic Intraocular Pressure
glaucoma	(C0017601, C0086543)	Exact/Approximate match	cataract	Cataract
ckd	(C1561643, C0022646)	Exact/Approximate match	renal	Kidney
ckd	(C1561643, C0022658)	Exact/Approximate match	kidney disease	Kidney Diseases
ckd	(C1561643, C0022661)	Exact/Approximate match	end-stage renal disease (ESRD)	Kidney Failure, Chronic
ckd	(C1561643, C0033687)	Exact/Approximate match	Proteinuria	Proteinuria
ckd	(C1561643, C0035078)	Exact/Approximate match	kidney failure	Kidney Failure
osteoarthritis	(C0029408, C0017718)	Exact/Approximate match	glucosamine	Glucosamine
osteoarthritis	(C0029408, C0019552)	Exact/Approximate match	hip	Hip structure
osteoarthritis	(C0029408, C0029410)	Exact/Approximate match	hip OA	Osteoarthritis of hip
anaemia	(C0002871, C0857322)	Exact/Approximate match	anaemic patients	Anemic
arthritis	(C0003864, C0022417)	Exact/Approximate match	joints	Joints
arthritis	(C0003864, C0029408)	Exact/Approximate match	Osteoarthritis	Degenerative polyarthritis
diabetes	(C0011849, C0017725)	Exact/Approximate match	glucose	Glucose
diabetes	(C0011849, C0241863)	Exact/Approximate match	diabetic	Diabetic
hypertension	(C0020538, C0005823)	Exact/Approximate match	Blood pressure	Blood Pressure
hypertension	(C0020538, C0857121)	Exact/Approximate match	hypertensive	Hypertensive (finding)
hypertension	(C0020538, C0242485)	Inexact match (hyponym/hypernym)	blood pressure measurement	Measurement
obesity	(C0028754, C0005910)	Exact/Approximate match	weight	Body Weight

The validation label from the third column denotes the relatedness of BMJ Best Practice term (the column previous to the last) to the UMLS Metathesaurus concept representing the focus concept for the candidate term (the last column). The rows with grey background correspond to UMLS Metathesaurus concept pairs that are retrieved by the SPARQL queries q1VU, or q2VU, or q3VU (see the cells with grey background in Table 11 for details)

The last column of Table 7 shows the UMLS Metathesaurus concept name representing the focus concepts for the candidate term. This column does not appear in the Additional file 3 as this information can be obtained with the UMLS API. The UMLS API can typically retrieve alternative names for a UMLS Metathesaurus concept in different terminologies, and thus, the Additional file 3 just provides the terminologies' identifiers (e.g. MeSH: D000068256) instead of the textual excerpts from terminological resources.

Table 8 displays the 11 UMLS Metathesaurus concept pairs that were identified by the domain experts as unrelated as they do not have an up-to-date clinically meaningful association. These 11 UMLS Metathesaurus concept pairs correspond to only five of the total of 11 medical conditions (i.e. target term) that are the subject of this study. For eight UMLS Metathesaurus concept pairs, the candidate concept could not be interpreted per se as clinically meaningful in connection with the medical condition, and thus, they have assigned the label "Unrelated: not clinically meaningful". For three UMLS Metathesaurus concept pairs, which corresponds to the medical condition epilepsy and glaucoma, the validation label "Unrelated: excluded" was assigned and the rationale behind is the following:

- UMLS Metathesaurus concept pair (C0014544|Epilepsy, C0071754|potassium bromide): in medicine "potassium bromide" is no longer a suitable treatment for epilepsy due to side effects [90] (see correspond excerpt in evalBMJ.xls).
- UMLS Metathesaurus concept pair (C0014544|Epilepsy, C3886490|Pexion): the active substance of pexion is imepitoin, which is a type of partial benzodiazepine site agonists. In medicine "benzodiazepine" is no longer a suitable treatment for epilepsy due to adverse effects [91] (see corresponding excerpt in evalBMJ.xls).
- UMLS Metathesaurus concept pair (C0017601|Glaucoma, C0014392|Enucleation procedure): As "malignant melanomas of the ciliary body" does not appear either explicitly or implicitly in BMJ Best Practice for open-angle glaucoma [92], which is the medical condition for which the "enucleation procedure" is recommended [93], the concept pair is judged as unrelated. As "enucleation of glaucomatous eyes" does appear in veterinary literature, another excerpt was added to capture this fact. For further details, see corresponding excerpts in evalBMJ.xls.

Table 8: The 11 UMLS Metathesaurus concept pairs identified by the domain experts as unrelated, as they do not have an up-to-date clinically meaningful association for human medicine

Neural language model (dataset)	Target term (n-gram) lower case	UMLS CUI pair (Target, Candidate)	Validation label	UMLS Metathesaurus concept name for focus concept of candidate term
CBOW(VetCN)	epilepsy	(C0014544, C0071754)	Unrelated: Excluded	potassium bromide
CBOW(VetCN)| Skip-gram(VetCN)	epilepsy	(C0014544, C3886490)	Unrelated: Excluded	Pexion
CBOW(PMSB)	glaucoma	(C0017601, C0023008)	Unrelated: Not clinically meaningful	Languages
CBOW(PMSB)	glaucoma	(C0017601, C1704324)	Unrelated: Not clinically meaningful	Scientific Publication
CBOW(VetCN)| Skip-gram(VetCN)	glaucoma	(C0017601, C0014392)	Unrelated: Excluded	Enucleation procedure
Skip-gram(PMSB)	osteoarthritis	(C0029408, C0392747)	Unrelated: Not clinically meaningful	Changing
CBOW(PMSB)	arthritis	(C0003864, C0442726)	Unrelated: Not clinically meaningful	Detected (finding)
Skip-gram(PMSB)	arthritis	(C0003864, C0558295)	Unrelated: Not clinically meaningful	preference
Skip-gram(VetCN)	obesity	(C0028754, C0686904)	Unrelated: Not clinically meaningful	Patient need for (contextual qualifier)
Skip-gram(VetCN)	obesity	(C0028754, C0849355)	Unrelated: Not clinically meaningful	Removed
CBOW(VetCN)	obesity	(C0028754, C3842589)	Unrelated: Not clinically meaningful	20%

The rows with grey background correspond to UMLS Metathesaurus concept pairs that are retrieved for some of the SPARQL queries q1 to q3 (see the cells with grey background in Table 9 for details)

As the above-mentioned three UMLS Metathesaurus concept pairs are up-to-date clinically meaningful associations in veterinary healthcare, they denote differences between human and veterinary healthcare.

The Additional file 3 contains a total of 342 unique 3-tuples considering both the VetCN and PMSB datasets, of which only 11 have UMLS Metathesaurus concept pairs that are considered as unrelated. The UMLS Metathesaurus concept pairs in Table 8 are considered False Positives (FP), and likewise the term pairs that have them as focus concepts. All other UMLS Metathesaurus concept pairs not included in Table 8 are considered as True Positives (TP), and likewise the term pairs that have them as focus concepts. Hence, it is feasible to calculate precision and estimate the performance of CBOW and Skip-gram for the semantic similarity and relatedness task using the 880 term pairs:

- For VetCN, precision is 98.18% for both CBOW and Skip-gram.
- For PMSB, the precision is 98.64% for CBOW and 99.09% for Skip-gram.

The BMJ health informatician considered it worthy to report to BMJ Best Practice editors that:

- Some concept pairs do not have candidate concepts that are easily found within BMJ Best Practice for the medical condition.

Hence, external evidence from systematic research was needed to validate them, and thus, the concept pairs have assigned the validation label "Relatedness by Inexact match (background knowledge)" instead of the expected label "Relatedness by exact/approximate match". For example, for the concept pair (C0004096|Asthma, C0006277|Bronchitis), bronchitis is not mentioned as a differential diagnosis in BMJ Best Practice for asthma, although BMJ Best Practice for acute bronchitis has asthma as a differential. Another example is the concept pair (C0004096|Asthma, C0232602|Retching), where retching does not currently appear along with coughing in the description of symptoms within BMJ Best Practice for asthma.

1. Some concept pairs have candidate concepts that may enhance the current content of BMJ Best Practice for the medical condition. For example, for the concept pair (C0017601|Glaucoma, C0020581|Hyphema), where hyphema appears as a common post-operative complication in patients with glaucoma that had trabeculotomy (a.k.a. glaucoma filter surgery). This concept pair has the validation label "Relatedness by Inexact match (background knowledge)" and can be interpreted as an example of transitivity: glaucoma has trabeculotomy as a surgical procedure, and trabeculotomy has hyphema as a common complication. Only by making "trabeculotomy" explicit does the relatedness for the concept pair (glaucoma, hyphema) become transparent. It should be noted that the relationship between "trabeculotomy" from BMJ and "C0020581|Hyphema" is a cause-effect relationship.

Step 6: formal representation of the knowledge acquired and validated

The third column of Table 1 shows OWL individuals for the main OWL Classes of the lemonEXT and OBANmod created programmatically applying the axiom patterns from the second column in Table 1. We use the reasoner FaCT++ to check that – once populated – the lemonEXT and OBANmod ontologies are consistent (i.e. no unsatisfiable classes).

Once the lemonEXT core ontology is populated with the term pairs (target term, candidate term) from VetCN, including their focus concepts from the UMLS Metathesaurus, it contains a total of 3477 axioms (class count: 324; individual count: 502) and its DL expressivity is ALEI. Once the lemonEXT is populated with the term pairs from PMSB, including their focus concepts, it contains a total of 3702 axioms (class count: 338; individual count: 549) and its DL expressivity is ALEI.

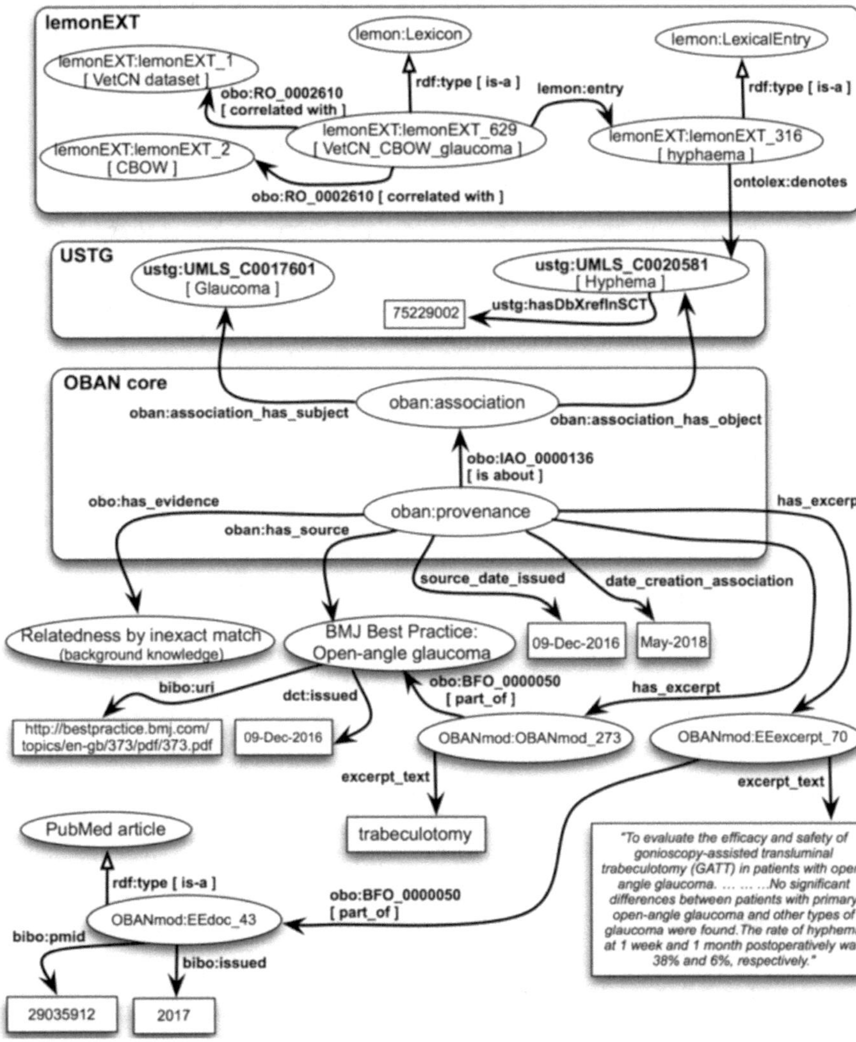

Figure 4: *Exemplifying the population of both the lemonEXT and OBANmod core ontologies* – the figure illustrates how the 3-tuple (C0017601|Glaucoma, C0020581|Hyphema, "Relatedness by Inexact match (background knowledge)") is represented as a modified version of the OBAN association. This 3-tuple has two bibo:excerpts: 1) the term "trabeculotomy" from BMJ Best Practice for open-angle glaucoma; and 2) few lines of text from the PubMed article with identifier PMID=29,035,912. The top of the figure shows the lex-

ical entry from the lemonEXT ontology corresponding to the focus concept C0020581|Hyphema.

Once the OBANmod core ontology is populated with the 192 unique 3-tuples from VetCN, it contains a total of 5893 axioms (class count: 347; individual count: 842) and its DL expressivity is ALEHI(D). Once the OBANmod is populated with the 185 unique 3-tuples from PMSB, it contains a total of 5545 axioms (class count: 361; individual count: 781) and its DL expressivity is ALEHI(D).

To illustrate the population of both the lemonEXT and OBANmod core ontologies, Fig. 4 takes as an example the term pair (glaucoma, hyphaema) from the neural embeddings created with CBOW using VetCN dataset. This term pair corresponds to the 3-tuple (C0017601|Glaucoma, C0020581|Hyphema, "Relatedness by Inexact match (background knowledge)"), and Fig. 4 displays how this 3-tuple is represented as a "sometimes true" association relationship in OWL with evidence-based provenance. The top of Fig. 4 shows the candidate term (i.e. an n-gram represented as a lexical entry from the lemonEXT) with a UMLS Metathesaurus concept as the focus concept with CUI=C0020581, which corresponds to the SNOMED CT concept with identifier=75,229,002.

Table 9 contains the number of UMLS Metathesaurus concept pairs retrieved from executing: a) the SPARQL SELECT queries q1 to q3 over the lemonEXT populated for each dataset VetCN and PMSB separately; and b) the SPARQL SELECT queries q1V to q3V over the OBANmod populated for each dataset VetCN and PMSB separately.

Table 9: *Results of the SPARQL SELECT queries performed over the lemon-EXT ontology and the OBANmod ontology* – **The table shows the number of UMLS** Metathesaurus concept pairs (target, candidate) retrieved for the SPARQL SELECT queries q1 to q3 as well as q1V to q3V

Target term (n-gram) lower case	Number of UMLS Metathesaurus concept pairs for VetCN dataset						Number of UMLS Metathesaurus concept pairs for PMSB dataset					
	lemonEXT ontology			OBANmod ontology			lemonEXT ontology			OBANmod ontology		
	q1	q2	q3	q1V	q2V	q3V	q1	q2	q3	q1V	q2V	q3V
heart_failure	2	3	3	2	3	3	2	2	3	2	2	3
asthma	0	4	1	0	4	1	2	4	3	2	4	3
epilepsy	0	3	4	0	3	2	0	4	3	0	4	3
glaucoma	4	0	2	3	0	2	2	3	0	2	3	0
ckd	0	2	1	0	2	1	2	2	0	2	2	0
osteoarthritis	2	2	2	2	2	2	3	1	3	3	1	3
anaemia	2	8	1	2	8	1	2	3	6	2	3	6
arthritis	0	2	2	0	2	2	1	2	3	1	1	3
diabetes	0	2	3	0	2	3	0	2	4	0	2	4
hypertension	1	1	1	1	1	1	2	3	3	2	3	3
obesity	0	2	0	0	1	0	0	6	0	0	6	0

The SPARQL SELECT queries appear within the Additional file 4 and the description of the queries appear within the Step 6 of the section Materials and methods. Each UMLS Metathesaurus concept pair represents the focus concepts of the term pair (target term, candidate term). The difference in the number of results between the query qi and the query qiV, with i = {1,2,3}, indicates that there are UMLS Metathesaurus concept pairs that have not passed the evaluation with BMJ Best Practice, and thus, the query qi (see cells with grey background) has a higher number of results than the query qiV

Extracting Locality-based Modules with SNOMED CT and Enabling one Health Queries

The worksheet "signatures" within the Additional file 3 contains the ontological signature (i.e. list of SNOMED CT identifiers) for each of the 11 medical conditions that are the subject of this study. We use the ModuleType BOT from the OWL API to create upper modules (a.k.a. locality-based modules). With a MacBook Pro Retina with 2.8 GHz Intel Core i7 and 16GB of RAM memory, the average time to create a locality-based module is 8 h.

All the locality-based modules created have a DL expressivity ALER, i.e. the same as the SNOMED CT ontology. Table 10 shows the details for each locality-based module extracted. From Table 10 it is easy to observe many axioms for the locality-based module extracted for diabetes mellitus. Although the guidelines introduced in Step 4 intend to avoid selection of focus concepts for the candidate terms that are too general for the medical condition (see the Additional file 2 for details), due to gaps in the terminological resources within the UMLS Metathesaurus, it may be that the focus concepts for some candidate terms (n-grams) are too broad in the biomedical or clinical meaning. Indeed, this is the case for diabetes mellitus. The UMLS Metathesaurus concept pair (C0011849|Diabetes Mellitus, C0012634|Disease) has the focus concepts for the term pair ("diabetes", "concurrent_diseases"), which was obtained with Skip-gram using VetCN. As the UMLS Metathesaurus concept "C0012634|Disease" was selected as the focus concept for the candidate term (n-gram) "concurrent_diseases", the SNOMED CT concept "64572001| Disease (disorder)" that maps the UMLS CUI=C0012634 is included in the ontological signature for diabetes mellitus. It should be noted that in the January 2017 version of SNOMED CT in OWL, the OWL Class for "64,572,001| Disease (disorder)" has 1810 OWL Classes as asserted descendants and 72,253 OWL Classes as asserted as well as inferred descendants using the reasoner FaCT++. Furthermore, one or more attribute-value pairs in an asserted or inferred descendant will trigger extraction of all axioms contributing the descendant's meaning.

Table 10: Locality-based modules extracted from the SNOMED CT ontology for the 11 well-known medical conditions

Target term (n-gram) lower case	**Number of SNOMED CT concept identifiers for the signature**	**Total number of axioms**	**Number of OWL Classes**	**Number of OWL object properties**	**Number of SubClassOf axioms**	**Number of Equivalent-Class axioms**
anaemia	34	105,205	22,227	46	7092	15,134
arthritis	33	74,398	16,180	31	7470	8709
asthma	37	97,647	20,804	39	7611	13,192
ckd	19	51,890	10,929	40	3899	7029
diabetes	29	463,437	100,119	44	49,818	50,300
epilepsy	19	10,072	2085	23	1055	1029
glaucoma	39	101,997	21,278	44	11,047	10,230
heart_failure	37	68,006	14,634	44	5490	9143
hypertension	31	152,283	32,270	48	16,425	15,844
obesity	28	90,224	18,994	40	12,376	6617
osteoarthritis	38	73,496	15,903	37	8922	6980

The second column just reports the total number of SNOMED CT identifiers for the ontological signature. The worksheet "signatures" within the Additional file 3 contains the list of SNOMED CT identifiers (as signature) for each target term. From the third to the last column ontology metrics information for the locality-based module created per target term is provided. The last two columns indicate the number of descriptions and definitions extracted from the SNOMED CT ontology for each locality-based module, respectively

Table 11 contains the results of executing the "One Health" queries proposed for this study. The worksheet "q One Health" within the Additional file 3 contains the number of UMLS CUI pairs validated with BMJ Best Practice content (i.e. human medicine) for each of the 27 UMLS Semantic Types that participates in the SPARQL SELECT query q1VU or q2VU or q3VU.

Table 11: *Results of the "One Health" queries performed that intend to acquire validated knowledge about the diagnosis and management of well-known medical conditions* – **The SPARQL SELECT queries appear within the Additional file 4 and the description of the queries appear within the subsection "Extracting locality-based modules with SNOMED CT and enabling One Health queries" of the section Materials and methods**

Target term (n-gram) lower case	Number of UMLS Metathesaurus concept pairs for both VetCN and PMSB datasets						Number of SNOMED CT concept pairs for both VetCN and PMSB datasets					
	OBANmod ontology						OBANmod ontology and locality-based modules extracted from SNOMED CT for the target term					
	q1VU	q2VU	q3VU	q1VM	q2VM	q3VM	q1VS	q2VS	q3VS	q1VR	q2VR	q3VR
heart_failure	4	5	6	3	1	5	3	1	9	248	28	1340
asthma	2	8	4	2	6	2	2	8	4	1	99	84
epilepsy	0	5	5	0	4	4	0	4	7	0	125	59
glaucoma	5	3	2	4	2	1	4	2	2	68	2	4
ckd	2	3	1	2	2	1	3	2	1	57	45	0
osteoarthritis	5	3	4	3	2	4	4	2	8	167	39	796
anaemia	4	10	7	3	6	5	4	6	10	890	9	778
arthritis	1	3	5	0	1	4	0	1	8	0	32	613
diabetes	0	3	6	0	1	4	0	1	7	0	23	261
hypertension	3	3	4	1	1	4	1	1	8	0	3	1359
obesity	0	7	0	0	6	0	0	8	0	0	37	0

Each query qiVU, with i={1,2,3}, is the union of the results obtained for the query qiV over VetCN dataset and the query qiV over PMSB dataset (see Table 9 for details). The cells with grey background indicate that there are common UMLS Metathesaurus concept pairs in both VetCN and PMSB datasets, and therefore, the total number of results for the query qiVU is lower that the summation of the results obtained for the query qiV in each dataset (see the rows with grey background in Table 7 for details of the common UMLS Metathesaurus concept pairs retrieved). As more than one SNOMED CT concept can map one UMLS Metathesaurus concept, the number of results for the query qiVM is equal to or lower than the number of results for the query qiVS, with i={1,2,3}. Each SPARQL SELECT query qiVR, with i={1,2,3}, retrieves the asserted and inferred descendants (with FaCT++) of those SNOMED CT concepts mapped to candidate concepts of the SNOMED CT pairs retrieved from the SPARQL SELECT query qiVS

According to Table 11:

- The SPARQL SELECT queries q1VU to q3VU obtain equal or more UMLS CUI pairs validated with BMJ Best Practice content (i.e. human medicine) than the queries q1V to q3V over VetCN and PMSB separately (Table 9). Hence, the higher number of up-to-date clinically meaningful associations for human medicine demonstrates that a "One Health" approach can provide added value by combining the UMLS CUI pairs (number of associations) from veterinary medicine (i.e. the SAVSNET veterinary clinical narratives) and from human medical science (i.e. the PubMed Systematic Reviews). The cells with a grey background indicate the existence of common UMLS CUI pairs in both datasets VetCN and PMSB.

- The query q2VU obtains UMLS CUI pairs validated with BMJ Best Practice content (i.e. human medicine) for all 11 target terms (i.e. the chosen medical conditions). By looking at worksheet "q One Health" within the Additional file 3 and examining the number of UMLS CUI pairs obtained for the UMLS Semantic Type "T033|Finding" alone, we can safely state that our SemDeep approach has captured "sometimes true" association relationships between each well-known medical condition and its clinical findings. Hence, "One Health" knowledge useful for the diagnosis of each of the 11 well-known medical conditions has been obtained.
- By combining the results of the queries q1VU and q3VU, it may be possible to obtain the UMLS CUI pairs validated with BMJ Best Practice content (i.e. human medicine) for the management of each well-known medical condition. We looked at worksheet "q One Health" within the Additional file 3 and examined the number of UMLS CUI pairs obtained for: a) the UMLS Semantic Type "T061|Therapeutic or Preventive Procedure" (included in query q1VU); and b) the UMLS Semantic Type "T121|Pharmacologic Substance", which belongs to the UMLS Semantic Group CHEM (included in query q3VU). Taking into account the number of UMLS CUI pairs for both UMLS Semantic Types, we can safely state that our SemDeep approach has captured "sometimes true" association relationships useful for the management of each medical condition with the only exception of obesity. According to BMJ Best Practice for obesity in adults [94], the first line of treatment for obesity is diet. There are candidate terms (n-grams) from VetCN dataset that have as focus concept the UMLS Metathesaurus "C0012155|Diet", which has the UMLS Semantic Type "T168|Food", and therefore, outside of the 27 UMLS Semantic Types involved in the SPARQL SELECT queries q1VU to q3VU. For some n-grams containing the token "diet" MetaMap provided also the UMLS Metathesaurus concept "C0012159|Diet (Diet therapy) "that has the UMLS Semantic Type "T058|Health

Care Activity", and thus, within the scope of query q1VU. The UMLS Metathesaurus concept "C0419178|Dietary regime", which has the UMLS Semantic Type "T061|Therapeutic or Preventive Procedure "was not provided by MetaMap for any n-gram with the token "diet" or "diets". Taking into account the two UMLS CUIs provided by MetaMap, the UMLS CUI=C0012155 was chosen instead of the UMLS CUI=C0012159 following the guidelines introduced in Step 4 of Materials and Method as the UMLS CUI=C0012155 is broader in meaning and it is mapped to SNOMED CT. Hence, after a closer inspection, as the UMLS CUI pair (C0028754|Obesity, C0012155|Diet) has passed the validation with the BMJ Best Practice content, we can safely state that our SemDeep approach has captured "sometimes true" association relationships useful for the management of each of the 11 well-known medical conditions including obesity.

- Looking at the results of the queries q1VS to q1VS and taking into account that the UMLS CUI pair (C0028754|Obesity, C0012155|Diet) is mapped to SNOMED CT, it seems reasonable to conclude that SNOMED CT concepts useful for the diagnosis/ management of the 11 medical conditions have been obtained. Hence, SemDeep can acquire "One Health" knowledge from veterinary medicine (VetCN) and medical sciences (PMSB) that can be reused and shared among organisations, i.e. there are sets of SNOMED CT concepts relevant for the diagnosis/management of each of the 11 well-known medical conditions (i.e. diseases or syndromes) under study.
- Looking at the results of the queries q1VR to q1VR, we can safely state that there are asserted and inferred descendants for some of the SNOMED CT concepts relevant for the diagnosis/ management of the 11 diseases that affect humans and animals. Hence, the knowledge within the locality-based modules extracted from the SNOMED CT ontology for each well-known medical condition can be used to enrich/expand the sets of SNOMED CT concepts mapped to the UMLS CUI pairs validated with BMJ Best Practice content (i.e. human medicine).

DISCUSSION

Semantic Deep Learning can be used to generate embeddings of nodes that contain both implicit and explicit information within biological knowledge graphs [95]. A knowledge graph is a "*graph-based representation of entities in the world and their interrelations*" [95]. In previous work, Jauhar et al. [96] adapted a neural language model (e.g. Skip-gram) to generate ontologically grounded sense vectors by adding latent variables representing senses and assuming a fixed ontology [96]. Other work combines both text corpora and ontologies, for example by requiring that the neural embeddings of two terms are similar if the two terms have a semantic relationship in the ontology [97]; this is complementary to retrofitting vectors (i.e. post-processing) to an ontology [96].

The work here transforms neural embeddings of n-grams into an augmented biomedical dataset of normalised and interlinked UMLS Metathesaurus concepts by reusing existing ontologies (i.e. lemon, OBAN, and SNOMED CT) that provide the foundation to acquire reusable One Health knowledge about well-known diseases in humans and animals. The approach is an external intervention in which neural embeddings are not modified but enhanced.

Associating vector representations to terms is the foundation of both classical distributional representations (e.g. LSA and LDA) and distributed representations (i.e. neural embeddings). Despite differences between these approaches (see [98] for details) a shared common problem is that most vector-space models represent a term with one vector, and thus, they do not capture homonymy and polysemy that require multiple vector representations per term [99]. In this study, assigning one or more focus concepts from UMLS Metathesaurus to neural embeddings of n-grams – where each UMLS Metathesaurus concept can have one or more senses (i.e. UMLS Semantic Types) – can be interpreted as a way to restore the homonymy and polysemy inherent in biomedical/clinical terms.

Inclusion of related terms is critical when querying PubMed or clinical narratives with the aim to help health professionals answer questions about patient care. Pakhomov et al. [25] acknowledge that when querying clinical narratives for a medical condition such as "heart failure" often requires additional terms, such as "pulmonary edema" or "shortness of breath", i.e. related terms that, without being synonyms to the underlying medical condition, denote different aspects of the same underlying medical condition. Similarly, the use of related terms in PubMed searches – either in the form of text words or MeSH terms – has been acknowledged as key to develop successful literature searches that yield meaningful results [100]. For example, to answer the focused clinical question "*in patients with sepsis (population), does treatment with steroids (intervention) compared with no steroids (comparison) alter mortality (outcome)?*" [101], the related term "septic shock" or "severe sepsis" can be used instead of the term "sepsis". Indeed, a PubMed search for systematic reviews of "severe sepsis" and "steroids" identified fewer useful articles than the more general search for "sepsis" and "steroids" [101]. It should be noted that a) narrowing a PubMed search to fewer useful articles is critical for retrieving a quantity of articles that is manageable; and b) the Cochrane Handbook for Systematic Reviews of Interventions [102] established PICO (Participants, Interventions, Comparisons and Outcomes) as the basis to prepare and maintain Cochrane systematic reviews, which lay the foundations for evidence-based medicine. PICO is typically applied when formulating focused clinical question as above.

Term pairs like ("heart failure", "pulmonary edema") or ("severe sepsis", "steroids") do not appear among the main four similarity and relatedness benchmarks [27] that are specific to the medical/clinical domain. Furthermore, in total there are fewer than 1 K term pairs in these benchmarks, whilst the 2018AA Metathesaurus release contains 3.67 M concepts and 14 M unique concept names [103]. McInnes et al. [104] developed the UMLS-Similarity that exploits both hierarchical and non-hierarchical information in the UMLS Metathesaurus (i.e. the MRREL table). UMLS-Similarity has two limitations. On the one hand, if the underlying terminologies in the UMLS Metathesaurus (e.g. SNOMED CT or MeSH) do not have pairwise relations, such as ("heart failure", "pulmonary edema"), they will not be included in the MRREL table. On the other hand, the semantic similarity measures included within UMLS-Similarity provide at most a numeric value indicating the relatedness of two UMLS CUIs with no further information. For example, using the UMLS-Similarity Web Interface [105] with the CUI pair (C0018801, C0034063) corresponding to the term pair ("heart failure", "pulmonary edema"), we obtain:

- "*The relatedness of Heart failure, NOS (C0018801) and Pulmonary edema, NOS (C0034063) using Adapted Lesk (lesk) is 4134*"; for further information about the Lesk measure see [106].
- "*The relatedness of Heart failure, NOS (C0018801) and Pulmonary edema, NOS (C0034063) using Vector Measure*

(vector) is 0.753"; for further information about vector measure see [106].

This paper uses evidence-based resources as a mechanism to validate the similarity and relatedness of term pairs such as ("heart failure", "pulmonary edema") while providing one or more textual excerpts of external evidence from systematic research or terminological resources. The focus concept pair (C0018801|Heart failure, C0034063| Pulmonary Edema) appears in the Additional file 1 within the worksheet "VetCN" and has the validation label "Relatedness by exact/approximate match" as the term "pulmonary oedema" appears within BMJ Best Practice for chronic congestive heart failure [85], and thus, there is current best evidence of "pulmonary edema" being clinically related to "heart failure" (see details in the Background section). An advantage of our proposal is providing evidence-based provenance for the association (semantic relatedness) where the evidence-based source is available for the clinicians to consult.

The validation with content from BMJ Best Practice is limited to only one document per medical condition. For example, for the "asthma" medical condition, BMJ Best Practice for "asthma in adults" [107] was chosen. However, if content of BMJ Best Practice for asthma was more broadly taken, for example including "asthma in children" [108], more 3-tuples (target concept, candidate concept, validation label) would have the validation label "Relatedness by exact/approximate match". Currently, the validation label "Relatedness by exact/approximate match" appears in: a) 206 of the total of 342 unique 3-tuples; b) 95 of the total 192 unique 3-tuples for VetCN; and c) 133 of the total 185 unique 3-tuples for PMSB. So overall, 60% of the 342 unique 3-tuples are "grounding" or "normalising" terms within the 11 documents of BMJ Best Practice. This percentage increases to 72% for the 3-tuples from the PMSB dataset, and drops to 49% for the 3-tuples from the VetCN dataset. As BMJ Best Practice documents need frequent updates from the literature, it is worth noticing the contribution to normalisation made from the PMSB dataset.

It has been acknowledged that "*MetaMap's greatest weakness is its reduced accuracy in the presence of ambiguity*" [109]. This is particularly so when short forms appear within the n-grams. This study implements a new short form detector, which deals with n-grams from both biomedical/clinical documents, and has demonstrated how precision, recall, and F measure of MetaMap can be improved if short form detection and expansion into long form is performed before applying MetaMap to n-grams (EXP-2). The difference in MetaMap performance between EXP-1 and EXP-2 is aligned with several studies [35,36,37] that have shown shortcomings of UMLS when dealing with short forms.

The assessment of MetaMap output has required close collaboration amongst three domain experts due to the large volume of MetaMap output [109]. In this study, MetaMap produced 2627 possible UMLS CUIs for the set of 613 n-grams. Another difficulty is a noticeable lack of mappings from MetaMap, such as:

- The n-gram "(HFpEF)" is a candidate term for the target term "heart_failure" from the neural embeddings created with Skip-gram using PMSB. "(HFpEF)" can be expanded into "(heart failure with preserved ejection fraction)" with Allie. Although the UMLS Metathesaurus concept "C3889077|Heart failure with preserved ejection fraction" exists in the UMLS releases of 2016, MetaMap could not do the mapping.
- The n-gram "macrovascular_disease" is a candidate term for the target term "diabetes" from the neural embeddings created with Skip-gram using PMSB. Although the UMLS Metathesaurus concept "C2609253|Macrovascular disease" exists in the UMLS releases of 2016, MetaMap could not do the mapping.

The column I of the worksheet "VetCN" in the Additional file 1 shows the value "WS" when there is a spelling error. For example, the candidate term "artritis" in row 292 and the candidate term "athritis" in row 321 are incorrect spellings of the term "arthritis". Candidate terms with wrong spellings are typically not recognised by MetaMap; this is an area for further work.

Aronson and Lang [109] note "*MetaMap's inability to perform in real-time situations*". Although this study presents detailed guidelines to decide if a UMLS CUI is a suitable focus concept for an n-gram, it is difficult to envision such guidelines working in real-time without human experts in the loop. Indeed, the SemDeep pipeline presented, whilst delivering promising results for both VetCN and PMSB, is time-consuming and labour-intensive from the domain experts perspective, e.g. four domain experts were needed to validate concept pairs with content from BMJ Best Practice and external resources (when necessary).

This study corroborates the observation made by Pakhomov et al. [60] that word embeddings from the biomedical literature have a performance alike to word embeddings from clinical notes for a semantic similarity and relatedness task. There are two main differences between Pakhomov et al. [60] study and this study:

- Pakhomov et al. [60] examined only the terms within the medical benchmark dataset [27], which contains 724 single-word pairs. Using single-word terms is a severe limitation as "*most medical terms consist of more than one word*" [60].
- This study has created neural embeddings from veterinary clinical notes instead of human clinical notes. Hence, this study demonstrated quantitatively the advantages of using a One Health approach instead of keeping the conventional division between veterinary medicine and medical science.

SNOMED CT is the world-leading clinical terminology. In UK, the National Health Service (NHS) has chosen SNOMED CT as the single terminology for the direct management of patient's care across all care settings in England [110]. A fundamental reason for uptaking a standardised terminology is enabling interoperability [111]. The proposed ontological representation supports:

- Storage of precoordinated SNOMED CT expressions, i.e. UMLS Metathesaurus concepts mapped to SNOMED CT concepts. Hence, this study supports the implementation level 1 (the second level of the three proposed, i.e. 0 to 2) of the SNOMED CT expression storage [112].
- Crafting subsets of SNOMED CT: we also have proposed a step-by-step methodology for SNOMED CT subset development, which has been acknowledged as an unmet need for implementing SNOMED CT in clinical settings [111]. Worldwide, there are three SNOMED CT datasets available to download through the UMLS [14]. In the UK, NHS has developed 80 SNOMED CT human-readable subsets [113]. None of the SNOMED CT datasets mentioned were developed for a well-known medical condition. For example, there is a NHS human-readable subset of SNOMED CT for "Ophthalmology" although not for glaucoma, and likewise, there is a NHS subset for "Respiratory medicine" although not for asthma.
- Building One Health SPARQL SELECT queries that successfully retrieved medical knowledge (validated with BMJ Best Practice content) about the diagnosis and management of the 11 well-known medical conditions that affect humans and animals. These One Health queries exploit knowledge within the locality-

based modules extracted from the SNOMED CT ontology and outside of SNOMED CT, i.e. within the OBAN "sometimes true" association relationships.

CONCLUSIONS

In this paper we demonstrated how a Semantic Deep Learning approach can transform neural embeddings of n-grams created from the unstructured text of 300K PubMed Systematic Reviews (medical science) and 2.5M veterinary clinical narratives (veterinary medicine) into augmented clinically meaningful biomedical datasets of normalised and interlinked concepts. This study applies the Semantic Web technologies and reuses existing ontologies to 1) encapsulate One Health knowledge about 11 well-known diseases in human and animals that is formal and computable (e.g. allowing a wide range of queries); and 2) separate a "sometimes true" association relationship between two biomedical concepts and its evidence-based provenance (i.e. BMJ Best Practice content). The main benefit of the Semantic Deep Learning approach proposed is in obtaining reliable and usable One Health knowledge (e.g. knowledge useful for public health) that enhances the world-leading clinical terminology SNOMED CT.

AVAILABILITY OF DATA AND MATERIALS

All data generated or analysed during this study are included in this article and its Additional files 1,2,3 and 4.

ACKNOWLEDGEMENTS

Thanks to Tim Furmston (University of Manchester) for help with software and e-infrastructure, and to the medical consultants from the Hospital do Salnés (Spain) for their support and valuable medical insights as well as to the anonymous reviewers for their useful comments. An initial version of this paper has been presented at the Healthcare Text Analytics Conference 2018 (HealTAC), in Manchester, UK in April 2018.

REFERENCES

1. WHO: One Health. September 2017. http://www.who.int/features/qa/one-health/en/.
2. Kahn LH. Perspective: the one-health way. Nature. 2017;543(7647):S47.
3. Stroud, C., Dmitriev, I., Kashentseva, E., Bryan, J.N., Curiel, D.T., Rindt, H., Reinero, C., Henry, C.J., Bergman, P.J., Mason, N.J. and Gnanandarajah, J.S., 2016, August. A One Health overview, facilitating advances in comparative medicine and translational research. In Clinical and translational medicine (Vol. 5, No. 1, p. 26). Springer Berlin Heidelberg.
4. Semantic Deep Learning. http://semdeep.iiia.csic.es.
5. Semantic Deep Learning. http://www.semantic-web-journal.net/blog/call-papers-special-issue-semantic-deep-learning. Accessed 25th April 2019.
6. LeCun Y, Bengio Y, Hinton G. Deep learning. Nature. 2015;521(7553):436–44.
7. Sackett DL, Rosenberg W, Gray JA, Haynes RB, Richardson WS. Evidence based medicine: what it is and what it isn't. Bmj. 1996;312(7023):71–2.
8. BMJ Best Practice. https://bestpractice.bmj.com.
9. DynaMed Plus. http://www.dynamed.com/.
10. UpToDate. https://www.uptodate.com/.
11. The National Institute for Health and Care Excellence (NICE). https://www.nice.org.uk/.
12. Field MJ, Lohr KN. Committee to advise the public health service on clinical practice guidelines IoM: clinical practice guidelines: directions for a new program. Washington, D.C.: National Academy Press; 1990.
13. Rebholz-Schuhmann D, Oellrich A, Hoehndorf R. Text-mining solutions for biomedical research: enabling integrative biology. Nature reviews. Genetics. 2012;13(12):829–39.
14. UMLS. https://www.nlm.nih.gov/research/umls/index.html.
15. MeSH. https://www.nlm.nih.gov/mesh/.
16. Huang M, Névéol A, Lu Z. Recommending MeSH terms for annotating biomedical articles. J Am Med Inform Assoc. 2011;18(5):660–7.
17. PubMed Systematic Reviews, https://www.nlm.nih.gov/bsd/pubmed_subsets/sysreviews_strategy.html.

18. One Health Initiative. http://www.onehealthinitiative.com.
19. SAVSNET. https://www.liverpool.ac.uk/savsnet/. Accessed 25th April 2019.
20. McCrae, J., Spohr, D. and Cimiano, P., 2011, May. Linking lexical resources and ontologies on the semantic web with lemon. In extended semantic web conference (pp. 245-259). Springer, Berlin, Heidelberg.
21. Sarntivijai S, Vasant D, Jupp S, Saunders G, Bento AP, Gonzalez D, Betts J, Hasan S, Koscielny G, Dunham I, Parkinson H, Malone J. Linking rare and common disease: mapping clinical disease-phenotypes to ontologies in therapeutic target validation. J Biomed Semantics. 2016;7(1):8.
22. Zhang Q-S, Zhu S-C. Visual interpretability for deep learning: a survey. Frontiers of Information Technology and Electronic Engineering. 2018;19(1):27–39.
23. Caviedes JE, Cimino JJ. Towards the development of a conceptual distance metric for the UMLS. J Biomed Inform. 2004;37(2):77–85.
24. Pedersen T, Pakhomov SV, Patwardhan S, Chute CG. Measures of semantic similarity and relatedness in the biomedical domain. J Biomed Inform. 2007;40(3):288–99.
25. Pakhomov SV, Pedersen T, McInnes B, Melton GB, Ruggieri A, Chute CG. Towards a framework for developing semantic relatedness reference standards. J Biomed Inform. 2011;44(2):251–65.
26. Pakhomov, S., McInnes, B., Adam, T., Liu, Y., Pedersen, T. and Melton, G.B., 2010. Semantic similarity and relatedness between clinical terms: an experimental study. In AMIA annual symposium proceedings (Vol. 2010, p. 572). American medical informatics association.
27. Semantic similarity and relatedness resources. http://rxinformatics.umn.edu/SemanticRelatednessResources.html.
28. Cohen T, Widdows D. Empirical distributional semantics: methods and biomedical applications. J Biomed Inform. 2009;42(2):390–405.
29. Landauer TK, Dumais ST. A solution to Plato's problem: the latent semantic analysis theory of acquisition, induction, and representation of knowledge. Psychol Rev. 1997;104(2):211.
30. Blei DM, Ng AY, Jordan MI. Latent dirichlet allocation. J Mach Learn Res. 2003;3(Jan):993–1022.
31. Faruqui, M., Tsvetkov, Y., Rastogi, P. and Dyer, C., 2016. Problems with evaluation of word Embeddings using word similarity tasks.

In proceedings of the 1st workshop on evaluating vector-space representations for NLP (pp. 30-35).

32. Hill F, Reichart R, Korhonen A. Simlex-999: evaluating semantic models with (genuine) similarity estimation. Computational Linguistics. 2015;41(4):665–95.
33. Gerz D, Vulić I, Hill F, Reichart R, Korhonen A. SimVerb-3500: a large-scale evaluation set of verb similarity. EMNLP 2016. Proceedings of the 2016 Conference on Empirical Methods in Natural Language Processing; 2016, pp. 2173–182.
34. Wren JD, Chang JT, Pustejovsky J, Adar E, Garner HR, Altman RB. Biomedical term mapping databases. Nucleic Acids Res. 2005;33(suppl_1):D289–93.
35. Liu, H., Lussier, Y.A. and Friedman, C., 2001. A study of abbreviations in the UMLS. In proceedings of the AMIA symposium (p. 393-7). American medical informatics association.
36. Xu, H., Stetson, P.D. and Friedman, C., 2007. A study of abbreviations in clinical notes. In AMIA annual symposium proceedings (Vol. 2007, p. 821-5). American medical informatics association.
37. Moon S, Pakhomov S, Liu N, Ryan JO, Melton GB. A sense inventory for clinical abbreviations and acronyms created using clinical notes and medical dictionary resources. J Am Med Inform Assoc. 2013;21(2):299–307.
38. Adar E. SaRAD: a simple and robust abbreviation dictionary. Bioinformatics. 2004;20(4):527–33.
39. Zhou W, Torvik VI, Smalheiser NR. ADAM: another database of abbreviations in MEDLINE. Bioinformatics. 2006;22(22):2813–8.
40. Yamamoto Y, Yamaguchi A, Bono H, Takagi T. Allie: a database and a search service of abbreviations and long forms. Database. 2011;2011.
41. Xu, H., Stetson, P.D. and Friedman, C., 2007. A study of abbreviations in clinical notes. In AMIA annual symposium proceedings (Vol. 2007, p. 821-825). American medical informatics association.
42. Xu H, Stetson PD, Friedman C. Methods for building sense inventories of abbreviations in clinical notes. J Am Med Inform Assoc. 2009;16(1):103–8.
43. Wu, Y., Denny, J.C., Rosenbloom, S.T., Miller, R.A., Giuse, D.A. and Xu, H., 2012. A comparative study of current clinical natural language processing systems on handling abbreviations in discharge summaries.

In AMIA annual symposium proceedings (Vol. 2012, p. 997-1003). American medical informatics association.

44. Arguello Casteleiro M., G. Demetriou, W.J. Read, M.J. Fernandez-Prieto, D. Maseda-Fernandez, G. Nenadic, J. Klein, J.A. Keane, R. Stevens. (Sept 2016). Deep Learning meets Semantic Web: A feasibility study with the Cardiovascular Disease Ontology and PubMed citations. In Proceedings of the 7th Workshop on Ontologies and Data in Life Sciences, organized by the GI Workgroup Ontologies in Biomedicine and Life Sciences (OBML). CEUR Vol. 1692.
45. Arguello Casteleiro M., M., Prieto, M.J.F., Demetriou, G., Maroto, N., Read, W.J., Maseda-Fernandez, D., Des Diz, J.J., Nenadic, G., Keane, J.A. and Stevens, R., 2016. Ontology Learning with Deep Learning: a Case Study on Patient Safety Using PubMed In SWAT4LS.
46. Arguello Casteleiro M., D. Maseda-Fernandez, G. Demetriou, W. Read, M.J. Fernandez-Prieto, J. Des-Diz, G. Nenadic, J. Keane, and R. Stevens (April 2017). A case study on Sepsis using PubMed and Deep Learning for Ontology Learning. In Proceedings of Informatics for Health 2017. In "Studies in Health Technology and Informatics" by IOS Press.
47. Arguello Casteleiro M., C. Martínez Costa, J. Des-Diz, M.J. Fernandez-Prieto, C. Wroe, D. Maseda-Fernandez, G. Demetriou, G. Nenadic, J. Keane, S. Schulz and R. Stevens (Dec 2017). Experiments to create ontology-based disease models for diabetic retinopathy from different biomedical resources. In proceedings of semantic web applications and tools for health care and life sciences (SWAT4HCLS 2017). CEUR Vol.
48. SPARQL query language. https://www.w3.org/TR/sparql11-query/.
49. Apache Jena ARQ. https://jena.apache.org/documentation/query/index.html.
50. VetSCT. https://www.nlm.nih.gov/research/umls/sourcereleasedocs/current/SNOMEDCT_VET/.
51. UMLS API. https://documentation.uts.nlm.nih.gov.
52. Horridge M, Bechhofer S. The owl api: a java api for owl ontologies. Semantic Web. 2011;2(1):11–21.
53. word2vec. http://code.google.com/p/word2vec/.
54. Mikolov T, Sutskever I, Chen K, Corrado GS, Dean J. Distributed representations of words and phrases and their compositionality.

InAdvances in neural information processing systems 2013; 3111–3119.

55. Arguello-Casteleiro, M., Demetriou, G., Read, W., Prieto, M.J.F., Maroto, N., Fernandez, D.M., Nenadic, G., Klein, J., Keane, J. and Stevens, R., 2018. Deep learning meets ontologies: experiments to anchor the cardiovascular disease ontology in the biomedical literature . J Biomed Semantics, 9(1), p.13
56. Pilehvar MT, Collier N. Improved semantic representation for domain-specific entities. ACL. 2016;2016:12.
57. Pyysalo, S., Ginter, F., Moen, H., Salakoski, T., & Ananiadou, S.: Distributional semantics resources for biomedical text pro-cessing. In Proc. of Languages in Biology and Medicine (2013).
58. Muneeb TH, Sahu SK, Anand A. Evaluating distributed word representations for capturing semantics of biomedical concepts: Proceedings of ACL-IJCNLP; 2015. p. 158.
59. Minarro-Giménez, J. A., Marín-Alonso, O., & Samwald, M.: Exploring the application of deep learning techniques on medical text corpora. In e-Health – for continuity of care, IOS Press, pp. 584–588 (2014).
60. Pakhomov SV, Finley G, McEwan R, Wang Y, Melton GB. Corpus domain effects on distributional semantic modeling of medical terms. Bioinformatics. 2016;32(23):3635–44.
61. Nadkarni PM, Ohno-Machado L, Chapman WW. Natural language processing: an introduction. J Am Med Inform Assoc. 2011;18(5):544–51.
62. SNOMED CT Compositional Grammar v2.3.1. http://snomed.org/scg.
63. SI unit. https://bitbucket.org/birkenfeld/ipython-physics/raw/default/physics.py .
64. Other non-SI units. https://www.bipm.org/utils/common/pdf/si_brochure_8_en.pdf.
65. Units of Length, Mass, and Liquid Volume. https://www.nist.gov/sites/default/files/documents/2017/04/28/AppC-12-hb44-final.pdf. Accessed 25th April 2019.
66. Word Frequencies in Written and Spoken English: based on the British National Corpus. http://ucrel.lancs.ac.uk/bncfreq/flists.html.
67. Manning CD, Schütze H. Foundations of statistical natural language processing: MIT press; 1999.

68. Pratt, W. and Yetisgen-Yildiz, M., 2003. A study of biomedical concept identification: MetaMap vs. people. In AMIA annual symposium proceedings (Vol. 2003, p. 529-533). American medical informatics association.

69. Smucker, M.D., Allan, J. and Carterette, B., 2007. A comparison of statistical significance tests for information retrieval evaluation. In proceedings of the sixteenth ACM conference on conference on information and knowledge management (pp. 623-632). ACM.

70. Box GE, Hunter WG, Hunter JS. Statistics for experimenters: John Wiley & Sons; 1978.

71. Pedregosa F, Varoquaux G, Gramfort A, Michel V, Thirion B, Grisel O, Blondel M, Prettenhofer P, Weiss R, Dubourg V, Vanderplas J. Scikit-learn: machine learning in python. J Mach Learn Res. 2011;12(Oct):2825–30.

72. Opal SM, Dellinger RP, Vincent JL, Masur H, Angus DC. The next generation of sepsis trials: What's next after the demise of recombinant human activated protein C ? critical care medicine. 2014;42(7):1714.

73. Extended lemon core ontology. http://semdeep.cs.man.ac.uk/inOWL/lemonEXT_core.owl.

74. Modified OBAN core ontology. http://semdeep.cs.man.ac.uk/inOWL/OBANmod_core.owl.

75. UMLS Semantic Types and Groups. https://metamap.nlm.nih.gov/Docs/SemGroups_2013.txt.

76. Basic Formal Ontology (BFO). http://www.obofoundry.org/ontology/bfo.html.

77. oboInOwl meta-model. http://www.geneontology.org/formats/oboInOwl.

78. Horridge, M., Drummond, N., Goodwin, J., Rector, A.L., Stevens, R. and Wang, H., 2006. The Manchester OWL syntax. In OWLed (Vol. 216).

79. Relations Ontology (RO). http://www.ontobee.org/ontology/RO.

80. Ontology Lexicon (Ontolex). https://www.w3.org/2016/05/ontolex/.

81. Cimiano, P., McCrae, J., Buitelaar, P. and Montiel-Ponsoda, E., 2013. On the role of senses in the ontology-lexicon. In new trends of research in ontologies and lexical resources (pp. 43-62). Springer, Berlin, Heidelberg.

82. UMLS Semantic Types. https://www.nlm.nih.gov/research/umls/META3_current_semantic_types.html .
83. Evidence & Conclusion Ontology (ECO). http://purl.obolibrary.org/obo/eco.owl .
84. Bibliographic Ontology Specification ontology (BIBO). http://purl.org/ontology/bibo/ .
85. BMJ Best Practice: Chronic congestive heart failure. http://bestpractice.bmj.com/topics/en-gb/61.
86. MedlinePlus. https://medlineplus.gov/ .
87. Grau, B.C., Horrocks, I., Kazakov, Y. and Sattler, U., 2007, May. Just the right amount: extracting modules from ontologies. In proceedings of the 16th international conference on world wide web (pp. 717-726). ACM.
88. Tsarkov, D. and Horrocks, I., 2006, August. FaCT++ description logic reasoner: system description. In international joint conference on automated reasoning (pp. 292-297). Springer, Berlin, Heidelberg.
89. Spelling Corrector. http://norvig.com/spell-correct.html.
90. Korinthenberg R, Burkart P, Woelfle C, Moenting JS, Ernst JP. Pharmacology, efficacy, and tolerability of potassium bromide in childhood epilepsy. J Child Neurol. 2007;22(4):414–8.
91. Rundfeldt C, Löscher W. The pharmacology of imepitoin: the first partial benzodiazepine receptor agonist developed for the treatment of epilepsy. CNS drugs. 2014;28(1):29–43.
92. BMJ Best Practice: Open-angle glaucoma. http://bestpractice.bmj.com/topics/en-gb/373.
93. Zhao M, Mu Y, Dang Y, Zhu Y. Secondary glaucoma as initial manifestation of ring melanoma: a case report and review of literature. Int J Clin Exp Pathol. 2014;7(11):8163.
94. BMJ Best Practice: Obesity in adults. http://bestpractice.bmj.com/topics/en-gb/211.
95. Alshahrani M, Khan MA, Maddouri O, Kinjo AR, Queralt-Rosinach N, Hoehndorf R. Neuro-symbolic representation learning on biological knowledge graphs. Bioinformatics. 2017;33(17):2723–30.
96. Jauhar, S.K., Dyer, C. and Hovy, E., 2015. Ontologically grounded multi-sense representation learning for semantic vector space models. In proceedings of the 2015 conference of the north American chapter

of the Association for Computational Linguistics: human language technologies (pp. 683-693).

97. Alsuhaibani M, Bollegala D, Maehara T, Kawarabayashi KI. Jointly learning word embeddings using a corpus and a knowledge base. PloS One. 2018;13(3):e0193094.
98. Turian, J., Ratinov, L. and Bengio, Y., 2010. Word representations: a simple and general method for semi-supervised learning. In proceedings of the 48th annual meeting of the association for computational linguistics (pp. 384-394). Association for Computational Linguistics.
99. Huang, E.H., Socher, R., Manning, C.D. and Ng, A.Y., 2012. Improving word representations via global context and multiple word prototypes. In proceedings of the 50th annual meeting of the Association for Computational Linguistics: long papers-volume 1 (pp. 873-882). Association for Computational Linguistics.
100. Hersh WR, Hickam DH. How well do physicians use electronic information retrieval systems?: a framework for investigation and systematic review. Jama. 1998;280(15):1347–52.
101. Doig GS, Simpson F. Efficient literature searching: a core skill for the practice of evidence-based medicine. Intensive Care Med. 2003;29(12):2119–27.
102. Cochrane Handbook for Systematic Reviews of Interventions. http://handbook-5-1.cochrane.org.
103. UMLS 2018AA. https://www.nlm.nih.gov/pubs/techbull/mj18/mj18_umls_2018aa_release.html.
104. McInnes, B.T., Pedersen, T. and Pakhomov, S.V., 2009. UMLS-Interface and UMLS-similarity: open source software for measuring paths and semantic similarity. In AMIA annual symposium proceedings (Vol. 2009, p. 431). American medical informatics association.
105. UMLS-Similarity Web Interface. http://maraca.d.umn.edu/cgi-bin/umls_similarity/umls_similarity.cgi.
106. UMLS-Similarity: Relatedness measures. http://maraca.d.umn.edu/umls_similarity/relatedness_measures.html.
107. BMJ Best Practice: Asthma in adults. http://bestpractice.bmj.com/topics/en-gb/44.
108. BMJ Best Practice: Asthma in children. http://bestpractice.bmj.com/topics/en-gb/782.

109. Aronson AR, Lang FM. An overview of MetaMap: historical perspective and recent advances. J Am Med Inform Assoc. 2010;17(3):229–36.
110. Personalised Health and Care 2020: A Framework for Action. https://www.gov.uk/government/publications/personalised-health-and-care-2020.
111. Lee D, Cornet R, Lau F, De Keizer N. A survey of SNOMED CT implementations. J Biomed Inform. 2013;46(1):87–96.
112. SNOMED CT® Technical Implementation Guide. January 2015 International Release. https://confluence.ihtsdotools.org/display/DOCTIG/Technical+Implementation+Guide.
113. NHS: SNOMED CT human-readable subsets. https://isd.digital.nhs.uk/trud3/user/guest/group/0/pack/40.

Section 3: Livestock

Cow

CHAPTER 13

The Burden of Mycobacterial Disease in Ethiopian Cattle: Implications for Public Health

Stefan Berg[1] , Rebuma Firdessa[2] , Meseret Habtamu[2] , Endalamaw Gadisa[2] , Araya Mengistu[2] , Lawrence Yamuah[2] , Gobena Ameni[2,3], Martin Vordermeier[1] , Brian D. Robertson[4] , Noel H. Smith[1] , Howard Engers[2] , Douglas Young[4] , R. Glyn Hewinson[1] , Abraham Aseffa[2] , Stephen V. Gordon[1,5,6,7,8]

[1] TB Research Group, Veterinary Laboratories Agency, Weybridge, New Haw, Addlestone, Surrey, United Kingdom

[2] Armauer Hansen Research Institute, Addis Ababa, Ethiopia

[3] Aklilu Lemma Institute of Pathobiology, Addis Ababa University, Addis Ababa, Ethiopia

[4] Department of Infectious Disease and Microbiology, Imperial College, South Kensington

Citation: Berg S, Firdessa R, Habtamu M, Gadisa E, Mengistu A, Yamuah L, et al. (2009) The Burden of Mycobacterial Disease in Ethiopian Cattle: Implications for Public Health. PLoS ONE 4(4): e5068. https://doi. org/10. 1371/journal. pone. 0005068.

Campus, London, United Kingdom

[5] School of Agriculture, Food Science and Veterinary Medicine, University College Dublin, Dublin, Ireland

[6] School of Medicine and Medical Science, University College Dublin, Dublin, Ireland

[7] School of Biomolecular and Biomedical Science, College of Life Sciences, University College Dublin, Dublin, Ireland

[8] UCD Conway Institute of Biomolecular and Biomedical Research, University College Dublin, Dublin, Ireland

ABSTRACT

Background

Bovine tuberculosis (bTB), caused by *Mycobacterium bovis*, is a debilitating disease of cattle. Ethiopia has one of the largest cattle populations in the world, with an economy highly dependent on its livestock. Furthermore, Ethiopia has one of the highest incidence rates of human extrapulmonary TB in the world, a clinical presentation that is often associated with transmission of *M. bovis* from cattle to humans.

Methodology/Principal Findings

Here we present a comprehensive investigation of the prevalence of bTB in Ethiopia based on cases identified at slaughterhouses. Out of approximately 32,800 inspected cattle, ~4. 7% showed suspect tuberculous lesions. Culture of suspect lesions yielded acid-fast bacilli in ~11% of cases, with *M. bovis* accounting for 58 of 171 acid-fast cultures, while 53 isolates were non-tuberculous mycobacteria. Strikingly, *M. tuberculosis* was isolated from eight cattle, an unusual finding that suggests human to animal transmission.

Conclusions/Significance

Our analysis has revealed that bTB is widely spread throughout Ethiopia, albeit at a low prevalence, and provides underpinning evidence for public health policy formulation.

INTRODUCTION

Bovine tuberculosis (bTB) is an infectious disease of cattle caused by *Mycobacterium bovis* and characterized by the formation of granulomatous lesions (tubercles) classically seen in the lungs and draining lymph nodes. Infection with *M. bovis* can be transmitted from cattle to humans, mainly through the consumption of contaminated milk and meat products; because of the route of infection, disease often manifests itself as extrapulmonary TB. Indeed, Kidane and colleagues concluded that among 35 PCR-positive cases of tuberculous lymphadenitis from southern Ethiopia, 29 (82. 9%) were caused by *M. tuberculosis* and six (17. 1%) were caused by *M. bovis* [1]. It is also noteworthy that 36% of incident TB cases in Ethiopia are extrapulmonary [2]. This is exceptionally high, considering that ~15% of incident TB is extrapulmonary in most high burden countries [2].

Ethiopia has ~40 million cattle, the largest cattle population in Africa and the seventh in size in the world (Ethiopian Ministry of Agriculture and Rural Development; http://www. moard. gov. et/statstical. htm). Approximately 80% of the labour-force works in agriculture, with this sector accounting for 47% of GDP and with an export value of agricultural goods of about USD $406 million in the year 2000 [3]. The livestock sector is of national importance and the Ethiopian government has set goals to improve productivity in this sector, such as intensification and the importation of "exotic" high performance cattle breeds such as Holstein.

Previous studies on bTB in Ethiopia have shown that prevalence varies depending on husbandry methods, with rural settings showing a lower prevalence compared to intensive dairy farms [4]. However, there have been no studies to date that have addressed the total burden of mycobacterial infection in cattle, nor do we know the molecular types of *M. bovis* that are circulating in Ethiopia. Furthermore, given the large cattle population and endemic bTB problem it is possible that some of the incidence of human extrapulmonary TB is caused by infection with *M. bovis*. While this latter connection has yet to be formally shown, knowledge of mycobacterial strains circulating in Ethiopian cattle is essential baseline data to inform public health policy.

Here we present an extensive study of the prevalence of mycobacterial infection from cattle slaughtered at abattoirs across different eco-epidemiological settings of Ethiopia. Suspect tuberculous lesions were processed for culture, with acid-fast isolates typed using molecular methods to provide definitive strain characterisation. The spatial distribution of strains and molecular types could then be mapped across the eco-epidemiological settings. We show that while infection with *M. bovis* is a significant cause of tuberculous lesions in cattle, other acid-fast bacilli (AFB), notably *M.*

tuberculosis and non-tuberculous mycobacteria (NTM), are also considerable causes of infection and disease.

MATERIALS AND METHODS

Study Sites

The investigated cattle specimens in this study were collected from abattoirs located in the rural areas of Gonder (abbrev. Go), Woldiya (Wo), Gimbi (Gi), Butajira (Bu), and Jinka (Ji) in Ethiopia (Figure 1). Each geographical area represents a different eco-epidemiological setting, comprising subsistence farming, agro-pastoral, and pastoral communities. People living in agro-pastoral settings are primarily dependent on cattle for farming their land, while pastoral communities are directly dependent on cattle to provide meat and milk products. To complement this study with samples from an urban setting, specimens from lesioned cattle were also collected at the "Enterprise" abattoir in the capital Addis Ababa (AA).

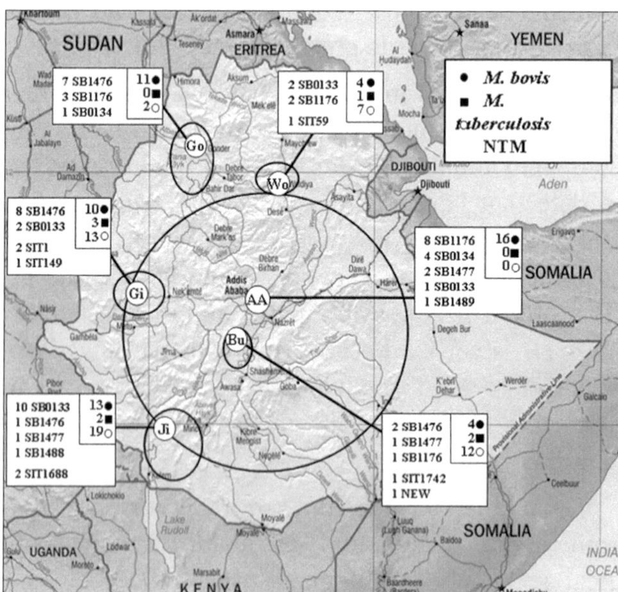

Figure 1: Geographical distribution of *Mycobacterium* **isolates from cattle in Ethiopia.**

The total number of *M. bovis* (•), *M. tuberculosis* (▪), and NTM (○), isolated from respective abattoir are indicated in respective box, as well as characterised spoligotype patterns. Approximate area coverage for each abattoir is shown by a solid circle.

https://doi. org/10. 1371/journal. pone. 0005068. g001

Identifying a precise geographical origin for each investigated animal was not possible. However, there were no nomads in our study areas, suggesting that any cattle movement was due to local, or possibly national, trading. The five rural abattoirs (Gonder, Woldiya, Gimbi, Butajira, and Jinka) documented the markets from where slaughtered animals were purchased by local butchers. Based on this information, and the assumption that no significant cattle movement over large distances was taking place, we determined the approximate area that each abattoir covers (Figure 1). The situation in Addis Ababa is different; the high demand for meat products in the capital necessitates larger supplies. Therefore, merchants buy cattle from a much larger geographical area and transport them to Addis Ababa for slaughter. The area covered by the Addis Ababa abattoir is therefore much greater (Figure 1).

Sample Collection, Processing, and Culturing

The study obtained ethical clearance from institutional (Armauer Hansen Research Institute (AHRI), All Africa Leprosy, Tuberculosis and Rehabilitation Training Centre (ALERT), Ethiopia; VLA, UK) and Ethiopian national ethical review committees. Trained staff at the six abattoirs performed ante-mortem examination (including sex, breed, and body condition *etc.*) of all or randomly selected cattle that were entering the abattoirs. All ante-mortem examinations were followed up by post-mortem inspections to look for suspect tuberculous lesions in lungs and a range of lymph nodes (hepatic, mesenteric, bronchial, mediastinal, mandibular, and medial retropharyngeal LNs). A description of the lesions (purulent, caseous, or calcified) was recorded. All sampled specimens from respective carcasses were pooled together and kept in phosphate buffer saline (pH 7. 2) at 4°C in universal containers and transported on ice to the AHRI laboratory in Addis Ababa within one week. Specimens were processed according to standard methods [5]. In brief, the tissue samples were dissected and manually homogenised using a pestle and mortar, followed by decontamination by shaking in an equal volume of 4% NaOH for 15 min and concentrated by centrifugation at 3,000×g for 15 min. The sediment was neutralized with 2 N HCl, using phenol red as an indicator, and then used to inoculate three different media slants: Löwenstein–Jensen (LJ) media with and without pyruvate, and Middlebrook 7H11 medium supplemented as previously described [6]. The slants were incubated at 37°C for six weeks; slants were examined on a weekly basis for the presence of mycobacterial colonies. Cultures were considered negative if no visible growth was detected after six weeks incubation. Microscopic examination of cultures using the Ziehl–Neelsen staining method was performed to select AFB positive isolates. Heat-killed cells of each isolate were prepared by mixing ~2 loopfuls of cells (≥20 µl cell pellet) in 200 µl dH_2O followed by incubation at 80°C for 1 hour. AFB positive cultures were prepared as 20% glycerol stocks and stored at −80°C.

Molecular Typing

Heat-killed AFB positive samples were investigated by multiplex PCR for the presence or absence of RD4, RD9, and TbD1 [7] using the primers listed in Table 1. Oligonucleotide primers used for deletion typing of RD105, RD142, RD150, and RD181 have been described previously [8]. Characterisation of NTM was performed by genus typing [9] and isolates positive for the *Mycobacterium* genus were sequenced at the 16S rDNA locus [10]. 16S rDNA sequences were used in BLAST searches [11] of databases at NCBI and RIDOM (http://rdna. ridom. de)[12]; particularly, the sequences of the hypervariable Region A and Region B [13] were considered when determining the *Mycobacterium* species. The PCR amplification mixtures used for RD and genus typing were as follows: reactions were performed in a total volume of 20 µl consisting of 10 µl HotStarTaq Master Mix (Qiagen, United Kingdom), 7. 1 µl distilled H_2O, 0. 3 µl of each oligonucleotide primer (100 µM), and 2 µl DNA template (heat-killed cells, see above). DNA sequencing was performed by The Sequencing Service (College of Life Sciences, University of Dundee, Scotland) using Applied Biosystems Big-Dye Ver 3. 1 chemistry on an Applied Biosystems model 3730 automated capillary DNA sequencer. Isolates genetically typed as belonging to the *M. tuberculosis* complex were spoligotyped as previously described [14].

Table 1: Oligonucleotide primers used for molecular typing of *Mycobacterium* **isolates and sizes of the expected PCR products**

https://doi. org/10. 1371/journal. pone. 0005068. t001

Locus	Primer name	Primer sequence	Present	Deleted
RD4	RD4_FlankF	CTCGTCGAAGGCCACTAAAG	335	446
	RD4_FlankR	AAGGCGAACAGATTCAGCAT		
	RD4_InternalF	ACACGCTGGCGAAGTATAGC		
RD9	RD9_FlankF	AACACGGTCACGTTGTCGTG	396	575
	RD9_FlankR	CAAACCAGCAGCTGTCGTTG		
	RD9_InternalR	TTGCTTCCCCGGTTCGTCTG		
TbD1	TbD1_FlankF	CTACCTCATCTTCCGGTCCA	909	486
	TbD1_FlankR	CATAGATCCCGGACATGGTG		
	TbD1_InternalR	AATCGAACTCGTGGAACACC		
16S rDNA	MYCGEN-F	AGAGTTTGATCCTGGCTCAG	1030	NA
	MYCGEN-R	TGCACACAGGCCACAAGGGA		
16S rDNA[#]	16S_AFB_F	GCGTGCTTAACACATGCAAGTC	660	NA
	16S_AFB_R	TCCTCCTGATATCTGCGCATTC		

[#]Primer set used for partial sequencing of the 16S rRNA locus [10].
NA, Not Applicable.
doi:10.1371/journal.pone.0005068.t001

RESULTS

Collection of Lesions

In total, about 32,800 cattle were investigated during the years 2006–2008 and the vast majority were of local *Bos indicus* breeds (*i. e.* Zebu). However, *Bos taurus* (*i. e.* Holstein-Frisian) and crossbreeds also occurred among the inspected cattle. At the Gonder, Woldiya, Gimbi and Butajira abattoirs, all of the slaughtered animals were also ante- and post-mortem examined during the period of 2006–2008, while in Jinka approximately 50% of the slaughtered animals were examined. From the end of 2007 to May 2008 the Addis Ababa abattoir investigated approximately 2,800 animals, corresponding to less than 2% of all cattle that passed through the abattoir during this period.

Culturing and ZN Staining Results

Approximately 1,500 samples from suspect TB lesions were processed and sown onto the three solid media: 7H11, LJ glycerol, and LJ pyruvate, as described in the Materials and Methods. From the inoculated slants, colonies could be collected at a higher frequency from 7H11 than LJ glycerol or LJ pyruvate (data not shown). In total, 171 AFB were cultured from 1524 lesioned samples (Table 2), with one animal from Gimbi showing dual infection of *M. tuberculosis* and *M. gordonae*.

Table 2: Summary of collected lesions and molecular typed AFB isolated from cattle slaughtered at six Ethiopian abattoirs

https://doi. org/10. 1371/journal. pone. 0005068. t002

Study Site	Animals investigated	Lesioned Animals	Culture and ZN positive	*M. bovis*	*M. tuberculosis*	NTM*	Total typed
Butajira	4606	281	26	4	2	12	18
Jinka	3471	278	54	13	2	19	34
Gimbi	3250	413	36	10	3	13	26
Gonder	14314	261	20	11	0	2	13
Woldiya	4338	240	19	4	1	7	12
Addis Ababa	2800	51	16	16	0	0	16
All abattoirs	32779	1524	171	58	8	53	119

*NTM = Typed as *Mycobacterium* species not from the *M. tuberculosis* complex.

doi:10.1371/journal.pone.0005068.t002

Characterisation of *Mycobacterium* Isolates by Molecular Typing

Of the 171 AFB isolated, 135 were subjected to a range of molecular typing techniques to determine their identity. To investigate if these isolates were *M. bovis*, deletion typing was performed to verify the specific deletion of the RD4 region in *M. bovis* strains (see Materials and Methods; Table 1). Fifty-eight samples generated a PCR product of 446 bp and were consequently identified as *M. bovis* (Table 2). In the cases where the RD4 region was shown to be intact, RD9 typing was performed to further identify the species. Eight isolates were shown to have both RD4 and RD9 present, suggesting that they were *M. tuberculosis*. Of the remaining AFB positive strains, 53 were typed as NTM, and 16 were not from the *Mycobacterium* genus. The frequency of NTM isolation varied between the collection sites (Table 2). Table 3 shows the NTM strains that were identified by DNA sequencing of the 16S rDNA gene as described in Materials and Methods. Overall, eleven different species were characterised of which *M. nonchromogenicum*, *M. gordonae*, *M. fortuitum*, and *M. peregrinum* were isolated from more than one abattoir.

Table 3: Non-tuberculous mycobacteria isolated from cattle in Ethiopia

https://doi. org/10. 1371/journal. pone. 0005068. t003

Species	Woldiya	Gimbi	Butajira	Jinka
M. acapulcensis		1		
M. colombiense			1	
M. engbaekii		1		
M. fortuitum	2			1
M. gordonae		3		2
M. intracellulare			2	
M. monacense		1		
M. mucogenicum				1
M. nonchromogenicum	1		7	
M. peregrinum		1	1	
M. vaccae			1	

doi:10.1371/journal.pone.0005068.t003

Strains of the *M. tuberculosis* Complex and Their Spoligotypes

Strains that were identified as *M. bovis* and *M. tuberculosis* from the six study sites were characterised by spoligotyping and their patterns are listed in Table 4. Strains of *M. bovis* were represented by seven different spoligotype patterns, of which four had not been registered before in the international *M. bovis* spoligotype database at www. Mbovis. org. Of the seven patterns SB0133, SB1476, and SB1176 were more prevalent. Interestingly, tendencies toward geographical clustering could be observed for these spoligotypes (Figure 1). First, SB0133 predominates in Jinka, but it is also represented in Addis Ababa, Gimbi, and Woldiya. The second, SB1476, is the most common pattern in both Gimbi and Gonder but it was found in Jinka and Butajira as well. SB1176 is most prevalent among the samples from Addis Ababa but can also be seen in Butajira, Gonder and Woldiya. The remaining four spoligotype patterns, SB0134, SB1488, SB1489, and SB1477, are all highly related to SB0133, and isolates with these patterns were mostly collected from the Addis Ababa or Jinka abattoirs. The largest diversity of *M. bovis* strains was found in Addis Ababa abattoir with five different spoligotypes, likely reflecting the wide geographical area from which cattle were sourced.

Table 4: Spoligotyping patterns of A) *M. bovis* **strains and B)** *M. tuberculosis* **strains isolated from cattle in Ethiopia**

https://doi. org/10. 1371/journal. pone. 0005068. t004

.	Spoligotype*	Spacers																																											Isolates per site						
		1	2	3	4	5	6	7	8	9	10	11	12	13	14	15	16	17	18	19	20	21	22	23	24	25	26	27	28	29	30	31	32	33	34	35	36	37	38	39	40	41	42	43	AA	Bu	Gi	Go	Ji	Wo	Total
A)	SB0134	■	■	·	·	·	■	■	■	·	■	■	■	■	■	■	·	■	■	■	■	■	■	■	■	■	■	■	■	■	■	■	■	■	■	■	■	■	■	·	·	·	·	·	4			1			5
	SB0133	■	■						■		■	■	■	■	■	■		■	■	■	■	■	■	■	■	■	■	■	■	■	■	■	■	■	■	■	■	■	■						1		2		10	2	15
	SB1488	■	■						■		■	■	■	■	■	■		■	■	■	■	■	■	■	■	■	■		■	■	■	■	■	■	■	■	■	■	■										1		1
	SB1489	■	■						■		■	■	■	■	■	■		■	■	■	■	■	■	■	■	■	■	■	■	■	■	■		■		■	■	■	■						1						1
	SB1477	■							■		■	■	■	■	■	■		■	■	■	■	■	■	■	■	■	■	■	■	■	■	■		■	■	■	■	■	■						2	1			1		4
	SB1476	■	■		■	■	■	■	■		■	■	■	■	■	■		■	■	■	■	■	■	■	■	■	■	■	■	■	■	■	■						■							2	8	7	1		18
	SB1176	■	■	·	·	·	·	·	■	·	■	■	■	■	■	■	·	■	■	■	■	■	■	■	·	·	·	■	·	·	·	·	·	·	·	·	·	·	■	·	·	·	·	·	8	1		3		2	14
B)	SIT1688	■	■	■	■	■	■	■	■	■	■	■	■	■	■	■	■	■	■	■							■	■	■	■	■	■	■					■	■	■	■	■	■	■					2		2
	NEW	■	■	■	■	■	■	■	■	■	■	■	■	■	■	■	■	■	■		■	■	■	■	■	■	■	■	■				■					■	■	■	■	■	■	■		1					1
	SIT1742	■	■	■	■	■	■	■		■	■	■	■	■	■	■	■	■	■	■	■	■	■	■	■	■							■					■	■	■	■	■	■	■		1					1
	SIT59	■	■	■	■	■	■	■	■	■	■	■	■	■	■	■	■	■	■	■	■					■	■					■	■					■	■	■	■	■	■	■						1	1
	SIT149	■	■	■	■	■	■	■	■	■											■	■	■	■	■	■	■	■	■	■	■	■	■					■	■	■	■	■	■	■			1				1
	SIT1 (Beijing)	·	·	·	·	·	·	·	·	·	·	·	·	·	·	·	·	·	·	·	·	·	·	·	·	·	·	·	·	·	·	·	·	·	·	■	■	■	■	■	■	■	■	■			2				2

*SB No = www.mbovis.org; SIT No = Spoligo-International-Typing.

doi:10.1371/journal.pone.0005068.t004

Eight *M. tuberculosis* strains were isolated from cattle, and they originated from four out of the six abattoirs. The TbD1 region [7] was deleted in all of these eight isolates. The spoligotype patterns of seven of these isolates had been registered previously in the Spoligo-International-Typing (SIT) database [15] and the majority were of the Euro-American lineage (Table 4). Two of the strains were of the Beijing lineage, based on the characteristic spoligotype pattern (SIT1) and by identification of the lineage specific deletions RD105 and RD181 [16]. However, the RD142 and RD150 regions that are frequently deleted in Beijing strains were present in these two cattle isolates.

DISCUSSION

The aim of this study was to obtain data on the prevalence and epidemiology of bovine TB in cattle in Ethiopia so as to understand the disease burden and to inform public health policies. Ethiopia has the largest cattle population in Africa, and practices such as the consumption of raw milk and meat mean that the opportunity for zoonotic transmission of bTB is high. However, accurate baseline data for the prevalence of bTB in Ethiopia, across a rage of eco-epidemiological settings, is needed to ensure that public health policies and disease control strategies are based on sound scientific data. We therefore carried out slaughterhouse surveys across major geographical areas of Ethiopia to determine the burden of mycobacterial infection, the prevalence of bTB, and the molecular types of *M. bovis* circulating in Ethiopia.

The Agricultural Settings

The societies in the highlands around the abattoirs of Gonder, Woldiya, Gimbi, and Butajira, are supported by farming in agro-pastoral settings. This is in contrast to the lowlands in the southern regions of Ethiopia, to the south of Jinka, where farming in pastoral settings is practised. In rural areas the local Zebu cattle are the dominant breeds, while exotic breeds are rare. However, closer to Addis Ababa, in the centre of Ethiopia, farming is more diverse and exotic breeds - such as Holstein-Frisian - are common in dairy farms. The cattle in dairy farms are usually subject to intensive farming methods as compared to husbandry practices among pastoralists who keep their animals grazing outdoors.

Lesion Prevalence

Results were based on post-mortem examinations of about 32,800 cattle (Table 2), making this one of the largest bovine TB studies of its kind in Africa. About 1,500 of these animals (~4. 7%) carried suspect TB lesions. In 1975 the prevalence of tuberculous lesions at meat inspections in slaughterhouses varied between 0. 02–1. 83% in different parts of Ethiopia [17], while data from reports in 2003 and 2004 showed an increase to 1. 48–5. 16% [18], [19]. No culturing of AFB was reported in these latter studies.

AFB Positives

The culture yield of AFB from lesions varied depending on the abattoir they were collected from, but the overall yield from visible lesions was calculated to be ~11%. This is a relatively low number when compared to other countries such as Great Britain, where typical TB lesions collected at slaughterhouses yield AFB in ~50% of cases [20]. This may reflect subjective differences in identifying tuberculous lesions across the study sites. From the 135 typed AFB, 58 were *M. bovis*, 8 *M. tuberculosis*, 53 NTM, and 16 could not be identified (Table 2).

Speciation of NTM

About one third of the isolated AFB positives that were typed using molecular methods in this study were confirmed as NTM. As a comparison, isolation of NTM from cattle at slaughterhouses in Great Britain runs at less than 1%, with the majority of isolates from the *M. avium* complex [21]. This may reflect a different population of NTM in Ethiopia as compared to Great Britain, and differences in husbandry practices. The 11 species of NTM listed in Table 3 contain both fast and slow growing mycobacteria. Most of these NTM have been collected before from cattle and wildlife animals in different parts of Africa, suggesting that they are widely spread in nature. For example, *M. engbaekii* and *M. vaccae* have been isolated in buffaloes in South Africa [22] and a study from Chad found *M. fortuitum*, *M. nonchromogenicum*, and strain of the *M. avium* complex to be common in cattle and humans [23]. *M. nonchromogenicum* was the most frequently isolated NTM in our study; it has previously been collected from cattle lymph nodes and nasal mucus [24]. Interestingly, *M. nonchromogenicum* has been linked to bacteremia in AIDS patients [25], suggesting a potential risk of transmission from cattle to HIV positive humans. Limited information is available for the three strains *M. acapulcensis*, *M. colombiense*, and *M. monacense*, but the latter two have previously been isolated from humans [26]–[28].

Isolation of *M. tuberculosis* from Cattle

Unexpectedly, 8 *M. tuberculosis* strains were collected from cattle from four different abattoirs, and isolated from lesions defined as purulent or calcified (Figure 1 and Table 2). Most of these cattle strains were of the Euro-American lineage but, interestingly, two of the strains collected from the Gimbi abattoir (but nine months apart) were of the *M. tuberculosis* Beijing lineage, and to our knowledge, this is the first description of this type in Ethiopia. *M. tuberculosis* has been sporadically isolated from cattle in other countries [29], [30] or zoo animals [31]. It is tempting to speculate that the isolation of *M. tuberculosis* from cattle is evidence of transmission from humans. The prevalence of *M. tuberculosis* in humans in Ethiopia is 643 cases per 100,000, the 7th highest in the world [2]; as rural dwellers live in close contact with their animals (*i. e.* sharing their house at night-time), many animals are at risk from human-to-cattle transmission. Future studies are needed to find out if isolation of *M. tuberculosis* from cattle is a "spill over" effect from humans, or if these *M. tuberculosis* strains are adapted to cattle. The latter also raises the question of the risk for cattle-to-cattle and cattle-to-human transmission and the impact it may have on public health.

Differentiation of *M. bovis* Strains by Spoligotyping

Fifty-eight *M. bovis* strains were spoligotyped and categorised into seven different types (Table 4) with five of these patterns so far apparently unique to Ethiopia. One of these spoligotypes SB1176 was first identified in a government farm in Holeta close to Addis Ababa [32]. Our data from the Addis Ababa abattoir suggests that SB1176 is the most common spoligotype in the central part of Ethiopia. However, it was also identified among strains isolated from the northern part of Ethiopia (Gonder and Woldiya). The cattle slaughtered in Addis Ababa are traded over a large area of Ethiopia, and as such the mycobacterial strains isolated from these cattle should provide a general picture of strain types in Ethiopia. This is supported by the large diversity of spoligotype patterns that was found at Addis Ababa abattoir.

As well as the five spoligotypes that appear to be unique to Ethiopia, we also identified patterns SB0133 and SB0134. The former is most frequent in Jinka in the southern part of the country. An interesting association is that SB0133 has been isolated from cattle in pastoral areas of nearby Uganda [33] and also from several areas of Tanzania [34]. However, SB0133 has so far not been reported outside Africa. It is therefore possible that SB0133 belongs to an *M. bovis* lineage that has spread over a large geographical area of Eastern Africa. SB0134, on the other hand, is a pattern that seems to be widespread in Europe where it has been found in countries such as Great Britain, France, Spain, and Belgium [35]–[37]. It has also been isolated in Africa; a recent publication from Mali suggests that SB0134 is one of the most common patterns in the country [38]. Further chromosomal analysis of strains from Ethiopia and neighbouring countries will help to resolve the population structure of *M. bovis* in this geographical region, and shed light on how *M. bovis* has evolved in this location.

CONCLUSION

Our analysis has revealed that bTB is widely spread throughout Ethiopia, although our rates of *M. bovis* culture-positives from suspect lesions indicate a low prevalence. Indeed, more than 40% of all culture-positives were NTM, underlining the importance of confirming the presence of *M. bovis* in suspect lesions by culture and molecular methods.

There are reported differences in the susceptibility of Zebu and Holstein cattle breeds to bTB [4]. Similarly, rates of bTB in Ethiopia have been shown to be directly related to husbandry, with intensive farming practices driving increased bTB rates [32]. It is possible therefore that the low prevalence of bTB in our study is at least in part explained by the fact that the vast majority of cattle investigated were of Zebu extraction and from agro-pastoral and pastoral settings (*i. e.* non-intensive). To gain a more complete picture of the impact of bTB in Ethiopia our current investigations are focussing on intensive farms in periurban areas.

In terms of the impact of bTB on public health, our results suggest that the high rates of extrapulmonary TB seen in humans in Ethiopia are likely due to a combination of factors that include, but are not solely a consequence of, *M. bovis* transmission from cattle to humans. However, isolation of *M. tuberculosis* from significant numbers of animals calls for further investigations to elucidate the importance of this finding to public health.

ACKNOWLEDGMENTS

This study would not have been possible to accomplish without the excellent work of our collaborators in the abattoirs of Gonder, Woldiya, Gimbi, Addis Ababa, Butajira, and Jinka. We thank the many veterinary students and research staff, at AHRI as well as at the VLA, who have taken part in this work in one way or another. We are also thankful to Dr. Rea Tschopp for her help throughout the project.

AUTHOR CONTRIBUTIONS

Conceived and designed the experiments: SB MV HE DBY GH AA SVG. Performed the experiments: SB RF MH EG AM GA. Analyzed the data: SB NHS GH AA SVG. Contributed reagents/materials/analysis tools: LY GA BDR. Wrote the paper: SB SVG.

REFERENCES

1. Kidane D, Olobo JO, Habte A, Negesse Y, Aseffa A, et al. (2002) Identification of the causative organism of tuberculous lymphadenitis in Ethiopia by PCR. J Clin Microbiol 40: 4230–4234.
2. World Health Organisation (2008) Global tuberculosis control - surveillance, planning, financing: WHO report 2008. WHO/HTM/TB/2008. 393: 1–304.
3. Halderman M (2004) The Political Economy of Pro-Poor Livestock Policy-Making in Ethiopia. Rome, Italy: FAO.
4. Ameni G, Aseffa A, Engers H, Young D, Gordon SV, et al. (2007) High prevalence and increased severity of pathology of bovine tuberculosis in Holsteins compared to zebu breeds under field cattle husbandry in central Ethiopia. Clin Vaccine Immunol 14: 1356–1361.
5. Roberts GD, Koneman EW, Kim YK (1991) *Mycobacterium*. In: Balow A, editor. Manual of Clinical Microbiology. Washington D. C. : pp. 304–339.
6. Gallagher J, Horwill DM (1977) A selective oleic acid albumin agar medium for the cultivation of *Mycobacterium bovis*. J Hyg (Lond) 79: 155–160.
7. Brosch R, Gordon SV, Marmiesse M, Brodin P, Buchrieser C, et al. (2002) A new evolutionary scenario for the *Mycobacterium tuberculosis* complex. Proc Natl Acad Sci U S A 99: 3684–3689.
8. Tsolaki AG, Hirsh AE, Deriemer K, Enciso JA, Wong MZ, et al. (2004) Functional and evolutionary genomics of *Mycobacterium tuberculosis*: insights from genomic deletions in 100 strains. Proc Natl Acad Sci U S A 101: 4865–4870.
9. Wilton S, Cousins D (1992) Detection and identification of multiple mycobacterial pathogens by DNA amplification in a single tube. PCR Methods Appl 1: 269–273.
10. Han XY, Pham AS, Tarrand JJ, Sood PK, Luthra R (2002) Rapid and accurate identification of mycobacteria by sequencing hypervariable regions of the 16S ribosomal RNA gene. Am J Clin Pathol 118: 796–801.
11. Altschul SF, Madden TL, Schaffer AA, Zhang J, Zhang Z, et al. (1997) Gapped BLAST and PSI-BLAST: a new generation of protein database search programs. Nucleic Acids Res 25: 3389–3402.

12. Harmsen D, Rothganger J, Frosch M, Albert J (2002) RIDOM: Ribosomal Differentiation of Medical Micro-organisms Database. Nucleic Acids Res 30: 416–417.

13. Kirschner P, Springer B, Vogel U, Meier A, Wrede A, et al. (1993) Genotypic identification of mycobacteria by nucleic acid sequence determination: report of a 2-year experience in a clinical laboratory. J Clin Microbiol 31: 2882–2889.

14. Kamerbeek J, Schouls L, Kolk A, van Agterveld M, van Soolingen D, et al. (1997) Simultaneous detection and strain differentiation of *Mycobacterium tuberculosis* for diagnosis and epidemiology. J Clin Microbiol 35: 907–914.

15. Brudey K, Driscoll JR, Rigouts L, Prodinger WM, Gori A, et al. (2006) *Mycobacterium tuberculosis* complex genetic diversity: mining the fourth international spoligotyping database (SpolDB4) for classification, population genetics and epidemiology. BMC Microbiol 6: 23.

16. Tsolaki AG, Gagneux S, Pym AS, Goguet de la Salmoniere YO, Kreiswirth BN, et al. (2005) Genomic deletions classify the Beijing/W strains as a distinct genetic lineage of *Mycobacterium tuberculosis.* J Clin Microbiol 43: 3185–3191.

17. Shitaye JE, Tsegaye W, Pavlik I (2007) Bovine tuberculosis infection in animal and human populations in Ethiopia: a review. Veterinarni Medicina 317–332.

18. Ameni G, Wudie A (2003) Preliminary study on bovine tuberculosis in Nazareth municipality abattoir of central Ethiopia. Bulletin of Animal Health and Production in Africa 51: 125–132.

19. Asseged B, Woldesenbet Z, Yimer E, Lemma E (2004) Evaluation of abattoir inspection for the diagnosis of *Mycobacterium bovis* infection in cattle at Addis Ababa abattoir. Tropical Animal Health and Production 36: 537–546.

20. Liebana E, Johnson L, Gough J, Durr P, Jahans K, et al. (2008) Pathology of naturally occurring bovine tuberculosis in England and Wales. Vet J 176: 354–360.

21. Vordermeier HM, Brown J, Cockle PJ, Franken WP, Drijfhout JW, et al. (2007) Assessment of cross-reactivity between *Mycobacterium bovis* and *M. kansasii* ESAT-6 and CFP-10 at the T-cell epitope level. Clin Vaccine Immunol 14: 1203–1209.

22. Michel AL, de Klerk LM, Gey van Pittius NC, Warren RM, van Helden PD (2007) Bovine tuberculosis in African buffaloes: observations regarding *Mycobacterium bovis* shedding into water and exposure to environmental mycobacteria. BMC Vet Res 3: 23.
23. Diguimbaye-Djaibe C, Vincent V, Schelling E, Hilty M, Ngandolo R, et al. (2006) Species identification of non-tuberculous mycobacteria from humans and cattle of Chad. Schweiz Arch Tierheilkd 148: 251–256.
24. McCorry TP, McCormick CM, Hughes MS, Pollock JM, Neill SD (2004) *Mycobacterium nonchromogenicum* in nasal mucus from cattle in a herd infected with bovine tuberculosis. Vet Microbiol 99: 281–285.
25. Mayo J, Collazos J, Martinez E (1998) *Mycobacterium nonchromogenicum* bacteremia in an AIDS patient. Emerg Infect Dis 4: 124–125.
26. Reischl U, Melzl H, Kroppenstedt RM, Miethke T, Naumann L, et al. (2006) *Mycobacterium monacense* sp. nov. Int J Syst Evol Microbiol 56: 2575–2578.
27. Hogardt M, Schreff AM, Naumann L, Reischl U, Sing A (2008) *Mycobacterium monacense* in a patient with a pulmonary tumor. Jpn J Infect Dis 61: 77–78.
28. Esparcia O, Navarro F, Quer M, Coll P (2008) Lymphadenopathy caused by *Mycobacterium colombiense*. J Clin Microbiol 46: 1885–1887.
29. Prasad HK, Singhal A, Mishra A, Shah NP, Katoch VM, et al. (2005) Bovine tuberculosis in India: potential basis for zoonosis. Tuberculosis (Edinb) 85: 421–428.
30. Ocepek M, Pate M, Zolnir-Dovc M, Poljak M (2005) Transmission of *Mycobacterium tuberculosis* from human to cattle. J Clin Microbiol 43: 3555–3557.
31. Lyashchenko KP, Greenwald R, Esfandiari J, Olsen JH, Ball R, et al. (2006) Tuberculosis in elephants: antibody responses to defined antigens of *Mycobacterium tuberculosis*, potential for early diagnosis, and monitoring of treatment. Clin Vaccine Immunol 13: 722–732.
32. Ameni G, Aseffa A, Sirak A, Engers H, Young DB, et al. (2007) Effect of skin testing and segregation on the prevalence of bovine tuberculosis, and molecular typing of *Mycobacterium bovis*, in Ethiopia. Vet Rec 161: 782–786.

33. Oloya J, Kazwala R, Lund A, Opuda-Asibo J, Demelash B, et al. (2007) Characterisation of mycobacteria isolated from slaughter cattle in pastoral regions of Uganda. BMC Microbiol 7: 95.
34. Kazwala RR, Kusiluka LJ, Sinclair K, Sharp JM, Daborn CJ (2006) The molecular epidemiology of *Mycobacterium bovis* infections in Tanzania. Vet Microbiol 112: 201–210.
35. Smith NH, Gordon SV, Rua-Domenech R, Clifton-Hadley RS, Hewinson RG (2006) Bottlenecks and broomsticks: the molecular evolution of *Mycobacterium bovis*. Nat Rev Microbiol 4: 670–681.
36. Aranaz A, Liebana E, Mateos A, Dominguez L, Vidal D, et al. (1996) Spacer oligonucleotide typing of *Mycobacterium bovis* strains from cattle and other animals: a tool for studying epidemiology of tuberculosis. J Clin Microbiol 34: 2734–2740.
37. Allix C, Walravens K, Saegerman C, Godfroid J, Supply P, et al. (2006) Evaluation of the epidemiological relevance of variable-number tandem-repeat genotyping of *Mycobacterium bovis* and comparison of the method with IS*6110* restriction fragment length polymorphism analysis and spoligotyping. J Clin Microbiol 44: 1951–1962.
38. Müller B, Steiner B, Bonfoh B, Fane A, Smith NH, et al. (2008) Molecular characterisation of *Mycobacterium bovis* isolated from cattle slaughtered at the Bamako abattoir in Mali. BMC Vet Res 4: 26.

CHAPTER 14

Correction: The Burden of Mycobacterial Disease in Ethiopian Cattle: Implications for Public Health

Stefan Berg, Rebuma Firdessa, Meseret Habtamu, Endalamaw Gadisa, Araya Mengistu, Lawrence Yamuah, Gobena Ameni, Martin Vordermeier, Brian D. Robertson, Noel H. Smith, Howard Engers, Douglas Young, R. Glyn Hewinson, Abraham Aseffa, Stephen V. Gordon

Citation: Berg S, Firdessa R, Habtamu M, Gadisa E, Mengistu A, Yamuah L, et al. (2009) Correction: The Burden of Mycobacterial Disease in Ethiopian Cattle: Implications for Public Health. PLoS ONE 4(4): 10.1371/annotation/f7240b30-f202-45c5-aacb-4fda2efb8b3d. https://doi.org/10.1371/annotation/f7240b30-f202-45c5-aacb-4fda2efb8b3d.

The black square symbols in Table 4 appear incorrectly. Please view the correct Table 4 here:

Table 4. Spoligotyping patterns of A) *M. bovis* strains and B) *M. tuberculosis* strains isolated from cattle in Ethiopia

		Spacers																					
	Spoligotype*	1	2	3	4	5	6	7	8	9	10	11	12	13	14	15	16	17	18	19	20	21	22
A)	SB0134	■	■				■	■	■		■	■	■	■	■	■		■	■	■	■	■	■
	SB0133	■	■						■		■	■	■	■	■	■		■	■	■	■	■	■
	SB1488	■	■						■		■	■	■	■	■	■		■	■	■	■	■	■
	SB1489	■	■						■		■	■	■	■	■	■		■	■	■	■	■	■
	SB1477	■							■		■	■	■	■	■	■		■	■	■	■	■	■
	SB1476	■	■		■	■	■	■	■		■	■	■	■	■	■		■	■	■	■	■	■
	SB1176	■	■						■		■	■	■	■	■	■		■	■	■	■	■	■
B)	SIT1688	■	■	■	■	■	■	■	■	■	■	■	■	■	■	■	■	■	■	■			
	NEW	■	■	■	■	■	■	■	■	■	■	■	■	■	■	■	■	■	■		■	■	■
	SIT1742	■	■	■	■	■	■	■		■	■	■	■	■	■	■	■	■	■	■	■	■	■
	SIT59	■	■	■	■	■	■	■	■	■	■	■	■	■	■	■	■	■	■	■	■		
	SIT149	■	■	■	■	■	■	■	■	■											■	■	■
	SIT1 (Beijing)																						

	Spacers																					Isolates per site						
Spoligotype*	23	24	25	26	27	28	29	30	31	32	33	34	35	36	37	38	39	40	41	42	43	AA	Bu	Gi	Go	Ji	Wo	Total
SB0134	■	■	■	■	■	■	■	■	■	■	■	■	■	■	■	■						4			1			5
SB0133	■	■	■	■	■	■	■	■	■	■	■	■	■	■	■	■						1		2		10	2	15
SB1488	■	■	■	■		■	■	■	■	■	■	■	■	■	■	■										1		1
SB1489	■	■	■	■	■	■	■	■	■		■		■	■	■	■						1						1
SB1477	■	■	■	■	■	■	■	■	■		■	■	■	■	■	■						2	1			1		4
SB1476	■	■	■	■	■	■	■	■	■	■						■							2	8	7	1		18
SB1176	■				■											■						8	1		3		2	14
SIT1688				■	■	■	■	■	■	■					■	■	■	■	■	■	■					2		2
NEW	■	■	■	■	■	■				■					■	■	■	■	■	■	■		1					1
SIT1742	■	■	■							■					■	■	■	■	■	■	■		1					1
SIT59			■	■					■	■					■	■	■	■	■	■	■						1	1
SIT149	■	■	■	■	■	■	■	■	■	■					■	■	■	■	■	■	■			1				1
SIT1 (Beijing)													■	■	■	■	■	■	■	■	■			2				2

* SB No = www.mbovis.org, SIT No = Spoligo-International-Typing

CHAPTER 15

Genotypic and Phenotypic Properties of Cattle-Associated Campylobacter and Their Implications to Public Health in the USA

Yasser M. Sanad[1,2], Issmat I. Kassem[1], Melanie Abley[2], Wondwossen Gebreyes[2], Jeffrey T. LeJeune[1,2], Gireesh Rajashekara[1,2]

[1]Food Animal Health Research Program, Ohio Agricultural Research and Development Center, The Ohio State University, Wooster, Ohio, United States of America

[2] Department of Veterinary Preventive Medicine, College of Veterinary Medicine, The Ohio State University, Columbus, Ohio, United States of America

Citation:Sanad YM, Kassem II, Abley M, Gebreyes W, LeJeune JT, Rajashekara G (2011) Genotypic and Phenotypic Properties of Cattle-Associated Campylobacter and Their Implications to Public Health in the USA. PLoS ONE 6(10): e25778. https://doi.org/10.1371/journal.pone.0025778.

ABSTRACT

Since cattle are a major source of food and the cattle industry engages people from farms to processing plants and meat markets, it is conceivable that beef-products contaminated with *Campylobacter* spp. would pose a significant public health concern. To better understand the epidemiology of cattle-associated *Campylobacter* spp. in the USA, we characterized the prevalence, genotypic and phenotypic properties of these pathogens. *Campylobacter* were detected in 181 (19.2%) out of 944 fecal samples. Specifically, 71 *C. jejuni*, 132 *C. coli*, and 10 other *Campylobacter* spp. were identified. The prevalence of *Campylobacter* varied regionally and was significantly ($P<0.05$) higher in fecal samples collected from the South (32.8%) as compared to those from the North (14.8%), Midwest (15.83%), and East (12%). Pulsed Field Gel Electrophoresis (PFGE) analysis showed that *C. jejuni* and *C. coli* isolates were genotypically diverse and certain genotypes were shared across two or more of the geographic locations. In addition, 13 new *C. jejuni* and two *C. coli* sequence types (STs) were detected by Multi Locus Sequence Typing (MLST). *C. jejuni* associated with clinically human health important sequence type, ST-61 which was not previously reported in the USA, was identified in the present study. Most frequently observed clonal complexes (CC) were CC ST-21, CC ST-42, and CC ST-61, which are also common in humans. Further, the cattle associated *C. jejuni* strains showed varying invasion and intracellular survival capacity; however, *C. coli* strains showed a lower invasion and intracellular survival potential compared to *C. jejuni* strains. Furthermore, many cattle associated *Campylobacter* isolates showed resistance to several antimicrobials including ciprofloxacin, erythromycin, and gentamicin. Taken together, our results highlight the importance of cattle as a potential reservoir for clinically important *Campylobacter*.

INTRODUCTION

Campylobacteriosis is estimated to affect over one million individuals in the United States annually, with symptoms ranging from mild diarrhea to more serious neuropathies [1]. Of major concern are *C. jejuni*, a species responsible for the majority of human campylobacteriosis, and *C. coli*, which has been exhibiting increased resistance to antimicrobials [2], [3]. Both *C. jejuni* and *C. coli* can readily and asymptomatically colonize major food-animals such as poultry and cattle, subsequently contaminating foods of animal origin including meats and dairy products [4]. Therefore, both species highlight the potential public health impact of *Campylobacter* contamination of food animals. Although it is known that chickens constitute a major reservoir for *Campylobacter* spp., the occurrence of these pathogens in other food animals such as cattle and its potential impact on human health remain largely uncharacterized. The latter can be partially attributed to the sporadic nature of *Campylobacter* infections and difficulties in isolating these fastidious pathogens. However, since cattle are a major source of food and the multi-faceted cattle industry engages people from farms to processing plants and meat markets, it is conceivable that both live cattle and contaminated cattle products could contribute significantly to *Campylobacter* infections in humans. Furthermore, cattle-associated *Campylobacter* also pose an additional indirect public health risk [5], [6]. For example, contamination of surface and ground water may occur with waste run-off from cattle farming and processing operations. Thus, it is important to further investigate the epidemiology of *Campylobacter* in the cattle population in order to assess associated risks to public health.

Recent studies have shown that the contributions of non-poultry associated *Campylobacter* to human infections were considerable and warrant investigation [7], [8]. For example, evidence collected using Multilocus Sequence Typing (MLST) showed that cases of human *Campylobacter* infections in Finland could be attributed equally to cattle and poultry [9]. Moreover, another study reported that cattle were the source of human infections in 35% of the cases examined in Lancashire, England [8]. This is not surprising since up to 80% of cattle herds and 40–60% of the individual animals shed *Campylobacter* [10], [11], [12]. Consequently, the role of cattle as reservoirs for these pathogens might be important for understanding the epidemiology of *Campylobacter* infections. However, particularly in the US, the prevalence of *Campylobacter* in cattle, their characteristics and relationship to isolates from humans have not been extensively described in the peer reviewed literature.

Host cell invasion and intracellular survival and resistance to antibiotics are important characteristics that affect *Campylobacter* infections in humans. However, the potential role of these properties in facilitating human infections with *Campylobacter* isolated from cattle is not clear. Although epidemiological studies deploy typing analysis to glean information about the relationships between cattle-associated *Campylobacter* and clinically-important human isolates, the capability of cattle-associated *Campylobacter* to invade and persist in the human host might require further assessment, especially when previously unknown sequence types are identified. This also applies to antimicrobial resistance properties, which have been posing a serious concern in *Campylobacter* collected from animal hosts along with the possibility of the transmission of these isolates to humans through the food chain. Therefore, epidemiological studies concerning cattle-associated *Campylobacter* would gain from attempts to amend molecular typing analysis with *in vitro* invasion studies using human intestinal cell lines and phenotypic assays for determining antibiotic resistance.

Since molecular typing data of cattle-associated *Campylobacter* in the United States are limited and little is known about their impact on human health, it is important to adopt a multiphasic approach to characterize cattle-associated *Campylobacter* by using a combination of molecular typing and *in vitro* assays. Therefore, in this study, we determined the genotypes of *C. jejuni* and *C. coli* isolated from cattle slaughtered for meat purpose in different geographical locations in the U.S. and investigated their antimicrobial susceptibility profiles as well as virulence associated phenotypes such as their potential for invasion and persistence in human intestinal epithelial cells.

MATERIALS AND METHODS

Distribution of Sampling Sites and Collection and Processing of Fecal Samples from Beef Cattle

A total of 944 fresh fecal samples (10 g each) were collected during the summer and early fall of 2008 from cattle presented to slaughter which included both feed-lot (n=482) as well as mature cows and bulls (n=462) culled from milking and breeding herds (Table 1). The samples were collected from colon on the conveyer belt during the normal harvest process from slaughtered cattle. To account for potential spatial heterogeneity between different beef processing plants, sampling efforts included seven plants that were distributed across four major geographical locations [North (N), East (E), Midwest (M), and South (S)] in the U.S. The time interval of sample collection per plant spanned two consecutive days and was divided into four periods during which at least four different lots of cattle were sampled in order to capture the genetic diversity of cattle-associated *Campylobacter*. To further limit sampling bias, the number of samples collected per plant was calculated in terms of the total number of animals expected for slaughter on each day, and a statistically calculated sample size to detect at least a single sequence type of *Campylobacter* should it be present in each group of animals at a prevalence of greater than 2.5% with a confidence of 95%. Samples were stored on ice immediately after collection and shipped overnight to the laboratory for *Campylobacter* isolation.

Table 1: ***Campylobacter*** **species isolated from cattle fecal samples that were collected from four major ge**ographical locations in the USA

https://doi.org/10.1371/journal.pone.0025778.t001

Regions	Number of samples	*Campylobacter* prevalence/region (%)	Number of *C. jejuni*	Number of *C. coli*
North	210	31/210 (14.8%)	13	16
Midwest	240	38/240 (15.83%)	36	12
South	244	80/244 (32.8%)	8	83
East	250	32/250 (12%)	14	21
Total	944	181/944 (19.2%)*	71	132

*A total of 213 *Campylobacter* isolates occurred in 181 fecal samples including 10 *Campylobacter* spp. other than *C. jejuni* and *C. coli*.
doi:10.1371/journal.pone.0025778.t001

Isolation and Identification of *Campylobacter* Species from Fecal Samples

To isolate *Campylobacter* spp., 1 g of each fecal sample was enriched in Preston broth for 48 h at 42°C under microaerobic conditions (5% O_2, 10% CO_2, and 85% N_2) [13]. From the enrichments, an inoculum (100 µl) was spread onto modified Cefoperazone Charcoal Deoxycholate Agar (mCCDA) plates, which were then incubated for an additional 48 h at 42°C under microaerobic conditions [14]. Three to five colonies suspected as *Campylobacter* were selected from each plate and sub-cultured onto Muller-Hinton agar plates, after which DNA was extracted from each isolate using the Genomic DNA Purification Kit (Epicenter, Madison, WI) as described by the manufacturer. The DNA was then quantified using the NanoDrop 1000 Spectrophotometer V3.7.1 (Fisher Scientific, Pittsburgh, PA) and subjected to species-specific PCR analysis to confirm the identity of the isolates as described elsewhere [15]. PCR analysis targeted a 16S rRNA gene fragment (F) 5′-ATCTAATGGCTTAACCATTAAAC-3′ and (R) 5′-GGACGGTAACTAGTTTAGTATT-3′, *mapA*, (F) 5′-CTATTTTATTTTTGAGTGCTTGTG-3′ and (R) 5′-GCTTTATTTGCCATTTGTTTTATTA-3′ and *ceu*E, (F) 5′ATTTGAAAATTGCTCCAACTATG-3′ and (R) 5′-TGATTTTATTATTTGTAGCAGCG-3′ which indicated the specific detection of *Campylobacter* spp., *C. jejuni*, and *C. coli*, respectively [16], [17]. All PCR products were resolved on a 1.5% agarose gel containing 0.5 µg/ml of ethidium bromide. The size of the PCR products was determined using a 1 Kb DNA ladder and detection was confirmed by comparison to

PCR products generated from *C. jejuni* 81–176 (wild-type strain) and *C. coli* (ATCC 33559), which were used as positive controls in all PCR analysis. Negative controls (reactions with no DNA templates) were included in all PCR analyses to ensure specific product amplification.

Pulsed Field Gel Electrophoresis (PFGE) of *Campylobacter* Isolates

To determine their genotypic relatedness, the *Campylobacter* isolated from fecal samples were analyzed using PFGE analysis as described in Ribot et al. [18]. Briefly, *C. jejuni* and *C. coli* isolates were harvested from MH agar plates and suspended to an OD_{610} of 1.4 in 1X PBS. To prepare agarose plugs, OD adjusted suspensions were gently mixed with 1% SeaKem Gold agarose (SKG, Fisher scientific, Pittsburgh, PA) that was pre-melted in TE (10 mM Tris, 1 mM EDTA, pH 8.0). The plugs were then incubated with shaking (200 rpm) in lysis buffer [50 mM Tris, 50 mM EDTA (pH 8.0), 1% sarcosine, 0.1 mg ml^{-1} of proteinase K] for 1 h at 55°C. After lysis the plugs were washed four times and suspended in 5 ml of fresh TE. The plugs were then sliced and digested overnight with *Sma*I at room temperature. The digested slices were loaded onto a 1% SKG agarose gel and DNA fragments were separated by electrophoresis for 20 h using the CHEF Mapper system (Bio-Rad, Hercules, CA) followed by post staining with ethidium bromide. The resulting PFGE patterns were documented and analyzed using the BioNumerics 5.1 software (Applied Maths Inc, Austin, TX). Similarity and clustering analysis of the PFGE patterns were performed using the Dice Coefficient and the unweighted pair-group method with arithmetic averages (UPGMA) with optimization of 1% and position tolerance of 1.5% [7], respectively. The PFGE analysis was also performed on *C. jejuni* 81–176 and *C. coli* (ATCC 33559), which were used as controls for facilitating gel to gel comparison. Furthermore, Lambda Ladder PFG Marker (50–1,000 kb, New England BioLabs, Ipswich, MA) was used as a molecular marker. A cut-off similarity value of 75%, was used to determine the sub-clusters of macro restriction profiles (MRPs).

Determining the Antimicrobial-resistance Properties of the *Campylobacter* Isolates

Minimal inhibitory concentrations (MIC) for *C. jejuni* and *C. coli* isolates were determined using commercially available 96-well plates containing antimicrobials (Sensititre Campy plates, TREK Diagnostic Systems Inc.,

Cleveland, OH, USA). These Sensititre Campy plates were used as described by the manufacturer and include *Campylobacter*-relevant antimicrobials such as azithromycin (AZI) (MIC for a resistant *Campylobacter* isolate: ≥8 μg ml^{-1}); ciprofloxacin (CIP) (MIC: ≥4 μg ml^{-1}); clindamycin (CLI) (MIC: ≥8 μg ml^{-1}); erythromycin (ERY) (MIC: ≥32 μg ml^{-1}); gentamicin (GEN) (MIC: ≥8 μg ml^{-1}); nalidixic acid (NAL) (MIC: ≥64 μg ml^{-1}); telithromycin (TEL) (MIC: ≥8 μg ml^{-1}); florfenicol (FEN) (MIC: ≥8 μg ml^{-1}); and tetracycline (TET) (MIC: ≥16 μg ml^{-1}). *Campylobacter* isolates were grown to mid-log phase and then suspended in Mueller-Hinton broth to achieve an OD_{600} of 0.05. For each isolate, one hundred microliter of the suspension were transferred to each well in the Sensititre Campy plates, including a control well that did not contain any antibiotics, while *C. jejuni* 81–176 and *C. coli* (ATCC 33559) were used for quality control. The plates were then incubated under microaerobic conditions at 42°C for 24 h after which the minimum inhibitory concentration (MIC) was measured. The MIC for each antibiotic was defined as the absence of bacterial growth in the well with the lowest concentration of the antibiotic and susceptibility was interpreted according to the Clinical and Laboratory Standards Institute (CLSI) [19].

Multilocus Sequence Typing (MLST) of *Campylobacter* Isolates

To investigate the clonality of the *Campylobacter* isolates and to assess similarity to strains associated with human infections, MLST analysis was performed on a total of 112 isolates (62 *C. jejuni* and 50 *C. coli*). Nine *C. jejuni* isolates were not typed because either they yielded incomplete profiles or last culturability upon storage. However, since the *C. coli* isolates showed limited genetic diversity using MLST, the tested number of isolates (n=50) was considered adequate to meet the objectives of this study. These isolates were selected to be representative of different PFGE clusters. MLST analysis was conducted as described by Dingle et al. [20]. Briefly, loci from seven housekeeping genes (*aspA*, *glnA*, *gltA*, *glyA*, *pgm*, *tkt*, and *uncA*) were amplified using gene specific primers by PCR and anticipated sizes of the amplicons were confirmed by agarose gel electrophoresis. The *Campylobacter* MLST oligonucleotides [20] were obtained from Integrated DNA Technologies (Coralville, IA). PCR products were purified (QIA quick 96 PCR purification kit, QIAGEN, Valencia, CA), sequenced in both directions as described previously [21], and, forward and reverse sequences were then aligned using ClustalW (www.ebi.ac.uk/clustalw). Allelic profiles were determined by performing BLAST analysis using the single-locus query function, while sequence types (STs) were assigned using the allelic profile query function available in the MLST *Campylobacter* database (http://pubmlst.org/campylobacter/). STs were then traced to their respective clonal complexes using BURST at http://pubmlst.org/.

Invasion and Intracellular Survival Potential of the *Campylobacter* Isolates in INT407 Cells

Representative *C. jejuni* (n=19) and *C. coli* (n=12) that belonged to clonal complexes associated with human infections as well unassigned clonal complexes including the newly identified STs were further tested for their virulence associated phenotypes by assessing their ability to invade and survive within human intestinal epithelial cells. The invasion studies were conducted as described in Konkel et al. [22] and Prasad et al. [23]. Briefly, 10^5 cells ml^{-1} of INT407 (human embryonic intestine, ATCC CCL 6) were seeded into each well of a 24-well tissue culture plates after suspension in Eagle›s Minimum Essential Medium (MEM, Fisher scientific, Pittsburgh, PA) supplemented with 10% fetal bovine serum (FBS, Fisher scientific, Pittsburgh, PA). The plates were then incubated at 37°C in a humidified incubator with 5% CO_2 until semi-confluent mono-layers were obtained. In preparation for infection with *Campylobacter*, the INT407 mono-layers were washed three times and covered in MEM supplemented with 1% FBS. Similarly, *Campylobacter* cultures were washed three times and suspended in MEM supplemented with 1% FBS to obtain 10^7 bacteria ml^{-1}. One ml of bacterial suspension was added to each well containing the INT407 cell monolayer, achieving a 1:100 multiplicity of infection (MOI) then incubated for 3 h. After 3 h of incubation with bacteria, cells were treated with gentamicin (150 μg/ml) and incubated for additional 2 h. The infected mono-layers were then washed three times with MEM, lysed using 0.1% Triton X-100 (Fisher scientific, Pittsburgh, PA), serially diluted (10-fold) in MEM and 100 μl of each dilution were spread on MH agar plates. The agar plates were then incubated for 48 h at 42°C under microaerobic conditions, after which colony forming units (CFU) were counted to determine the number of *Campylobacter* that invaded the monolayers. Each isolate was tested in duplicate per assay, and the experiment was repeated three times on separate occasions. *C. jejuni* 81–176 (highly invasive) [24], *C. jejuni* NCTC11168 (poorly invasive) [24] and *C. coli* (ATCC 33559) were used as controls in all invasion assays.

To assess intracellular survival [22], *Campylobacter* cultures and the INT407 cells were processed as described above. However, after treatment with gentamicin, MEM containing 3% FBS and bacteriostatic concentration of gentamicin (10 μg ml^{-1}) were added to each well to allow strict quantitation of intracellular bacteria. After further incubation at 37°C for 24 h, the monolayers were washed three times, lysed and serially diluted in MEM as described earlier and plated on MH agar to determine CFU of *Campylobacter* surviving within the intestinal cells. In addition, we also cultured the supernatant of gentamicin treated monolayers to ensure the quality of the gentamicin protection assay. The experiment was repeated three times on separate occasions. *C. jejuni* 81–176, *C. jejuni* NCTC11168, and *C. coli* (ATCC 33559) were used as controls in all intracellular survival assays.

Statistical Analysis

A Chi square test was used to evaluate data collected from the *Campylobacter* prevalence analysis. The data derived from cell culture assays were analyzed using one-way ANOVA followed by Tukey's Multiple Comparison Test. A P value of <0.05 was considered statistically significant for all experiments. Measurements expressed as mean ± SE (standard error) were averages of at least three replicas.

RESULTS

Occurrence and Distribution of *Campylobacter* spp. in Feces Sampled from Cattle

The occurrence and distribution of *Campylobacter* in cattle feces was investigated in order to determine the role of cattle as a potential reservoir for these bacteria. *Campylobacter* were detected in 181 (19.2%) out of 944 fecal samples, while a total of 213 isolates were retrieved from the *Campylobacter*-positive fecal samples. The *C. jejuni* and *C. coli* constituted the majority of the isolates and were detected in 71 (7.5%) and 132 (14%) of the samples, respectively (Table 1). There was no significant difference (P<0.12) in the number of *C. jejuni* isolated from fecal samples obtained from; either feed lot (6.2%) or the cull cattle (8.9%), while *C. coli* was more frequently isolated from cull cattle (17%) as compared to feed lot cattle (11%) (P<0.01). Furthermore, the occurrence of *Campylobacter* in cattle feces varied according to geographic location of the sampling sites.

Specifically, *Campylobacter* species were retrieved from 32.8% of the fecal samples collected from the South, which was significantly higher ($P<0.05$) than those from the North (14.8%), Midwest (15.83%), and East (12%) (Table 1).

C. jejuni and *C. coli* Strains were Genotypically Diverse across Different Geographic Regions

PFGE analysis was performed on *C. jejuni* and *C. coli* isolated from feces in order to determine the genetic diversity of these bacteria and the relationship between the isolates retrieved from different geographic locations. PFGE analysis was successfully performed on 67 out of 71 *C. jejuni*, while 4 isolates could not be typed using the aforementioned method. Regardless, analysis of the PFGE profiles of the *C. jejuni* suggested that the isolates possessed diverse genotypes, especially when comparing profiles of isolates belonging to different geographic locations (Fig. 1A). Furthermore, there were 7 main clusters (Fig. 1A) and by using a cut-off similarity value of 75%, profiles of the *C. jejuni* were classified to15 sub-clusters. With the exception of 4 clusters that included isolates that were collected from 2 to 3 different geographic locations, *C. jejuni* profiles were mostly found to form geographically homogenous groupings. For example, one cluster was composed of three isolates collected from the East, the Midwest, and the South, respectively, while others included 4 isolates, 3 of which were isolated from the Midwest and one from the South (see arrows in Fig. 1A). Overall, it was interesting to observe that profiles of the isolates from the North did not group with those collected from the East and the South.

Similarity analysis was performed using the Dice coefficient, and clustering was performed by the unweighted pair-group method with arithmetic averages UPGMA (optimization, 1% and position tolerance, 1.5%). Clustering cut-off was 75% similarity, which was represented by dashed line. Numbers on bootstraps represent Cophenetic correlations. N=North, S=South, M=Midwest, and E=East. Cj=*C. jejuni*, CC=*C. coli*.

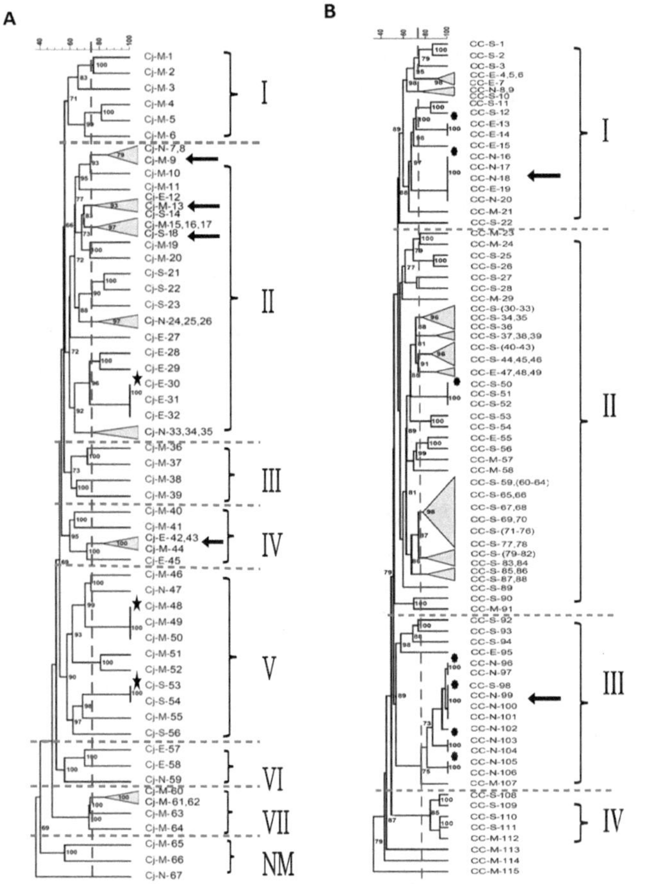

Figure 1: Dendrogram showing the Pulsed-field gel electrophoresis patterns for *Sma*I restricted cattle-associated *C. jejuni* **(A) and** *C. coli* **(B).**

https://doi.org/10.1371/journal.pone.0025778.g001

Similarly, PFGE analysis was successfully performed on 115 out of 132 *C. coli* isolates, while the remaining isolates could not be typed using the aforementioned method (Fig. 1B). Unlike the *C. jejuni*, PFGE profiles of certain *C. coli* isolated from all four geographic locations were found to share a 100% similarity. Specifically, analysis grouped the *C. coli* isolates into 4 main clusters and by using a cut-off similarity value of 75%, 21 sub-clusters were identified, 2 of which included identical profiles (100% similar) for isolates from different geographic locations (See arrows Fig. 1B). For example, a cluster was composed of 5 isolates (4 from the North and 1 from the East) that displayed identical profiles, while another cluster similarly contained identical profiles of 4 isolates (3 from the North and 1 from the South). However, other clusters contained profiles of *C. coli* isolates from different geographic locations that were not identical, yet exhibited high similarity (between 79 to 95%) than that of the cut-off (75%). Similar to PFGE analysis of *C. jejuni*, many *C. coli* isolated from the same geographic location showed profiles that tend to cluster together.

Antimicrobial Susceptibility of *C. jejuni* and *C. coli* Isolates

To better assess the potential public health impact of the *Campylobacter* spp. associated with cattle, the isolates were assayed for their potential to resist antibiotics that are of both clinical and veterinary importance. Antimicrobial susceptibility was assessed on 66 (the 5 remaining isolates lost cultivability after storage) *C. jejuni* isolates using commercially available Sensititre Campy plates. The *C. jejuni* isolates were resistant to different antimicrobials including, ciprofloxacin (MIC: 4–64 μg ml^{-1}), erythromycin (MIC: 32 μg ml^{-1}), tetracycline (MIC: 16–64 μg ml^{-1}), and clindamycin (MIC: 8–16 μg ml^{-1}) (Fig. 2A). However, resistance to tetracycline was observed for the majority of the tested *C. jejuni* (72.7%), while resistance to nalidixic acid was observed for only 27.3% of the isolates (Fig. 2A). Furthermore, 10 (15.1%) tested *C. jejuni* were resistant only to tetracycline, while 34 (51.5%) isolates were resistant to 3 or more antimicrobials, including ciprofloxacin, tetracycline, nalidixic acid, erythromycin, and clindamycin. Overall, a higher percentage of *C. jejuni* isolates from the South region showed decreased resistance to most of the antimicrobials (See asterisk Table 2). The MRPs and antimicrobial resistance profiles of the *C. jejuni* isolates analyzed by MLST are summarized.

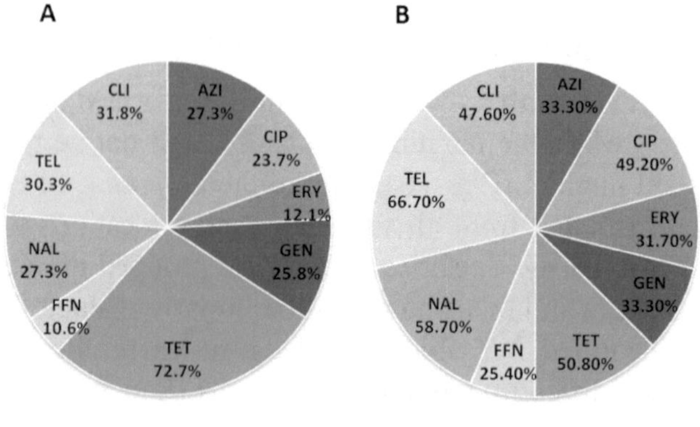

Figure 2: Antimicrobial resistance of *Campylobacter* **isolated from cattle.**

C. jejuni (A) and *C. coli* (B) isolates obtained from four different locations across the US. Percentage of isolates resistant to different antimicrobials used in this study is shown. *C. jejuni* 81–176 and *C. coli* ATCC 33559 strains were used for quality control.

https://doi.org/10.1371/journal.pone.0025778.g002

Table 2: Antimicrobials resistance of *C. jejuni* **and** *C. coli* **isolate**d from four different locations

https://doi.org/10.1371/journal.pone.0025778.t002

		Number of resistant isolates (%) and range of MICs in μg/ml								
Isolates		AZI	CIP	ERY	GEN	TET	FFN	NAL	TEL	CLI
CJ	N	5(7.6%) 8–32[a]	6(9.1%) 4–64	1(1.5%) 32	3(4.5%) 8–16	9(13.6%) 16–64	3(4.5%) 8–16	8(12.1%) 64	7(10.6%) 8	4(6.1%) 16
	S*	0	1(1.5%) 64	0	0	2(3.0%) 16–64	0	2(3.0%) 64	0	0
	M	6(9.1%) 8–16	4(6.1%) 16–32	4(6.1%) 32	8(12.1%) 8–16	26(39.4%) 16–64	4(6.1%) 8–16	4(6.1%) 64	9(13.6%) 8	11(16.7%) 8–16
	E	7(10.6%) 8–32	4(6.1%) 16–32	3(4.5%) 32	6(9.1%) 8–16	11(16.7%) 16–64	0	4(6.1%) 64	4(6.1%) 8	6(9.1%) 8–16
CC	N	0	2(3.2%) 4–16	0	1(1.6%) 8	1(1.6%) 16	0	6(9.5%) 64	4(6.3%) 8	2(3.2%) 8–16
	S	10(15.9%) 8–64	24(38.1%) 8–64	10(15.9%) 32–64	13(20.6%) 8–32	20(31.7%) 16–64	10(15.9%) 8–32	23(36.5%) 64	26(41.3%) 8	16(25.4%) 8–16
	M	3(4.8%) 64	2(3.2%) 8–32	2(3.2%) 32–64	4(6.3%) 8–16	3(4.8%) 16–64	0	2(3.2%) 64	5(7.9%) 8	3(4.8%) 16
	E	8(12.7%) 16–64	3(4.8%) 4–8	8(12.7%) 32–64	3(4.8%) 8–16	8(12.7%) 16–32	6(9.5%) 8–64	6(9.5%) 64	7(11.1%) 8	10(15.9%) 8–16

[a]resistance MIC range of antimicrobials.
*decreased resistance to most of antimicrobials.
doi:10.1371/journal.pone.0025778.t002

Since, *C. coli* isolates belonging to different PFGE clusters were highly similar, we tested representative isolates from each cluster for a total of 63 isolates. Although the range of MIC to disparate antibiotics was at instances different from those observed for *C. jejuni*, the *C. coli* isolates also exhibited resistance to the several of the antimicrobials tested (Fig. 2B). For example, *C. coli* isolates were resistant to ciprofloxacin (MIC: 4–64 μg ml^{-1}), erythromycin (MIC: 32–64 μg ml^{-1}), tetracycline (MIC: 16–64 μg ml^{-1}), and clindamycin (MIC: 16 μg ml^{-1}) (Table 3). Furthermore, resistance to telethromycin was observed for the majority of the tested *C. coli* (66.7%), while resistance to flourofenicol was observed for only 25.4% of the isolates (Fig. 2A). Of the 63 tested *C. coli*, 3 (4.8%), 9 (14.3%) and 48 (76.2%) isolates were resistant to only one (e.g. nalidixic acid), two (e.g. ciprofloxacin and nalidixic acid), and three or more antimicrobials, respectively. Furthermore, a higher percentage of *C. coli* isolates from the South showed increased resistance to all the antimicrobials (Table 2).

Table 3: Distribution of multilocus sequence types among cattle *C. jejuni* **and** *C. coli*

https://doi.org/10.1371/journal.pone.0025778.t003

ST*	*C. jejuni* Isolates	Isolates/Region				ST- CC
		N	E	M	S	
590	3	0	1	2	0	**ST-42**
1013	11	2	0	7	2	
459	5	0	0	5	0	
797	7	0	1	5	1	**ST-21**
4924	5	0	0	3	2	
2876	2	0	0	2	0	
4026	2	1	0	1	0	
4930	1	0	1	0	0	
4922	1	0	0	1	0	
4923	1	0	0	1	0	
239	1	0	0	1	0	
61	2	0	2	0	0	**ST-61**
500	1	0	0	1	0	
3091	1	1	0	0	0	**ST-45**
3084	1	0	0	0	1	**ST-48**
4929	2	0	2	0	0	**Unassigned**
4925	1	0	0	1	0	
4926	1	0	0	1	0	
4927	1	1	0	0	0	
4928	1	0	1	0	0	
4931	1	0	0	1	0	
4932	1	0	0	0	1	
922	1	0	1	0	0	
5447	2	0	0	2	0	
5448	1		1			
ST	***C. coli* Isolates**	**Isolates/Region**				**ST- CC**
		N	**E**	**M**	**S**	
1068	40	8	2	7	23	**ST-828**
902	2	0	0	2	0	
2501	2	1	0	0	1	
1110	2	0	2	0	0	
3866	1	0	0	0	1	
5372	1	0	0	0	1	
4933	1	0	0	0	1	
5446	1	0	0	1	0	

*Complete allelic profiles for six *C.jejuni* isolates could not be ascertained.

doi:10.1371/journal.pone.0025778.t003

MLST Analysis Identifies STs belonging to Human Clonal Complexes

To determine the genotypic properties of *Campylobacter* spp. isolated from cattle as well as their relationship to those of human origin, MLST analysis was performed on selected *C. jejuni* and *C. coli* isolates. A total of 112 isolates, 62 *C. jejuni* and 50 *C. coli* that represented the main PFGE clusters and different geographic locations, were selected for MLST analysis. Thirteen new *C. jejuni* STs were identified and designated as STs 4922 to 4932, 5447, and 5448 (Table 3). These new STs were recognized following query with the PubMLST database. Furthermore, three new alleles, *gly*A; 416 (Cj-M-46), *pgm*; 537 (Cj-M-63) and *tkt*; 436 (Cj-M-64) were detected for 3 STs out of 13 new *C. jejuni* STs. Both *pgm* and *tkt* alleles had only a single nucleotide polymorphism, while *gly*A had a two nucleotide polymorphism.

In general *C. jejuni* isolates showed high genetic diversity by MLST analysis (Table 3). A total of 25 different STs were identified. Twenty isolates with 8 STs belonged to Clonal Complex (CC) ST-21, while 3 of these isolates were assigned to new STs. Additional isolates (n=19) were grouped into 3 STs that belonged to CC ST-42, while five, three, and 11 isolates were identified as ST-459 which belonged to CC ST-42, ST-590, and ST-1013, respectively. Finally, one isolate belonged to CC ST-45, one belonged to CC ST-48, and 9 belonged to undefined CC (Table 3). Interestingly, we also detected in our study three isolates belonging to clonal complex ST-61, and two of them were identified as ST-61, which is clinically important and has not been reported previously in cattle in the USA according to the PubMLST database. The most frequently observed clonal complexes (CC) were CC ST-21, CC ST-42, and CC ST-61, which are also common in humans. Six isolates in our collection had incomplete allelic profiles and could not be assigned to any ST.

In contrast to *C. jejuni* isolates, *C. coli* isolates showed very limited diversity (Table 3). Of the 50 *C. coli* isolates analyzed (Table 3), a total of 8 STs were identified with 1 isolate belonging to new ST-4933 and ST-5446. Interestingly, 40 (80%) isolates were identified as ST-1068. Twenty three of those 40 isolates were isolated from the South as well as 8 from the North, 7 from the Midwest and 2 from the East. Additionally, seven different STs were identified from the remaining 10 strains that were isolated from all four regions. Two of these isolates were identified as ST-902 (from Midwest), while the remaining two isolates (from the North and the South, respectively) were identified as ST-2501. Additional isolates (n=2) were identified as ST-1110 (East), while three more isolates were identified as ST-3866, ST-5372, and ST-4933, respectively. Additionally, one isolate was identified as a new ST-5446 (Midwest). All 50 *C. coli* isolates belonged to CC ST-828 (Table 3).

Invasion and Intracellular Survival Potential of Cattle Associated *C. jejuni and C. coli*

To investigate the virulence-associated potential of the cattle-associated *C. jejuni* and *C. coli*, we tested selected isolates for their ability to invade and survive in human INT407 intestinal epithelial cells. *C. jejuni* and *C. coli* isolates that belonged to all the detected CCs were selected for these assays; in addition isolates that exhibited disparate antibiotic resistance profiles and/or originated from different geographic locations were also included in this analysis. All tested *C. jejuni* isolates (n=19) invaded INT407 (Fig. 3A); however, the invasion potential varied between isolates, ranging from an average of 2.75×10^1 to 8.0×10^4 CFU ml^{-1} (Fig. 3A). The invasion capability of 17 of the 19 isolates tested was significantly lower than that of the highly invasive strain *C. jejuni* 81–176 (*P<0.01*); however, 11 of these isolates invaded the INT407 cells with higher numbers as compared to the poorly invasive *C. jejuni* NCTC11168 (Fig. 3A). Furthermore, two *C. jejuni* isolates Cj-M-13 and Cj-N-33 exhibited high invasion potential that matched that of *C. jejuni* 81–176 (Figure 3A). Interestingly, all tested *C. jejuni* isolates were capable of intracellular survival, albeit the potential varied between isolates, ranging from an average of 8.0×10^1 to 1.3×10^4 CFU ml^{-1}(Fig. 3B). Additionally, the same aforementioned two isolates showed a significantly higher ($P<0.01$) potential for intracellular survival as compared to that of *C. jejuni* 81–176.

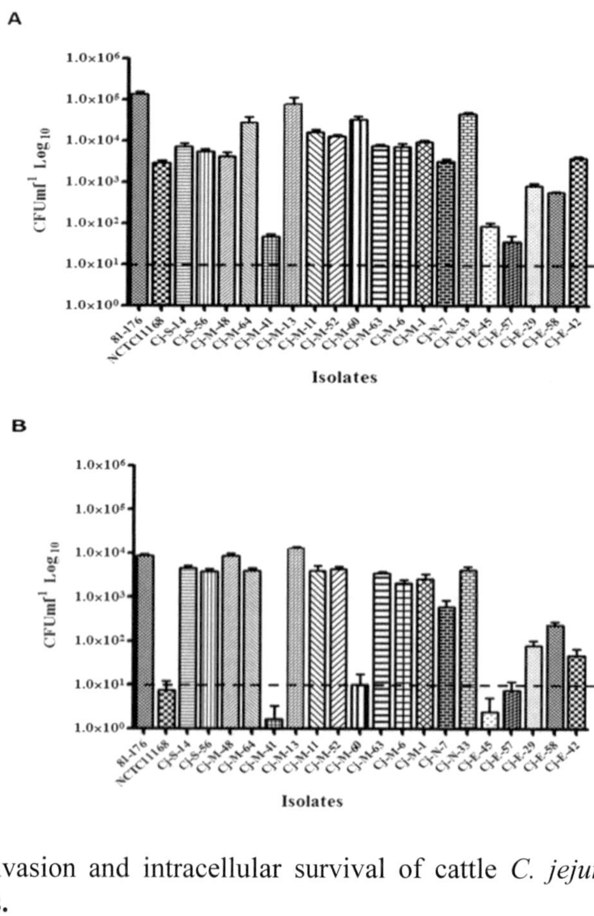

Figure 3: Invasion and intracellular survival of cattle *C. jejuni* **isolates in INT407 cells.**

A. CFU ml^{-1} representing the number of the internalized bacteria which could be retrieved after treatment of cells with gentamicin. B. Intracellular survival of *C. jejuni* isolates in INT407cells. CFU ml^{-1} representing the numbers of internalized bacteria retrieved after 24 h of incubation following the gentamicin treatment. The INT407 were infected with 1:100 MOI of *C. jejuni* strains. *C. jejuni* 81–176 and NCTC11168 were used as controls. The detection limit of the assay is represented by the dashed line. Each bar represents the mean ± SE of three independent experiments done in duplicates for each sample ($P<0.01$).

https://doi.org/10.1371/journal.pone.0025778.g003

On the other hand, only nine *C. coli* isolates could invade the INT407 cells with average numbers ranging between 2.75×10^1 and 1.34×10^4 CFU ml^{-1} (Fig. 4A). There was a significant difference in the invasion capabilities between the *C.coli* isolates ($P<0.05$). Furthermore, only seven isolates were capable of intracellular survival (Fig. 4B). Three isolates that did not invade were not tested for intracellular survival. In general *C. coli* isolates were less invasive and displayed reduced intracellular survival compared to *C. jejuni*.

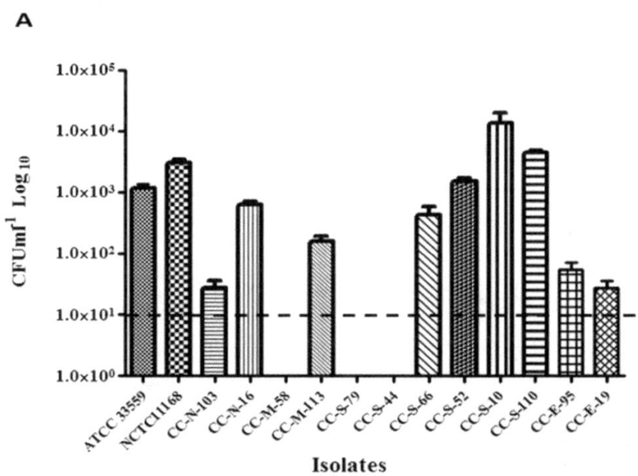

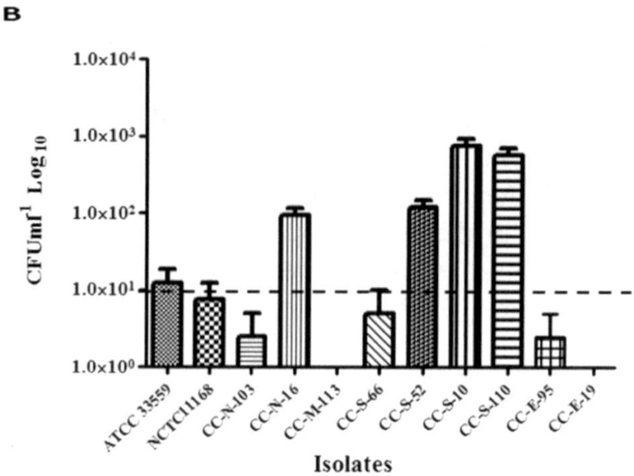

Figure 4: Invasion and intracellular survival of cattle *C. coli* **isolates in INT407 cells.**

A. CFU ml^{-1} representing the number of the internalized bacteria which could be retrieved after treatment of cells with gentamicin. B. Intracellular survival of *C. coli* isolates in INT407cells. CFU ml^{-1} representing the numbers of internalized bacteria retrieved after 24 h of incubation following the gentamicin treatment. The INT407 cells were infected with 1:100 MOI of *C. coli* strains. *C. coli* (ATCC 33559) and *C. jejuni* NCTC11168 were used as controls. The detection limit of the assay is represented by the dashed line. Each bar represents the mean ± SE of three independent experiments done in duplicates for each sample ($P<0.01$).

https://doi.org/10.1371/journal.pone.0025778.g004

Our data also show that invasive *Campylobacter* possessed a variable antimicrobial phenotype that ranged from complete susceptibility to resistance to 7 different antimicrobials (Table 4). However, it was notable that *C. jejuni* isolates with high invasion potential (Cj-M-13 and Cj-N-33) were also resistant to multiple antimicrobials (see asterisks Table 4; Fig. 3A).

Table 4: The properties of *C. jejuni* **and** *C. coli* isolatesthat were tested for invasion and intracellular survival in INT407 cell line

https://doi.org/10.1371/journal.pone.0025778.t004

C.jejuni Isolates	Invasion[a]	Survival	Antimicrobial Resistance Profile	ST	CC ST	MRP cluster
Cj-M-6	Moderate	High	Susceptible	797	CC ST21	II
Cj-M-11	Moderate	High	GEN, TET, NAL	4922	CC ST21	II
Cj-M-52	Moderate	High	CIP, TET, TEL, CLI	4924	CC ST21	V
Cj-M-13*	High	High	AZI, CIP, ERY, GEN, FFN, TEL, CLI	2876	CC ST21	II
Cj-N-33*	High	High	AZI, CIP, GEN, TET, NAL, TEL, CLI	4026	CC ST21	II
Cj-M-48	Moderate	High	ERY, CLI	4923	CC ST21	V
Cj-M-41	Poor	Poor	TET, FFN, NAL	239	CC ST21	IV
Cj-E-29	Poor	Moderate	CIP, TET, TEL	4930	CC ST21	II
Cj-M-60	Moderate	Poor	TET	459	CC ST42	VII
Cj-S-14	Moderate	High	Susceptible	4924	CC ST21	II
Cj-E-58	Poor	Moderate	AZI, GEN, TET, NAL, TEL	4929	UA	VI
Cj-M-63	Moderate-High	High	TET	4925	UA	VII
Cj-M-1	Moderate	High	TET, TEL, CLI	4931	UA	I
Cj-M-64	Moderate-High	High	AZI, CIP, ERY, GEN, TET, CLI	4926	UA	VII
Cj-N-7	Poor	Moderate-High	AZI, ERY, FFN, TEL, CLI	4927	UA	II
Cj-E-57	Poor	Poor	CIP, ERY, TET	922	UA	VI
Cj-E-42	Poor	Moderate	CIP, GEN, TET, NAL, CLI	4928	UA	IV
Cj-S-56	Moderate	High	TET	4932	UA	V
Cj-E-45	Moderate	Poor	AZI, ERY, GEN, CLI	61	CC ST61	IV
***C.coli* Isolates**	**Invasion**	**Survival**	**Antimicrobial Resistance Profile**	**ST**	**CC ST**	**MRP cluster**
CC-N-103	Poor	Poor	NAL	1068	CC ST828	III
CC-N-16	Poor	Moderate	NAL	1068	CC ST828	I
CC-M-58	Non	N/A	AZI, GEN, TEL, CLI	902	CC ST828	II
CC-M-113	Poor	Non	CIP, TET, NAL, TEL	1068	CC ST828	IV
CC-S-79	Non	N/A	CIP, TET, NAL, TEL	1068	CC ST828	II
CC-S-44	Non	N/A	CIP, TET, NAL, TEL	1068	CC ST828	II
CC-S-66	Poor	Poor	CIP, TET, TEL	1068	CC ST828	II
CC-S-52	Poor	Moderate	AZI, ERY, GEN, FFN, TEL, CLI	2501	CC ST828	II
CC-S-10	Moderate	Moderate- high	AZI, ERY, GEN, TET, TEL, CLI	5372	CC ST828	I
CC-S-110	Poor	Moderate- high	CIP, TET, NAL	4933	CC ST828	IV
CC-E-95	Poor	Poor	AZI, ERY, GEN, TET, FFN, NAL, TEL, CLI	1110	CC ST828	III
CC-E-19	Poor	Non	AZI, ERY, GEN, CLI	1068	CC ST828	I

[a]Classification index: High = higher or comparable to *C. jejuni* 81-176. Moderate-High: Significantly lower than *C. jejuni* 81-176 but significantly higher than moderate. Moderate: Significantly higher than *C. jejuni* NCTC11168. Poor: significantly lower or comparable to *C. jejuni* NCTC11168.
doi:10.1371/journal.pone.0025778.t004

DISCUSSION

We investigated the occurrence of two important *Campylobacter* spp. (*C. jejuni* and *C. coli*) in cattle that are slaughtered for meat production across 4 major geographic locations in the USA. This is in contrast to previous studies of cattle-associated *Campylobacter* in the USA, which have been mostly geographically confined and/or limited to investigating the prevalence of certain species as well as other phenotypic characteristics such as antibiotic resistance [12], [25], [26]. Therefore, our approach was unique as it included analysis of genotypic diversity and antimicrobial resistance properties as well as attempts to understand the potential role of cattle-associated *Campylobacter* in virulence related phenotypes that might impact human health. Significantly, our findings highlight the importance of cattle as a reservoir for genotypically diverse, antimicrobial resistant, and potentially virulent *Campylobacter* in the USA.

Although the association of *Campylobacter* spp. with cattle in the USA has been investigated previously, available literature reported differing prevalence numbers for these pathogens. For example, Hoar et al [26] isolated *Campylobacter* from 5% of the fecal samples collected from cattle, while Sato et al [27] and Gharst et al [28] showed that these bacteria occurred in 27.9% and 23.4% of their samples, respectively. These differences in the prevalence of cattle-associated *Campylobacter* can be attributed to several factors, including methods for isolation, sample size and type (e.g. dairy versus feedlot), seasonal variations, and geographical location [2]. However, the latter factor, geographical location, was a prominent difference in the aforementioned prevalence studies, which were conducted in California, Wisconsin and Southeastern region of the USA, respectively [26], [27], [28]. Since the prevalence of cattle-associated *Campylobacter* in the USA seemed to vary according to the geographic location, it was necessary in this study to investigate this assumption as well as the overall occurrence of these bacteria in cattle. Our results showed that the occurrence of *Campylobacter* in Southern USA (prevalence of 32.8%) was significantly ($P<0.05$) higher than those observed in Northern (14.8%), Midwestern (15.83%), and Eastern (12%) locations. The precise reasons that contribute to the observed differences in geographic distribution of the *Campylobacter* are not clear. Nevertheless, previous studies have suggested that environmental conditions such as temperature, humidity, and sunlight might contribute to increases in human infections with *Campylobacter* [29]. Subsequently, the climate might be a driver of the relatively higher prevalence of cattle-associated *Campylobacter* in the Southern USA, which is generally warmer as compared to the other locations. Interestingly, Stanley et al. [30] suggested that indirect-temperature dependent factors such as migratory animals might impact the occurrence of *Campylobacter* in dairy cattle. Additionally, *C. coli* have been previously reported to predominantly occur in samples collected from broiler flocks during warmer months [15]. Strikingly higher prevalence of *C. coli* (83 isolates) in our samples from the South (Table 1) further support a role for climate in affecting the geographic prevalence of *Campylobacter*. Regardless, the overall prevalence of cattle-associated *Campylobacter* in this study was 19.2%, falling between the percentages (0.8 to 46.7%) reported for cattle from other countries [31]. However, it is important to note that samples in our study were collected only over a two-day period from a limited number of locations in each region. Furthermore, to emphasize a public health relevance of cattle as potential reservoir for *Campylobacter* (discussed below), our sampling efforts focused on the seasons that normally witness an increase in *Campylobacter* incidences [8]. As such our prevalence data don›t account for possible seasonal and temporal variations and, from this study alone, the extent to which differences between geographic locations contributed to *Campylobacter* prevalence in cattle cannot be conclusively ascertained. Nevertheless, our prevalence data provide evidence that like in other countries cattle in the USA might constitute a considerable reservoir for these pathogens, promoting their potential persistence in cattle-derived products and environment, which might pose a risk for human consumers and cattle-handlers.

The genus *Campylobacter* contains species that are known for their adaptation capabilities, readily acquiring genetic material necessary for survival and persistence [32]. High genotypic diversity has been reported in *Campylobacter* isolated from humans, chickens, sheep, pigs, and cattle [33], [34], [35], [36]. This is important since the genotypic diversity of these bacteria might serve as an indicator for the value of a reservoir in providing a suitable niche for the persistence of these microorganisms and their evolution [36]. As elegantly demonstrated by Wilson et al. [36], the broiler gastrointestinal tract constitutes such a niche, allowing multiplication of *C. jejuni* and enhancing its genetic diversity. Subsequently, it was important to investigate the genetic diversity of *Campylobacter* in cattle in order to assess the potential attributes of this reservoir. Our PFGE analysis of the cattle-associated *Campylobacter* revealed that all *C. jejuni* and *C. coli* isolates were generally grouped into 7 and 4 major clusters, respectively (Fig. 1). Remarkably, some *C. jejuni* isolates possessed genotypes that were 100% similar on 3 different occasions (see stars in Fig. 1A). Similarly, several *C. coli* exhibited 100% similar genotypes and were encountered on seven occasions (see asterisks in Fig. 1B). Taken together, these observations confirm a high genetic diversity within the cattle-associated *C. jejuni* and *C. coli* isolated in the USA, which also corroborates the findings of studies conducted on cattle in other countries, including Spain [35], the United Kingdom [37], Denmark [38], Finland [9], and Turkey [39]. Additionally, our PFGE analysis revealed clustering of some *C. jejuni* isolates (more than 75% similarity) that originated from different geographic locations. Likewise, some clusters included *C. coli* isolates with identical genotypes originating from different geographic locations (see arrows Fig. 1). This indicated that isolates with identical genotypes can occur in different herds, which highlights the possibility for certain strains to be transmitted among disparate herds/individuals. However, the latter requires further investigation to elucidate potential routes and mechanisms of transmission. Our PFGE analysis shows that cattle can be considered a potential reservoir for genetically diverse *Campylobacter* that is worth investigating in order to determine its relevance to public health in the USA.

An outstanding feature in livestock associated *Campylobacter* is its ability to resist antibiotics (reviewed in Luangtongkum et al. [40], an increasingly important emerging threat to public health [41]. Furthermore, previous studies have demonstrated that cattle-associated *Campylobacter* in the USA can potentially resist front-line therapeutic drugs for treating human infections. For example, 31.8% and 44% of *C. coli* collected from calves and feedlot cattle were resistant to erythromycin and ciprofloxacin, respectively [31]. Additionally, 47.7% and 49.1% of *C. jejuni* isolated from dairy and feedlot cattle, respectively, were resistant to tetracycline [25], [42]. In agreement with these studies, our antibiotic resistance analysis showed that a high percentage of the cattle-associated strains were resistant to erythromycin, ciprofloxacin, tetracycline and others (Fig 2A and Table 2). Remarkably, 23.7% and 12.1% of the cattle-associated *C. jejuni* that were analyzed in our study exhibited resistance to ciprofloxacin and erythromycin, respectively (Fig. 2A). This was an important and marked difference when compared to previous studies in the US that reported a relatively lower frequency of resistance to ciprofloxacin (1.8% to 5%) and erythromycin (0.4% to 2.9%) in *C. jejuni* isolated from different cattle operations [25], [31], [42]. Our data further support previous findings and predictions that livestock-associated *Campylobacter* are becoming increasingly resistant to important antibiotics [41], [43]. These observations highlight the need for rigorous surveillance of antibiotics used in cattle operations in order to facilitate interventions and curb further emergence of antibiotic resistant *Campylobacter*.

Understanding the contributions of cattle to human infections with *Campylobacter* relies heavily on: i) assessing the relationship between the cattle-associated *Campylobacter* and those isolated from- or previously implicated in human infections and ii) directly examining the potential of these isolates to invade and persist in the human intestine. However, these two criteria were rarely investigated together in previous research that focused on cattle-associated *Campylobacter* and are, to our knowledge, absent in similar studies conducted in the USA. To meet the aforementioned criteria, we typed a pool of the cattle-associated *Campylobacter* using MLST and further analyzed a subset of isolates for their properties in an *in vitro* surrogate for the human host (i.e. INT407 intestinal cell line). Since interpretation of the MLST typing depends on pre-existing online data bases (PubMLST) to assign sequence types (STs) to the isolates, it is possible to compare the cattle-associated *Campylobacter* from this study to others from different sources (e.g. humans) and locations, including other countries. Subsequently, our MLST analysis confirmed that cattle-associated *C. jejuni* (n=62) were diverse and belonged to disparate STs (n=25; Table 3), while only 8 STs were assigned to *C. coli* (n=50) (Table 3). Furthermore, we identified 12 *C. jejuni* that belonged to unassigned clonal complexes (n=10), which along with newly discovered alleles (n=3) and STs (n=13) (Table 3) emphasize the genetic diversity of this bacterium in cattle, possibly indicating that certain *C. jejuni* might be highly associated with this potential source of infection. The latter is supported by reports showing that CC ST-21 and ST-61 were predominant in cattle samples from other countries, including the UK, Canada, New Zealand, and Finland [9], [33], [44], [45]. The detection of CC ST-61 and CC ST-21 (Table 3) in this study was of particular interest, since these complexes constitute two of six CCs that harbor 60% of *C. jejuni* associated with human disease [20], [37]. Although all tested cattle-associated *C. jejuni* were capable of invading the human INT407 cells with variable efficiency, two isolate (Cj-M-13) and (Cj-N-33) that belong to CC ST-21 not only exhibited an invasive potential that was comparable to the hyper-invasive strain *C. jejuni* 81–176 [46] but also showed a higher capability to survive in the intracellular milieu (Fig. 3). Furthermore, these two isolates were resistant to 7 out of 9 tested antimicrobials, including ciprofloxacin, erythromycin, and gentamicin (Table 4). Along the same lines, 11 out of 19 *C. jejuni* with varying invasive capacities were able to either match or outperform the intracellular survival potential of *C. jejuni* 81–176 (Fig. 3). No specific association was observed with either invasion or intracellular survival of the tested *C. jejuni* isolates and their MRPs, antimicrobial resistance properties or their geographical distributions. Since many of the tested *C. jejuni* (13 out of 19) were moderate to highly invasive (Table 4) and given this bacterium›s aforementioned capability for natural transformation, the occurrence of invasive strains in a reservoir that is witnessing a rise in antibiotic-resistant *Campylobacter* constitutes a cause for caution and increased surveillance. Since invasion and intracellular survival of pathogens facilitates evading the immune system of the host and allows for persistent and recurrent infection [47], it can be argued that the cattle-associated *C. jejuni* would pose a plausible risk to human health.

Furthermore, MLST data and invasion studies also showed relatively lower clinical relevance of selected cattle-associated *C. coli* isolates, which all belonged to CC ST-828, a complex that include strains which are mainly isolated from agricultural and environmental sources and some from clinical cases [48]. Although the invasion and intracellular survival capacities of *C. coli* were overall lower than those of *C. jejuni*, one cattle associated *C. coli* (CC-S-10) invaded the cells with approximately 10^4 CFU per ml, while also surviving inside the cells in relatively high numbers (Fig. 4). Taken together, our data confirm the observations of Sheppard et al. [48] that ruminants also are among the most likely sources of human infections with *C. coli*.

We conclude that cattle in the USA constitute a suitable niche for the persistence of *Campylobacter* and a potentially important source for human infections with these pathogens. The potential of cattle-associated *Campylobacter* to resist antibiotics, invade and persist in human cells, and their genetic classification into clinically important clonal complexes warrant highlighting the contribution of cattle to the epidemiology of these pathogens in the USA.

AUTHOR CONTRIBUTIONS

Conceived and designed the experiments: GR JL YS. Performed the experiments: YS MA. Analyzed the data: YS IK GR JL WG. Wrote the paper: YS IK GR.

REFERENCES

1. Scallan E, Griffin PM, Angulo FJ, Tauxe RV, Hoekstra RM (2011) Foodborne illness acquired in the United States–unspecified agents. Emerg Infect Dis 17: 16–22.
2. Allos BM (2001) *Campylobacter jejuni* infections: update on emerging issues and trends. Clin Infect Dis 32: 1201–1206.
3. Saenz Y, Zarazaga M, Lantero M, Gastanares MJ, Baquero F, et al. (2000) Antibiotic resistance in *Campylobacter* strains isolated from animals, foods, and humans in Spain in 1997–1998. Antimicrob Agents Chemother 44: 267–271.
4. Miller WG, Mandrell RE (2005) Prevalence of *Campylobacter* in the food and water supply: incidence, outbreaks, isolation and detection. In: Ketley JM, Konkel ME, editors. *Campylobacter*: Molecular and Cell Biology. Norfolk, UK: Horizon Biosciences. pp. 101–63.
5. Piddock IJN, Ricci V, Stanley K, Jones K (2000) Activity of antibiotics used in human and medicine for *Campylobacter jejuni* isolated from farm animals and their environment in Lancashire, UK. J Antimicrob Chemother 46: 303–306.
6. Vandeplas S, Marcq C, Dauphin RD, Beckers Y, Thonart P, et al. (2008) Contamination of poultry flocks by the human pathogen *Campylobacter* spp. and strategies to reduce its prevalence at the farm level. Biotechnol Agron Soc Environ 12: 317–334.
7. Ragimbeau C, Schneider F, Losch S, Even J, Mossong J (2008) Multilocus sequence typing, pulsed-field gel electrophoresis, and fla short variable region typing of clonal complexes of *Campylobacter jejuni* strains of human, bovine, and poultry origins in Luxembourg. Appl Environ Microbiol 74: 7715–7722.
8. Wilson DJ, Gabriel E, Leatherbarrow AJH, Cheesbrough J, Gee S, et al. (2008) Tracing the source of campylobacteriosis. PLoS Genet 4: e1000203.
9. de Haan CPA, Kivistö RI, Hakkinen M, Corander J, Hänninen M-L (2010) Multilocus sequence types of Finnish bovine *Campylobacter jejuni* isolates and their attribution to human infections. BMC Microbiol 10: 200.
10. Besser TE, Lejeune JT, Rice DH, Berg J, Stilborn RP, et al. (2005) Increasing prevalence of *Campylobacter jejuni* in feedlot cattle through the feeding period. Appl Environ Microbiol 71: 5752–5758.

11. Milnes AS, Stewart I, Clifton-Hadley FA, Davies RH, Newell DG, et al. (2007) Intestinal carriage of verocytotoxigenic *Escherichia coli O157*, *Salmonella*, thermophilic *Campylobacter* and *Yersinia enterocolitica*, in cattle, sheep and pigs at slaughter in Great Britain during 2003. Epidemiol Infect 1–13:

12. Wesley IV, Wells SJ, Harmon KM, Green A, Schroeder-Tucker L, et al. (2000) Fecal shedding of *Campylobacter* and *Arcobacter* spp. in dairy cattle. Appl Environ Microbiol 66: 1994–2000.

13. Krause M, Josefsen MH, Lund M, Jacobsen NR, Brorsen L, et al. (2006) Comparative, collaborative, and on-site validation of a TaqMan PCR method as a tool for certified production of fresh, Campylobacter-free chickens. Appl Environ Microbiol 72: 5463–5468.

14. Engberg J, On SLW, Harrington CS, Gerner-Smidt P (2000) Prevalence of *Campylobacter*, Arcobacter, Helicobacter, and Sutterella spp. In human fecal samples as estimated by a reevaluation of isolation methods for campylobacters. J Clin Microbiol 38: 286–291.

15. Denis M, Rose V, Huneau-Salaün A, Balaine L, Salvat G (2008) Diversity of pulsed-field gel electrophoresis profiles of *Campylobacter jejuni* and *Campylobacter coli* from broiler chickens in France. Poult Sci 87: 1662–1671.

16. Denis M, Soumet C, Rivoal K, Ermel G, Blivet D, et al. (1999) Development of a m-PCR for simultaneous identification of *Campylobacter jejuni* and *C. coli*. Lett Appl Microbiol 29: 406–410.

17. Linton D, Lawson AJ, Owen RJ, Stanley J (1997) PCR detection, identification to species level, and fingerprinting of *Campylobacter jejuni* and *Campylobacter coli* direct from diarrheic samples. J Clin Microbiol 35: 2568–2572.

18. Ribot ER, Fitzgerald C, Kubota K, Swaminathan B, Barrett D (2001) Rapid pulsed-field gel electrophoresis protocol for subtyping of *Campylobacter jejuni*. J Clin Microbiol 39: 1889–1894.

19. Clinical and Laboratory Standards Institute (2006) Performance standards for antimicrobial disk susceptibility tests, approved standard, 9th ed. Clinical and Laboratory Standards document M2-A9. Clinical and Laboratory Standards Institute, Wayne, PA.

20. Dingle KE, Colles FM, Wareing DRA, Ure R, Fox AJ, et al. (2001) Multilocus sequence typing system for *Campylobacter jejuni*. J Clin Microbiol 39: 14–23.

21. Thakur S, Gebreyes W (2005) Prevalence and antimicrobial resistance of *Campylobacter* in antimicrobial-free and conventional pig production systems. J Food Prot 68: 2402–2410.
22. Konkel ME, Cieplak W Jr (1992) Altered synthetic response of *Campylobacter jejuni* to cocultivation with human epithelial cells is associated with enhanced internalization. Infect Immun 60: 4945–4949.
23. Prasad KN, Dhole TN, Ayyagari A (1996) Adherence, invasion and cytotoxin assay of *Campylobacter jejuni* in HeLa and HEp-2 cells. J Diarrhoeal Dis Res 14: 255–259.
24. Bacon DJ, Alm RA, Burr DH, Hu L, Kopecko DJ, et al. (2000) Involvement of a plasmid in virulence of *Campylobacter jejuni* 81–176. Infect Immun 68: 4384–4390.
25. Englen MD, Hill AE, Dargatz DA, Ladely SR, Fedorka-Cray PJ (2007) Prevalence and antimicrobial resistance of *Campylobacter* in US dairy cattle. J Appl Microbiol 102: 1570–1577.
26. Hoar BR, Atwill ER, Elmi C, Farver TB (2001) An examination of risk factors associated with beef cattle shedding pathogens of potential zoonotic concern. Epidemiol Infect 127: 147–155.
27. Sato K, Bartlett PC, Kaneene JB, Downes FP (2004) Comparison of Prevalence and Antimicrobial Susceptibilities of *Campylobacter* spp. Isolates from Organic and Conventional Dairy Herds in Wisconsin. Appl Environ Microbiol 1442–1447.
28. Gharst G, Hanson D, Kathariou S (2006) Effect of direct culture versus selective enrichment on the isolation of thermophilic *Campylobacter* from feces of mature cattle at harvest. J Food Prot 69: 1024–1027.
29. Patrick ME, Christiansen LE, Waino M, Ethelberg S, Madsen H, et al. (2004) Effects of climate on incidence of *Campylobacter* spp. in humans and prevalence in broiler flocks in Denmark. Appl Environ Microbiol 70: 7474–7480.
30. Stanley KN, Wallace JS, Currie JE, Diggle PJ, Jones K (1998) Seasonal variation of thermophilic *Campylobacters* in lambs at slaughter. J Appl Microbiol 84: 1111–1116.
31. Bae W, Kaya KN, Hancock DD, Call DR, Park YH, et al. (2005) Prevalence and antimicrobial resistance of thermophilic *Campylobacter* spp. from cattle farms in Washington State. Appl Environ Microbiol 71: 169–74.

32. Meinersmann RJ, Patton CM, Evins GM, Wachsmuth IK, Fields PI (2002) Genetic diversity and relationships of *Campylobacter* species and subspecies. Int J Syst Evol Microbiol 52: 1789–1797.
33. Levesque S, Frost E, Arbeit RD, Michaud S (2008) Multilocus sequence typing of *Campylobacter jejuni* isolates from humans, chickens, raw milk, and environmental water in Quebec, Canada. J Clin Microbiol 46: 3404–11.
34. Oporto B, Barandika JF, Hurtado A, Moreno B, Aduriz G, et al. (2006) Incidence of ovine abortion by *Coxiella burnetii* in Northern Spain. Ann N Y Acad Sci 1078: 498–501.
35. Oporto B, Esteban JI, Aduriz G, Juste RA, Hurtado A (2007) Prevalence and strain diversity of thermophilic campylobacters in cattle, sheep and swine farms. J Appl Microbiol 103: 977–984.
36. Wilson DL, Rathinam VA, Qi W, Wick LM, Landgraf J (2010) Genetic diversity in *Campylobacter jejuni* is associated with differential colonization of broiler chickens and C57BL/6J IL10-deficient mice. Microbiol 156: 2046–2057.
37. Grove-White DH, Leatherbarrow AJH, Cripps PJ, Diggle PJ, French NP (2010) Molecular epidemiology and genetic diversity of *Campylobacter jejuni* in ruminants. Epidemiol Infect 7: 1–11.
38. Litrup E, Torpdahl M, Nielsen EM (2007) Multilocus sequence typing performed on *Campylobacter coli* isolates from humans, broilers, pigs and cattle originating in Denmark. J Appl Microbiol 103: 210–218.
39. Açik MN, Çetinkaya B (2006) Heterogeneity of *Campylobacter jejuni* and *Campylobacter coli* strains from healthy sheep. Vet Microbiol 115: 370–375.
40. Luangtongkum T, Jeon B, Han J, Plummer P, Logue CM, et al. (2009) Antibiotic resistance in *Campylobacter*: emergence, transmission and persistence. Future Microbiol 4: 189–200.
41. Inglis GD, McConville M, de Jong A (2006) Atypical *Helicobacter canadensis* strains associated with swine. Appl Environ Microbiol 72: 4464–4471.
42. Englen MD, Fedorka-Cray PJ, Ladely SR, Dargatz DA (2005) Antimicrobial resistance patterns of *Campylobacter* from feedlot cattle. J Appl Microbiol 99: 285–291.
43. Châtre P, Haenni M, Le meunier D, Botrel M, Calavas D, et al. (2010) Prevalence and Antimicrobial Resistance of *Campylobacter jejuni* and

Campylobacter coli Isolated from Cattle between 2002 and 2006 in France. J Food Prot 73: 825–831.

44. Dingle KE, Colles FM, Ure R, Wagenaar JA, Duim B, et al. (2002) Molecular characterisation of *Campylobacter jejuni* clones : a basis for epidemiologic investigation. Emerg Infect Dis 8: 949–955.
45. Kwan PS, Birtles A, Bolton FJ, French NP, Robinson SE, et al. (2008) Longitudinal study of the molecular epidemiology of *Campylobacter jejuni* in cattle on dairy farms. Appl Environ Microbiol 74: 3626–33.
46. Hu L, Kopecko DJ (1999) *Campylobacter jejuni* 81–176 associates with microtubules and dynein during invasion of human intestinal cells. Infect Immun 67: 4171–4182.
47. Day WA Jr, Sajecki JL, Pitts TM, Joens LA (2000) Role of catalase in *Campylobacter jejuni* intracellular survival. Infect Immun 68: 6337–6345.
48. Sheppard SK, Dallas JF, Wilson DJ, Strachan NC, McCarthy ND, et al. (2010) Evolution of an Agriculture-Associated Disease Causing *Campylobacter coli* Clade: Evidence from National Surveillance Data in Scotland. Appl Environ Microbiol 76: 5269–5277.

CHAPTER 16

High Prevalence of Bovine Tuberculosis in Dairy Cattle in Central Ethiopia: Implications for the Dairy Industry and Public Health

Rebuma Firdessa[1], Rea Tschopp[1,4,6]., Alehegne Wubete[2]., Melaku Sombo[2], Elena Hailu[1], Girume Erenso[1], Teklu Kiros[1], Lawrence Yamuah[1], Martin Vordermeier[5], R. Glyn Hewinson[5], Douglas Young[4], Stephen V. Gordon[3], Mesfin Sahile[2], Abraham Aseffa[1], Stefan Berg[5]

[1]Armauer Hansen Research Institute, Addis Ababa, Ethiopia

[2] National Animal Health Diagnostic and Investigation Center, Sebeta, Addis Ababa, Ethiopia

[3] School of Veterinary Medicine, University College Dublin, Dublin, Republic of Ireland

Citation:Firdessa R, Tschopp R, Wubete A, Sombo M, Hailu E, Erenso G, et al. (2012) High Prevalence of Bovine Tuberculosis in Dairy Cattle in Central Ethiopia: Implications for the Dairy Industry and Public Health. PLoS ONE 7(12): e52851. https://doi.org/10.1371/journal.pone.0052851

[4] Centre for Molecular Microbiology and Infection, Imperial College London, London, United Kingdom

[5] Department for Bovine Tuberculosis, Animal Health and Veterinary Laboratories Agency, Weybridge, Surrey, United Kingdom

[6] Swiss Tropical and Public Health, Basel, Switzerland

ABSTRACT

Background

Ethiopia has the largest cattle population in Africa. The vast majority of the national herd is of indigenous zebu cattle maintained in rural areas under extensive husbandry systems. However, in response to the increasing demand for milk products and the Ethiopian government's efforts to improve productivity in the livestock sector, recent years have seen increased intensive husbandry settings holding exotic and cross breeds. This drive for increased productivity is however threatened by animal diseases that thrive under intensive settings, such as bovine tuberculosis (BTB), a disease that is already endemic in Ethiopia.

Methodology/Principal Findings

An extensive study was conducted to: estimate the prevalence of BTB in intensive dairy farms in central Ethiopia; identify associated risk factors; and characterize circulating strains of the causative agent, *Mycobacterium bovis*. The comparative intradermal tuberculin test (CIDT), questionnaire survey, post-mortem examination, bacteriology, and molecular typing were used to get a better understanding of the BTB prevalence among dairy farms in the study area. Based on the CIDT, our findings showed that around 30% of 2956 tested dairy cattle from 88 herds were positive for BTB while the herd prevalence was over 50%. Post-mortem examination revealed gross tuberculous lesions in 34/36 CIDT positive cattle and acid-fast bacilli were recovered from 31 animals. Molecular typing identified all isolates as *M. bovis* and further characterization by spoligotyping and MIRU-VNTR typing indicated low strain diversity within the study area.

Conclusions/Significance

This study showed an overall BTB herd prevalence of 50% in intensive dairy farms in Addis Ababa and surroundings, signalling an urgent need for intervention to control the disease and prevent zoonotic transmission of *M. bovis* to human populations consuming dairy products coming from these farms. It is suggested that government and policy makers should work together with stakeholders to design methods for the control of BTB in intensive farms in Ethiopia.

INTRODUCTION

The population of Ethiopia has increased dramatically in the last two decades, growing from approximately 55 million people in 1992 to a current estimate of around 85 million [1]. Increased population size has led to an inexorable increase in demand for food, putting pressure on the agricultural sector in which 85% of the work force is employed. Ethiopia has the largest livestock population in Africa, including an estimated ~52 million cattle [2] that contributes to the livelihoods of 60–70% of the population [3]. The vast majority of the cattle are indigenous zebu (*Bos indicus*) managed under traditional husbandry systems (grazing in the field) in rural areas. However, in recent years the number of dairy cattle of highly productive exotic (*Bos taurus*, mainly Holstein-Friesian) and cross breeds has been on the rise, particularly in urban and peri-urban areas in response to the increasing demand for milk products and the Ethiopian government's effort to improve productivity in the livestock sector. The population of dairy cows accounts for 6.3 million animals (around 12% of the total cattle population) and the estimated total national milk production per year is 2.6 billion litres [4] of which the urban and peri-urban dairy farmers produce 2%.

In a country such as Ethiopia, where livestock are extremely important for people's livelihood, animal diseases can be a real threat to animal productivity and thus negatively impact on the agricultural sector and economic development. Bovine tuberculosis (BTB), caused by *Mycobacterium bovis*, is a chronic and contagious disease of cattle and other domestic and wild animals [5], [6]. BTB is prevalent worldwide but prevalence data is scarce in most developing countries due to lack of active control programmes. Several studies conducted since 2006 have confirmed that BTB is endemic in Ethiopia with prevalence rates varying from 0.8% to around 10% in extensive rural farming systems [7]–[11], while higher prevalence rates have been reported from regions in Ethiopia where intensive husbandry systems are more common [12]–[14]. As well as causing a high morbidity, BTB can also be a financial burden to farmers owning infected cattle; it has been suggested that cattle with BTB have a reduced productivity affecting milk yield and carcass value [15] as well as through reduced pulling power in traditional farming system [16]. BTB has also zoonotic potential [17], [18] - mainly through consumption of unpasteurised milk products - and its prevalence in Ethiopian cattle can therefore be a contributing factor to the human burden of TB in Ethiopia that currently is ranked as the 7th highest in the world [19], [20].

This study was designed to get a better understanding of the prevalence of BTB among intensive dairy farms in the urban and peri-urban belt of Addis Ababa (central Ethiopia), areas from where the vast majority of commercial dairy products in the country are produced. Furthermore, we show molecular typing results of *M. bovis* strains circulating in this region of Ethiopia and discuss the importance of our findings for informing BTB control policies in Ethiopia.

MATERIALS AND METHODS

Study Sites and Ethical Considerations

Statistics from the Ethiopian Ministry of Agriculture and Rural Development (MoARD) as well as data from regional agricultural offices recognised a high density of intensive dairy farms in a 50 km radius region from Addis Ababa city centre. Most dairy farms in this region had access to one of the five main roads that lead into the capital. Therefore, the region was divided into six geographically defined study areas, including Addis Ababa city, Debre Zeit, Sebeta, Holeta, Sululta and Sendafa (Figure 1). The recruitment of dairy farms within each study area was based on the existence of high numbers of Holstein-Friesian and cross breed cattle, as well as variation in herd size, with the latter being categorized into three groups; small herds (1–10 cattle), medium herds (11–50 cattle), and large herds (>50 cattle). This study was approved by the MoARD and was based on voluntary participation, and consent to participate was given from all farm owners after being informed about the study by local agricultural office authorities as well as by study representatives. Ethical approval for the study was given by the institutional ethical committee at the Armauer Hansen Research Institute (AHRI) and the All Africa Leprosy, Tuberculosis and Rehabilitation Training centre (ALERT). As part of the commitment of this study, one week training was given to 63 of the farm owners and/or their staff by professionals from National Animal Health Diagnostic and Investigation Center (NAHDIC), AHRI, the National Artificial Insemination Center (NAIC), and the Ministry of MoARD Extension Directorate and Oromia Special Zone Surrounding Finfinne Livestock Development & Marketing Agency Office. The training was done in the Amharic language and included topics such as mastitis, abortion, dystocia, calcium deficiency, BTB, brucellosis, heat detection, artificial insemination, vaccination, and herd management (e.g. annual planning, dairy management).

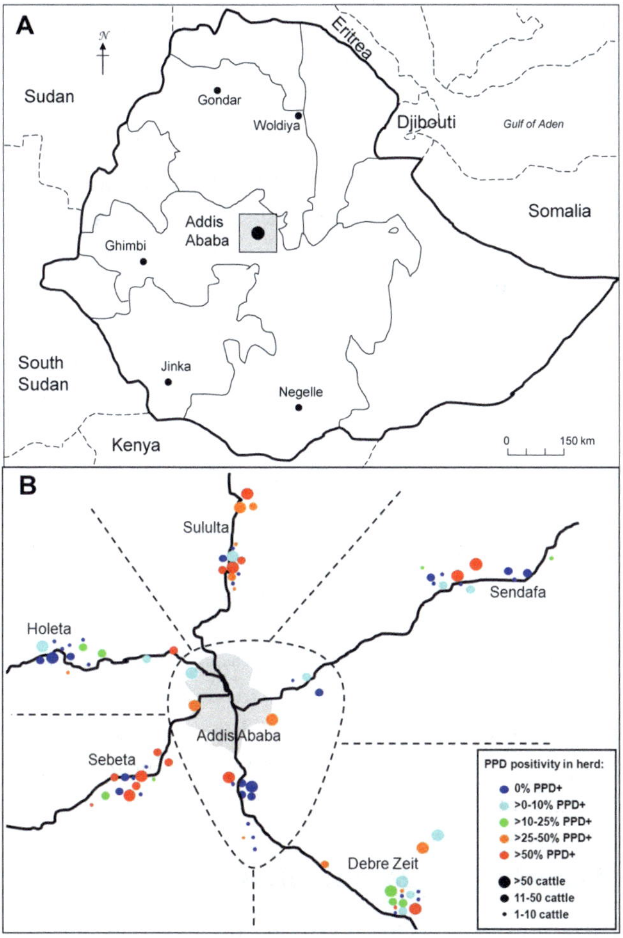

Figure 1: Mapping of (A) the study area (marked in grey) and other sites in Ethiopia where M. bovis **isolates used in this study were collected, and (B) BTB prevalence in investigated farms in the five study areas.**

PPD positivity and farm size category are defined by colour and size of respective circle.

https://doi.org/10.1371/journal.pone.0052851.g001

Comparative Intradermal Test and Questionnaire

All cattle older than six months within a herd, except clinically sick animals and cows one month pre-and post partum, were tested by CIDT: Two sites, 12 cm apart, horizontally of the mid-neck of the animal were shaved and the skin thickness was measured with a calliper. In calves with short neck, the sites were shaven vertically. Aliquots of 0.1 ml of 2500 IU/ml bovine purified protein derivative (PPD) and 0.1 ml of 2500 IU/ml avian PPD (Synbiotics Corporation, Lyon, France) were injected separately into the respective shaved site. The thickness of the skin at each injection site was measured again after 72 hours. The test results were interpreted in line with OIE recommendations (2004): if the difference between skin thickness at the bovine site of injection and the avian site of injection was <2 mm, between 2 mm and ≤ 4 mm, or >4 mm, the animal was classified as negative, doubtful (inconclusive), or positive for BTB, respectively. A herd with at least one positive reactor was considered as "PPD positive". *M. avium* positivity was defined as previously described [9] with the avian PPD injection site showing a reaction of >4 mm between the testing day and the reading day. In addition, we calculated how many animals showed an avian skin reaction of >4 mm as compared to a bovine skin reaction [(A2-A1) - (B2-B1)>4 mm]. Information on farm structure and management were collected from 80 out of the 88 investigated farms using a standardised questionnaire with closed questions. The owner or manager of each farm was interviewed in the local language at the time the CIDT was performed. Sanitation status was judged as poor, medium (satisfactory), or good, based on aspects such as odour, waste drainage, cleanness of floor and animals, barn ventilation and light source, and animal stocking.

Ante and Post Mortem Examination

All cattle tested by CIDT were recorded for age, sex, breed, and given a body condition score (poor, medium, or good). Selected CIDT positive cattle from farms of all six study areas were purchased and subjected to post mortem examination performed by an experienced veterinarian. The selection criteria for these purchased animals were based on strong PPD response, unproductive animal, and willingness of farmer to sell the animal. The pathology scoring system was based on the semi-quantitative procedure developed by Vordermeier et al. [21]. Each lobe of the lung was inspected externally and then sliced into 2 cm-thick slices to facilitate detection of any typical TB lesion. Similarly, liver, spleen, and kidney as well as other lymph nodes (mandibular, medial retropharyngeal, bronchial, mediastinal, and mesenteric) were sliced into thin sections and inspected for the presence of typical TB lesions. The severity scores of the gross lesions and the pathology scores for lymph nodes, lungs, and other organs were added up to determine the total pathology score per animal. Furthermore, lesion type (caseous or calcified) and TB stage (localised or generalized) were documented.

Sample Collection, Processing and Culturing of Mycobacteria

During post-mortem, up to seven suspected visible TB-like lesions were collected per animal in individual 50 ml sterile universal tubes (containing sterile PBS buffer) and transported at 4°C to the AHRI for further processing. At AHRI, samples were stored at 4°C when processed within 48 hours or at −20°C if processed later than two days after sample collection. All samples were processed and cultured for mycobacteria as previously described [7]. Samples were inoculated on three different media, including two Löwenstein-Jensen (LJ) media (supplemented with either glycerol or pyruvate) and a modified Middlebrook 7H11 medium optimised for the culture of *M. bovis* [22].

Identification and Molecular Typing of Mycobacteria

Bacterial colonies from culture positive samples were Ziehl-Neelsen stained according to standardized protocol to identify Acid-Fast Bacilli (AFB). Slants with no growth at week 8 were considered negative. Heat-killed cells of each AFB positive isolate were prepared by mixing ~2 loopfuls of cells (~20 µl cell pellet) in 200 µl dH_2O followed by incubation at 80°C for 1 hour and then stored at 4°C until used for molecular typing. AFB positive isolates were also prepared as 20% glycerol stocks and stored at −80°C. Heat-killed AFB positive isolates were investigated by multiplex PCR for the presence or absence of RD4 [23], a chromosomal deletion characteristic of *M. bovis*. PCR amplification for RD4 typing was as follows: reactions were performed in a total volume of 20 µl consisting of 10 µl HotStarTaq Master Mix (Qiagen, United Kingdom), 7.1 µl distilled H_2O, 0.3 µl of each oligonucleotide primer (100 µM), and 2 µl DNA template (heat-killed cells, see above). Isolates genetically typed as *M. bovis* were spoligotyped as previously described [24]. Selected *M. bovis* isolates were typed for 24-loci MIRU-VNTR by Genoscreen, France [25], [26].

Data Analysis

All collected field and laboratory data were double entered into Microsoft Access database. Data validation was done with EpiInfo software ("Data compare" package). Data analysis was performed with STATA Version 10.1 (StataCorp, Texas, USA). We used logistic regression models for the analysis of BTB prevalence and the univariate analysis of parameters associated with positive farms. A GEE model (Generalized Estimating Equations) with binary outcome was used to analyse potential risk factors for individual reactors, in which farms were used as random effect and farm size as fixed effect with adjusted odds ratio (OR). In both cases, results were presented with 95% confidence interval for OR and p-values.

RESULTS

Study Areas and Farm Recruitment

As shown in Figure 1, six study areas were chosen in the region of Addis Ababa in central Ethiopia; these were Addis Ababa, Sendafa, Holeta, Sululta, Debre Zeit and Sebeta.

Table 1: Individual and herd prevalence of BTB based on CIDT and individual PPD-A response, both stratified by farm size and study site

https://doi.org/10.1371/journal.pone.0052851.t001

Farm size		Addis Ababa	Debre Zeit	Sebeta	Holeta	Sululta	Sendafa	Total
Small#	Farm number	6	6	5	6	5	6	34
	Animal number	40	25	37	36	34	46	218
	Indiv. prev. (%)	0	16	16.2	13.9	11.8	2.2	9.2
	Herd prev. (%)	0	33	40	16.7	40	17	23.5
	PPD-A resp. (%)*	0 (0)	11 (0)	2.7 (0)	0 (0)	6.2 (0)	2.2 (0)	3.2 (0)
Medium#	Farm number	5	5	7	6	5	5	33
	Animal number	187	106	177	181	154	137	942
	Indiv. prev. (%)	0.5	20.7	38.4	14.9	47.4	1.5	20.5
	Herd prev. (%)	20	100	71	67	80	40	63.6
	PPD-A resp. (%)*	3.2 (0.5)	1.9 (0)	8.5 (1.1)	2.8 (0)	22.7 (0.6)	0.7 (0)	6.8 (0.4)
Large#	Farm number	5	5	2	2	4	3	21
	Animal number	531	518	235	144	210	158	1796
	Indiv. prev. (%)	36.2	23.5	72.8	1.4	53.8	89.9	41.3
	Herd prev. (%)	80	100	100	100	100	67	90.5
	PPD-A resp. (%)*	17.1 (0.2)	11 (1.3)	11.1 (0)	9 (0.7)	10 (0.5)	24.7 (0.6)	13.8 (0.6)

#Farm size definition: Small, 1–10 animals; Medium, 11–50 animals; Large, >50 animals.

*The percent of animals with response to PPD-A alone [(A2-A1)>4 mm] regardless of the PPD-B response. In bracket, percentage of animals with a skin reaction defined as [(A2-A1) – (B2-B1)>4 mm].

doi:10.1371/journal.pone.0052851.t001

A total of 88 farms were included in the study, with a distribution of 14–16 farms per study area (Table 1). At least five small farms (1–10 cattle) and five medium size farms (11–50 cattle) were included from each study area, while the number of large dairy farms (>50 cattle) varied between two and five due to limited large herds in the Holeta (3), Sebeta (2), and Sendafa (2) study areas.

Dairy Farm Management

The vast majority (95%) of the investigated farms were privately owned while the remaining 5% were owned by a cooperative or the government. Seventeen of the 88 farms were owned by women. In our survey 80 out of these 88 farms answered our questionnaire. These 80 farms were intensive dairy farms with similar husbandry practice and nearly all cattle herds were kept in indoor barns; only three small herds shared the same house as the owners and one farm kept its animals outdoors. Herds were strictly kept separate from others at 78 out of 80 farms while owners of the remaining two farms allowed mixing with animals from other farms. Sanitation status was assessed for the barns on each farm. The sanitation was mostly poor (20%) to medium (65%), while only 12 farms (15%) showed good sanitation standards. All farms included in this study used mainly their own bull or artificial insemination for restocking the herd. However, about 1/3 of the farms purchased animals during 2008 and 2009 and they did so within the area where this study was conducted (Figure 1). Fifty of the farm owners sold cattle during these two years of which 13 found a buyer in other locations of Ethiopia (outside the described study area) such as Tigray in the north, Ambo and Jimma in the west, and Nazareth, Hawassa, Woleita, Hosana, and Shashemene southwards of Addis Ababa. More than half of the farms that sold animals to regions outside the study area were medium-sized farms, whereas only ~1/4 of both small and large farms had such behaviour. Animals were sold particularly when they showed low productivity (N=50/80; 62%). Other motives for selling animals included being old (5%), weak and emaciated (5%), or being diseased (6.3%). Seven owners (8.9%) reported to have sold while still at high productivity.

Trading and Consumption of Milk

The vast majority of milk produced in the investigated farms was either sold (N=36/80; 45%), or partly sold and partly used for own consumption (50%). The remaining 5% of the farms produced milk for their own consumption only. Thirty-five farms sold their milk to processing plants, 23 to cafés and restaurants, 19 to intermediate caterers, and 16 directly to private individuals. The number of people living on a farm varied between one and 53. However, the majority (81%) of farms accommodated less than ten people while half had only between one and four people living on the farm. The majority (86%) of the farm owners did not drink raw milk and 71% knew that BTB is an animal and a zoonotic disease.

Tuberculin Testing

Over a period of 13 months (January 2009 to January 2010), a total of 2956 cattle from 88 dairy farms were tuberculin tested by CIDT. The vast majority (94%) of these cattle were female and either of Holstein-Frisian breed (37%) or cross breeds thereof (62%) with local zebu cattle (Table 2). The remaining tested animals were of either zebu or Jersey breed, confined to a few farms in Holeta, Sendafa, Sebeta, and Debre Zeit.

Table 2: Univariate analysis of risk factors for having positive reactors using a GEE model with binary outcome

https://doi.org/10.1371/journal.pone.0052851.t002

Variable	Category	Animal No.*	BTB reactors*	OR (95%CI)	p-value
Herd size	Small	216 (7.3)	20 (2.1)		
	Medium	942 (31.9)	193 (20.2)	2.5 (0.91; 6.85)	0.07
	Large	1796 (60.8)	742 (77.7)	5.9 (2.06; 16.84)	0.001
Breed	Crossbreed	1837 (62.8)	351 (37.2)		
	Holstein	1011 (34.6)	582 (61.8)	1.2 (0.69; 2.23)	0.47
	Zebu	45 (1.5)	3 (0.3)	1.1 (0.89; 1.33)	0.39
	Jersey	33 (1)	7 (0.7)	1.1 (0.73; 1.71)	0.59
Sex	Female	2783 (94)	919 (96.3)		
	Male	170 (6)	35 (3.7)	0.9 (0.76; 1.23)	0.8
Body condition score	Poor	212 (7.2)	62 (7)		
	Medium	1454 (49.5)	409 (43)	0.8 (0.57; 1.06)	0.1
	Good	1272 (43.3)	481 (50)	0.8 (0.58; 1.13)	0.2
Age class	<12 months	139 (5.2)	27 (3)		
	≥12 mo - 3 yrs	823 (31)	225 (24.3)	2.2 (0.91; 5.51)	0.07
	>3 yrs - 10 yrs	1642 (61.8)	656 (71)	3.6 (1.43; 9.30)	0.007
	>10 yrs	54 (2)	16 (1.7)	4.1 (1.59; 10.56)	0.003
PPD-A status	Negative	2636 (89.2)	755 (79)		
	Positive	318 (10.8)	200 (21)	1.6 (1.01; 2.60)	0.04

*A number in brackets shows percentage of animals recorded in a category of one variable. The total number of animals recorded for in one variable may vary as some data are missing.
doi:10.1371/journal.pone.0052851.t002

The result of the tuberculin testing is presented in Table 1 as prevalence per study area and herd size. The CIDT result shows that BTB was prevalent in all six study areas. It also demonstrates that the prevalence increased with farm size category. The average individual prevalence of small, medium, and large farms was 9.2%, 20.5%, and 41.3%, respectively, and a similar trend was seen in herd prevalence with the extreme of over 90% infected herds among the large farms (Table 1). The overall number of herds with at least one reactor animal was 48 (55%); however, 38 farms (43%) had ≥10% of its animals infected (Figure 1B).

More detailed analysis of small and medium sized farms indicated that the areas of Addis Ababa and Sendafa (Figure 1) had the lowest individual animal prevalence; only one out of 86 tested animals from small farms in those two areas combined was a reactor, and in the medium-sized farms only three out of 324 animals were positive according to the CIDT. Small and medium farms tested in the remaining four study areas had significantly higher individual prevalence, ranging between 12–16% in small farms while it varied more in medium farms (between 15–47.4%); elevated prevalence was seen in medium farms in Sebeta and Sululta. The distribution of tuberculin reactors between farms is reflected in the herd prevalence (Table 1). More than 1/3 of small farms and over 2/3 of medium farms were infected in Debre Zeit, Sebeta, and Sululta. The latter was valid also for the medium farms in Holeta. We conclude that BTB was widely distributed among the tested small and medium farms in at least Debre Zeit, Sebeta, and Sululta.

The results obtained from the analysis of large dairy farms showed a somewhat different picture. The individual BTB prevalence in any tested large farm varied from 0% and up to 93%. By study area, Holeta showed the lowest prevalence with only two CIDT positive animals between the two large farms tested. The individual prevalence in all other study areas varied between 23.5–90%, (Table 1). Overall, 18 out of the 21 large farms were infected with BTB (overall herd prevalence: 85.7%) of which 11 had an animal prevalence of over 30%. The crude overall individual prevalence was 32.3% (30.6%–34.0%) (not considering stratification of farm size or study sites).

Sensitisation to Environmental Mycobacteria

Although the CIDT is used to detect BTB infection with high specificity, reactivity to the avian PPD indicates infection with or exposure to species of the *Mycobacterium avium* complex (MAC) or other environmental mycobacteria. In this study 318 out of the 2956 tested animals (~11%) reacted positive to PPD-A. The PPD-A response was lowest in Holeta (5%; 95%CI: 3.1; 7.8) and highest in Sululta (14.6%; 95%CI: 11.5; 18.5) (Table 1), regardless of farm size. However, 200 out of the 318 PPD-A positive animals reacted positive also to PPD-B (Table 2) and the PPD-A responses could be down to cross reactivity between *M. bovis* antigens and antigens from environmental mycobacteria. Interestingly, PPD-A responders with a stronger response than to PPD-B were few (between 0 and 1.3%; Table 1).

Factors Associated with BTB

The univariate analysis (Table 2) showed that reactor animals differed statistically according to age categories. Adult animals between 3 and 10 years old were over threefold more at risk of being tuberculin positive (with adults older than 10 years fourfold more at risk) as compared to young animals (OR=3.6, p-value=0.007 and OR=4.1, p-value=0.003, respectively). Medium and large farms had a significant risk increase of having positive reactors (OR=2.5 and OR=5.9 respectively, p-value: 0.001). Positive PPD-A reactors had an OR of 1.6 for being BTB reactors (p-value: 0.04).

Table 3 shows the univariate analysis of parameters potentially affecting the BTB herd status: Herd size and education level of the owners were factors associated with having positive farms; medium herds had a statistically significant fivefold increase of tuberculin positivity (OR=5.4; p-value: 0.002) compared to small herds. Higher education of owner was linked with BTB positive farms (secondary school with OR=4.8, p-value: 0.04 and diploma/degree with OR=4.5, p-value: 0.05). The majority of the owners (65.4%) had an education above junior secondary school (secondary school, diploma, degree). However, 87.5% of the owners having large farms had gone through secondary school and/or higher education (diploma, degree). On the other hand 50% of the owners with no education at all or primary school attendance governed over small farms. When stratified by farm size, there was no statistical difference between level of education and BTB farm status (data not shown).

Table 3: Univariate analysis for potential risk factors being a positive farm (logistic regression with likelihood ratio test)

https://doi.org/10.1371/journal.pone.0052851.t003

Farm size		Addis Ababa	Debre Zeit	Sebeta	Holeta	Sululta	Sendafa	Total
Small#	Farm number	6	6	5	6	5	6	34
	Animal number	40	25	37	36	34	46	218
	Indiv. prev. (%)	0	16	16.2	13.9	11.8	2.2	9.2
	Herd prev. (%)	0	33	40	16.7	40	17	23.5
	PPD-A resp. (%)*	0 (0)	11 (0)	2.7 (0)	0 (0)	6.2 (0)	2.2 (0)	3.2 (0)
Medium#	Farm number	5	5	7	6	5	5	33
	Animal number	187	106	177	181	154	137	942
	Indiv. prev. (%)	0.5	20.7	38.4	14.9	47.4	1.5	20.5
	Herd prev. (%)	20	100	71	67	80	40	63.6
	PPD-A resp. (%)*	3.2 (0.5)	1.9 (0)	8.5 (1.1)	2.8 (0)	22.7 (0.6)	0.7 (0)	6.8 (0.4)
Large#	Farm number	5	5	2	2	4	3	21
	Animal number	531	518	235	144	210	158	1796
	Indiv. prev. (%)	36.2	23.5	72.8	1.4	53.8	89.9	41.3
	Herd prev. (%)	80	100	100	100	100	67	90.5
	PPD-A resp. (%)*	17.1 (0.2)	11 (1.3)	11.1 (0)	9 (0.7)	10 (0.5)	24.7 (0.6)	13.8 (0.6)

#Farm size definition: Small, 1–10 animals; Medium, 11–50 animals; Large, >50 animals.

*The percent of animals with response to PPD-A alone [(A2-A1)>4 mm] regardless of the PPD-B response. In bracket, percentage of animals with a skin reaction defined as [(A2-A1) – (B2-B1)>4 mm]

There was no significant difference in BTB status amongst farms with different sanitation status and amongst the type of cattle housing (data not shown).

Ante- and Post-mortem Investigation

Over 90% of all investigated animals had medium or good body condition. However, no significant difference of the body condition score was observed between tuberculin positive and negative cattle (Table 2). To further confirm that the recorded high rates of BTB were due to *M. bovis* and to get a better understanding of the population structure of the causative agent, we purchased and slaughtered a minimum of three PPD positive animals per study area. In total 36 cattle from 19 farms (up to three animals per farm) were purchased and investigated by post-mortem and each animal was given a pathology score based on disease progression (Table 4). A trend of inverted correlation between pathology score and PPD response was observed; animals with strong response to the tuberculin test had in general a lower than average pathology score. Suspect TB lesions from up to seven infected sites per animal were collected from 34 animals. In addition, samples were taken from mediastinal and bronchial lymph nodes from two animals with non-visible lesions. Thirteen out of the 34 (32%) animals with lesions had generalised disease while the remaining 21 had lesions localised to one or a few infection sites. Figure 2 shows the distribution of lesions found in the 34 animals. Visible lesions were most frequent in mediastinal lymph nodes (N=30; 88%), lungs (74%), bronchial lymph nodes (62%), and head lymph nodes (50%), but prevalent also in mesenteric and hepatic lymph nodes, and in liver. Calcification was the most commonly observed stage of lesion (Table 4).

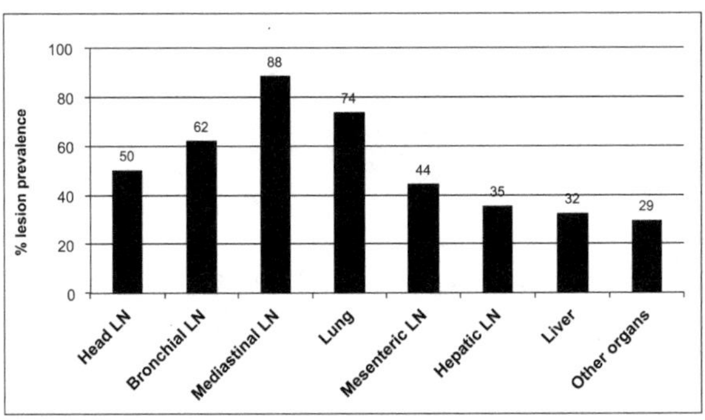

Figure 2: Organ association of suspect TB lesions in 34 cattle with visible lesions (LN, lymph node).

https://doi.org/10.1371/journal.pone.0052851.g002

Table 4: Post-mortem results of 36 animals from 19 farms

https://doi.org/10.1371/journal.pone.0052851.t004

Study area	Farm ID: Animal ID	CIDT (Δmm)	PM score	Type of TB	Type of lesion	AFB	Strain ID	*M. bovis* Spoligotype	24-loci MIRU-VNTR type*
Addis Ababa	1:A	24	4	localized	caseous	Yes	BTB-2269	SB0134	35223 40554 21415 24222 2XXX
Addis Ababa	1:B	26	15	localized	calcified	No	-	-	-
Addis Ababa	2:A	22	23	generalized	calcified/caseous	Yes	BTB-2850	SB1477	36223 32454 21445 24222 3213
Addis Ababa	2:B	18	23	generalized	calcified	Yes	BTB-2840	SB1176	35223 32554 21435 24222 3323
Holeta	3:A	30	10	generalized	caseous	Yes	BTB-2352	SB1176	35223 32554 21435 24222 3323
Holeta	3:B	26	0	-	-	No	-	-	-
Holeta	3:C	12	29	generalized	calcified	Yes	BTB-2400	SB1176	35223 32554 21435 24222 3323
Holeta	4:A	29	5	generalized	calcified	Yes	BTB-2349	SB0134	-
Holeta	5:A	11	20	localized	calcified	Yes	BTB-2350	SB0134	35223 40554 21415 24222 2332
Sululta	6:A	16	14	localized	calcified/caseous	Yes	BTB-2762	SB0133	23223 32454 21425 24222 3323
Sululta	6:B	18	21	generalized	calcified	Yes	BTB-2822	SB0133	23X23 32454 21425 24222 3323
Sululta	7:A	15	46	generalized	calcified	Yes	BTB-2466	SB1176	35223 32554 21435 24222 3323
Sululta	7:B	28	12	localized	calcified	Yes	BTB-2785	SB1176	35223 32554 21435 24222 3323
Sululta	7:C	14	43	generalized	calcified	Yes	BTB-2469	SB1176	35223 32554 21435 24222 3323
Sululta	8:A	38	5	localized	calcified	Yes	BTB-2465	SB1176	35223 32554 21435 24222 3323
Sululta	8:B	24	1	localized	calcified	Yes	BTB-2461	SB0134	35223 40554 21415 24222 2332
Sendafa	9:A	21	14	generalized	calcified/caseous	Yes	BTB-2847	SB0134	35223 40554 21415 24222 2332
Sendafa	9:B	13	8	localized	caseous	Yes	BTB-2849	SB0134	35223 40554 21415 24222 2332
Sendafa	10:A	17	2	localized	caseous	Yes	BTB-2862	SB0133	-
Sendafa	10:B	41	4	localized	calcified/caseous	Yes	BTB-2861	SB0133	23223 32454 2X425 24222 3X23
Debre Zeit	11:A	17	20	generalized	calcified	Yes	BTB-2266	SB1176	-
Debre Zeit	11:B	16	19	generalized	calcified	Yes	BTB-2261	SB0134	35223 40554 21415 24222 2332
Debre Zeit	12:A	4	9	localized	calcified	Yes	BTB-2260	SB1176	35223 32552 21435 24222 3323
Debre Zeit	13:A	23	19	generalized	calcified	No	-	-	-
Debre Zeit	13:B	11	8	localized	calcified	Yes	BTB-2257	SB0133	23223 32454 21425 24222 3323
Sebeta	14:A	6	2	localized	calcified	No	-	-	-
Sebeta	14:B	20	12	localized	caseous	Yes	BTB-2813	SB1176	3522X 32554 2X435 24X22 3323
Sebeta	15:A	18	37	generalized	calcified	Yes	BTB-2803	SB1176	35223 32552 21435 24222 3323
Sebeta	15:B	18	6	localized	caseous	Yes	BTB-2805	SB1176	35223 32552 21435 24222 3323
Sebeta	16:A	20	0	-	-	No	-	-	-
Sebeta	16:B	22	4	localized	calcified	Yes	BTB-2768	SB1176	-
Sebeta	17:A	21	8	localized	calcified	Yes	BTB-2838	SB0134	35223 40554 21415 24222 2332
Sebeta	17:B	16	8	localized	calcified/caseous	Yes	BTB-2765	SB0134	35223 40554 21415 24222 2332
Sebeta	18:A	13	6	localized	caseous	Yes	BTB-2757	SB0134	35223 40554 21415 24222 2332
Sebeta	19:A	22	9	localized	calcified	Yes	BTB-2773	SB0134	35223 40554 21415 24222 2332
Sebeta	19:B	20	13	localized	calcified/caseous	Yes	BTB-2766	SB0134	35223 40554 21415 24222 2332

The average CIDT value and post-mortem score [21] were 20 mm and 13, respectively.

*Consecutive order of the 24-loci MIRU-VNTR: 2165; 2461; 577; 580; 3192; 154; 960; 1644; 2059; 2531; 2687; 2996; 3007; 4348; 802; 424; 1955; 2163b; 2347; 2401;3171; 3690; 4052; 4156; X = Typing failed.

doi:10.1371/journal.pone.0052851.t004

Isolation and Molecular Typing of Mycobacteria

Mycobacterial culturing on selected media yielded in total 107 AFB isolates from samples corresponding to 31 animals with visible lesion. No AFB were cultivated from the remaining three animals with suspect TB lesions or from samples of the two animals with non-visible lesions (Table 4).

Heat-killed cells of 67 AFB isolates originating from 31 animals were enrolled in a typing scheme initiated by RD4 deletion typing. The result showed that all isolates could be identified as *M. bovis* as they had the RD4 region deleted [23]. Spoligotyping was then employed to further differentiate these isolates and the outcome is shown in Table 4. Overall, four different spoligotype patterns were recognized among these *M. bovis* isolates and defined as of type SB0134, SB1176, SB0133, and SB1477 at the international spoligotyping database www.mbovis.org. None of the 31 animals were found to be infected with more than one *M. bovis* spoligotype suggesting no dual infection among these animals. On the other hand, *M. bovis* isolates of two different spoligotype patterns were collected from three farms (out of 19). The geographical distribution of these four spoligotypes suggested three main clusters; SB0134 was the most dispersed and found in all six study areas, SB1176 was represented in all areas but Sendafa, while SB0133 was only found in Sululta, Sendafa and Debre Zeit. The one isolate of type SB1477 was collected in a farm from Addis Ababa.

In attempts to find possible transmission links between dairy farms included in this study we performed 24-loci MIRU-VNTR typing to further discriminate the collected *M. bovis* isolates. MIRU-VNTR results were generated for 27 isolates of the four different spoligotypes (Table 4). All 11 isolates of type SB0134 had the same genotype (spoligotype plus MIRU-VNTR type) and that was also the case for four isolates of SB0133. Only the collected isolates of type SB1176 showed diversity; two different genotypes were identified among 11 isolates, however the two types differed in only one out of the 24 loci (Locus 2531; Table 4). Overall, this genotyping result suggested very low diversity within each spoligotype cluster among the collected strains.

In a previous study on BTB in Ethiopia [7] we isolated and identified *M. bovis* from several abattoirs in the country. Here we performed 24-loci MIRU-VNTR typing of 15 *M. bovis* isolates from that study with spoligotypes SB0133, SB1176, SB0134, the three most frequent types isolated in the present study (Table 4) with the aim to investigate genotype diversity among strains collected from different parts of Ethiopia. Table 5 summarises the different *M. bovis* genotypes and their frequency from each collection site. Two distinct MIRU-VNTR features were seen among strains within spoligotype SB0133 as well as within SB1176, while strains compared within SB0134 showed no diversity.

Table 5: Geographical diversity of *Mycobacterium bovis* **genotypes in Ethiopia**

https://doi.org/10.1371/journal.pone.0052851.t005

Spoligotype	Region	Collection site	24 loci MIRU-VNTR type*	Frequency
SB0133	North Ethiopia	Woldiya	23223 32454 21425 24222 3323	1
	Central Ethiopia	This study	23223 32454 21425 24222 3323	3
	Central Ethiopia	Addis Ababa	23223 32454 214**3**5 24222 3323	1
	South-East Ethiopia	Negelle	23223 32454 21425 24222 3**2**23	1
	South-East Ethiopia	Negelle	23223 32454 21425 24222 3323	2
	South Ethiopia	Jinka	**36**223 **2253**4 214**36** 24222 **4**323	1
	South Ethiopia	Jinka	**36**223 **2254**4 214**36** 24222 **4**323	4
	West Ethiopia	Ghimbi	**36**223 **2254**4 214**36** 24222 **4**323	1
SB1176	North Ethiopia	Woldiya	35223 32554 2143**4** 24222 3323	2
	Central Ethiopia	This study	35223 32554 21435 24222 3323	8
	Central Ethiopia	This study	35223 3255**2** 21435 24222 3323	3
	South Ethiopia	Jinka	35223 3**3**5**7**4 **3**1**223** 24222 33**7**3	1
SB0134	North Ethiopia	Gondar	35223 40554 21415 24222 2332	1
	Central Ethiopia	This study	35223 40554 21415 24222 2332	11

Repeats of locus highlighted in bold diverge between isolates of the same spoligotype. All collection sites are shown in Figure 1.
*Consecutive order of the 24-loci MIRU-VNTR: 2165; 2461; 577; 580; 3192; 154; 960; 1644; 2059; 2531; 2687; 2996; 3007; 4348; 802; 424; 1955; 2163b; 2347; 2401; 3171; 3690; 4052; 4156.
doi:10.1371/journal.pone.0052851.t005

DISCUSSION

With the recent shift towards larger intensive farming systems, mainly in urban areas, it is clear that the farming practices are becoming more favourable for BTB transmission. In this study we explored the prevalence of BTB in cattle in intensively managed dairy farms located in and around Addis Ababa and that supply milk to the capital. Based on the CIDT test, all six surveyed areas (Figure 1) recorded a high rate of the disease and we concluded that, overall, approximately one third of the tested animals and more than half of the investigated herds were infected with BTB (if stratification or random effect is not taken into account). However, the prevalence fluctuated significantly between the three investigated farm size categories. Both the overall individual and overall herd prevalence of medium and large farms were double and fourfold, respectively, as compared to the corresponding numbers of small farms (Table 1). Since as many as 20 out of 21 large herds (>50 animals/herd) were infected, we concluded that BTB had a strong-hold in dairy farms in this investigated region of central Ethiopia.

Risk Factors

Several past and recent studies have shown that susceptibility to BTB can vary between cattle breeds [12], [27] with suggestions that cattle of *Bos indicus* (Zebu) breeds are more resistant to BTB than *Bos taurus* (European breeds). The vast majority of the dairy cattle included in this study were cross breeds (62%) and Holstein-Frisian (34%), however the BTB prevalence did not differ significantly between these two categories (Table 2). The number of cattle recorded as Jersey and Zebu breeds in this study were too few to make relevant comparisons on susceptibility. On the other hand, Ameni et al. [28] have demonstrated that animal husbandry conditions are a major influence on the prevalence of BTB. Cattle kept under high-intensity conditions showed significantly higher skin-test prevalence as compared to cattle kept under extensive conditions. Intensification, stressed animals, and overcrowding are all possible explanations for such relationship. The main routes of BTB transmission are through aerosol as gross lesions usually involve the lungs and thoracic lymph nodes [29], and therefore BTB transmission benefits from overcrowded herds. The post-mortem data collected in this study support this as most animals had TB lesions in lungs and/or lung associated lymph nodes (Figure 2). Once an animal is infected with BTB, it usually takes several months or longer for this chronic disease to develop clinical signs. It is also likely that infected cattle, similarly to humans, enter an asymptomatic phase after infection with *M. bovis*, in which no clinical signs are developed for years [30]. Our ante- and post-mortem investigation also confirmed this common attribute of BTB. The body condition scoring of the nearly 3,000 animals suggested no significant differences between tuberculin reactors and non-reactors (Table 2) and a poor correlation was also seen among the 36 slaughtered animals; despite that the majority of these animals were highly diseased with severe pathology, little or no clinical signs typical for BTB were observed. Thus, if BTB gets into a herd where no diagnostic measure is in place (such as CIDT), then it can be difficult to detect infected animals on a visual basis solely and to prevent further transmission.

Most owners of medium and large farms had higher educational level while illiterate farmers tended to have smaller farms. Although this leads to an association in our dataset between higher education and higher prevalence of BTB in the herds (Table 3), this is more likely to be an expression of higher educated people engaging in dairy farm business as an investment, hence larger farms being associated with higher education.

Farmers participating in the current study said that the main means of restocking their herd was through artificial insemination. However, frequently animals were purchased or sold between farms within the study areas. Trading of animals beyond the study area occurred too (mainly by medium sized farms) but seemed unidirectional; dairy cattle were not purchased but only sold to other regions across Ethiopia. The response from farmers in this survey also suggested that some animals were sold because they were weak or diseased. As no official test-and-movement policy is in place, selling of infected animals from this highly affected region of Ethiopia may consequently contribute to spreading BTB to other parts of the country.

The Causative Agent

The mycobacterial strains isolated from cattle in this study were solely *M. bovis* and the CIDT test was confirmed by culture in 86% of the slaughtered cattle. This culturing yield can be compared with a 90% yield normally seen in slaughterhouse cases in the UK from animals with visible lesions and of which the vast majority were caused by *M. bovis* [31]. No strain of *M. tuberculosis* was isolated in our study in contrast to several other Ethiopian studies [7], [13], [32]. It can be argued though that only ~3% of all cattle that were tuberculin positive in this study were investigated by post-mortem, leaving out a large group of animals that could have been infected with *M. tuberculosis* as CIDT testing do not distinguish between *M. tuberculosis* and *M. bovis*. However, the dairy cattle of this study were managed under intensive husbandry systems with animals staying in separate barns, as compared to management systems common in the rural areas where domestic animals often sleep in the same house as people. It has previously been suggested that human-to-cattle transmission of *M. tuberculosis* is possible in settings where farm members in high-burden TB areas are in close contact with their animals [32].

Molecular typing of the 31 *M. bovis* strains isolated from the slaughtered animals identified four different spoligotype patterns which previously have been shown as common in several regions of Ethiopia [7]. Three of these four spoligotypes (SB1176, SB1477, and SB0133) carry a specific spoligotype feature (spacers 4–7 missing) typical for members of a clonal complex identified as *M. bovis* African 2 (Af2) so far only found in East Africa [33]. The fourth spoligotype (SB0134) do not belong to the Af2 clonal complex and is therefore likely to have a different epidemiological history. *M. bovis* isolates with spoligotype SB0134 has been found also in Europe. However, the phylogenetic relationship between the African and European strains of SB0134 is not known. Future chromosomal sequencing may shed light over their relationship.

Further discrimination of the *M. bovis* isolates performed by MIRU-VNTR typing suggested low diversity within each isolated spoligotype. This is in agreement with other studies as in high prevalence settings the genetic diversity of sampled strains is usually low, reflecting on-going local transmission events (due to high transmission rate) [34]. However, in comparison with genotypes of *M. bovis* strains that were collected from other parts of Ethiopia about 2–3 years earlier [7], some variations were observed and possible epidemiological links between specific regions of the country can be suggested (Table 5). Two distinct MIRU-VNTR features were seen within spoligotype SB0133 of which the one identified in central Ethiopia (this study) was also seen in sites along the Rift Valley, from Woldiya in the north to Negelle in the south. The other genotype of SB0133 was isolated from Jinka (south-west Ethiopia) and Ghimbi (west Ethiopia). Similarly, the single strain of type SB1176 collected in Jinka deviated in the MIRU-VNTR pattern from those seen in Woldiya and in central Ethiopia. However, no genotype diversity was seen between the one isolate of SB0134 from Gondar (North-west) when compared to the 11 *M. bovis* strains of SB0134 isolated in central Ethiopia (this study). Overall, based on these molecular typing results of our *M. bovis* collection, epidemiological links can be suggested for strains collected within the North/Central Ethiopian highlands and through the rift valley down to Negelle, while strains isolated from Jinka and Ghimbi in the south/west of Ethiopia diverged. This may reflect on past and ongoing cattle movements and trading patterns in the country. The low strain frequency from some collection sites means that these results should be interpreted with care. Nevertheless, this genotype comparison gives an indication of the diversity and the epidemiology of *M. bovis* in Ethiopia. Future studies may give better clarification on this subject. The 24-loci MIRU-VNTR typing employed for this study showed no diversity in at least 14 of the 24 loci (variation seen between spoligotypes; Table 5). This is in agreement with other studies [35], which has shown that VNTR typing of only a selective set of loci is informative for generating a better resolution between *M. bovis* isolates. This suggests that any future MIRU-VNTR typing of *M. bovis* strains from Ethiopia could consider including only loci that show diversity in its local strain population, however, a more comprehensive evaluation of what loci to select is recommended.

Reasons for Controlling BTB

At least three important reasons for controlling BTB should be considered: animal welfare, the financial burden to farmers with diseased animals, and the risk of zoonotic transmission. This chronic disease can take years to develop. Animals infected with BTB can therefore slowly develop symptoms and may, if not carefully observed by animal keepers, suffer unnecessarily. Depending on which organ or associated lymph node is affected, animals can display a wide range of symptoms e.g. coughing, dyspnoea, gastrointestinal problems, bone deformation, and emaciation [36].

Few thorough studies have been conducted to investigate loss of productivity due to BTB, but it is probable that there is a productivity impact in animals with BTB due to the nature of the disease and thus this is likely linked to economical losses [37]. It has previously been suggested that reduced milk production, the food value of the carcass, and reproduction (or fertility) are factors that affects the animal productivity [38].

The zoonotic risk of BTB is often associated with consumption (ingestion) of dairy products based on unpasteurised milk infected with *M. bovis*. However, aerosol transmission (inhalation) from cattle-to-human should also be considered as a potential risk factor. Ethiopian milk consumers generally prefer raw milk (as compared to treated milk) because of its taste, availability and lower price. Therefore the demand for fresh whole milk in rural areas is and will probably remain high, and in general, the demand from these consumers is likely to be satisfied by home production or by purchasing from neighbouring producers. On the other hand, the markets for surplus milk are mainly found in the urban population, of which currently ~65% is accounted for by the capital Addis Ababa and its neighbouring districts [39].

The annual total milk supply to Addis Ababa has been estimated at 65 million litres and the major sources are from small private farms and smallholder urban dairies in and around the city that own upgraded breeds [40]. These over 5,000 dairy farms produce around 35 million litres of milk per year. Grossly 10% is used to feed calves, 10% is used for home consumption, 8% is processed into butter and cheese, while the remaining ~2/3 is sold to the market as un-boiled or unpasteurised milk [41]. Other important sources of milk to the capital city are medium to large dairy plants run privately or as enterprises/cooperatives. Today there are around eight major dairy plants in central Ethiopia with a total production capacity of around 100,000 litres per day (~35 million litres per year) (survey by Alehegne Wubete, unpublished), but the actual yearly production is likely to be less as animal product consumption can vary through the year due to cultural and religious habits such as fasting. These plants collect milk from farmers within a 150 km radius around Addis Ababa and provide standardized pasteurized milk in plastic sachets to the markets [39]. Our survey suggests that many dairy farms from the investigated area sell their milk to restaurants and catering companies, but it is difficult to estimate how much of that milk is pasteurised or boiled before consumption. Overall, the available information suggests that a significant volume of milk may be distributed without being pasteurised.

In a recent Ethiopian study of the rate of *M. bovis* causing TB in humans [42] we reported that the rate of TB in humans due to *M. bovis* is currently below 1%. However, we concluded that areas where BTB prevalence in cattle is high (such as the "hot-spot" area of central Ethiopia shown in this study) could generate an increase of BTB in the human population in such areas. It is important to note that existing human TB control methods relying on the directly observed treatment short course (DOTS) strategy will not directly impact on BTB transmission. The high rate of BTB in the dairy farms of this study is therefore likely to pose a serious risk to public health and deserves a targeted intervention as early as possible.

Conclusion and Control Strategies

This study suggests that the overall prevalence of BTB in intensive dairy farms in Addis Ababa and its surroundings is very high and requires urgent intervention to control the disease. To comprehensively control BTB regardless of the method chosen, a robust system of individual animal and herd identification is required. The powers to restrict movements to prevent disease dissemination and the ability to trace movements to catch up with translocation are essential components of an eradication strategy. Most developed countries where BTB is a problem have introduced a test-and-slaughter policy to control/eradicate the disease. The CIDT, although not 100% sensitive, is the standard test widely used in developed countries for this purpose. Most importantly, infected animals have to be detected and removed at an early stage to minimise further transmission. As such policy is considered costly it has rarely been implemented in developing countries. No compensation scheme for elimination of infected animals is currently in place in Ethiopia and due to financial constraints changes toward such policy might not be feasible in the near future. However, a test-and-segregation policy of tuberculin positive animals can be suggested and pursued and the milk of infected animals strictly and consequently pasteurized before selling and consumption. Additional interventions by the government authorities could be to encourage farmers to regularly test their animals for BTB and create incentives to keep their herds free from BTB. It could be compulsory for dairy farms that are supplying milk to the public to test their cattle for BTB, and create incentives to TB free herds – "good milk certificate" – that could lead to better economic values for farmers. Also dairy plants should be engaged and could e.g. pay a higher rate for milk from herds certified as being BTB free. Pasteurization at a bigger scale has to be promoted, either directly at farm level before selling the milk or at household level through boiling the milk before consumption (e.g. awareness campaigns).

As the dairy industry in Ethiopia has expanded in recent years and is expected to continue doing so, significant number of high productive exotic and cross breed animals are likely to be traded from the urban areas around the capital to the rural areas where dairy cattle numbers are still relatively low. Such movement is linked with increasing financial well-being of farmers and amelioration of artificial insemination technique/delivery. However, without any control strategy the risk of spreading BTB by such movements is high and may create new hot-spots of BTB in other parts of the country. Several recommendations can be made to minimize the risk of spreading the disease between farms and regions. First and most important is to spread knowledge about BTB and make farmers and people involved in trading of cattle aware of the risks involved in trading animals. On a smaller scale, our field work and specific training to the farmers aimed on this, informing them in how to handle cattle infected with BTB and other diseases, and encouraging them not to trade with diseased animals. Another way to decrease risk of transmission is to encourage farmers who are restocking by purchasing animals to request that the animals should be tested for BTB. The CIDT test is relatively inexpensive and would reduce the risk to the farmer of purchasing an infected animal. A possibility for Ethiopia to in part meet the demand for cattle of upgraded breeds could be to establish farms with dairy cattle free of BTB and from which farmers could restock their farms without the risk of introducing the disease into their herds. This might imply scaling up of existing artificial insemination centres to improve productivity in the national dairy herd and possibly purchase of new stocks of exotic breeds from abroad.

ACKNOWLEDGMENTS

We thank all farm owners and staff that participated in this study, and we appreciate laboratory and database support from AHRI staff.

AUTHOR CONTRIBUTIONS

Conceived and designed the experiments: RF SB AW M. Sombo GE RT AA SVG DY MV M. Sahile RGH. Performed the experiments: RF AW M. Sombo EH GE TK RT. Analyzed the data: RT RF SB AW MV. Contributed reagents/materials/analysis tools: AW RT LY AA M.Sahile. Wrote the paper: SB RT RF AW. Revised the manuscript: M. Sombo AA SVG DY MV.

REFERENCES

1. Bureau PR (2010) World Population Data Sheet. Available: http:// www. prb. org/Publications/Datasheets/2010/2010wpds. aspx.
2. CSA (2011) Report on Livestock and Livestock Characteristics. Agricultural Sample Survey 2011/2012. Addis Ababa: Central Statistical Agency of Ethiopia.
3. Halderman M (2004) The Political Economy of Pro-Poor Livestock Policy-Making in Ethiopia. In: Initiative P-PLP, editor. Rome, Italy. pp. 1–59.
4. CSA (2007) Report on Livestock and Livestock Characteristics. Agricultural Sample Survey 2006/2007. Addis Ababa: Central Statistical Agency of Ethiopia.
5. Cosivi O, Grange JM, Daborn CJ, Raviglione MC, Fujikura T, et al. (1998) Zoonotic tuberculosis due to *Mycobacterium bovis* in developing countries. Emerg Infect Dis 4: 59–70.
6. Ayele WY, Neill SD, Zinsstag J, Weiss MG, Pavlik I (2004) Bovine tuberculosis: an old disease but a new threat to Africa. Int J Tuberc Lung Dis 8: 924–937.
7. Berg S, Firdessa R, Habtamu M, Gadisa E, Mengistu A, et al. (2009) The burden of mycobacterial disease in ethiopian cattle: implications for public health. PLoS One 4: e5068.
8. Tschopp R, Schelling E, Hattendorf J, Aseffa A, Zinsstag J (2009) Risk factors of bovine tuberculosis in cattle in rural livestock production systems of Ethiopia. Prev Vet Med 89: 205–211.
9. Tschopp R, Schelling E, Hattendorf J, Young D, Aseffa A, et al. (2010) Repeated cross-sectional skin testing for bovine tuberculosis in cattle kept in a traditional husbandry system in Ethiopia. Vet Rec 167: 250–256.
10. Gumi B, Schelling E, Firdessa R, Aseffa A, Tschopp R, et al. (2011) Prevalence of bovine tuberculosis in pastoral cattle herds in the Oromia region, southern Ethiopia. Trop Anim Health Prod 43: 1081–1087.
11. Demelash B, Inangolet F, Oloya J, Asseged B, Badaso M, et al. (2009) Prevalence of bovine tuberculosis in Ethiopian slaughter cattle based on post-mortem examination. Trop Anim Health Prod 41: 755–765.
12. Ameni G, Aseffa A, Engers H, Young D, Gordon S, et al. (2007) High prevalence and increased severity of pathology of bovine tuberculosis

in Holsteins compared to zebu breeds under field cattle husbandry in central Ethiopia. Clin Vaccine Immunol 14: 1356–1361.

13. Tsegaye A, Aseffa A, Mache A, Mengistu Y, Berg S, et al. (2010) Conventional and Molecular Epidemiology of bovine tuberculosis in dairy farms in Addis Ababa City, the capital of Ethiopia. J Appl Res Vet Med 8: 143–151.
14. Elias K, Hussein D, Asseged B, Wondwossen T, Gebeyehu M (2008) Status of bovine tuberculosis in Addis Ababa dairy farms. Rev Sci Tech 27: 915–923.
15. Meisinger G (1970) Economic effects of the elimination of bovine tuberculosis on the productivity of cattle herds. 2. Effect on meat production. Monatsh Veterinarmed 25: 7–13.
16. Tschopp R, Aseffa A, Schelling E, Zinsstag J (2010) Farmers' perceptions of livestock, agriculture, and natural resourses in the rural Ethiopian highlands. Mountain Research and Development (MRD) 30: 381–390.
17. Grange JM (2001) *Mycobacterium bovis* infection in human beings. Tuberculosis (Edinb) 81: 71–77.
18. de la Rua-Domenech R (2006) Human *Mycobacterium bovis* infection in the United Kingdom: Incidence, risks, control measures and review of the zoonotic aspects of bovine tuberculosis. Tuberculosis (Edinb) 86: 77–109.
19. Gumi B, Schelling E, Berg S, Firdessa R, Erenso G, et al. (2012) Zoonotic Transmission of Tuberculosis Between Pastoralists and Their Livestock in South-East Ethiopia. Ecohealth 9: 139–149.
20. WHO (2011) Global tuberculosis control: World Health Organisation report 2011. pp. 1–258.
21. Vordermeier HM, Chambers MA, Cockle PJ, Whelan AO, Simmons J, et al. (2002) Correlation of ESAT-6-specific gamma interferon production with pathology in cattle following *Mycobacterium bovis* BCG vaccination against experimental bovine tuberculosis. Infect Immun 70: 3026–3032.
22. Gallagher J, Horwill DM (1977) A selective oleic acid albumin agar medium for the cultivation of *Mycobacterium bovis*. J Hyg (Lond) 79: 155–160.

23. Brosch R, Gordon SV, Marmiesse M, Brodin P, Buchrieser C, et al. (2002) A new evolutionary scenario for the *Mycobacterium tuberculosis* complex. Proc Natl Acad Sci U S A 99: 3684–3689.
24. Kamerbeek J, Schouls L, Kolk A, van Agterveld M, van Soolingen D, et al. (1997) Simultaneous detection and strain differentiation of *Mycobacterium tuberculosis* for diagnosis and epidemiology. J Clin Microbiol 35: 907–914.
25. Allix-Beguec C, Supply P, Wanlin M, Bifani P, Fauville-Dufaux M (2008) Standardised PCR-based molecular epidemiology of tuberculosis. Eur Respir J 31: 1077–1084.
26. Allix-Beguec C, Fauville-Dufaux M, Supply P (2008) Three-year population-based evaluation of standardized mycobacterial interspersed repetitive-unit-variable-number tandem-repeat typing of *Mycobacterium tuberculosis*. J Clin Microbiol 46: 1398–1406.
27. Carmichael J (1939) Bovine tuberculosis in the tropics with special reference to Uganda, part 1. J Comp Pathol Therap 52: 322–335.
28. Ameni G, Aseffa A, Engers H, Young D, Hewinson G, et al. (2006) Cattle husbandry in Ethiopia is a predominant factor affecting the pathology of bovine tuberculosis and gamma interferon responses to mycobacterial antigens. Clin Vaccine Immunol 13: 1030–1036.
29. Thoen CO, Bloom BR (1995) Pathogenesis of *Mycobacterium bovis* In: Thoen CO, Steele JH, editors. *Mycobacterium bovis* Infection in Animals and Humans. First ed. Ames, Iowa: Iowa State University Press. pp. 3–14.
30. Phillips CJ, Foster CR, Morris PA, Teverson R (2003) The transmission of *Mycobacterium bovis* infection to cattle. Res Vet Sci 74: 1–15.
31. Liebana E, Johnson L, Gough J, Durr P, Jahans K, et al. (2008) Pathology of naturally occurring bovine tuberculosis in England and Wales. Vet J 176: 354–360.
32. Ameni G, Vordermeier M, Firdessa R, Aseffa A, Hewinson G, et al. (2011) *Mycobacterium tuberculosis* infection in grazing cattle in central Ethiopia. Vet J 188: 359–361.
33. Berg S, Garcia-Pelayo MC, Muller B, Hailu E, Asiimwe B, et al. (2011) African 2, a clonal complex of *Mycobacterium bovis* epidemiologically important in East Africa. J Bacteriol 193: 670–678.
34. Supply P, Warren RM, Banuls AL, Lesjean S, Van Der Spuy GD, et al. (2003) Linkage disequilibrium between minisatellite loci supports

clonal evolution of *Mycobacterium tuberculosis* in a high tuberculosis incidence area. Mol Microbiol 47: 529–538.

35. Le Fleche P, Fabre M, Denoeud F, Koeck JL, Vergnaud G (2002) High resolution, on-line identification of strains from the *Mycobacterium tuberculosis* complex based on tandem repeat typing. BMC Microbiol 2: 37.
36. OIE (2009) Bovine tuberculosis. Terrestrial manual: Office International des Epizootie
37. Tschopp R, Hattendorf J, Roth F, Choudhoury A, Shaw A, et al. (2012) Cost Estimate of Bovine Tuberculosis to Ethiopia. Curr Top Microbiol Immunol DOI: 10. 1007/82_2012_245.
38. Thoen CO, Steele JH, editors (1995) *Mycobacterium bovis* infection in animals and humans. 1 ed. Ames, Iowa: Iowa State University Press. 355 p.
39. SNV (2008) Dairy Investment Opportunities in Ethiopia. Addis Ababa: The Netherlands Development Organisation. pp. 1–48.
40. Abreha T (2006) Dairy production in Addis Ababa. Paper presented at a regional workshop organized by ASARECA, Kampala, Uganda.
41. Tegegne A, Sileshi Z, Tadesse M, Alemayhu M (2002) Scoping Study of Urban and Peri-urban Livestock Keepers in Addis Ababa: Part I. Literature Review on Dairy Production, Management and Marketing. ILRI and EARO: Addis Ababa.
42. Firdessa R, Berg S, Hailu E, Schelling E, Gumi B, et al. (2013) Mycobacterial Lineages Causing Pulmonary and Extrapulmonary TB in Ethiopia. Emerging Infectious Diseases In press.

CHAPTER 17

Survey on Antimicrobial Usage in Local Dairy Cows in North-central Nigeria: Drivers for Misuse and Public Health Threats

Nma Bida Alhaji[1], Mohammed Baba Aliyu[1] , Ibrahim Ghali-Mohammed[2] , Ismail Ayoade Odetokun[2]

[1] Department of Public Health and Epidemiology, Niger State Ministry of Livestock and Fisheries, Minna, Nigeria
[2] Department of Veterinary Public Health and Preventive Medicine, University of Ilorin, Ilorin, Nigeria

Citation: Alhaji NB, Aliyu MB, Ghali-Mohammed I, Odetokun IA (2019) Survey on antimicrobial usage in local dairy cows in North-central Nigeria: Drivers for misuse and public health threats. PLoS ONE 14(12): e0224949. https://doi.org/10.1371/journal.pone.0224949.

ABSTRACT

Background

Antimicrobials are used as a measure to maintain good health and productivity of lactating cows. This study assessed pastoralists' knowledge and practices regarding AMU in lactating cows; and risk pathways for AMR dissemination from cow milk to humans.

Methods

Interview questionnaire-based cross-sectional study was conducted in Fulani pastoral communities. Frequencies and proportions were used for descriptive statistics. Chi-square test and multivariable logistic regressions were used for analytic statistics at 95% confidence level.

Results

All recruited 384 pastoral households participated. About 11% of participants indicated antimicrobials misuse as when given at under-dose, while 58.9% had no knowledge of what antimicrobial misuse entailed. Most participants (51.6%) were unaware about effects of improper AMU. Most respondents (61.7%) reported self-prescription of antimicrobials used on cows. Also, 67.4% of respondents reported arbitrary applications of antimicrobials used in cows, while 15% used antimicrobials to increase milk yield. Frequently used antimicrobials were: tetracycline (98.7%), penicillin (96.6%), streptomycin (95.8%) and sulfonamides (95.3%). Consumption of raw milk and milk products ($p = 0.010$); contacts with contaminated udder ($p = 0.002$); and aerosols of discarded contaminated milk $P = 0.001$) were perceived risk pathways for spread of antimicrobial resistance from cow milk. Improper AMU ($p<0.001$), non-enforcement of regulating laws ($p<0.001$), weak financial status ($p<0.001$), and low education and expertise ($p<0.001$) influenced antimicrobials misuse in lactating cows.

Conclusions

This study highlighted low levels of knowledge, risk perceptions and practices regarding AMU and AMR among survey pastoralists. This calls for education of the vulnerable populations on promotion of prudent AMU in lactating cows through 'One Health' approach, to assure food safety, food security, and public and environmental health.

INTRODUCTION

Milk is an important source of nutrients and most popular natural health food to both humans and animals [1]. Dairy products provide many highly bioavailable essential nutrients, which are especially important in human diets, and even small amounts of dairy products can improve the nutritional status of those living in low-income households [2]. Dairy production is rapidly expanding in developing countries due to population growth [3]. In Nigeria, the annual average national milk consumption is about 20 liters, which is four times less than World Health Organization (WHO) minimum recommendation [4]. One of the major problems of local cattle milk production worldwide is bovine mastitis, which has significant impact on the economy of milk production, due to high cost of antimicrobial agents [5,6].

Annual antimicrobial usage (AMU) in food animals was globally estimated at 63,000 tons in 2015, with projected increase of about 67% by 2030 [7]. Apart from the top consumers of antimicrobials that include China, United States and Brazil, the largest relative increase of more than 200% is projected to occur in developing countries, with Myanmar, Indonesia and Nigeria taking the lead [7]. Improper AMU in food animals contributes to development of antimicrobial resistance (AMR), currently a global health problem and can spread between countries [8–10]. Improper AMU in food animals creates selective evolutionary pressure that enables antimicrobial resistant pathogens to emergence and increase in numbers more rapidly than antimicrobial susceptible pathogens, posing a serious health threat [8].

Fulani pastoralists in Nigeria live in rural areas, herding about 90% of ruminants, relied on the livestock for social and economic well-being and practice nomadic or seasonal transhumance grazing system [11]. These pastoralists commonly use antimicrobials on lactating cows primarily to treat or prevent mastitis and increase milk production or improve feed efficiency. However, because they are domiciled in remote areas, conventional veterinary services for the animals are poor and basic information on AMU is not readily available. [12].

Due to public health and socio-economic impacts of AMR, investigation into risks associated with AMU dynamics in lactating cows becomes imperative. However, the risk of AMR emergence and dissemination through milk to humans can better be assessed within the premise of "Perceived Susceptibility" construct of the Health Belief Model, which postulates that perceived personal risk or susceptibility prompt people to adopt healthier behaviours. The greater the risk perception about a health threat, the greater the likelihood of engaging in behavours that will decrease its risk [13]. Exploration of pastoralists' local knowledge and practices regarding AMU in lactating cows is, therefore, crucial for the development of effective AMR surveillance, control and prevention. Study objectives were to: 1) assess pastoralists' knowledge and practices regarding AMU and AMR in lactating cows; and 2) assess risk pathways for AMR dissemination from lactating cows to humans. In addition, we sought to explore socio-demographic determinants of knowledge on AMU and resistance, and socio-economic activities that influence antimicrobials misuse and resistance emergence in lactating cows.

MATERIALS AND METHODS

Structure of Target Population and Livelihood

Target populations were households in Fulani pastoral communities, who are seasonally mobile with herds of local breeds of cattle, domiciled in the study area during the survey. Each pastoral community had an average 28 households deriving their livelihoods mainly from sales of milk and milk products. A Fulani pastoral household constituted a herd managed by herd head or owner. Average number of cattle in a herd was 82 animals, with ratio of 1:10 bull to cows. Study eligibility was based on a participant being a household head and at least 20 years of age. They are expected at these ages to possess existing veterinary knowledge on livestock health and production management because of long time intimate relationship with the animals.

Study Design, Sample Size and Sampling Procedure

An interview questionnaire-based cross-sectional survey was conducted in randomly selected pastoral households in Fulani nomadic pastoral communities of North-central Nigeria in 2017. The sample size was calculated using the Open Source Epidemiologic Statistics for Public Health (OpenEpi) 2.3.1 software [14] for percentage frequency in a population (random sample). The following assumptions were used: finite population because the target population was not large and mobile in nature, 50% expected frequency of respondents that use antimicrobials on cows, a desired absolute precision set at 5%, and 95% confidence interval. Based on these assumptions, a sample size of 384 pastoralists' households was obtained. A two-stage sampling method was used. In the first stage, 30

Fulani pastoral communities were purposively selected across the study area. In the second stage, systematic random sampling was used to select all targeted households.

Questionnaire Design, Pretesting and Data Collection

A structured questionnaire that contained mostly close-ended questions, to ease data processing and improve response precision [15], was designed based on experts' opinions. It consisted of five sections: i) pastoralist's socio-demographic characteristics (6 questions); ii) knowledge about AMU in lactating cows: antimicrobials misuse, AMR and its effects in animals and humans (13 questions); (iii) practices of AMU in cows (13 questions); (iv) risk pathways for antimicrobial resistance dissemination (8 questions); and (v) factors that influence antimicrobials misuse and resistance emergence in cows (6 questions). The questionnaire was designed in English and verbally translated into local *Hausa* languages during administration, for those without formal education. They were asked questions in *Hausa* and responses translated to English during recording. Six animal health officials were trained to administer the questionnaires, pre-tested on households in a community before final administration, to identify problems and eliminate them for adequate data delivery.

Respondents were provided with verbal information on objectives of the study. Their informed consent was verbally obtained with signatures and thumbs printing on a sheet before questionnaire administration and none declined to participate. They were assured of voluntary participation, confidentiality of responses and the opportunity to withdraw at any time without prejudice in line with the World Medical Association Declaration of Helsinki [16]. Verbal information and informed consent were deemed necessary because of low literacy levels among participants. Advocacy visits were made to each pastoral community a week prior to the proposed interview and necessary permission obtained from '*Ardos*' (leaders).

Ethics Statement

The Research Ethics Committee of Niger State Ministry of Livestock and Fisheries, Minna, Nigeria approved the research proposal (Ref #. MLF/NGS/693).

Data Management and Statistical Analysis

Participants' responses were first summarized into Microsoft Excel 7 (Microsoft Corporation, Redmond, WA, USA) spreadsheets. Descriptive statistics were calculated for all variables in forms of frequencies and proportions, and associations between variables were determined by Chi-square test or Fischer' exact test where appropriate, and by multivariable logistic regressions models.

The participants' level of knowledge was determined according to outcome criteria previously identified by Alhaji et al. [17]: the word "very low" represented a proportion of respondents with "know" knowledge that ranged from 1% to 24%; "low" represented proportion with "know" knowledge that ranged between 25% and 49%; "high" represented proportion with "know" knowledge that ranged between 50% and 74%; and "very high" represented proportion with "know" knowledge that ranged from 75% to 100%. Similar approach was used to identify levels of practices and perceptions. To assess association, independent (explanatory) variables were created from the socio-demographic characteristics and socio-economic factors that influence antimicrobial misuse, while respondents' overall response levels constituted the dependent (outcome) variables. The outcome variables were coded by a unique scoring system, with a response score that ranged between 1 and 20 points and converted to 100%. The score range was further categorized into 'poor' or 'satisfactory' to keep them in categorical forms. Response scores within 1–10 points were considered 'poor' (≤49%), and those within 11–20 points were considered 'satisfactory' (≥50%). Associations between explanatory and outcome variables were first subjected to univariable analysis using Chi-square tests [18]. Factors found to be statistically significant at this analysis were finally subjected to likelihood stepwise backward multivariable logistic regressions models to control for confounding and test for effect modification. All statistical analyses were conducted using EpiInfo 3.4.3 (CDC, Atlanta, GA, USA) and OpenEpi version 2.3.1 [14]. A $p<0.05$ was considered statistically significant in all analyses.

RESULTS

Socio-demographic Characteristics of Participants

All recruited 384 Fulani pastoralists participated in the study. Most of the participants (25.0%, 96/384) were in age group 50–59 years. Majority the respondents were males (84.4%, 328/384) and married 84.4%, while 8.6% (33/384) were single and 7.0% (27/384) widows. The majority of pastoralists (65.2%, 250/384)) had no formal education and only 8.3% (32/384) had tertiary education (Fig 1).

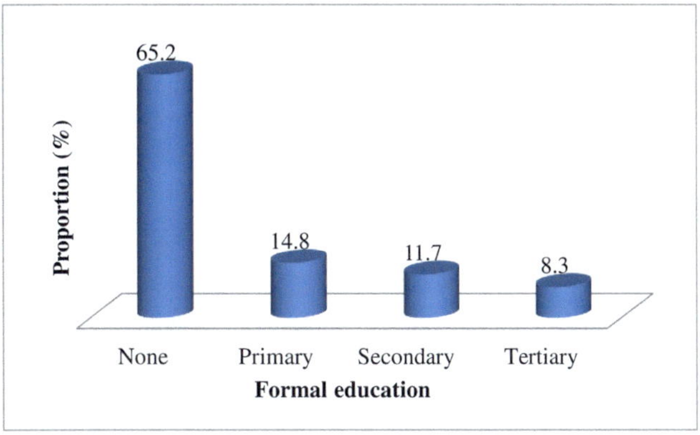

Figure 1: Formal educational status of Fulani pastoralists in North-central Nigeria.

https://doi.org/10.1371/journal.pone.0224949.g001

Knowledge about AMU and Antimicrobial Resistance in Lactating Cows

Most respondents (53.4%) believe that antimicrobials are used to treat mastitis in lactating cows, 24.2% to prevent mastitis in lactating cows, and 15.4% to promote production of milk in lactating cows, while 7.0% of participants indicated knowing antimicrobials to be used in cows for all of the above.

Also, only 10.9% of participants knew antimicrobials misuse in cows to be when given as under-dose, 10.2% mentioned when given at over-dose and 58.9% did not know what misuse entailed. Regarding whether antimicrobials misuse in cows can predispose to the emergence of resistance, 26.8% of pastoralists agreed and 41.1% disagreed. On AMR effects, 51.6% and 60.4% of participants did not know the consequences in lactating cows and humans, respectively. Regarding dissemination of AMR from lactating cows to humans, only 13.0% of respondents mentioned drinking raw milk, 12.2% reported drinking fermented milk (*nono*), while 8.9% indicated milking cows and 60.4% had no idea (Table 1).

Table 1: Fulani pastoralists' knowledge about AMU and resistance in lactating cows in North-central Nigeria

https://doi.org/10.1371/journal.pone.0224949.t001

Variable	Frequency (n)	Proportion (%)	Confidence interval
Know antimicrobials to be used			
To treat mastitis in lactating cows	205	53.4	48.3858.34
To prevent mastitis in lactating cows	93	24.2	20.13, 28.69
To promote milk yield in lactating cows	59	15.4	12.02, 19.23
All of the above	27	7.0	4.78, 9.93
Antimicrobials misuse in lactating cows is when			
Administered under-dose	42	10.9	8.10, 14.36
Administered over-dose	39	10.2	7.43, 13.49
Administered in normal dose	77	20.0	16.27, 24.28
Don't know	226	58.9	53.87, 63.7
Effects of antimicrobials misuse on lactating cows			
Non response to bacterial infection treatment	121	31.5	27.01, 36.29
Extra costs on treatment of bacterial infection	65	16.9	13.42, 20.93
Don't know	198	51.6	46.56, 56.54
Antimicrobials misuse in lactating cows can predispose to resistance emergence			
Agree	103	26.8	22.57, 31.42
Disagree	158	41.1	36.3, 46.13
Don't know	123	32.1	27.51, 36.83
Antimicrobial resistance can be passed from lactating cows to humans through			
Drinking raw milk	50	13.0	9.93, 16.67
Drinking fermented milk (*nono*)	47	12.2	9.24, 15.81
Eating raw cheese (*wara*)	21	5.5	3.51, 8.10
Milking cows	34	8.9	6.31, 12.02
Don't know	232	60.4	55.46, 65.22
Effects of antimicrobial resistance in humans			
Non response to bacterial infection treatment	56	14.6	11.32, 18.38
Extra costs on treatment of bacterial infection	41	10.7	7.88, 14.07
Longer duration of illness and treatment	55	14.3	11.08, 18.1
Don't know	232	60.4	55.46, 65.22

https://doi.org/10.1371/journal.pone.0224949.t001

Practices of AMU on Lactating Cows

On personnel that prescribed antimicrobials for usage in lactating cows, 28.1% of the pastoralists reported animal health officials, while 61.7% engaged in self-prescription. More than one-third (34.9%) of the participants purchased antimicrobials used on cows from veterinary drug shops; 10.7% obtained the drugs from human drug shops, while 54.4% patronized animal drug hawkers. Also, 65.4% of participants practiced self-administration of antimicrobials used on cows, and 32.6% engaged services of animal health officials. Regarding rate of AMU on lactating cows with mastitis, 19.3% of respondents reported following prescribed instructions, 27.1% administered antimicrobials only once on sick cows until they recovered, while 53.6% administered antimicrobials once daily on sick animals until recovered.

On dosage determination before usage on cows, 34.6% of the pastoralists reported following instructions on labels, and 65.4% mentioned arbitrary applications. Majority of the pastoralists (57.6%) frequently administered antimicrobials by injection and very few of them (3.6%) applied the drugs through feed. However, 18.7% of respondents observed withdrawal periods after AMU on cows, and most of them (81.3%) reported non-compliance with withdrawal periods. On purpose for AMU on lactating cows, most respondents (53.4%) mentioned that they used antimicrobials to treat mastitis in lactating cows, 24.2% to prevent mastitis in lactating cows, and 15.4% to promote milk production in lactating cows (Table 2).

Table 2: Practices of antimicrobial usage on lactating cows by Fulani pastoralists in North-central Nigeria

https://doi.org/10.1371/journal.pone.0224949.t002

Practice	Frequency (n)	Proportion (%)	Confidence interval
Personnel that prescribe antimicrobials for usage in lactating cows			
Animal health officials	108	28.1	23.8, 32.78
Self-prescription	237	61.7	56.78, 66.48
Friends and relations	39	10.2	7.43, 13.49
Purchasing places of antimicrobials			
Veterinary drug shops	134	34.9	30.25, 39.77
Human drug shops	41	10.7	7.88, 14.07
Animal drug hawkers	209	54.4	49.42, 59.37
Who administer antimicrobials on cows			
Self-administered	259	67.4	62.64, 72.00
Animal health officials	125	32.6	28.00, 37.36
Frequency of antimicrobial usage on lactating cows with mastitis			
As prescribed	74	19.3	15.56, 23.45
Only once	104	27.1	22.82, 31.7
Once daily until recovered	206	53.6	48.64, 58.6
Dosage determination before antimicrobials use			
From instructions on the label	133	34.6	30.00, 39.50
Arbitrary	251	65.4	60.50, 70.00
Frequently used route of administration			
Injection	221	57.6	52.56, 62.43
Mouth (POS)	101	26.3	22.08, 30.88
On the skin (topical)	48	12.5	9.47, 16.10
In feed	14	3.6	2.09, 5.90
Observe about withdrawal periods			
Yes	72	18.7	15.08, 22.89
No	312	81.3	77.11, 84.92
Purpose for antimicrobial usage			
Treatment of mastitis in lactating cows	205	53.4	48.38, 58.34
Prevention of mastitis in lactating cows	93	24.2	20.13, 28.69
Promotion of milk yield in lactating cows	59	15.4	12.02, 19.23
All of the above	27	7.0	4.78, 9.93

https://doi.org/10.1371/journal.pone.0224949.t002

Antimicrobials Frequently used on Lactating Cows

Pastoralists recounted usage of a range of different classes of antimicrobials on lactating cows. Most frequently used antimicrobials were: tetracycline (98.7%), penicillin (96.6%), streptomycin (95.8%), sulfonamides (95.3%), gentamicin (94.3%), ciprofloxacin (86.2%) and neomycin (51.8%) (Fig 2).

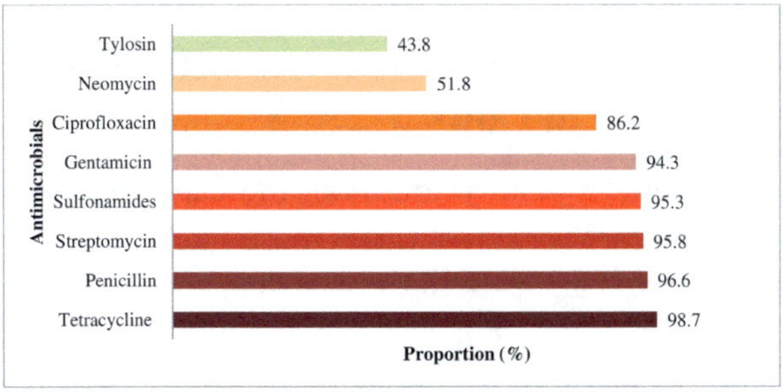

Figure 2: Antimicrobials frequently used by Fulani pastoralists on lactating cows in North-central Nigeria.

https://doi.org/10.1371/journal.pone.0224949.g002

Perceptions on Risk Pathways for Antimicrobial Resistance Dissemination from Cow Milk to Humans

Fulani pastoralists significantly perceived variably low, moderate and high risks associated with pathways for transmission of antimicrobial resistance from cow milk to humans. Few pastoralists perceived consumption of contaminated raw milk (14.8%), fermented milk (10.1%), and raw cheese (9.3%) to be of high risk for spread of antimicrobial resistance to humans. Also, very few of the participants perceived contacts with contaminated udder and milk (13.3%), and fomites (6.2%) to be of high risk for dissemination of resistance from cow milk to humans. Furthermore, very few participants perceived aerosols of discarded contaminated milk (18.0%), milking of cows in the herd site (10.4%), and flies attracted to the contaminated milk in the site (7.5%) to be high risk pathway for dissemination of AMR from cow milk to humans through environment (Table 3).

Table 3: Pastoralists' perceptions on risk pathways for antimicrobial resistance dissemination from cow milk to humans in North-central Nigeria

https://doi.org/10.1371/journal.pone.0224949.t003

Risk pathway	Low risk n (%)	Moderate risk n (%)	High risk n (%)	Chi-square	P-value
Consumption of raw milk and milk products				**11.98**	**0.010**
Raw milk	225 (58.6)	102 (26.6)	57 (14.8)		
Fermented milk (*nono*)	221 (57.6)	124 (32.3)	39 (10.1)		
Raw cheese (*wara*)	251 (65.4)	97 (25.3)	36 (9.3)		
Direct or indirect contacts				**11.66**	**0.002**
Contacts with contaminated udder and milk	243 (63.3)	90 (23.4)	51 (13.3)		
Contacts with contaminated fomites	251 (65.4)	109 (28.4)	24 (6.2)		
Environment				**21.22**	**0.001**
Aerosols of discarded contaminated milk	229 (59.6)	86 (22.4)	27 (18.0)		
Aerosols during milking of cows	253 (65.9)	120 (31.2)	11 (2.9)		
Flies attracted to contaminated milk	279 (72.7)	76 (19.8)	29 (7.5)		

Low risk (< 35%); Moderate risk (35–65%); High risk (> 66%); Significant at $p<0.05$

https://doi.org/10.1371/journal.pone.0224949.t003

Socio-demographic Characteristics Associated with Knowledge on AMU in Cows

Results of multivariable logistic regressions indicated that Fulani pastoralists in age group 70–79 years were five times more likely to possess satisfactory knowledge about AMU in lactating cows than those in age group 20–29 years (OR = 4.56; 95% CI: 1.89, 11.02; p = 0.001). Also, male pastoralists were more likely to possess satisfactory knowledge about AMU in cows than females (OR = 3.10; 95% CI: 1.69, 5.65; p = 0.001). Meanwhile, pastoralists with tertiary education were four times more likely to possess satisfactory knowledge about AMU in lactating cows than those without formal education (OR = 4.21; 95% CI: 1.82, 9.74; p = 0.001) (Table 4).

Table 4: Socio-demographic characteristics of Fulani pastoralists associated with knowledge regarding AMU on lactating cows in North-central Nigeria

https://doi.org/10.1371/journal.pone.0224949.t004

Variable	Number of respondents n (%)	Poor knowledge n (row %)	Satisfactory knowledge n (row %)	Odds ratio (OR)	95% Confidence interval	P-value
Age (in years)						
20–29	44 (11.5)	29 (65.9)	15 (34.1)	1.00		
30–39	56 (14.5)	33 (58.9)	23 (41.1)	1.35	0.59, 3.06	0.485
40–49	69 (18.0)	32 (46.4)	37 (53.6)	2.24	1.02, 4.89	0.040
50–59	96 (25.0)	40 (41.7)	56 (58.3)	2.71	1.29, 5.69	0.008
60–69	72 (18.8)	26 (36.1)	46 (63.9)	3.42	1.56, 7.52	0.002
70–79	47 (12.2)	14 (29.8)	33 (70.2)	4.56	1.89, 11.02	0.001
Gender						
Female	56 (14.6)	38 (67.9)	18 (32.1)	1.00		
Male	328 (84.4)	133 (40.5)	195 (59.5)	3.10	1.69, 5.65	0.001
Formal education						
None	250 (62.2)	146 (58.4)	104 (41.6)	1.00		
Primary	57 (14.8)	22 (38.6)	35 (61.4)	2.23	1.24, 4.03	0.007
Secondary	45 (11.7)	19 (42.2)	26 (57.8)	1.92	1.01, 3.65	0.040
Tertiary	32 (8.3)	8 (25.0)	24 (75.0)	4.21	1.82, 9.74	0.001

Statistically significant at $p<0.05$

https://doi.org/10.1371/journal.pone.0224949.t004

Socio-economic Factors that Influence Antimicrobials Misuse and Resistance Emergence in Lactating Cows

Many factors were identified to influence antimicrobials misuse on lactating cows, and which can predispose to emergence and dissemination of antimicrobial resistant pathogens through cow milk to humans. On multivariable logistic regressions models, improper AMU was more likely to satisfactorily influenced antimicrobials misuse on lactating cows (OR: 26.20; 95% CI: 13.79, 49.59; p<0.001). Also, non-enforcement of laws regulating AMU was more likely to influenced antimicrobials misuse in cows (OR: 5.32; 95% CI: 3.35, 8.25; p<0.001). Pastoralists' weak financial status (OR: 6.73; 95% CI: 4.20, 10.62; p<0.001) and low education and expertise (OR: 10.41; 95% CI: 6.37, 16.87; p<0.001) were more likely to satisfactorily influenced their misuse of antimicrobials on lactating cows. Furthermore, pastoralists' nomadic culture (OR: 9.75; 95% CI: 5.47, 17.38; p<0.001) and extensive management system of herds (OR: 18.40; 95% CI: 10.97, 30.86; p<0.001) were more likely to influenced antimicrobials misuse on lactating cows (Table 5).

Table 5: Socio-economic factors that influence antimicrobials misuse by pastoralists and resistance emergence in lactating cows in North-central Nigeria

https://doi.org/10.1371/journal.pone.0224949.t005

Factor	Poor influence (%)	Satisfactory influence (%)	Odds ratio (OR)	95% CI	P-value
Improper antimicrobial usage					
No	111 (63.4)	64 (36.6)	1.00		
Yes	13 (6.2)	196 (93.8)	26.20	13.79, 49.59	<0.001
Non-enforcement of laws regulating antimicrobial usage					
No	98 (57.6)	72 (42.4)	1.00		
Yes	44 (20.6)	170 (79.4)	5.32	3.35, 8.25	<0.001
Weak financial status of pastoralists					
No	93 (60.0)	62 (40.0)	1.00		
Yes	42 (18.3)	187 (81.7)	6.73	4.20, 10.62	<0.001
Low education and expertise of pastoralists					
No	97 (66.9)	48 (33.1)	1.00		
Yes	39 (16.3)	200 (83.7)	10.41	6.37, 16.87	<0.001
Nomadic and transhumance culture of pastoralists					
No	55 (73.3)	20 (26.7)	1.00	5.47, 17.38	<0.001
Yes	68 (22.0)	241 (78.0	9.75		
Extensive husbandry system					
No	151 (77.0)	45 (23.0)	1.00		
Yes	29 (15.4)	159 (85.6)	18.4	10.97, 30.86	<0.001

Statistically significant at $p<0.05$

https://doi.org/10.1371/journal.pone.0224949.t005

DISCUSSION

This study, to our knowledge, was the first to survey AMU and AMR in lactating cows, and consequent dissemination from cow milk to humans in rural pastoral communities in Nigeria. It provides critical information regarding antimicrobials stewardship among local smallholders' dairy farmers under extensive husbandry system. Antimicrobials misuse and resistance in food animals and humans have been recognized as emerging global threat to food safety, food security and public health [19].

This study identified low level of knowledge about proper AMU on lactating cows among surveyed Fulani pastoralists. This could be due to their low literacy level, since 65.2% of them do not possessed formal education. Lack of formal education could influence antimicrobials misuse, which can in turns predispose to residues in milk following treatment of clinical mastitis, a major public health threat with adverse effects on safety of milk for human consumption. Human consumption of milk contaminated with antimicrobials can results in allergic responses, carcinogens, changes in intestinal flora and development of antimicrobial resistant pathogens [20,21]. Knowledge about patterns of AMU is very vital to understanding animal health management in dairy farms. Bridging the knowledge gap through sensitization will increase pastoralists' risk perceptions associated with AMR in lactating cows.

We found the majority of pastoralists practicing self-prescription (61.7%) and self-administration (67.4%) of antimicrobials on lactating cows without professionals' inputs. This could be due to gross inadequate veterinary services associated with hard-to-reach terrains of Fulani pastoral communities. In Nigeria, the use of non-prescribed antimicrobials in food animals has been related to unavailability of veterinary services and extra cost of veterinary services [22]. Most of the pastoralists (54.4%) purchase antimicrobials used on cows from animal drugs hawkers without prescriptions, thus facilitating antimicrobials misuse and resistance emergence. Also, antimicrobials were found to be used for therapeutic and preventive purposes as well as for milk yield promotion in lactating cows. This is consistent with findings that reported similar uses in food animals [23,24,25]. Also, improper AMU in food animals has been found to be significant contributor to emergence and dissemination of antimicrobial resistance to humans [26,27]. Noteworthy is our finding on practices of low dose AMU in lactating cows. Antimicrobials at low dosages (subtherapeutic dosages) have been reported to predispose to the resistance emergence in exposed bacteria due to selection bias from promotion of genetic and phenotypic variability [28,29].

This study found some antimicrobials to be frequently used on lactating cow, which include: tetracycline, penicillin, streptomycin, sulfonamides, gentamicin, ciprofloxacin, neomycin and tylosine. This is consistent with previous studies that reported these antimicrobials to be frequently used in food animals in Nigeria [30,31]. About half of the world's antimicrobials productions have been reported to be frequently used in food animals [32,33]. Frequent use of antimicrobials in food animals has been reported in other African countries (South Africa and Tanzania), where tetracycline, sulfonamides, penicillin and streptomycin were mostly used in livestock [32,34]. Similar levels of AMU have been reported in India, where penicillin and tetracycline were frequently used in food animals [35]. Mastitis in lactating cows in pastoral communities was commonly treated with oxytetracycline and gentamicin, which are critically important antimicrobials for human medicine but not approved for use in dairy cattle in the United States due to public health implications [36].

Very few surveyed pastoralists perceived pathways for dissemination of antimicrobial resistance from cow milk to humans to be of high risks. These were through: consumption of contaminated raw milk and milk products (cheese, fermented milk); contacts with contaminated udder and fomites; and contaminated environment (aerosols of contaminated milk). This finding is in consonance with the results that reported transmission of antimicrobial resistance from food animals to humans through food consumption, contact with contaminated animals, and associated wastes spilled into the environment [37,38]. Also, reports have indicated that drug-resistant strains of food animal origin can spread to humans either through food supply chain (dairy products); direct animal contact; or through environment [39,40,41].

We found socio-demographic characteristics of age, gender and education to have significant influence on antimicrobials misuse and emergence of resistance in lactating cows. Health education of pastoralists through mass media, especially radio, is imperative for modification of behaviours and effective social change towards proper AMU on lactating cows in hard-to-reach smallholder pastoral rural settlements in developing economies, especially in Africa.

The study found some socio-economic activities of improper AMU, non-enforcement of laws regulating AMU, weak financial status, low education and expertise, nomadic culture, and extensive husbandry system of Fulani pastoralists to significantly influenced antimicrobials misuse and AMR emergence on lactating cows. Improper AMU has been reported as the major driver of AMR in food animals due to selection pressure on animal microbiota [42]. Improper AMU and lack of enforcement of regulating laws are major factors contributing to increase AMR [43]. Although most of the antimicrobials used by pastoralists were within the categories of the OIE List allowed [44], the major problems were low levels of knowledge and practices on usage. These observed interconnected factors that driven antimicrobials misuse could create complex challenges associated with AMR in cows with public health consequences.

Major limitation of this study was the use of questionnaire tool for data collection. In this case, we pre-tested the questionnaire for quality control and ensured that no information was lost in the process. Also, we were limited by lack of full adjustments for community clustering in the designed random sampling. However, the used of central tendency measures, would be valuable enough to tolerate the likely imperfections in the confidence intervals.

CONCLUSIONS

This study highlighted low levels of knowledge, risk perceptions and practices regarding AMU and AMR among surveyed pastoralists. These challenging gaps, influenced by socio-demographic factors, call for education of the vulnerable populations on promotion of prudent AMU in lactating cows. Consideration of the risk pathways for antimicrobial resistance dissemination is crucial for the development of AMR surveillance system, control and prevention. Gradual reform of socio-economic activities that influence antimicrobials misuse and resistance emergence through collaborations, in line with 'One Health' approach, will assure food safety, food security, and public health.

ACKNOWLEDGMENTS

The authors wish to thank all Area Animal Health Officers in the Niger State Ministry of Livestock and Fisheries who assisted in identifying Fulani pastoral settlements and data collection. Great thanks are extended to the pastoralists, especially their '*Ardos*' who voluntarily participated in the study.

REFERENCES

1. Sharma C, Rokana N, Chandra M, Singh BP, Gulhane RD, Gill JPS, et al. Antimicrobial Resistance: Its Surveillance, Impact, and Alternative Management Strategies in Dairy Animals. Front Vet Sci. 2018; 4:237. pmid:29359135
2. FAO. World Livestock 2011: Livestock in food security. Food and Agriculture Organization of the United Nations, Rome, Italy; 2011. Available: http://www. fao. org/3/i2373e/i2373e. pdf Accessed 18 October 2019.
3. Hemme T, Otte J. Status and Prospects for Smallholder Milk Production: A Global Perspective. In: Pro-Poor Livestock Policy Initiative, editor. A Living from Livestock. Food and Agriculture Organization of the United Nations, Rome, Italy; 2010. pp. 1–12. Available: http://www. fao. org/docrep/012/i1522e/i1522e00. pdf. Accessed 18 October 2019.
4. Annatte I. Nigeria- Milk production Fact Sheet. In: IFCN Dairy Report 2009, International Farm Comparison Network, Dairy Research Center, Kiel, Germany 2010; Vol. 3. 63 (129).
5. Gussmann MK, Steeneveld W, Kirkeby CT, Hogeveen H, Farre M, Halasa T. Economic and epidemiological impact of different intervention strategies for subclinical and clinical mastitis. 2019; Prev Vet Med. 166:78–85. pmid:30935508
6. Stevens M, Piepers S, De Vliegher S. The effect of mastitis management input and implementation of mastitis management on udder health, milk quality, and antimicrobial consumption in dairy herds. J Dairy Sci. 2019; 102(3):2401–2415. pmid:30692005
7. Van Boeckel TP, Brower C, Gilbert M, Grenfella BT, Levina SA, Robinson TP, et al. Global trends in antimicrobial use in food animals. Proc Natl Acad Sci USA (PNAS). 2015; 12(18):5649–5654. http://dx. doi. org/10. 1073/pnas. 1503141112
8. WHO. Antimicrobial resistance. World Health Organization Fact Sheet, No. 194, 2016. Available: http://www. who. int/mediacentre/ factsheets/fs194/en. Accessed 22 November 2018.
9. Littmann J, Viens AM. The ethical significance of antimicrobial resistance. Pub Health Ethics. 2015; 8(3):209–24. https://doi. org/10. 1093/phe/phv025

10. Robinson TP, Bu DP, Carrique-Mas J, Fevre EM, Gilbert M, Grace D, et al. Antibiotic resistance is the quintessential One Health issue. Trans R Soc Trop Med Hyg. 2016; 110(7):377–380. pmid:27475987

11. Suleiman H. Policy issues in agropastoral development in Nigeria. Proceedings of the National Conference on Pastoralism in Nigeria, Ahmadu Bello University Zaria, Nigeria; 1988.

12. de Leeuw PN, McDermott JJ, Lebbie SHB. Monitoring of livestock health and production in sub-Saharan Africa. Prev Vet Med. 1995; 25:195–212.

13. Hochbaum GM. Public Participation in Medical Screening Programs: A Socio-psychological Study, Public Health Service Publication No. 572, Washington DC; 1958.

14. Dean AG, Sullivan KM, Soe MM. Open Source Epidemiologic Statistics for Public Health (OpenEpi), Version 2. 3. 1; 2009.

15. Thrusfield M. Veterinary Epidemiology. 3rd ed, Blackwell Science Ltd, a Blackwell Publishing company, 9600 Garsington Road, Oxford OX4 2DQ, UK, 2009; p. 228–238.

16. WMADH. World Medical Association Declaration of Helsinki. Ethical principles for medical research involving human subjects. Bull World Health Organ. 2001; 373–374. pmid:11357217

17. Alhaji NB, Babalobi OO, Isola TO. A quantitative exploration of nomadic pastoralists' knowledge and practices towards Rift Valley fever in Niger State, North-central Nigeria: The associated sociocultural drivers. One Health, 2018; 6(1):16–22. https://doi. org/10. 1016/j. onehlt. 2018. 09. 001

18. Dohoo I, Martin W, Studahl H. Measures of Association. In: McPike SM, editor. Veterinary Epidemiologic Research, 2nd ed. , University of Prince Edward Island, Charlottetown, PE, Canada; 2009; 136–148.

19. Aminov R. I, Mackie RI. Evolution and ecology of antibiotic resistance genes, FEMS Microbiol Lett. 2007; 271:147–161. pmid:17490428

20. More SJ, Clegg TA, O'Grady L. Insights into udder health and intramammary antibiotic usage on Irish dairy farms during 2003–2010. Irish Vet J. 2012; 65:7. https://doi. org/10. 1186/2046-0481-65-7

21. Asredie T, Engdaw TA. 2015. Antimicrobial Residues in Cow Milk and its Public Health Significance. World J Dairy Food Sci. 2015; 10(2):147–153. https://doi. org/10. 5829/idosi. wjdfs. 2015. 10. 2. 96244

22. Okoli IC, Nwosu CI, Okoli GC, Okeudo NJ, Ibekwe V. Drug management of anti-microbial resistance in avian bacterial pathogen in Nigeria, Intern J Environ Health Human Dev. 2002; 3(1):39–48.

23. Sirdar MM, Picard J, Bisschop S, Gummow BA. A questionnaire survey of poultry layer farmers in Khartoum State, Sudan, to study their antimicrobial awareness and usage patterns. Onderstepoort J Vet Res. 2012; 79(1):1–8. https://doi. org/10. 4102/ojvr. v79i1. 361

24. Oluwasile BB, Agbaje M, Ojo OE, Dipeolu MA. Antibiotic usage pattern in selected poultry farms in Ogun state. Sokoto J Vet Sci. 2014; 12(1):45–50. https://doi. org/10. 4314/sokjvs. v12i1. 7

25. Alhaji NB, Haruna AE, Muhammad B, Lawan MK, Isola TO. Antimicrobials usage assessments in commercial poultry and local birds in North-central Nigeria: Associated pathways and factors for resistance emergence and spread. Prev Vet Med. 2018; 154:139–147. pmid:29685438

26. Morgan DJ, Okeke LN, Laxminarayan R, Perencevich EN, Weisenberg S. Nonprescription antimicrobial use worldwide: a systematic review. Lancet Infect. Dis. 2011; 11:692–701. pmid:21659004

27. Locatelli C, Cremonesi P, Caprioli A, Carfora V, Ianzano A, Barberio A, Morandi S, et al. , Occurrence of methicillin-resistant Staphylococcus aureus in dairy cattle herds, related swine farms, and humans in contact with herds. J Dairy Sci. 2017; 100(1):608–619. pmid:27865508

28. Martinez JL. Antibiotics and antibiotic resistance genes in natural environments. Science. 2008; 321:365–367. pmid:18635792

29. You Y, Silbergeld EK. Learning from agriculture: understanding low-dose antimicrobials as drivers of resistome expansion, Front Microbiol. 2014; 5:284. pmid:24959164

30. Adesokan HK, Akanbi IO, Akanbi IM, Obaweda RA. Pattern of antimicrobial usage in livestock animals in south-western Nigeria: the need for alternative plans. Onderstepoort J Vet Res. 2015; 82(1):Art #816, 6 pages. http://dx. doi. org/10. 4102/ojvr. v82i1. 816

31. Alhaji NB, Isola TO. Antimicrobial usage by pastoralists in food animals in North-central Nigeria: The associated socio-cultural drivers for antimicrobials misuse and public health implications. One Health. 2018; 6(1):41–47. https://doi. org/10. 1016/j. onehlt. 2018. 11. 001

32. Katakweba AAS, Mtambo MMA, Olsen JE, Muhairwa AP. Awareness of human health risks associated with the use of antibiotics among

livestock keepers and factors that contribute to selection of antibiotic resistance bacteria within livestock in Tanzania. Livest Res Rural Dev. 2012; 24:170. Art #.

33. Landers TF, Cohen B, Wittum TE, Larson EL. A review of antibiotic use in food animals: perspective, policy, and potential. Pub Health Rep. 2012; 127(1):4–22. https://doi. org/10. 1177/003335491212700103
34. Eagar H, Swan G, Van Vuuren M. A survey of antimicrobial usage in animals in South Africa with specific reference to food animals. J South Afr Vet Assoc. 2012; 83(1):Art #16. 8 pages. http://dx. doi. org/10. 4102/jsava. v83i1. 16
35. Sahoo KC, Tamhankar AJ, Johansson E, Lundborg CS. Antibiotic use, resistance development and environmental factors: a qualitative study among healthcare professionals in Orissa, India. BMC Pub Health. 2010; 10(629):1–10. http://dx. doi. org/10. 1186/1471-2458-10-629
36. Smith GW, Gehring R, Craigmill AL, Webb AI, Riviere JE. Extra-label intramammary use of drugs in dairy cattle. J Am Vet Med Assoc. 2005; 226:1994–1996. pmid:15989181
37. Woolridge M, Evidence for the circulation of antimicrobial-resistant strains and genes in nature and especially between humans and animals, Rev Sci Techn Off Int Epiz. 2012; 31:231–247.
38. Mceachran AD, Blackwell BR, Hanson JD, Wooten KJ, Mayer GD, Cox SB et al. Antibiotics, bacteria, and antibiotic resistance genes: aerial transport from cattle feed yards via particulate matter. Environ Health Perspect. 2015; 123:337–343. pmid:25633846
39. Lhermie G, Grohn YT, Raboisson D. Addressing antimicrobial resistance: an overview of priority actions to prevent suboptimal antimicrobial use in food animal production. Front Microbiol. 2016; 7:2114. pmid:28111568
40. Marshall BM, Levy SB. Food animals and antimicrobials: impacts on human health. Clin Microbiol Rev. 2011; 24:718–33. pmid:21976606
41. Chang Q, Wang W, Regev-Yochay G, Lipsitch M, Hanage WP. Antibiotics in agriculture and the risk to human health: how worried should we be? Evol Appl. 2015; 8:240–5. pmid:25861382
42. McEwen SA. Antibiotic use in animal agriculture: what have we learned and where are we going? Anim Biotechnol. 2006; 17:239–250. pmid:17127534

43. Mtenga A, Emanuel M, Mabula J, Peter R. Consumer and Practitioner Education: Status of Antibiotic Resistance, Alliance for the Prudent Use of Antibiotics (APUA), 136 Harrison Avenue, Boston, 2011.
44. OIE. List of Antimicrobial Agents of Veterinary Importance. The World Organisation for Animal Health, Paris, France, 2015.

CHAPTER 18

Coxiella Burnetii in Slaughterhouses in Brazil: A Public Health Concern

Mateus de Souza Ribeiro Mioni[1] , Francisco Borges Costa[2] , Bruna Letı´cia Devide´ Ribeiro[1] , Wanderson Sirley Reis Teixeira[1,3], Vanessa Cristina Pelicia[1] , Marcelo Bahia Labruna[4] , E´lodie Rousset[5] , Karim Sidi-Boumedine[5] , Richard Thie´ry[5] , Jane Megid[1]

[1] Departamento de Higiene Veterina´ria e Sau´de Pu´blica, Universidade Estadual Paulista "Ju´lio de Mesquita Filho", Botucatu, São Paulo, Brazil

[2] Departamento de Patologia, Universidade Estadual do Maranhão, São Luı´s, Maranhão, Brazil

[3] Instituto de Quı´mica, Universidade de São Paulo, São Paulo, São Paulo, Brazil

Citation: Mioni MdSR, Costa FB, Ribeiro BLD, Teixeira WSR, Pelicia VC, Labruna MB, et al. (2020) Coxiella burnetii in slaughterhouses in Brazil: A public health concern. PLoS ONE 15(10): e0241246. https://doi.org/10.1371/journal.pone.0241246.

[4] Departamento de Medicina Veterina´ria Preventiva e Sau´de Animal, Universidade de São Paulo, São Paulo, São Paulo, Brazil

[5] Anses, French Agency for Food, Environmental and Occupational Health Safety, Sophia Antipolis laboratory, Animal Q Fever Unit, Sophia Antipolis, France

ABSTRACT

Q fever is an important zoonosis, yet it is often neglected and can present large outbreaks, as observed in the Netherlands. In the past few years, cases of Q fever have been described in Brazil; however, the epidemiological situation of Q fever in ruminants, the main reservoir of the pathogen, is unknown in this country. Our study aimed to estimate the prevalence of *C. burnetii* in cattle sent to slaughterhouses using an immunofluorescence assay (IFA) and quantitative real-time PCR (qPCR). From 1515 cattle serum samples collected from nine slaughterhouses, 23.8% (360/1515) were serologically positive by IFA (cutoff titer>1:64), indicating past or recent exposure to *C. burnetii* infection. Among the 54 cities sampled during the study, 83.3% (45/54) had at least one seropositive animal. Subsequently, all seropositive samples were submitted to qPCR for *C. burnetii* DNA, and 12.2% (44/360) of the sera were qPCR positive, which indicates bacteremia and suggests active or recent infection. The results highlight the risk for abattoir workers that results from exposure to contaminated aerosols produced during slaughter procedures. Moreover, the heat maps that were construction from the positive samples demonstrate the widespread distribution of *C. burnetii* in the State of São Paulo, Brazil and denotes the need for surveillance and preventive measures to reduce the prevalence in cattle.

INTRODUCTION

Coxiella burnetii is the causative agent of Q fever in humans and animals [1, 2], and this condition is distributed throughout the world, with the exception of New Zealand [1, 3]. It is an obligatory intracellular, gram-negative bacterium that can persist in the environment for long periods, due to its highly infectious spore-like form [3].

In humans, Q fever is mainly asymptomatic. However, an acute form with flu-like syndromes, acute hepatitis, pneumonia, and a chronic and more severe form with persistent focalized infection, such as endocarditis or vascular infections, are also possible [2, 4]. *C. burnetii* is mainly transmitted by the inhalation of aerosol particles in the vicinity of infected animals, their reproductive tissues, or other animal products [1]. However, people who live far from livestock can also acquire the disease through the inhalation of particles transported by wind [5, 6]. Additionally, *C. burnetii* can be transmitted orally through the consumption of milk and dairy products, although this route has a minor importance in the epidemiology of the disease [1, 7]. Moreover, despite the fact that *C. burnetii* has been detected in ticks the possibility of transmission to humans through the bite of these arthropods is low. Nevertheless, ticks may play a role in transmission to animals due to ecological features (e.g. host preference and bite rate) [8].

Coxiellosis is most often latent in animals, without clinical manifestation, but may cause bacteria sheeding into the environment via feces or parturition products [1]. *C. burnetii* can infect a diverse group of species that range from arthropods and cold blooded animals to birds and mammals [1, 3, 8], and domestic ruminants are considered the main reservoirs for human disease [1, 9]. In cows, ewes, and goats, the disease is associated with late abortion. Reproductive disorders, such as metritis and infertility, have also been associated with the disease [1].

The worldwide prevalence of coxiellosis is somewhat higher in cattle than in small ruminants [10], although, in European countries, goats, followed by ewes, are the species with the highest infection rates [11, 12]. In Brazil, only a few epidemiological surveys have aimed to establish seroprevalence in animals. In 1955, Valle et al. [13] found 14% (24/171) positivity in dairy cows. Later, Riemann et al.[14] reported a seroprevalence of 29% (45/156) in beef cattle. Recently, a survey using a commercial ELISA found 55.1% (172/312) seropositivity in a goat herd with reproductive disorders [15], and Mioni et al. [16] reported the presence of *C. burnetii* in dairy cattle and their fetuses in the state of Goiás [17].

Lipopolysaccharide (LPS) is the main *C. burnetii* antigen and is found in two different phases on the bacterium surface. The phase I antigen is the complete form of LPS and is linked to the virulence. The phase II antigen is the truncated form that is obtained after several passages in cell culture, and it is the result of gene deletion and consequent loss of sugars from LPS [2, 9]. Serological antigens are based on these two antigenic forms of *C. burnetii* [1, 2]. Phase II is imunodominant and is inserted in phase I LPS. For this reason, antibodies against phase II are always present during the course of infection, and this feature is important for epidemiological studies where the detection of phase II antibodies is preferable [18].

In humans, the serodiagnostic is based on the different titers of IgG to phase I and phase II antigens. Antibodies against phase II are the first to arise in primo infections; and for this reason, such antibodies are linked to acute Q fever, whereas the increase in the antibody response to phase I (IgG≥1:800 when using immunofluorescence assay–IFA) is related to persistent focalize infections [2]. In animals, although the antibody response to the infection has the same pattern, the interpretation of the antibody response to phase I and phase II antigens has not been validated for diagnosing acute and chronic Q fever [1]. Instead, commercially available kits can detect mixtures of anti-phase I and II antibodies to reflect the total immunoglobulin content in the sera [1].

Moreover, for acute Q fever in humans, the use of real-time PCR (qPCR) in the early stages of the disease is necessary due to the lack of antibodies in the first two weeks of symptoms and evidentiate bacteremia [2, 19]. This approach is essential for the diagnosis of clinical manifestations after Q fever primo-infection and the identification of new cases during outbreaks, in association with IgM rates [19, 20]. However, only a few studies have reported molecular diagnoses using serum samples [21–25]. Moreover, due to a lack of data, it is difficult to establish its significance in animal disease.

Q fever is considered an occupational hazard for farmers, veterinarians, and abattoir workers [26]. Since its description in Australia in 1937, epidemic Q fever was predominantly associated with abattoirs in that country, until 1980. In some regions, 92% of cases of the disease were among slaughterhouse workers [27, 28]. In France, one slaughterhouse was the source of a Q fever outbreak that affected 29 people [29]. In Brazil, a survey reported 29% (41/144) of seroprevalence for *C. burnetii* among cattle slaughterhouse workers, and the positivity was higher for those who worked in corrals (40%) and the killing floor (36%) [14]. In 2015, an outbreak related to cattle slaughterhouse workers was reported in the city of Barbosa (21°160″S 49°5654″W), which is located in the state of São Paulo [30]. Apart from slaughterhouse scenarios, human Q fever in Brazil has been confirmed in patients with endocarditis [31–33], patients suspected of having dengue [34, 35], and a family cluster case [36]. Additionally, the most recently reported case of Q fever involved six cadets after a military training where they had contact with a goat and capybaras [37, 38]. The state of Rio de Janeiro is the region of Brazil with the most reports of Q fever.

Although the risk of human outbreaks is usually linked to small ruminants; in Brazil, the Q fever can be due to a different scenario, since the country is a primary producer of bovine meat, with a herd of more than 214 million animals. In comparison, the flocks of small ruminants combined are limited to 28 million animals [39].

The public health link to Q fever in Brazil is unknown, as is the epidemiological status regarding the prevalence of *C. burnetii* in reservoirs and the risk of acquiring Q fever during the slaughter of animals. Therefore, this study aimed to establish the prevalence of *C. burnetii* in beef cattle in the State of São Paulo, Brazil, using serological and molecular diagnostic methods.

MATERIALS AND METHODS

The Ethics Committee on Animal Use (CEUA-FMVZ-UNESP/Botucatu, State of Sao Paulo, Brazil) approved this study (protocol numbers 141/2013 and 0203/2016). CEUA-FMVZ-UNESP/Botucatu is registered in the National Council for Animal Control and Experimentation (CONCEA), under registration number CIAEP/CONCEA 01.0115.2014. The CONCEA is regulated by LAW No. 11.794, October 8th, 2008, which establishes procedures for the scientific use of animals. Moreover, the CONCEA is responsible for the periodic review of standards for use and care of animals for teaching or scientific research proposes, following the international conventions of which Brazil is a participant.

The Ministry of Agriculture, Livestock, and Food Supply (MAPA), the Secretariat of Agriculture and Supply of the State of São Paulo, and the municipal authorities of the cities involved, as well as those responsible for each establishment, provided permission to sample animals in slaughterhouses.

Samples

Serum samples were obtained from nine slaughterhouses in the state of São Paulo, Brazil, from October 2014 to March 2015. To calculate the sample size, we used an Openepi proportion test [40]. To estimate the number of bovines to be tested through IFA, a slaughter population of 3.5 million bovines per year in the State of São Paulo was estimated [41]. As the real situation of the disease is unknown, we used an apparent coxiellosis prevalence of 50%, with 3% precision and a confidence interval of 95%. Using these parameters, the minimum sample size needed to estimate the prevalence in the slaughter population was 1066 samples. To avoid sample bias, we collected 1515 cattle (*Bos taurus* and *Bos indicus*) samples, which were divided among the nine slaughterhouses in the state of São Paulo, with a minimum of 157 samples per facility. All facilities in the experiment were sampled only after authorization by the authority of the inspection system of the establishment. Animals were sampled without breed and sex discrimination. Blood samples were collected at the bleeding step after desensitization and vessel sectioning. Blood samples were collected in tubes (BD Vacutainer Serum Plus Blood collection tubes with a clot activator, 10 mL; BD, Franklin Lakes, NJ), and serum was obtained by centrifugation (2,000g /10 min) within 12 h. All serum samples were stored at– 80 °C until the time of analysis.

Immunofluorescence Assay

Serum samples were defrosted, diluted two-fold, and analyzed by an IFA, according to Horta *et al.* [42], with a cutoff of 1:64. For the IFA, we used slides that contained a mixture of phase I and phase II *C. burnetii* antigens, and the *C. burnetii* were purified after production on Vero cells, to detect total immunoglobulin (IgM and IgG). The in-house antigen was produced from the Argentinian *C. burnetii* strain At12, which was isolated from *Amblyomma tigrinum* [43]. Positive samples were subjected to real-time PCR for the detection of *C. burnetii* in the bloodstream.

For serological analysis, all slides used during the assay were assembled as follows: 20 μL of serum samples were placed in the wells that were previously prepared with the antigen and incubated for 30 min at 37 °C. The slides were then washed with washing buffer (WB) and placed for 10 min in a vat that contained WB solution, and this operation was repeated twice. Subsequently, the slides were dried at 37 °C for 10 min. The material was organized in a humid chamber, and 20 μL of the bovine conjugate was added and incubated for 30 min at 37°C. After incubation, slides were washed with WB and deposited in a vat that contained WB solution and 1.5 mL of Evans solution, for 10 min (step repeated twice). The slides were dried and read under a fluorescence microscope. In each slide, positive and negative sera were used as controls. Sera that showed reactivity following a 1:64 dilution were considered positive.

DNA Extraction

The extraction of serum samples was carried out with a DNeasy Blood and Tissue Kit (Qiagen, Hilden, Germany) according to the manufacturer's protocol.

Real-time PCR

All samples that showed a positive IFA reaction (n = 360) were screened in Brazil through real-time PCR using a melt analysis protocol. For quantification, qPCR was performed using specific probes for *C. burnetii* as well as reference materials prepared from the Nine Mile strain. The procedure was performed at the French Agency for Food, Environmental, and Occupational Health Safety (ANSES, Sophia Antipolis, France), which is mandated as the French National Reference Laboratory and a World Organization for Animal Health (O.I.E.) reference laboratory for Q fever.

The melt-analysis protocol was performed according to Vaidya *et al.* [44] with primers Trans 3 (5'-GTAACGATGCGCAGGCGAT-3') and Trans 4 (5'-CCACCGCTTCGCTCGCTA-3'), as previously described [45] for the detection of the *IS1111a* element. The qPCR assay was performed in a Real-Time Applied Biosystems Step-One (model ABI 7500 Fast), using the standard cycle. As a positive control, we used DNA extracted from *C. burnetii* strain At12, which was isolated from ticks [43]. The positive control dissociation curve was recorded a melting temperature (Tm) of 86.91°C.

For quantifying *C. burnetii* in the samples, qPCR was used with specific primers and probes developed by the reference center. Briefly, the PCR reaction was performed in a total volume of 25 μL that contained 320 nM of each forward and reverse primer, 200 nM of the specific probe (FAM-TAMRA), 1X × TaqMan Universal PCR Master Mix (Applied Biosystems), 5.7 μL nuclease-free PCR grade water, and 5 μL of DNA template. As a control, primers (320 nM) targeting the glyceraldehyde-3-phosphate dehydrogenase (GAPDH) gene of ruminants and the VIC-TAMRA probe at a concentration of 200 nM were used. All assays were performed using CFX96™ equipment (Bio-Rad Laboratories, Inc.). Cycling conditions were as follows: 50°C for 2 min, 95°C for 10 min, and 40 cycles of 95 / 15 s, 60°C / 1 min. To quantify the samples, in each assay, a standard curve that ranged from 2×10^2 to 2×10^7 genome equivalent (GE) bacteria/mL was prepared using a tenfold dilution of purified genomic DNA from a calibrated *C. burnetii* Nine Mile strain. The theoretical number of GEs was calculated based on the measured DNA concentration and the length of the *C. burnetii* Nine Mile genome sequence (1,995,275 bp) published by Seshadri et al. [46]. As a quality control for the assay, two sentinel samples that contained Nine Mile isolates at a concentration of 10^4 were used to check the accuracy and reproducibility of the quantification.

Sequencing

To confirm the detection of *C. burnetii*, we sequenced seven positive samples by qPCR. They were analyzed by 1% agarose gel, purified with Agencourt AMPure XP (Beckman Coulter, Inc.), and subjected to automatic sequencing by capillary electrophoresis on an ABI3500 Genetic Analyzer (Applied Biosystems). The nucleotide sequences that were produced were aligned with the reference sequences deposited in GenBank. The PCR product was also confirmed by 1.5% agarose gel electrophoresis with ethidium bromide, to visualize a 243 bp product. Seven sequence analyses that were produced during this survey were deposited in GenBank (accession numbers MK125016, MK125017, MK125018, MK125019, MK125020, MK125021, and MK125022).

Phylogenetic Analysis

To confirm the detection of *C. burnetii*, we performed a phylogenic analysis using the online tool phylogeny.fr [47–49] with the seven sequences generated here, sixteen *IS1111a C. burnetii* strain sequences from different countries and hosts, and an *IS1111*-like sequence from a Coxiella-like bacteria (CLB) with 90% identity with the insertion element of *C. burnetii* strains [50]. All GenBank accession numbers of the sequences used in this study are available in the phylogenetic tree.

Georeferencing

All animals evaluated were spatially distributed in the cities of origin, sampled during work, and analyzed through thematic density maps with ArcGIS 10.0 [51]. We created three sample distribution thematic maps that showed i) the total number of samples, ii) the number of serologically-positive samples by IFA, and iii) the number of serum samples that were positive in the real-time PCR analysis. All heat maps represent the origin of the bovines that were sampled in the different slaughterhouses.

The true prevalence was calculated considering the sensibility and specificity values that were previously reported to IFA [52] in ruminants. All statistical analyses took into account a confidence interval of 95% and were performed with "epiR" [53] in the R environment [54].

RESULTS

The apparent and true prevalence observed by IFA was 23.8% (CI95% = 21.7%– 25.9%) (360/1515) and 20.0% (CI95% = 17.8%– 22.3%), respectively. Among seropositive samples, the presence of *C. burnetii* DNA was observed by qPCR in 12.2% of the cattle (44/360). When we analyzed the seroprevalence and detection by slaughterhouse, we observed homogenous *C. burnetii* positivity, since all facilities had positive samples in both diagnostic methods (Table 1). Moreover, the chi-squared test showed no statistical difference ($p>0.05$) in seroprevalence among the slaughterhouses.

Table 1: Prevalence for each slaughterhouse evaluated during the study by IFA and qPCR

https://doi.org/10.1371/journal.pone.0241246.t001

Slaughterhouse	Samples analyzed (n)	Positives (n)	Prevalence (%)	Samples analyzed (n)	Positives (n)	Positivity (%)
1	170	42	24.7%	42	8	19.0%
2	166	38	22.9%	38	7	18.4%
3	170	29	17.0%	29	3	10.3%
4	170	49	28.8%	49	5	10.2%
5	181	64	35.3%	64	2	3.1%
6	158	27	17.1%	27	15	55.6%
7	162	42	25.9%	42	3	7.1%
8	172	21	12.2%	21	0	0.0%
9	166	48	28.9%	48	1	2.1%
Total	1515	360	23.8%	360	44	12.2%

https://doi.org/10.1371/journal.pone.0241246.t001

Quantification of the positive serum samples showed an average of 144 bacteria/mL in the serum, with a minimum of 10 bacteria and a maximum of 814 bacteria/mL. For the average estimation, we used only samples with Ct < 36.

All *IS 1111a* sequences presented 100% identity with the *C. burnetii* strains available in GenBank. The phylogenetic analysis confirmed the diagnosis of *C. burnetii*, as all *C. burnetii* strains were clustered together, and the *IS 1111-like* sequence of the CLB detected on a tick was placed in a different branch.

Regarding the positive sample distribution, density maps were constructed, and using the herd location of the slaughtered animals allowed revealed that *C. burnetii* circulates across the state of São Paulo, since we observed positive serological and molecular diagnoses in all sampled regions (Figs 1–3). Using slaughterhouses as a tool for Q fever surveillance, we were able to sample animals derived from 54 cities across the state of São Paulo. We observed that 83% (45/54) of the cities had at least one seropositive animal.

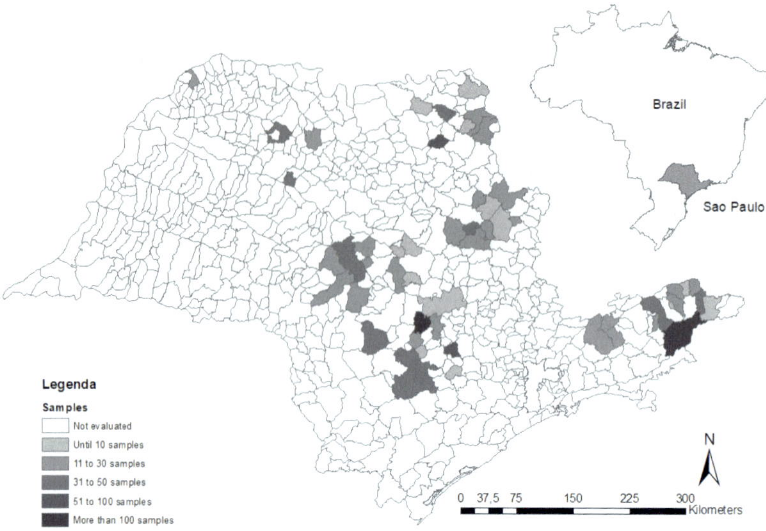

Figure 1: Distribution of collected serum samples by municipalities.

https://doi.org/10.1371/journal.pone.0241246.g001

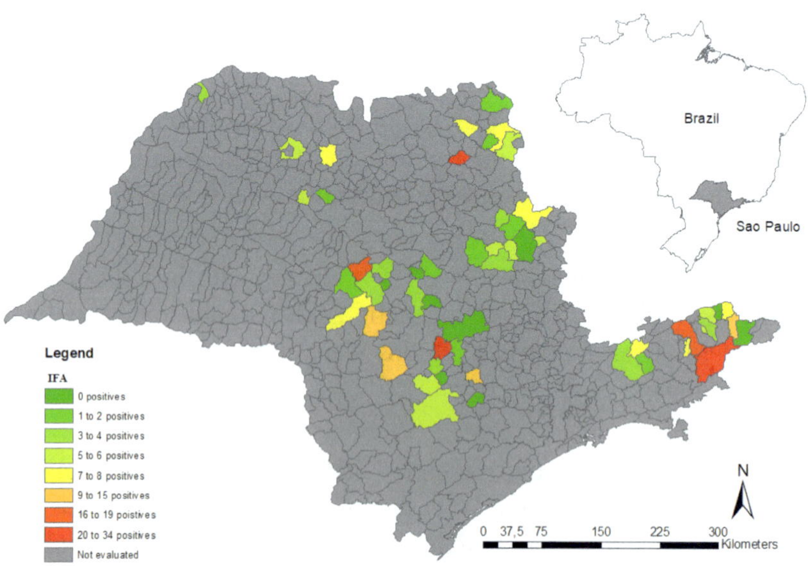

Figure 2: Distribution of seropositive *C. burnetii* **animals based on the immunofluorescence assay (IFA)**.

https://doi.org/10.1371/journal.pone.0241246.g002

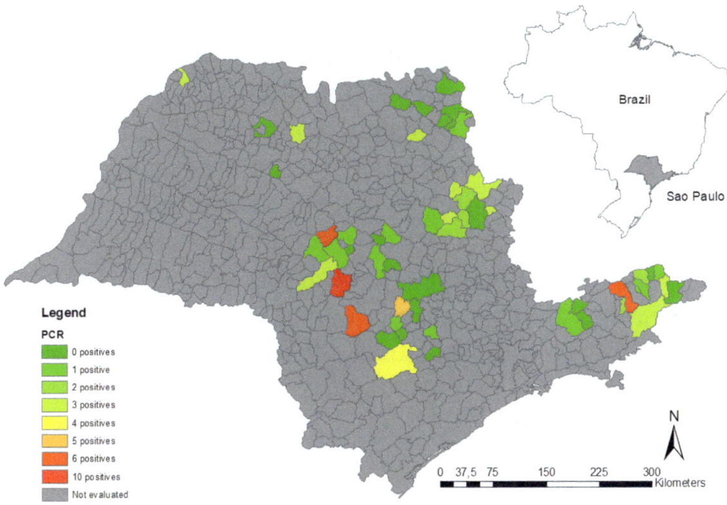

Figure 3: Detection of *Coxiella burnetii* **DNA in cattle from different municipalities.**

https://doi.org/10.1371/journal.pone.0241246.g003

DISCUSSION

During the study, we observed *C. burnetii* positivity in beef cattle sent to slaughter using serological and molecular methods. Cattle exposure to the bacterium could be verified through IFA in 360 of the 1515 samples (23.8%), which indicates a past or current infection. The presence of the pathogen in the bloodstream of cattles at the time of slaughter was confirmed using qPCR in 12.2% (44/360) of the seropositive animals, and this finding is suggestive of bacteremia.

In several countries, the use of serological diagnoses in slaughterhouses is the primary method of zoonotic disease surveillance. The efficiency of this approach relies on a traceability system, which is useful for detecting and monitoring pathogens in regions with relatively high prevalence [55]. Sampling in abattoirs has some advantages when the prevalence of a disease is unknown, such as the sampling of a large number of animals from different cities, regions, and producers in a short time. However, some pathogens have specific infection characteristics, such as a preference for sex, age, or type of production (extensive versus intensive) [56]. Therefore, the results obtained from samples taken at slaughterhouses can be adjusted to estimate the prevalence in the living population, if necessary; for example, when differences in prevalence are expected in different age or sex cohorts. The slaughter animal population is mainly composed of healthy males. Thus, our results estimate *C. burnetii* infections in beef cattle, rather than the prevalence of coxiellosis disease, since most animals are asymptomatic [1, 57, 58]. A recent study conducted in small ruminants in slaughterhouses suggested a lack of correlation between sex and Q fever positivity in serodiagnoses, and similar prevalence rates were observed among males and females [59]. However, the evaluation of females from herds that experience abortions due to Q fever can lead to higher seroprevalence rates, as observed in northeast Brazil, where 55.1% (172/312) of dairy goats were positive for *C. burnetii* [15].

The median prevalence reported worldwide for cattle is 37.7% [10]. Although lower than the reported average, the true prevalence of 20.0% that was found in the IFA serological analysis is considered high. Our results are similar to those of a previous study of slaughterhouses in Brazil, where the prevalence assessed using the complement fixation test was 29% [14]. The serology diagnosis outcome can influence the antigen used [60]. Notably, studies that compared antigens produced from the Nine Mile strain, a bovine strain, and a ruminant strain reported better efficiency in sensibility and specificity bovine and ruminant strains [11, 61]. Additionally, phase II LPS is an immunodominant antigen that is inserted in phase I LPS; therefore, antibodies against phase II are always present during the infection and have a higher prevalence in epidemiological studies [18, 62]. In our study, we used an in-house antigen produced from a *C. burnetii* strain that was isolated from ticks in Argentina. Since it was a low-passage Vero cell culture strain, it is possible that it is mainly composed of phase I cells. Therefore, the estimated prevalence can be even higher because antibodies against phase II might be underestimated. Furthermore, the use of a tick strain from Argentina can also impact the study outcome. Serological evidence of *C. burnetii* can reflect a recent or past exposure to the pathogen; therefore, it cannot infer an actual hazard for slaughterhouse workers. However, qPCR detection in 12.2% (44/360) of the seropositive samples revealed presence of the pathogen at the moment of slaughter.

In humans, the diagnosis of acute Q fever using qPCR in sera is indispensable for diagnosing bacteremia. The use of molecular detection in sera was previously deployed to identify the source of a human infection in Japan, where two patients, two dogs, and three cats were positive by nested

PCR. Moreover, the investigators were able to isolate *C. burnetii* from the serum samples, which demonstrates the presence of viable bacteria in the bloodstream of the animals [21]. Recently, *C. burnetiid* was isolated from the serum of a patient with acute Q fever in Australia using cell culture [63], and this finding corroborates the presence of viable bacteria in the sera and blood of infected hosts. Our results suggest that bacteremia is present in a significant proportion of tested animals and most likely represents active infections. Due to the lack of data that correlates bacteria detection in sera with clinical disease or shedding in animals, we can only wonder if bacteremia is associated with primary infection exclusively or with chronic phases that result in an active infection.

The molecular detection of *C. burnetii* in the bloodstream of cattle was confirmed by phylogenetic analysis of *IS1111a* sequences. All seven sequences that were investigated in our study were clustered together, along the 16 *IS1111a* of *C. burnetii* strain sequences that were used in the comparison. Moreover, the *IS1111*-like CLB sequence appeared in a different branch, which demonstrates the specificity of the Trans 3 and Trans 4 primers that were used in this study. *Coxiella*-like bacteria have a symbiont relationship with ticks [64] and are widespread in these arthropods, with a 100% prevalence in some species [8]. This can lead to the transmission of CLB to animals, as reported in birds [65, 66] and horses [67]. Additionally, the existence of an *IS1111*-like CLB with 90% identity to *IS1111* has been found in *C. burnetii* [50] and can impact Q fever diagnosis if these species are misidentified.

The lack of discrimination between *C. burnetii* strains using the *IS1111a* marker is probably due to the small amplicon size (243 bp) and its location in a conserved region of the gene. Notably, studies that used Trans 1 and Trans 2 primers, which produce a 600 bp amplicon, identified differences between *C. burnetii* strains derived from different hosts and countries [68, 69]. Nevertheless, recent multi spacer typing (MST) and multi-locus variable tandem repeat analysis (MLVA) investigations in Brazil have identified evidence of unique genotypes [17], and this emphasizes the need of epidemiological and virulence studies that aim to better understand the disease in this region. MST-74, detected in Brazil, belongs to the same phylogenetic branch as MST 20 and MST 61 [16], which places the strains circulating in Brazil in genomic group III of Hendrix et al. [70–72]. In a recent virulence study, it was shown that strains belonging to genomic groups I and III were more virulent, with increased pathogenic potential compared to *C. burnetii* strains from other genomic groups [73].

The main route of Q fever transmission is through the inhalation of infectious particles. Quantification of *C. burnetii* in the bloodstream through qPCR showed a concentration that ranged from 10 to 814 bacteria / mL (average of 144 cells / mL). As *C. burnetii* is an intracellular pathogen [3], there are more bacteria per milliliter in blood than in the sera of slaughtered cattle. During bleeding, cattle eliminate approximately 4.5 liters of blood, which can result in the dissemination of approximately $6x10^4$ bacteria, at the abovementioned concentration. Thus, the results presented here evidentiate the risk of infection for slaughterhouse workers due to aerosol formation during slaughter procedures, such as bleeding and sawing of half-carcasses, among others [74]. These facts emphasize the risk of acquiring Q fever due to the contact with blood from infected animals, a possibility that increases inside abattoirs due to the formation of aerosols from slaughter procedures and the main infective respiratory route of *C. burnetii* in humans [74, 75].

The lack of human Q fever surveillance in Brazil, in addition to the high animal prevalence reported here and by other authors in Brazil [13–15], suggests that Q fever must be a neglected disease due to misdiagnosis in Brazil. Active infection also represents a threat for people who live in areas that surround livestock once asymptomatic and symptomatic animals shed bacteria into the environment, leading to possible *C. burnetii* dissemination to urban areas, through the wind over long distances [5, 6]. Additionally, the possibility of transmission through consumption of cattle milk and dairy products should not be neglected, as the presence of *C. burnetii* has been evaluated in such products in Brazil [16, 76].

CONCLUSION

The high prevalence of *C. burnetii* infection in cattle sent to slaughter and the wide dispersion observed in the density maps indicates that the pathogen is widely prevalent in herds of the State of São Paulo and is not restricted to a single region. These results indicate a risk for public health, since *C. burnetii* is an airborne pathogen that can be spread by wind over long distances and cause Q fever outbreaks in cities close to livestock production. In addition, the occupational hazard for farmers, veterinarians, and especially, abattoir workers, has to be highlighted due to the potential accumulation of *C. burnetii* contamination in their environments. Surveillance programs for humans and animals should be implemented in Brazil to monitor cases of Q fever and establish preventive and control measures.

ACKNOWLEDGMENTS

For the Ministry of Agriculture and Livestock Supply of Brazil, the Secretary of Agriculture of the State of São Paulo, and all the prefectures that collaborated for this work.

REFERENCES

1. OIE. Q Fever. Manual of Diagnostic Tests and Vaccines for Terrestrial Animals. Paris: World Organisation for Animal Health; 2018. pp. 1–18. http://www. oie. int/standard-setting/terrestrial-manual/access-online/
2. Anderson A, Bijlmer H, Fournier P-E, Graves S, Hartzell J, Kersh GJ, et al. Diagnosis and management of Q fever—United States, 2013: recommendations from CDC and the Q Fever Working Group. MMWR Recomm reports Morb Mortal Wkly report Recomm reports. 2013;62: 1–30. Available: http://www. ncbi. nlm. nih. gov/pubmed/23535757
3. Marrie TJ, Minnick MF, Textoris J, Capo C, Mege J-L. Coxiella. In: Tang Y-W, Sussman M, Liu D, Poxton I, Schwartzman J, editors. Molecular Medical Microbiology. Second. Elsevier; 2015. pp. 1941–1972.
4. Raoult D. Rickettsial Infections. In: Goldman L, Schafer AI, editors. Goldman's Cecil Medicine. 24th ed. Elsevier; 2012. pp. 1954–1964.
5. Tissot-Dupont H, Amadei M, Nezri M, Raoult D. Wind in November, Q Fever in December. Emerg Infect Dis. 2004;10: 1264–1269. pmid:15324547
6. Hawker JI, Ayres JG, Blair I, Evans MR, Smith DL, Smith EG, et al. A large outbreak of Q fever in the West Midlands: windborne spread into a metropolitan area? Commun Dis public Heal. 1998;1: 180–7. Available: https://www. researchgate. net/profile/Jon_Ayres/publication/13502962_A_large_outbreak_of_Q_fever_in_the_West_Midlands_windborne_spread_into_a_metropolitan_area/links/09e4150c05597897c9000000/A-large-outbreak-of-Q-fever-in-the-West-Midlands-windborne-spread
7. Anderson A, Bijlmer H, Fournier P-E, Graves S, Hartzell J, Kersh GJ, et al. Diagnosis and management of Q fever—United States, 2013: recommendations from CDC and the Q Fever Working Group. MMWR Recomm reports Morb Mortal Wkly report Recomm reports. 2013;62: 1–30. Available: http://www. ncbi. nlm. nih. gov/pubmed/23535757
8. Duron O, Sidi-Boumedine K, Rousset E, Moutailler S, Jourdain E. The Importance of Ticks in Q Fever Transmission: What Has (and Has Not) Been Demonstrated? Trends Parasitol. Elsevier Ltd; 2015;31: 536–552. pmid:26458781
9. Toman R, Heinzen RA, Samuel JE, Mege J-L. Coxiella burnetii: Recent Advances and New Perspectives in Research of the Q Fever Bacterium

[Internet]. 1st ed. Toman R, Heinzen RA, Samuel JE, Mege J-L, editors. Advances in Experimental Medicine and Biology. Dordrecht: Springer Netherlands; 2012.

10. Guatteo R, Seegers H, Taurel A-F, Joly A, Beaudeau F. Prevalence of Coxiella burnetii infection in domestic ruminants: A critical review. Vet Microbiol. 2011;149: 1–16. pmid:21115308
11. EFSA. European Food Safety Authority panel on animal health and welfare: 2010, Scientific opinion on Q fever. EFSA J. 2010;8: 1–114.
12. Gache K, Rousset E, Perrin JB, De Cremoux R, Hosteing S, Jourdain E, et al. Estimation of the frequency of Q fever in sheep, goat and cattle herds in France: results of a 3-year study of the seroprevalence of Q fever and excretion level of Coxiella burnetii in abortive episodes. Epidemiol Infect. 2017;145: 3131–3142. pmid:29039279
13. Valle LAR, Brandão H, Cristovão D de A, D'Apice M. Investigações sobre a Febre Q em São Paulo. 2. Estudos em tratadores de gado e em bovinos. Arq Fac Hig Saude Publica Univ Sao Paulo. 1955;9: 167–180.
14. Riemann HP, Brant PC, Behymer DE, Franti CE. Toxoplasma gondii and Coxiella burnetii antibodies among Brazilian slaughterhouse employees. Am J Epidemiol. 1975;102: 386–393. Available: http:// www. ncbi. nlm. nih. gov/entrez/query. fcgi?cmd=Retrieve&db=Pub Med&dopt=Citation&list_uids=1200023 pmid:1200023
15. de Oliveira JMB, Rozental T, de Lemos ERS, Forneas D, Ortega-Mora LM, Porto WJN, et al. Coxiella burnetii in dairy goats with a history of reproductive disorders in Brazil. Acta Trop. Elsevier; 2018;183: 19–22. pmid:29621535
16. Mioni M, Ribeiro BLD, Peres MG, Teixeira WSR, Pelícia VC, Motta RG, et al. Real-time quantitative PCR-based detection of Coxiella burnetii in unpasteurized cow's milk sold for human consumption. Zoonoses Public Health. 2019; zph. 12609. pmid:31173477
17. Mioni M de SR, Sidi-Boumedine K, Dalanezi FM, Joaquim SF, Denadai R, Teixeira WSR, et al. New Genotypes of Coxiella burnetii Circulating in Brazil and Argentina. Pathogens. 2020;9: 30. pmid:31905637
18. Rousset É, Cremoux R de, Bronner A, Jourdain E, Touratier A, Sidi-Boumedine K. La fièvre Q. Bull des GTV—Hors Série "Zoonose Mal bactériennes. " 2012; 53–67. https://www. sngtv. org/4DACTION/NS2013_INDEX/5/V1131#

19. Schneeberger PM, Hermans MHA, van Hannen EJ, Schellekens JJA, Leenders ACAP, Wever PC. Real-time PCR with serum samples is indispensable for early diagnosis of acute Q fever. Clin Vaccine Immunol. 2010;17: 286–90. pmid:20032219
20. Wielders CCH, Wijnbergen PCA, Renders NHM, Schellekens JJA, Schneeberger PM, Wever PC, et al. High Coxiella burnetii DNA load in serum during acute Q fever is associated with progression to a serologic profile indicative of chronic Q fever. J Clin Microbiol. 2013;51: 3192–8. pmid:23863573
21. Komiya T, Sadamasu K, Toriniwa H, Kato K, Arashima Y, Fukushi H, et al. Epidemiological survey on the route of Coxiella burnetii infection in an animal hospital. J Infect Chemother. Elsevier; 2003;9: 151–5. pmid:12825114
22. Kirkan Ş, Kaya O, Tekbiyik S, Parin U. Detection of Coxiella burnetii in cattle by PCR. Turkish J Vet Anim Sci. 2008;32: 215–220.
23. To H, Htwe KK, Kako N, Kim HJ, Yamaguchi T, Fukushi H, et al. Prevalence of Coxiella burnetii infection in dairy cattle with reproductive disorders. J Vet Med Sci. 1998;60: 859–61. pmid:9713817
24. Emery MP, Ostlund EN, Ait Ichou M, Ballin JD, McFarling D, McGonigle L. Coxiella burnetii serology assays in goat abortion storm. J Vet Diagn Invest. 2014;26: 141–5. pmid:24532695
25. Roest H-J, van Gelderen B, Dinkla A, Frangoulidis D, van Zijderveld F, Rebel J, et al. Q Fever in Pregnant Goats: Pathogenesis and Excretion of Coxiella burnetii. R RM II, editor. PLoS One. 2012;7: e48949. pmid:23152826
26. Maurin M, Raoult D. Q Fever. Clin Microbiol Rev. 1999;12: 518–553. pmid:10515901
27. Spelman DW. Q fever: A study of 111 consecutive cases. Med J Aust. 1982;1: 547–553. pmid:7110006
28. Derrick EH. The course of infection with Coxiella burnetii. Med J Aust. 1973;1: 1051–1057. pmid:4736969
29. Armengaud A, Kessalis N, Desenclos JC, Maillot E, Brousse P, Brouqui P, et al. Urban outbreak of Q fever, Briançon, France, March to June 1996. Eurosurveillance. European Centre for Disease Prevention and Control (ECDC); 1997;2: pii = 137. http://www. eurosurveillance. org/ ViewArticle. aspx?ArticleId=137

30. CIEVS. Informe Epidemiológico Cievs—Paraná. Semana Epidemiológica. Curitiba: Centro de Informações e Respostas Estratégicas de Vigilância em Saúde—CIEVS; 2015. www. saude. pr. gov. br
31. Siciliano RF, Strabelli TM, Zeigler R, Rodrigues C, Castelli JB, Grinberg M, et al. Infective endocarditis due to Bartonella spp. and Coxiella burnetii: experience at a cardiology hospital in Sao Paulo, Brazil. Ann N Y Acad Sci. 2006;1078: 215–22. pmid:17114712
32. Siciliano RF, Ribeiro HB, Furtado RHDM, Castelli JB, Sampaio RO, Dos Santos FCP, et al. Endocardite por Coxiella burnetii (febre Q). Doença rara ou pouco diagnosticada? Relato de caso. Rev Soc Bras Med Trop. 2008;41: 409–412. pmid:18853017
33. Siciliano RF, Castelli JB, Mansur AJ, Pereira dos Santos F, Colombo S, do Nascimento EM, et al. Bartonella spp. and Coxiella burnetii Associated with Community-Acquired, Culture-Negative Endocarditis, Brazil. Emerg Infect Dis. 2015;21: 1429–1432. pmid:26197233
34. da Costa PSG, Brigatte ME, Greco DB. Questing one Brazilian query: reporting 16 cases of Q fever from Minas Gerais, Brazil. Rev do Inst Med Trop São Paulo. 2006;48: 5–9. Available: http://www. ncbi. nlm. nih. gov/pubmed/16547572
35. Mares-Guia MAMM, Rozental T, Guterres A, Ferreira MDS, Botticini RDG, Terra AKC, et al. Molecular Identification of Q Fever in Patients with a Suspected Diagnosis of Dengue in Brazil in 2013–2014. Am J Trop Med Hyg. 2016;94: 1090–4. pmid:26928831
36. Lemos ERS, Rozental T, Mares-Guia MAM, Almeida DNP, Moreira N, Silva RG, et al. Q fever as a cause of fever of unknown origin and thrombocytosis: first molecular evidence of Coxiella burnetii in Brazil. Vector Borne Zoonotic Dis. 2011;11: 85–87. pmid:20569012
37. Lemos ERS, Rozental T, Siqueira BN, Júnior AAP, Joaquim TE, da Silva RG, et al. Q Fever in Military Firefighters during Cadet Training in Brazil. Am J Trop Med Hyg. 2018;99: 303–305. pmid:29943714
38. von Ranke FM, Clemente Pessoa FM, Afonso FB, Gomes JB, Borghi DP, Alves de Melo AS, et al. Acute Q fever pneumonia: high-resolution computed tomographic findings in six patients. Br J Radiol. 2019;92: 20180292. pmid:30608178
39. IBGE. IBGE | Portal do IBGE [Internet]. [cited 26 Feb 2019]. https:// www. ibge. gov. br/

40. Dean AG, Sullivan KM, Soe MM. OpenEpi: Open Source Epidemiologic Statistics for Public Health. In: Versão. www. OpenEpi. com [Internet]. 2013 [cited 23 Aug 2016]. http://openepi. com/Menu/OE_Menu. htm
41. Bueno CRF. Produção Animal: Estado de São Paulo e Brasil, 2005 a 2014. Análises e indicadores do Agronegócio. 2016;11: 1–6. Available: http:// www. iea. sp. gov. br/out/LerTexto. php?codTexto=14035
42. Horta MC, Labruna MB, Sangioni LA, Vianna MCB, Gennari SM, Galvão MAM, et al. Prevalence of antibodies to spotted fever group rickettsiae in humans and domestic animals in a Brazilian spotted fever-endemic area in the state of São Paulo, Brazil: serologic evidence for infection by rickettsia rickettsii and another spotted fever gro. Am J Trop Med Hyg. 2004;71: 93–97. pmid:15238696
43. Pacheco RC, Echaide IE, Alves RN, Beletti ME, Nava S, Labruna MB. Coxiella burnetii in ticks, Argentina. Emerg Infect Dis. 2013;19: 344–6. pmid:23343647
44. Vaidya VM, Malik SVS, Bhilegaonkar KN, Rathore RS, Kaur S, Barbuddhe SB. Prevalence of Q fever in domestic animals with reproductive disorders. Comp Immunol Microbiol Infect Dis. Elsevier Ltd; 2010;33: 307–321. pmid:19101035
45. Hoover TA, Vodkin MH, Williams JCA. A *Coxiella burnetii* repeated DNA element resembling a bacterial insertion sequence. J Bacteriol. 1992;174: 5540–8. pmid:1324903
46. Seshadri R, Paulsen IT, Eisen JA, Read TD, Nelson KE, Nelson WC, et al. Complete genome sequence of the Q-fever pathogen Coxiella burnetii. Proc Natl Acad Sci. 2003;100: 5455–5460. pmid:12704232
47. Dereeper A, Guignon V, Blanc G, Audic S, Buffet S, Chevenet F, et al. Phylogeny. fr: robust phylogenetic analysis for the non-specialist. Nucleic Acids Res. 2008;36: W465–W469. pmid:18424797
48. Chevenet F, Brun C, Bañuls A-L, Jacq B, Christen R. TreeDyn: towards dynamic graphics and annotations for analyses of trees. BMC Bioinformatics. 2006;7: 439. pmid:17032440
49. Guindon S, Dufayard J-F, Lefort V, Anisimova M, Hordijk W, Gascuel O. New Algorithms and Methods to Estimate Maximum-Likelihood Phylogenies: Assessing the Performance of PhyML 3. 0. Syst Biol. 2010;59: 307–321. pmid:20525638
50. Vilcins I-ME, Old JM, Deane E. Molecular detection of Rickettsia, Coxiella, and Rickettsiella DNA in three native Australian tick species. Exp Appl Acarol. 2009;49: 229–242. pmid:19296229

51. ESRI. ArcGIS Desktop: Release 10. Redlands, CA: Environmental Systems Research Institute. Redlands CA. 2011.
52. Slabá K, Skultéty L, Toman R. Efficiency of various serological techniques for diagnosing Coxiella burnetii infection. Acta Virol. 2005;49: 123–7. pmid:16047740
53. Stevenson M, Nunes T, Heuer C, Marshall J, Sanchez J, Thornton R, et al. epiR: Tools for the Analysis of Epidemiological Data. R package version 0. 9–96. [Internet]. 2018. https://cran. r-project. org/web/ packages/epiR/index. html
54. R Core Team. R: The R Project for Statistical Computing. In: R Foundation for Statistical Computing, Vienna, Austria. [Internet]. 2017. https://www. r-project. org/
55. Nielsen K, Duncan JR. Animal Brucellosis [Internet]. Second. Nielsen K, Duncan JR, editors. Florida: CRC Press; 1990. https://books. google. fr/books?id=QQFDHBpNnTQC&printsec=frontcover&source=gbs_ge_summary_r&cad=0#v=onepage&q&f=false
56. May RM, Anderson RM. Population biology of infectious diseases: Part II. Nature. 1979;280: 455–61. pmid:460424
57. Bildfell RJ, Thomson GW, Haines DM, McEwen BJ, Smart N. Coxiella Burnetii Infection is Associated with Placentitis in Cases of Bovine Abortion. J Vet Diagnostic Investig. 2000;12: 419–425. pmid:11021428
58. Agerholm JS. Coxiella burnetii associated reproductive disorders in domestic animals-a critical review. Acta Vet Scand. 2013;55: 13. pmid:23419216
59. Folitse RD, Opoku-Agyemang T, Amemor E, Opoku ED, Bentum KE, Emikpe BO. Serological evidence of Coxiella burnetii infection in slaughtered sheep and goats at Kumasi Abattoir, Ghana. J Immunoass Immunochem. Taylor & Francis; 2020;41: 152–157. pmid:32028862
60. Cooper A, Hedlefs R, McGowan M, Ketheesan N, Govan B. Serological evidence of Coxiella burnetii infection in beef cattle in Queensland. Aust Vet J. 2011;89: 260–264. pmid:21696375
61. Horigan MW, Bell MM, Pollard TR, Sayers AR, Pritchard GC. Q fever diagnosis in domestic ruminants: comparison between complement fixation and commercial enzyme-linked immunosorbent assays. J Vet Diagn Invest. 2011;23: 924–31. pmid:21908348
62. O'Neill TJ, Sargeant JM, Poljak Z. A Systematic Review and Meta-Analysis of Phase I Inactivated Vaccines to Reduce Shedding of

Coxiella burnetii From Sheep and Goats From Routes of Public Health Importance. Zoonoses Public Health. 2013;61: n/a-n/a. pmid:24251777

63. Vincent GA, Graves SR, Robson JM, Nguyen C, Hussain-Yusuf H, Islam A, et al. Isolation of Coxiella burnetii from serum of patients with acute Q fever. J Microbiol Methods. Elsevier B. V. ; 2015;119: 74–8. pmid:26462766
64. Duron O, Noël V, McCoy KD, Bonazzi M, Sidi-Boumedine K, Morel O, et al. The Recent Evolution of a Maternally-Inherited Endosymbiont of Ticks Led to the Emergence of the Q Fever Pathogen, Coxiella burnetii. Rasgon JL, editor. PLOS Pathog. 2015;11: e1004892. pmid:25978383
65. Vapniarsky N, Barr BC, Murphy B. Systemic Coxiella-like Infection With Myocarditis and Hepatitis in an Eclectus Parrot (Eclectus roratus). Vet Pathol. 2012; pmid:21712515
66. Shivaprasad HL, Cadenas MB, Diab SS, Nordhausen R, Bradway D, Crespo R, et al. Coxiella-like infection in psittacines and a toucan. Avian Dis. 2008; pmid:18939630
67. Seo M-G, Lee S-H, VanBik D, Ouh I-O, Yun S-H, Choi E, et al. Detection and Genotyping of Coxiella burnetii and Coxiella-Like Bacteria in Horses in South Korea. Samuel JE, editor. PLoS One. 2016;11: e0156710. pmid:27244230
68. Ghoneim NH, Abdel-Moein KA, Zaher HM, Abuowarda MM. Investigation of Ixodidae ticks infesting camels at slaughterhouse and its potential role in transmitting Coxiella burnetii in Egypt. Small Rumin Res. 2020;
69. Shabbir MZ, Akram S, Hassan Z ul, Hanif K, Rabbani M, Muhammad J, et al. Evidence of Coxiella burnetii in Punjab province, Pakistan. Acta Trop. 2016;163: 61–69. pmid:27456937
70. Hendrix L, Samuel J, Mallavia L. Differentiation of Coxiella burnetii isolates by analysis of restriction-endonuclease-digested DNA separated by SDS-PAGE. J Gen Microbiol. 1990;137: 269–276. pmid:1673152
71. Hornstra HM, Priestley RA, Georgia SM, Kachur S, Birdsell DN, Hilsabeck R, et al. Rapid typing of Coxiella burnetii. PLoS One. 2011;6. pmid:22073151
72. Glazunova O, Roux V, Freylikman O, Sekeyova Z, Fournous G, Tyczka J, et al. Coxiella burnetii genotyping. Emerg Infect Dis. 2005;11: 1211–7. pmid:16102309

73. Long CM, Beare PA, Cockrell DC, Larson CL, Heinzen RA. Comparative virulence of diverse Coxiella burnetii strains. Virulence. 2019; pmid:30782062

74. Kaufmann AF, Fox MD, Boyce JM, Anderson DC, Potter ME, Martone WJ, et al. Airborne spread of brucellosis. Ann N Y Acad Sci. 1980;353: 105–114. pmid:6939379

75. Jones RM, Nicas M, Hubbard AE, Reingold AL. The Infectious Dose of Coxiella Burnetii (Q Fever). Appl Biosaf. 2006;11: 32–41.

76. Rozental T, de Faria LS, Silva MR, Ribeiro JB, Araújo FR, da Costa RR, et al. Ocorrência de Coxiella burnetii em queijo Minas artesanal de leite cru: resultados preliminares de um preocupante problema de saúde pública. Rev Med Minas Gerais. 2018;28: 85–91. Available: https://linkinghub. elsevier. com/retrieve/pii/S0001706X19304140

Section 3: Livestock

Pigs

CHAPTER 19

Reassortant between Human-Like H3N2 and Avian H5 Subtype Influenza A Viruses in Pigs: A Potential Public Health Risk

Yanlong Cong[1], Guangmei Wang[1], Zhenhong Guan[2], Shuang Chang[1], Quanpeng Zhang[1], Guilian Yang[3], Weili Wang[4], Qingfeng Meng[4], Weiming Ren[3], Chunfeng Wang[3], Zhuang Ding[1]

[1]College of Animal Science and Veterinary Medicine, Jilin University, Changchun, Jilin, China

[2] Institute of Zoonoses, Jilin University, Changchun, Jilin, China

[3] College of Animal Science and Technology, Jilin Agricultural University, Changchun, Jilin, China

[4] Jilin Entry-Exit Inspection and Quarantine Bureau, Changchun, Jilin, China

Citation: Cong Y, Wang G, Guan Z, Chang S, Zhang Q, Yang G, et al. (2010) Reassortant between Human-Like H3N2 and Avian H5 Subtype Influenza A Viruses in Pigs: A Potential Public Health Risk. PLoS ONE 5(9): e12591. https://doi.org/10.1371/journal.pone.0012591.

ABSTRACT

Background

Human-like H3N2 influenza viruses have repeatedly been transmitted to domestic pigs in different regions of the world, but it is still uncertain whether any of these variants could become established in pig populations. The fact that different subtypes of influenza viruses have been detected in pigs makes them an ideal candidate for the genesis of a possible reassortant virus with both human and avian origins. However, the determination of whether pigs can act as a "mixing vessel" for a possible future pandemic virus is still pending an answer. This prompted us to gather the epidemiological information and investigate the genetic evolution of swine influenza viruses in Jilin, China.

Methods

Nasopharyngeal swabs were collected from pigs with respiratory illness in Jilin province, China from July 2007 to October 2008. All samples were screened for influenza A viruses. Three H3N2 swine influenza virus isolates were analyzed genetically and phylogenetically.

Results

Influenza surveillance of pigs in Jilin province, China revealed that H3N2 influenza viruses were regularly detected from domestic pigs during 2007 to 2008. Phylogenetic analysis revealed that two distinguishable groups of H3N2 influenza viruses were present in pigs: the wholly contemporary human-like H3N2 viruses (represented by the Moscow/10/99-like sublineage) and double-reassortant viruses containing genes from contemporary human H3N2 viruses and avian H5 viruses, both co-circulating in pig populations.

Conclusions

The present study reports for the first time the coexistence of wholly human-like H3N2 viruses and double-reassortant viruses that have emerged in pigs in Jilin, China. It provides updated information on the role of pigs in interspecies transmission and genetic reassortment of influenza viruses.

INTRODUCTION

The pig is considered to be an important host of influenza A viruses as it might be associated with the generation of human pandemic influenza strains [1]. Historically, two human pandemic viruses, H1N1 in 1918 and H3N2 in 1968, were almost simultaneously detected both in humans and pigs [2], [3]. Although the pandemic H3N2 virus was initially detected in humans and the full genome of the 1918 H1N1 pandemic virus was recently decoded [4], it is still unknown whether these viruses were first introduced into humans or into pigs before they became human pandemic strains.

Currently, H1N1, H1N2 and H3N2 influenza subtype viruses co-circulate in pigs widely throughout the world. All of these viruses were the result of either interspecies transmission or reassortment events [5]–[7]. The Sydney-like H3N2 variants from pigs in the U.S.A. in 1998 were double and triple reassortants containing viral genes of human, swine and avian origin [8]. This highlights the complex and dynamic influenza ecology in pig populations.

Here we present the results of the genetic and phylogenetic characterization of swine H3N2 influenza viruses isolated from 2007 to 2008 in Jilin province of China. Genetic analysis showed that wholly contemporary human-like H3N2 viruses and double-reassortant viruses containing genes from contemporary human (PB2, PB1, PA, HA, NP, and NA) and avian H5 (M and NS) viruses were co-circulating in pig populations. This is the first description of an instance of reassortment between mammal H3N2 and avian H5 influenza viruses. The coexistence of entirely contemporary human-like viruses and double-reassortant viruses provides further evidence that pigs serve as intermediate hosts or mixing vessels. This emphasizes the importance of reinforcing swine influenza virus surveillance.

RESULTS

HA1 Amino Acid Analysis

To investigate all of the detailed genetic characteristics, we compared the deduced amino acid sequences of the hemagglutinin 1 (HA1) gene from the three swine H3N2 isolates against the representatives of five lineages (avian, European swine, earliest human, early human and contemporary human) available in GenBank. For the five lineages, the three H3N2 isolates showed a close relationship to HA1 genes from the contemporary human lineage with a sequence similarity of 94.8~97.6% to Moscow/10/99, while there was 80.9~87.2% similarity compared with the representatives of four other lineages (Dk/Hong Kong/7/75 (ABB88256.1), Sw/Italy/1461/96

(CAC40048.1), Hong Kong/1/68 (ACU79871.1), Port Chalmers/1/73 (AAC78096.1), and Victoria/3/75 (CAA24270.1)).

Analysis of amino acid variations of the proposed antigenic sites [9]–[11], receptor-binding sites [12], and potential glycosylation sites was conducted and variations are shown in Figure 1. The HA1 domain of HA, the major antigenic protein of influenza A viruses, contains all the antigenic sites of HA and is under continual immune-driven selection. For H3N2 viruses, the antigenic sites A~E have been described [9]–[11]. All variations were accumulated at the antigenic sites A and B, while C, D and E were relatively conserved for the contemporary human-like H3N2 lineage. Concerning the three swine H3N2 isolates, two or four amino acid substitutions were observed at the major antigenic sites (A and B) of the HA1 molecule compared with Moscow/10/99, the representative for contemporary human lineage (Figure 1 and 2).

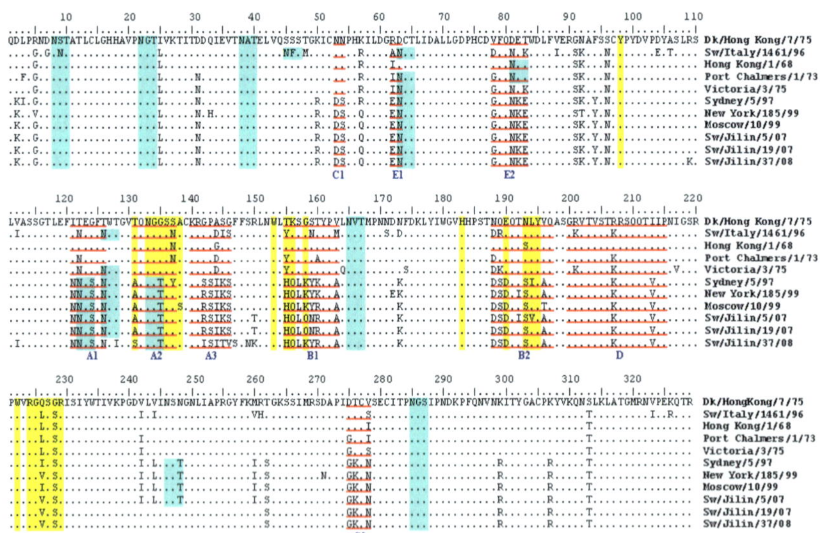

Figure 1: Alignment of the HA1 amino acid sequences of three swine H3N2 influenza virus isolates and the representatives of five H3N2 lineages.

The underlined residues represent the antigenic sites (lowercase letters indicate discrete antigenic sites), residues in green represent the potential glycosylation sites, and residues in yellow shade denote the receptor-binding sites.

https://doi.org/10.1371/journal.pone.0012591.g001

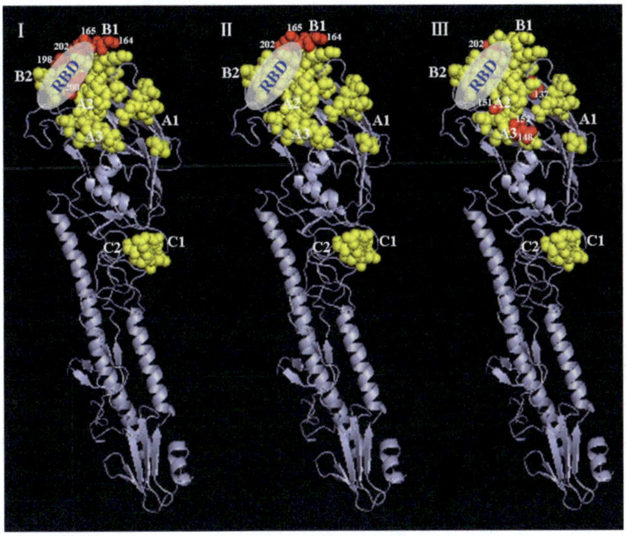

Figure 2: Cartoon diagram representing the amino acid changes at the HA1 molecule of H3 subtype influenza viruses.

RBD: receptor-binding domain. A, B and C: major antigenic sites at the HA molecules. Yellow color: conserved amino acids and red color: changed amino acids. (|) Sw/Jilin/5/07 vs. Moscow/10/99, (||) Sw/Jilin/19/07 vs. Moscow/10/99, (|||) Sw/Jilin/37/08 vs. Moscow/10/99.

https://doi.org/10.1371/journal.pone.0012591.g002

Amino acids at the receptor-binding sites of the HA1 protein are associated with the differences in the receptor-binding specificity [13]. All of the H3N2 viruses had relatively conserved receptor-binding sites at Y98, G134, S136, W153, H183, Y195, and R224. Eight obvious mutations (A/S131T, T135G, H155T/Y, Q156K, K/Q158G, D190E, S193N, I/V194L) occurred between the contemporary human lineage and other lineages; these mutations were unique to the contemporary human lineage. Residues responsible for the sialic acid-α2,6-galactose (SAα2,6Gal) of H3N2 are L226 and S228 [14]. The viruses of avian lineage had Q at position 226, while the viruses of European swine, earliest human, and early human lineages had L, and the viruses of contemporary human lineage had I/V. The viruses of avian lineage had G at position 228, with other four lineages being S. In this study, all of the swine H3N2 isolates contained V instead of L at position 226. It has been considered that carbohydrate side chains might affect receptor-binding capacity and antigenicity [11], [15], [16]. Analysis of potential glycosylation sites in the HAs of the three H3N2 isolates revealed eight common sites (N8, 22, 38, 63, 122, 133, 165 and 285, respectively) with the NXT/S motif (in which X may be any amino acid except aspartic acid and proline). Furthermore, as shown in Figure 1, both Sw/Jilin/5/07 and Sw/Jilin/19/07 possessed an additional glycosylation site at position 126, and Sw/Jilin/5/07 acquired another one at position 246.

Phylogenetic Analysis

Phylogenetic analysis and identification of antigenically different strains are necessary to monitor the evolution of influenza virus. The phylogenetic relationships among swine H3N2 viruses prevalent in Jilin province compared to the selected reference strains available in the GenBank were estimated from the nucleotide sequences of each viral gene. The phylograms for all genes are shown in Figures 3a~3h.

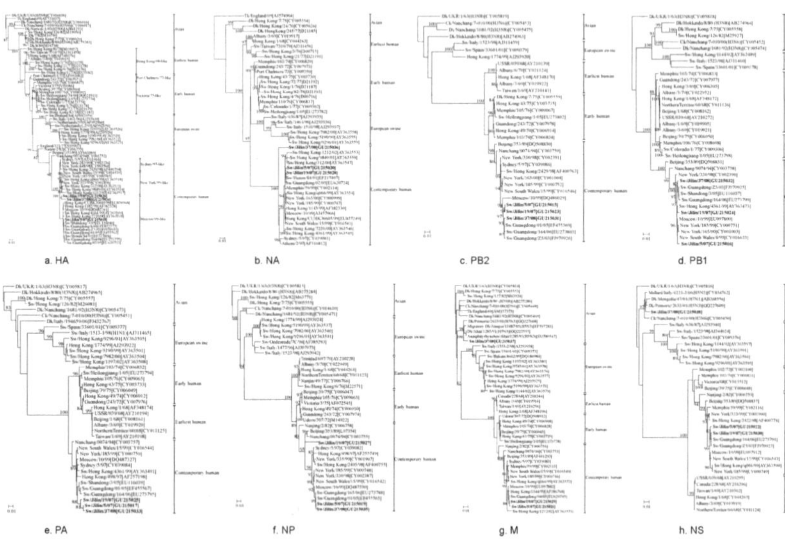

Figure 3: Phylogenetic trees of the eight gene segments of the influenza viruses.

The tree was created by the neighbor-joining method and bootstrapped with 1,000 replicates. The bootstrap numbers are given for each node. Only bootstrap values above 80 are shown. Viruses isolated in the present study are in bold. Abbreviations used in virus designations are as follows: Ck, chicken; Dk, duck; Sw, swine; Tk, turkey.

https://doi.org/10.1371/journal.pone.0012591.g003

Phylogenetic analysis of H3 HAs showed that H3N2 viruses could be segregated into five distinct lineages, including avian strains, European swine strains, earliest human strains, early human strains, and contemporary human strains (Figure 3a). The swine viruses of earliest human lineage seemed to be derived from the human strain A/Hong Kong/1/68. The early-human-derived swine viruses were closely related to A/Victoria/75-like viruses. The swine viruses of European swine lineage were probably derived from European H3N2 swine influenza viruses, for which the HA genes originated from A/Victoria/3/75. The contemporary human lineage included three sublineages (represented by Sydney/97-like, New York/99-like and Moscow/99-like viruses). The HAs of the three swine H3N2 virus isolates tested in this study clustered into the Moscow/99-like virus sublineage.

The phylogeny of the NA and internal genes of H3N2 viruses paralleled to that of the HA genes, in which five different lineages were defined. The NA, PB2, PB1, PA, and NP genes of the three isolates in this study belonged to the contemporary human lineage. Additionally, in the NA tree, Sw/Jilin/37/08 was equally closely related to the early and contemporary H3 strains as well as an intermediate swine virus between the early lineage and the contemporary human lineage. In the M and NS trees, Sw/Jilin/5/07 and Sw/Jilin/19/07 still clustered in the contemporary human lineage, whereas Sw/Jilin/37/08 was incorporated into the avian lineage (Figures 3b~3h).

The results of the published data in GenBank revealed that in recent years H3N2 influenza viruses were cocirculating with H5 influenza viruses, raising the possibility of genetic exchange between these viruses. The evolutionary trees revealed that the M gene of Sw/Jilin/37/08 formed a sublineage with H5N3 viruses (A/Dk/Altai/1285/1991, A/Dk/Primorie/2633/2001, A/Anasplatyrhynchos/Altai/1285/1991, and A/Migratory Dk/Jiangxi/13487/2005 (Figure 3g)). The NS gene of Sw/Jilin/37/08 had a close relationship with A/Mallard/Italy/4223-2/2006 (H5N2) and A/Dk/Primorie/2633/2001 (H5N3) (Figure 3h).

DISCUSSION

The rapid evolution of influenza viruses occurs both clonally and non-clonally through a variety of genetic mechanisms and selection pressure. Three mechanisms on the non-clonal evolution of influenza viruses have been proposed: (i) relatively frequent reassortment among gene segments of multiple host types, (ii) possibly but rarely non-homologous recombination, in which short regions of sequence are transferred among different segments, and (iii) controversial and extremely rare homologous recombination within segments [17]. Reassortment, therefore, plays an important role in the non-clonal evolution of the influenza viruses.

Infection of pigs with H1, H3, H5 and H9 subtype influenza viruses has occurred frequently on multiple occasions [5], [18]–[23]. As pigs can serve as intermediate hosts or mixing vessels for the reassortment of human and avian influenza viruses, swine influenza infections have been the focus of increasing attention. It often appeared that multiple reassortments occurred among the same subtype or different subtypes of human, swine and avian viruses in pig populations. The best examples are the swine H1N2 isolates and the pandemic novel H1N1 influenza virus that emerged in the human population. Peiris et al [24] were the first to confirm that cocirculation of contemporary human H3N2 viruses and avian H9N2 in pigs had occurred in southeastern China. Xu et al [25] reported that the avian H9N2 viruses caused pig disease and death in clinics, and deduced that it probably originated from a reassortant of avian H5 influenza virus and H9N2 viruses. All of these situations indicated that the cocirculation of H1, H3, H5 and H9 viruses in pigs would provide an opportunity for genetic reassortment, leading to the emergence of viruses with pandemic potential. Avian influenza viruses, especially when avian H5 subtype viruses reassorted with suitable surface glycoproteins, can infect humans [26]. Yu et al [27] described the coexistence of wholly human-like H3N2 viruses, double-reassortant and triple-reassortant H3N2 viruses in pigs in China from 1970 to 2006. However, reassortment was not found between mammal H3N2 viruses and avian H5 viruses. From July 2007 to October 2008, we collected 279 nasopharyngeal swabs from pigs diagnosed with respiratory tract illness on different pig farms in Jilin province. The H3N2 influenza virus infections in the pigs were diagnosed by virus isolation. The results showed that one Moscow/10/99-like reassortant isolate, A/swine/Jilin/37/2008, contained avian H5-like M and NS genes, suggesting that after introduction to pigs, avian H5 viruses further reassorted with contemporary human-like H3N2 viruses. To our knowledge, such reassortment between mammal H3N2 viruses and avian H5 viruses has not been described in existing publications.

A comparison of the amino acid sequences of the HA1 region of our isolates with the representatives of avian lineage (Dk/Hong Kong/7/75), European swine lineage (Sw/Italy/1461/96), earliest human lineage (Hong Kong/1/68), early human lineage (Victoria/3/75 and Port Chalmers/1/73), and contemporary human lineage (Sydney/5/97, New York/185/99, and

Moscow/10/99) showed that the three H3N2 swine isolates were more closely related to Moscow/10/99 (94.8~97.6% amino acid similarity), indicating that the HAs of them seemed to be derived from those of the contemporary human lineage. Phylogenetic analysis of the HAs also suggested that the three H3N2 isolates had been introduced into pigs at several points in time from the human side.

The comparison of antigenic sites of HA1 regions revealed evolution by antigenic drift of their HA genes. Codons under positive selection were associated with antigenic site A and B [28]. Sw/Jilin/19/07 had two variations at the antigenic sites A and B, while Sw/Jilin/5/07 and Sw/Jilin/37/08 accumulated four variations at the antigenic sites A and B. The gradual mutations of HA might generate new antigenic strains. Wilson and Cox [29] proposed that a drift variant with ≥4 amino acid changes at ≥2 out of 5 antigenic sites would be of epidemiologic importance. It has been also observed that new antigenic variations are created either when ≥2 variations occur in antigenic sites or when one variation occurs in an antigenic site and one in a sialic acid receptor-binding site [30].

Although the molecular basis of host-range restrictions is not completely defined, the compatibility between the HA protein of the virus and its corresponding receptor, sialic acid, on the host cell is thought to contribute in part to the infection of the virus in a specific host [31], [32]. Pigs, unlike humans, seem to be readily infected by most, if not all, mammalian and avian influenza viruses. The susceptibility of pigs to both mammalian and avian viruses is due to the presence of receptors for both lineages of virus in the pig trachea [33]. For H3N2 viruses, residues 226 and 228 on the receptor-binding domain of the HA1 molecule were shown to play a critical role in determining receptor specificity [14], [34], [35]. The three H3N2 isolates possess V226 and S228, which are the same as those of both turkey and swine triple reassortants. While L/I226 and S228 are usually expressed in human viruses [36], Q226 and G228 are usually found in avian viruses [14]. V, L and I are neutral non-polar amino acids, and substitutions between them most likely maintain the hydrophobic interactions and the proper 3D conformation at the binding domain [37].

Carbohydrate side chains are important for the structure and stability of glycoproteins [38], [39]. Changes in carbohydrate side chains are an important mechanism in the structural variation underlying antigenic drift. In this study, the three swine H3N2 isolates had eight to ten glycosylation sites, although all of them may not be used. Of these sites, positions 122, 133, and 246 were unique to the contemporary human lineage. In addition, the lineages of European swine, earliest human, early human, and contemporary human had more glycosylation sites than the avian lineage. The addition of new carbohydrate side chains to HA may have provided the viruses with the ability to evade antibody pressure by changing the antigenicity and by having an increased ability to prevail. For instance, human virus variants which cocirculated in the epidemic area with a higher number of glycosylation sites appeared to prevail at the end of the outbreak [40]. Conclusively, molecular analysis of the hemagglutinin gene showed that the 2007–2008 H3N2 influenza viruses circulating in swine populations in Jilin province accumulate variations at the antigenic sites A and B, receptor-binding sites, as well as glycosylation sites.

Phylogenies of the whole genome of the swine H3N2 influenza viruses in Jilin province, China during 2007 to 2008 provided evidence of the persistence of both the contemporary human lineage and the avian lineage in pig populations. These results also revealed multiple interspecies transmissions of influenza viruses from human and avian to pig and subsequent reassortment events, particularly together with the participation of internal genes from the avian H5 lineage in this region. This is the first report on reassortment between mammal H3N2 and avian H5 viruses.

Experimental infection suggested that introduction of certain H5 viral segments into circulating human H3N2 viruses may increase their virulence for mice and perhaps other mammalian species [41]. Further research is needed to determine whether Sw/Jilin/37/08 has higher virulence than Sw/Jilin/5/07 and Sw/Jilin/19/07. The principal evolutionary mechanism of influenza virus is by antigenic drift, creating small progressive antigenic changes in the hemagglutinin and neuraminidase surface antigens [42]. However, genetic reassortment readily occurs between influenza viruses and may also contribute to the evolution of new strains. Therefore Sw/Jilin/37/08 may represent a potential threat in the emergence of new human viruses.

There have been numerous descriptions of human infection with swine influenza viruses [5]. The swine populations have become a reservoir of a much more diverse array of influenza viruses. The replicative gene constellation of H3N2 viruses has the capacity to reassort among avian, swine and human viruses. This means that activity in swine virus reservoirs is of concern for human health. The above study provides additional evidence for continuing interspecies transmission and reassortment events occurring in pigs, which naturally increased the possibility of pigs as an important host for the emergence of novel reassortants with genes adapted for replication in pigs or even humans. The coexistence of reassortant viruses, especially reassortants of H3 and H5 viruses, emphasizes that genetic reassortment is an important factor in the evolution of H3N2 viruses and a formal surveillance system is needed for swine and avian influenza. It is only through such a system that cross-interspecies transmission and novel reassortment events will be identified in a timely fashion.

MATERIALS AND METHODS

Viruses

A total of 279 nasopharyngeal swabs were collected in Jilin province, China from July 2007 to October 2008. The initial isolation of the viruses was performed in Madin-Darby canine kidney (MDCK) cells. The viruses were grown in Eagle minimal essential medium (GIBCO/BRL) supplemented with 5% fetal bovine serum (GIBCO/BRL), penicillin-streptomycin (GIBCO/BRL), amphotericin B (Fungizone; GIBCO/BRL) and tolylsulfonyl phenylalanyl chloromethyl ketone-treated trypsin (1 μg/ml; Worthington Biochemical Corporation, Lakewood, N.J.). Subtype identification of these viruses were determined by standard hemagglutination inhibition tests and neuraminidase inhibition tests with a panel of reference antisera recommended by the World Health Organization (http://www.who.Int/csr/resources/publications/en/#influenza). Viral fluids were harvested for MDCK-passaged viruses and used as stock for sequence analysis. The three H3N2 virus isolates obtained in this study were named as follows: A/swine/Jilin/5/2007 (Sw/Jilin/5/07), A/swine/Jilin/19/2007 (Sw/Jilin/19/07), and A/swine/Jilin/37/2008 (Sw/Jilin/37/08).

Gene Sequencing and Phylogenetic Analysis

Viral RNA was extracted using Trizol reagents (GIBCO/BRL) and reverse transcription was performed using oligonucleotide influenza universal primer Uni12: 5′-AGC AAA AGC AGG-3′ [43]. After reverse transcription, PCR was done as described by Shu [44] using primers (Table 1) specific for each of the eight RNA segments. PCR products were purified with the QIA quick PCR purification kit (Qiagen). The purified PCR products were then partially sequenced using an Amersham ET Dye terminator kit and analyzed with an ABI 3730 DNA sequencer (Perkin-Elmer Appllied Biosystems, Foster City, CA, USA).

Table 1: The primer sequences used for PCR amplification of H3N2 influenza virus genes

https://doi.org/10.1371/journal.pone.0012591.t001

Gene	Forward primer 5′→3′	Reverse primer 5′→3′	Expected size (bp)
PB2	GCTGATAGTGAGTGGAAGAGACGAACA	AGTAGAAAAAGGTCGTTTTTAAACTATTC	1166
PB1	TGCGAGCTGACTGATTCAATCTGGATA	AGTAGAAACAAGGCATTTTTTCATGAA	2321
PA	TGCGAGCTGACTGATTCAATCTGGATA	AGTAGAAACAAGGTACTTTTTTGGACA	967
HA	ATGAAGACTATCATTGCTTTGAGCTAC	TCAAATGCAAATGTTGCACCTAATG	1701
NP	AGCAAAAGCAGGGTAGATAATCACTCA	AGTAGAAACAAGGGTATTTTTCTTTAA	1565
NA	AGCAAAAGCAGGAGTAAAGATGAAT	AAGCTTATATAGGCATGAGATTGAGG	1433
M	ATATTGAAAGATGAGCCTTCTAACCG	ACTCCAACTCTATGCTGACAAAATGAC	990
NS	AGCAAAAGCAGGGTGACAAAGACATAA	AGTAGAAACAAGGGTGTTTTTTATTAT	890

doi:10.1371/journal.pone.0012591.t001

Assembly of sequences, translation of nucleotide sequences into protein sequences, and initial multiple sequence alignments were performed with the Clustal V method using MegAlign software version 1.03 (SNAStar Inc., Madison, WI).

The reference strains selected for phylogenetic analysis are based on the following criteria: 1. Using a blast search (http://blast.ncbi.nlm.nih.gov/Blast.cgi), the most genetically closest segment sequence is selected. 2. The selected strains are well-characterized phylogenetically, so that they can represent their lineage and host origin, such as avian, pig or human. 3. The general topology of the phylogenetic tree constructed using the selected reference strains is consistent with previously well-recognized evolutionary analysis. Bootstrap support for tree topologies was accomplished using the Neighbor-joining (NJ) methods implemented in MEGA 4.0 with 1,000 iterations [45]. Genetic distances based on NJ phylogenetic trees were calculated applying Kimura's two-parameter method. In this study, the nucleotide sequences used for the phylogenetic analysis are as follows: PB2 1185-2306, PB1 64-2241, PA 1266-1579, HA 78-1064, NP 46-1504, NA 89-1399, M 35-990, and NS 27-822.

Molecular Graphic Visualization

Amino acid changes at the five antigenic sites (A~E) [9]–[11] of the HA1 molecule were determined by amino acid sequence alignment using the MegAlign program. Changes at the major antigenic sites (A, B and C) of the HA monomer were located using PyMOL software (v0.99) (DeLano Scientific LLC, South San Francisco, California, U.S.A.) based on the HA structure of the H5 subtype influenza virus, A/duck/Singapore/3/97 [46],

(1JSM) downloaded from the Protein Data Bank website (http://www.rcsb.org/pdb/home/home.do). The H5 structure was used because it is available as a monomer structure and it would be clearer to visualize the amino acid changes on it than the H3 structure, which is only available as a trimer structure.

Nucleotide Sequence Accession Numbers

The nucleotide sequences for all H3N2 influenza virus isolates analyzed in this study are available from GenBank under accession numbers GU215015 to GU215038 (Table 2).

Table 2: The accession numbers of protein and nucleotide acid of swine H3N2 influenza viruses

https://doi.org/10.1371/journal.pone.0012591.t002

	A/swine/Jilin/5/2007				A/swine/Jilin/19/2007				A/swine/Jilin/37/2008			
	Protein		Nucleotide acid		Protein		Nucleotide acid		Protein		Nucleotide acid	
	Accession	Length (aa)	Accession	Length (bp)	Accession	Length (aa)	Accession	Length (bp)	Accession	Length (aa)	Accession	Length (bp)
PB2	ACZ53952	373	GU215015	1121	ACZ53963	373	GU215023	1121	ACZ53974	373	GU215031	1121
PB1	ACZ53953	757	GU215016	2317	ACZ53964	757	GU215024	2317	ACZ53975	757	GU215032	2317
PA	ACZ53955	302	GU215017	930	ACZ53966	302	GU215025	930	ACZ53977	302	GU215033	930
HA	ACZ53956	566	GU215018	1701	ACZ53967	566	GU215026	1701	ACZ53978	566	GU215034	1701
NP	ACZ53957	498	GU215019	1541	ACZ53968	498	GU215027	1541	ACZ53979	498	GU215035	1541
NA	ACZ53958	469	GU215020	1414	ACZ53969	469	GU215028	1414	ACZ53980	469	GU215036	1414
M1	ACZ53959	249	GU215021	957	ACZ53970	249	GU215029	957	ACZ53981	249	GU215037	957
NS1	ACZ53961	230	GU215022	890	ACZ53972	230	GU215030	890	ACZ53983	230	GU215038	890

doi:10.1371/journal.pone.0012591.t002

AUTHOR CONTRIBUTIONS

Conceived and designed the experiments: YC ZD. Performed the experiments: YC ZG SC QZ GY. Analyzed the data: YC. Contributed reagents/materials/analysis tools: YC GW ZG WW QM WR CW. Wrote the paper: YC.

REFERENCES

1. Ludwig S, Stitz L, Planz O, Van H, Fitch WM, et al. (1995) European swine virus as a possible source for the next influenza pandemic? Virology 212: 555–561.
2. Kundin WD (1970) Hong Kong A-2 influenza virus infection among swine during a human epidemic in Taiwan. Nature 228: 857.
3. Shortridge KF, Webster RG, Butterfield WK, Campbell CH (1977) Persistence of Hong Kong influenza virus variants in pigs. Science 196: 1454–1455.
4. Reid AH, Fanning TG, Hultin JV, Taubenberger JK (1999) Origin and evolution of the 1918 "Spanish" influenza virus hemagglutinin gene. Proc Natl Acad Sci U S A 96: 1651–1656.
5. Brown IH (2000) The epidemiology and evolution of influenza viruses in pigs. Vet Microbiol 74: 29–46.
6. Brown IH, Harris PA, McCauley JW, Alexander DJ (1998) Multiple genetic reassortment of avian and human influenza A viruses in European pigs, resulting in the emergence of an H1N2 virus of novel genotype. J Gen Virol 79 (Pt 12): 2947–2955.
7. Campitelli L, Donatelli I, Foni E, Castrucci MR, Fabiani C, et al. (1997) Continued evolution of H1N1 and H3N2 influenza viruses in pigs in Italy. Virology 232: 310–318.
8. Karasin AI, Schutten MM, Cooper LA, Smith CB, Subbarao K, et al. (2000) Genetic characterization of H3N2 influenza viruses isolated from pigs in North America, 1977-1999: evidence for wholly human and reassortant virus genotypes. Virus Res 68: 71–85.
9. Caton AJ, Brownlee GG, Yewdell JW, Gerhard W (1982) The antigenic structure of the influenza virus A/PR/8/34 hemagglutinin (H1 subtype). Cell 31: 417–427.
10. Nakajima K, Nobusawa E, Tonegawa K, Nakajima S (2003) Restriction of amino acid change in influenza A virus H3HA: comparison of amino acid changes observed in nature and in vitro. J Virol 77: 10088–10098.
11. Wiley DC, Wilson IA, Skehel JJ (1981) Structural identification of the antibody-binding sites of Hong Kong influenza haemagglutinin and their involvement in antigenic variation. Nature 289: 373–378.
12. Nobusawa E, Aoyama T, Kato H, Suzuki Y, Tateno Y, et al. (1991) Comparison of complete amino acid sequences and receptor-binding

properties among 13 serotypes of hemagglutinins of influenza A viruses. Virology 182: 475–485.

13. Weis W, Brown JH, Cusack S, Paulson JC, Skehel JJ, et al. (1988) Structure of the influenza virus haemagglutinin complexed with its receptor, sialic acid. Nature 333: 426–431.
14. Vines A, Wells K, Matrosovich M, Castrucci MR, Ito T, et al. (1998) The role of influenza A virus hemagglutinin residues 226 and 228 in receptor specificity and host range restriction. J Virol 72: 7626–7631.
15. Ohuchi M, Ohuchi R, Feldmann A, Klenk HD (1997) Regulation of receptor binding affinity of influenza virus hemagglutinin by its carbohydrate moiety. J Virol 71: 8377–8384.
16. Zhou NN, Senne DA, Landgraf JS, Swenson SL, Erickson G, et al. (1999) Genetic reassortment of avian, swine, and human influenza A viruses in American pigs. J Virol 73: 8851–8856.
17. Boni MF, de Jong MD, van Doorn HR, Holmes EC (2010) Guidelines for Identifying Homologous Recombination Events in Influenza A Virus. Plos ONE 5: e10434.
18. Brown IH (2001) The pig as an intermediate host for influenza A viruses between birds and humans. Int Congr Ser 1219: 173–178.
19. Cong YL, Pu J, Liu QF, Wang S, Zhang GZ, et al. (2007) Antigenic and genetic characterization of H9N2 swine influenza viruses in China. J Gen Virol 88: 2035–2041.
20. Marozin S, Gregory V, Cameron K, Bennett M, Valette M, et al. (2002) Antigenic and genetic diversity among swine influenza A H1N1 and H1N2 viruses in Europe. J Gen Virol 83: 735–745.
21. Reeth KV, Brown I, Essen S, Pensaert M (2004) Genetic relationships, serological cross-reaction and cross-protection between H1N2 and other influenza A virus subtypes endemic in European pigs. Virus Res 103: 115–124.
22. Yu H, Hua RH, Wei TC, Zhou YJ, Tian ZJ, et al. (2008) Isolation and genetic characterization of avian origin H9N2 influenza viruses from pigs in China. Vet Microbiol 131: 82–92.
23. Bi YH, Fu GH, Chen J, Peng JS, Sun YP, et al. (2010) Novel Swine Influenza Virus Reassortants in Pigs, China. Emerg Infect Dis 16: 1162–1164.
24. Peiris JS, Guan Y, Markwell D, Ghose P, Webster RG, et al. (2001) Cocirculation of avian H9N2 and contemporary "human" H3N2

influenza A viruses in pigs in southeastern China: potential for genetic reassortment? J Virol 75: 9679–9686.

25. Xu C, Fan W, Wei R, Zhao H (2004) Isolation and identification of swine influenza recombinant A/Swine/Shandong/1/2003(H9N2) virus. Microbes Infect 6: 919–925.
26. Claas EC, Kawaoka Y, de Jong JC, Masurel N, Webster RG (1994) Infection of children with avian-human reassortant influenza virus from pigs in Europe. Virology 204: 453–457.
27. Yu H, Hua RH, Zhang Q, Liu TQ, Liu HL, et al. (2008) Genetic evolution of swine influenza A (H3N2) viruses in China from 1970 to 2006. J Clin Microbiol 46: 1067–1075.
28. Bush RM, Bender CA, Subbarao K, Cox NJ, Fitch WM (1999) Predicting the evolution of human influenza A. Science 286: 1921–1925.
29. Wilson IA, Cox NJ (1990) Structural basis of immune recognition of influenza virus hemagglutinin. Annu Rev Immunol 8: 737–771.
30. Shih AC, Hsiao TC, Ho MS, Li WH (2007) Simultaneous amino acid substitutions at antigenic sites drive influenza A hemagglutinin evolution. Proc Natl Acad Sci U S A 104: 6283–6288.
31. Ito T (2000) Interspecies transmission and receptor recognition of influenza A viruses. Microbiol Immunol 44: 423–430.
32. Ito T, Couceiro JN, Kelm S, Baum LG, Krauss S, et al. (1998) Molecular basis for the generation in pigs of influenza A viruses with pandemic potential. J Virol 72: 7367–7373.
33. Webby RJ, Rossow K, Erickson G, Sims Y, Webster R (2004) Multiple lineages of antigenically and genetically diverse influenza A virus co-circulate in the United States swine population. Virus Res 103: 67–73.
34. Tumpey TM, Maines TR, Van Hoeven N, Glaser L, Solorzano A, et al. (2007) A two-amino acid change in the hemagglutinin of the 1918 influenza virus abolishes transmission. Science 315: 655–659.
35. Rogers GN, Paulson JC, Daniels RS, Skehel JJ, Wilson IA, et al. (1983) Single amino acid substitutions in influenza haemagglutinin change receptor binding specificity. Nature 304: 76–78.
36. Lindstrom S, Sugita S, Endo A, Ishida M, Huang P, et al. (1996) Evolutionary characterization of recent human H3N2 influenza A isolates from Japan and China: novel changes in the receptor binding domain. Arch Virol 141: 1349–1355.

37. Yassine HM, Lee CW, Suarez DL, Saif YM (2008) Genetic and antigenic relatedness of H3 subtype influenza A viruses isolated from avian and mammalian species. Vaccine 26: 966–977.
38. Gallagher PJ, Henneberry JM, Sambrook JF, Gething MJ (1992) Glycosylation requirements for intracellular transport and function of the hemagglutinin of influenza virus. J Virol 66: 7136–7145.
39. Olden K, Parent JB, White SL (1982) Carbohydrate moieties of glycoproteins. A re-evaluation of their function. Biochim Biophys Acta 650: 209–232.
40. Seidel W, Kunkel F, Geisler B, Garten W, Herrmann B, et al. (1991) Intraepidemic variants of influenza virus H3 hemagglutinin differing in the number of carbohydrate side chains. Arch Virol 120: 289–296.
41. Li MC, Todd Davis C, Zhou H, Nancy JCox, Donis RO (2008) Genetic Compatibility and Virulence of Reassortants Derived from Contemporary Avian H5N1 and Human H3N2 Influenza A Viruses. Plos Phathogens 4: 1–10.
42. Daly JM, Wood JM, Robertson JS (1998) Co-circulation and divergence ofhumaninfluenza viruses. In: Nicholson KG, Webster RG, Hay AJ, editors. Textbook of influenzaOxford: Backwell Science. pp. 168–177.
43. Hoffmann E, Stech J, Guan Y, Webster RG, Perez DR (2001) Universal primer set for the full-length amplification of all influenza A viruses. Arch Virol 146: 2275–2289.
44. Shu LL, Lin YP, Wright SM, Shortridge KF, Webster RG (1994) Evidence for interspecies transmission and reassortment of influenza A viruses in pigs in southern China. Virology 202: 825–833.
45. Tamura K, Dudley J, Nei M, Kumar S (2007) MEGA4: Molecular Evolutionary Genetics Analysis (MEGA) software version 4. 0. Mol Biol Evol 24: 1596–1599.
46. Ha Y, Stevens DJ, Skehel JJ, Wiley DC (2002) H5 avian and H9 swine influenza virus haemagglutinin structures: possible origin of influenza subtypes. Embo J 21: 865–875.

CHAPTER 20

Regional Variation in Pig Farmer Awareness and Actions Regarding Japanese Encephalitis in Nepal: Implications for Public Health Education

Santosh Dhakal[1] , Durga Datt Joshi[1] , Anita Ale[1] , Minu Sharma[1] , Meena Dahal[1] , Yogendra Shah[1] , Dhan Kumar Pant[1] , Craig Stephen[2]

[1] National Zoonoses and Food Hygiene Research Centre, Kathmandu, Nepal

[2] Centre for Coastal Health, Nanaimo British Columbia, Canada and Faculty of Veterinary Medicine and Faculty of Medicine, University of Calgary, Calgary, Alberta, Canada

Citation: Dhakal S, Joshi DD, Ale A, Sharma M, Dahal M, Shah Y, et al. (2014) Regional Variation in Pig Farmer Awareness and Actions Regarding Japanese Encephalitis in Nepal: Implications for Public Health Education. PLoS ONE 9(1): e85399. https://doi.org/10.1371/journal.pone.0085399.

ABSTRACT

Japanese encephalitis (JE) is a mosquito-borne zoonotic disease that has pigs as the major amplifying hosts. It is the most important cause of viral encephalitis in people in Nepal and is spreading in its geographic distribution in that country. Pig farming is increasing in Nepal due to reducing cultural biases against pigs and government programs to support pig farming for poverty alleviation. Major strategies for JE prevention and control include education, vector control, and immunization of people and pigs. This study used a survey of 400 pig farmers in 4 areas of Nepal with different JE and pig farming histories to explore regional variations in farmer awareness and actions towards JE, the association of awareness and actions with farm and farmer variables, and the implications of these associations for public health education. Exposure to JE risk factors was common across pig farms and pig farming districts but there were significant district level differences in knowledge and practices related to on-farm JE risk reduction. Social factors such as literacy, gender, and cultural practices were associated with farmer attitudes, knowledge and practices for JE control. JE vaccine uptake was almost non-existent and mosquito control steps were inconsistently applied across all 4 districts. Income was not a determining factor of the differences, but all farmers were very poor. The low uptake of vaccine and lack of infrastructure or financial capacity to house pigs indoors or away from people suggest that farmer personal protection should be a priority target for education in Nepal. This study re-enforces the need to attack root causes of people's personal disease prevention behaviours and take into account local variation in needs and capacities when designing health or agriculture education programs.

INTRODUCTION

Japanese encephalitis (JE) is the most important viral encephalitis in Nepal and Asia [1], [2]. It is a mosquito borne disease caused by a flavivirus that cycles between birds, pigs and people [3]. Its distribution has, in recent years expanded and is expected to spread more widely geographically with anticipated changes in climate, land use (particularly rice and pig farming and urbanization), and access to health care [4], [5]. The high case fatality rate, high rate of severe long lasting neurological symptoms and the majority of deaths occurring in children make JE a major public health problem [6]. JE is not homogenously distributed in Nepal. The disease was first confirmed in western Nepal in 1978 [7] in the lowland plains of Terai which borders

India, and has regularly occurred since then with epidemic peaks every 2–5 years [8]. JE has now been documented in 54 of the 75 districts of Nepal, including 24 hill and mountain districts, and is considered to be endemic in the Kathmandu Valley as well as in 24 other districts [8], [9], [10].

The World Health Organization Regional Office for South-East Asia (SEARO/WHO) has pointed out four major strategies for JE prevention and control; health education, vector control, immunization of people and pigs, and epidemic preparedness and response [12]. Other countries like Japan, South Korea and Taiwan have successfully controlled JE by human and pig vaccination, modernization of pig farms, change in agricultural practices and improved living standards [4], [11]. The investments required to achieve these changes is beyond the current economic capacity of Nepal. Even regular human vaccination is not affordable or sustainable at the present time. Low-cost alternatives need to be found to begin to address the SEARO/WHO recommendations.

Pig farming is increasing in Nepal due to reduced cultural biases against pigs and new government programs to support pig farming as a low cost means to create income for poverty alleviation and to improve food security. A 48% increase in the Nepal pig population in 14 years [13] shows how fast this industry is growing. Pig farms are not homogeneously distributed in Nepal. Pig farming is most prevalent in the eastern and central regions of the country where the pig-raising ethnic community predominates and is also more common in peri-urban areas because of an increased demand of pig meat in urban areas. Because pigs are a major JE virus amplifying hosts [14], ensuring pig farmers are aware of and able to use means for personal and community risk reduction should be an important part of a Nepalese JE prevention strategy. Our previous study in the Kathmandu district [15] found that the pig farmers had high occupational exposures to JE risk factors but they also had low levels of awareness and adoption of preventive measures against JE. Knowledge, beliefs and practices can affect how individuals or groups apply control efforts against disease [16], [17]. Because we found socio-economic associations with knowledge about JE and the use of preventive practices in the Kathmandu district, in this study we sought to replicate Dhakal et. al. [15] in four districts with different experiences with both JE and pig farming to see if there were regional variations that may influence how farmer JE education programs might need to be adapted to local needs and situations. In this study we used a survey methodology to ascertain information on what pig farmers knew about JE, its risk factors and/or its control in; a long term endemic region (>30yrs) that is the source of most human JE cases in the country (Rupandehi district); a neighbouring endemic district with lower numbers of outbreaks (Kapilvastu near the Indian border); an area of endemic JE with the highest level of pig amplifying hosts (Morang district); and a region of comparatively recent JE emergence in the hill region (Kathmandu valley).Our objective was to determine if there was regional variation in JE knowledge about the disease and practices within pig farm communities to help public health and agriculture extension planners tailor educational material to local needs.

METHODS

This research was conducted from September 2011 to June 2012 in four districts of Nepal as described above. Within each district 2–4 communities known for active pig farming were selected in order to recruit 100 participants per district. The same procedure used in Dhakal et. al., 2012 was used for all districts. One hundred farms were selected by simple random sampling based on a sampling frame derived from local agriculture agency registry of pig farms. Before visiting the farms, the survey team went to the community and introduced the project to the para-veterinarians servicing the area and to community leaders. Farmers were provided the incentive to participate by being offered free physical examinations of the pigs owned by the farmers and treatment of any ill animals by a veterinarian. Local para-veterinarians were informed about any health issues or animal treatments after each visit to ensure follow up. Farmers were also offered a short training session on JE and its preventive measures after they completed the survey. This recruitment process resulted in 100% compliance with all requests to participate.

Ethics Statement

Participation in the survey was voluntary and all farmers were informed of their right to withdraw at any time in the survey and were required to give their consent to participate. Verbal consent to participate was requested because the high anticipated rate of illiteracy would preclude many from providing written consent. Consent was noted by the field investigator. All responses were recorded anonymously and without any identifying information. Ethical approval was provided for this procedure, including the use of verbal consent, by the ethics committee of the National Zoonoses and Food Hygiene Research Centre (approval number NZFHRC 22/10/7/2010/2011).

Participating farmers were verbally asked questions from a standardized survey and their oral response recorded by the interviewer. There were six main parts to the questionnaire; (i) farmer attributes such as gender, education, income, experience and training, (ii) farm attributes such as the number of pigs raised, management, and source and marketing of pigs, (iii) farmer proximity to known risk factors such as rice fields, pig barns, and standing water, (iv) farmer knowledge about JE and what can be done to prevent the disease, (v) JE specific practices used by the farmer for themselves, their pigs and their family including vaccination and mosquito avoidance and (vi) information on clinically compatible signs or past diagnosis of JE in people or animals on the farm. As this project was focused on farmer knowledge and exposure to known risks, we did not undertake a sero-survey of people or animals to document the prevalence of JE, but instead relied on self-reported information.

The survey contained both open- and closed-ended questions. Open-ended questions were grouped based on their similarities and turned into closed-ended replies after a thematic review of the answers. All answers were coded and entered into the SPSS software version 19 (http://www-01.ibm.com/software/analytics/spss/). Descriptive statistics used for analysis included frequencies, sums, ranges and means, and the chi-squared test was used as a test of association with $p< 0.05$ selected as the level for statistical significance. Chi-square was estimated to identify association between JE knowledge and predictors; mosquitoes avoiding practice and predictors. Univariate analysis was undertaken to calculate the odds ratio with 95% confidence interval. Significant variables were entered in to logistical model. Logistic regression was used for multivariate analysis of potential predictors of farmer's knowledge about JE; with $p<0.05$ used as a threshold for statistically significant interactions. The default "enter" method was used in SPSS for variable selection in which all variables in a block were entered in a single step.

RESULTS

All farmer houses in Kathmandu, Morang and Kapilvastu and all except one in Rupandehi were within 500 meters of the pig pens. Most pig farms were small family operations. The average number of pigs in Kathmandu was 26, followed by 8 in Kapilvastu, 6 in Rupandehi and 4 in Morang study area. Free range or open-air pens were common types of housing in all practices. Eighty-seven percent (346/400) of houses overall were within 1 km of rice paddy fields (range of 73–95%) and 91% (362/400) of farm houses within 1 Km of a standing water sources (range 80–99%) (table 1). There were regional differences in proximity to potential breeding sites for mosquitoes (paddy field [p<0.001]; standing water sources (p<0.001)]. All pig farmers had encountered mosquitoes on their farm and all reported mosquito biting their pigs and themselves. All farmers reported encountering wild birds on their farms. Among the 400 farmers, 62 (15.5%) also had domestic ducks and 193 (57%) had duck farms within 1 Km distance (ranging from 100%, 78%, 40% and 37% respectively, in Kathmandu, Morang, Kapilvastu and Rupandehi).

Table 1: Farmer and farm characteristics for 400 pig farmers and farms in 4 districts of Nepal

https://doi.org/10.1371/journal.pone.0085399.t001

Parameters	Kathmandu (n = 100)	Morang (n = 100)	Rupandehi (n = 100)	Kapilvastu (n = 100)	P value
Male: Female respondents	50:50	58:42	48:52	79:21	<0.001
Number of farmers self-declared as illiterate	39	55	23	78	<0.001
Farms with monthly income≤ 10000 NRS from pig farming	70	95	82	94	<0.001
Pig farming as sole occupation	73	18	27	25	<0.001
Number of farms who owned the land used for pig farming	15	65	90	87	<0.001
Number of farmers with ≤3 years' experience pig farming	27	16	21	40	0.001
Farm houses located ≤ 500 m from the pig farm	100	100	99	100	NT
Farm houses located ≤ 1 km from rice field	95	88	73	90	<0.001
Farm houses located ≤ 1 km from standing water bodies	99	91	92	80	<0.001
Number of farmers who knew people can get diseases from pigs	72	39	67	12	<0.001
Number of farmers who heard about JE	42	25	38	15	<0.001
Number of farmers vaccinating pigs against at least one disease	87	13	44	6	<0.001

doi:10.1371/journal.pone.0085399.t001

There were district level differences in the ratio of male:female respondents (p<0.001) mainly due to Kapilvastu, where females were generally reluctant to respond to interviewers from outside of their district. Across the four study districts we found significant differences in literacy rate (p<0.001), monthly income (p<0.001), pig farming being the sole income source (p<0.001), land ownership (p<0.001) and experience as a pig farmer (p=0.001) (table 1). Out of 400 pig farmers 195 (49%) were illiterate and the monthly income of 85% of farm families was not more than 10,000 NRS. Pig farming was sole source of income for 73% of pig farmers in Kathmandu study area but overall only 36% (143/400) farmers had this as sole occupation. The other occupations however, were low income generating type. Seventy-four percent (296/400) of pig farmers had 3 or more years' experience in pig farming. Many pig farms in Kathmandu were mobile due to being located on leasehold land but in other districts they were more permanent. Except for few farms in Rupandehi, all pig farms were of poor sanitary conditions and lacked biosecurity measures.

Out of 400 pig farmers, only 36 (9%) had received formal training on pig husbandry management or disease. Training was provided to these 36 individuals from government (n=21,58%) followed by farmers groups (n=9, 25%) and non-governmental organizations.

(n=6, 17%). Of the 364 farmers who had not received training, 198 (54%) said they did not know training was available, 146 (40%) said they did not know where to go for training and 20 (6%) said they couldn't afford training. Seventy-three percentages of pig farmers (292/400) said they learned about pig diseases through their own experience or through pig farming friends and the community. Only 27% (108/400) said they got information from veterinary sources. We found field veterinarians were the most trusted source of information on immunization practices for pigs in all districts: 83% of farmers relied on veterinarians followed by the pig farming community members (17%) for making immunization decisions.

Less than half (190/400) of the pig farmers interviewed knew they could acquire a disease from pigs but only 10% (40) could name a pig associated zoonosis. They named swine flu (17/40), JE (21/40), and neurocysticercosis (2/40). However, 30% (120/400) of the farmers were aware of JE. Eighteen percent (73/400) knew signs of JE in people, 7% (29/400) knew JE signs for pigs, 17% (66/400) knew JE was transmitted by mosquitoes, 9% (34/400) knew it can be prevented by vaccine in pigs and 15% (59/400) knew it is vaccine preventable in people. There was, however regional variation in the proportion of farmers who were aware of JE; Kathmandu 42%, Rupandehi 38%; Morang 25% and Kapilvastu 15%. Farmers who were aware of JE were more likely to know that people could acquire diseases from pigs ($p<0.001$) (table 1). Of the 120 farmers who were aware of JE, 53% learned about the disease through media sources, 28% from friends and community members, 9% from health care providers; 7% from training events and 4% through academic study. There was a significant difference in where farmers in different districts learned about JE ($p=0.008$). For example, 20% (20/100) of farmers in Rupandehi heard about JE through the media compared to only 8% (8/100) in Kalivastu and 20% (20/100) of farmers in Katmandu heard about JE from friends or the community compared to 2% (2/100) in Morang.

Only one family in Rupandehi from our survey group reported JE being diagnosed in a family member. Reports of clinically compatible signs like high fever, unconsciousness, severe headache, neck rigidity, convulsion and/or paralysis were infrequently reported in farm family members (table 2). Severe headache were more commonly reported in Kapilvastu families (p<0.001). There were significant differences between districts for JE compatible signs in in the pig health including; abortion (p<0.001), weak piglets (p<0.001) and convulsion (p<0.001).

Table 2: Pig and human clinical signs compatible with Japanese Encephalitis as reported by 400 pig farmers in four districts of Nepal

https://doi.org/10.1371/journal.pone.0085399.t002

	Kathmandu (n = 100)	Morang (n = 100)	Rupandehi (n = 100)	Kapilvastu (n = 100)	P-Value
Pig Health Disorders (n)					
Abortion	36	8	10	15	<0.001
False Pregnancy	12	5	6	7	0.24
Weak piglets	36	5	15	66	<0.001
Convulsions	20	5	7	3	<0.001
Hydrocephalus	2	1	2	1	NT
Swollen testicles	2	1	0	2	NT
Human Health Disorders (n)					
High fever	8	14	17	21	0.07
Severe headache	12	7	7	46	<0.001
Unconsciousness	2	3	1	2	NT
Neck rigidity	1	1	0	0	NT
Convulsion	1	2	1	0	NT
Paralysis	0	2	2	0	NT

NT indicates associations were not tested because of the magnitude of difference.
doi:10.1371/journal.pone.0085399.t002

Education status of the farmers, study district, farmer's sex and whether or not the farmer had raised pigs for more than 3 years were each found to be significantly associated with whether or not the farmer was aware of JE (table 3). Only one farmer reported that family members had received JE vaccine. This same family also reported the death of a family member due to JE and explained this was the motivation to have the rest of the family immunized. Nearly 38% (150/400) had vaccinated their pigs against various diseases (120 against classical swine fever, 3 against foot and mouth disease, 25 against both classical swine fever and foot and mouth disease, and 2 against hemorrhagic septicemia). Vaccines against JE were not commercially available for pigs in Nepal. There was a significant difference in pig vaccination status in four districts for other pig diseases (p<0.001): 87% of pig farmers in Kathmandu had vaccinated pigs against at least one disease followed by 44 in Rupandehi, 13% in Morang and only 6% in Kapilvastu (table 1). Reasons for not vaccinating pigs varied and included; they didn't know pigs needed vaccine (117/250), they didn't have problems in pigs that needed vaccines (94/250), they couldn't afford to purchase vaccines (13/250) or they didn't think vaccines worked (3/250). The remaining 23/250 couldn't say why they were not vaccinating pigs.

Table 3: Associations between farmer attributes and practices and awareness of Japanese Encephalitis (JE) among Nepalese pig farmers

https://doi.org/10.1371/journal.pone.0085399.t003

Association tested	P value	Odds ratio	95%CI
Literacy and awareness of JE	<0.001	3.62	(2.27–5.76)
Know people can get disease from pigs and awareness of JE	<0.001	4.29	(2.61–7.05)
Gender and awareness of JE	0.004	1.89	(1.20–2.97)
Income ≤ 10000 NRS and attendance at pig farming training sessions	<0.001	5.19	(2.49–10.82)
Literacy and use of at least one mosquito avoiding practices	0.01	4.41	(1.22–15.89)
Time period of raising pig and awareness of JE	0.02	1.69	(1.06–2.70)
Awareness of JE and use of at least one mosquito avoiding practices	0.03	6.26	(0.87–44.99)
Attendance at pig farming training sessions and awareness of JE	0.08	1.76	(0.88–3.52)
Income ≤ 10000 NRS and awareness of JE	0.12	0.68	(0.38–1.20)
Attendance at pig farming training sessions and use of at least one mosquito avoiding practices	0.24	3.24	(0.20–53.20)
Income ≤ 10000 NRS and use of at least one mosquito avoiding practices	0.32	0.40	(0.05–3.07)

doi:10.1371/journal.pone.0085399.t003

Ninety-six percent (385/400) of the farmers knew at least one method of preventing mosquito bites in their family. The remaining 4% (15/400) were unaware of means to prevent or avoid mosquito bites. The various techniques used included use of window screen, use of repellents, use of mosquito coils, staying indoor at dawn/dusk, wearing clothes that cover full body, improving drainage and use of mosquito nets (table 4). There were significant differences across districts in the use of various mosquito avoiding techniques (table 4). There was also variation in the frequency or intensity of their use. For example, reporting use of mosquito nets could mean that several family members used nets or only the children used them. Twenty-four percent (97/400) of the farmers reported they practiced mosquito avoidance practices in their pig sheds. The practices included spraying chemicals (47/97), maintaining cleanliness (7/97), using smoke from a fire (42/97) and using repellents (1/97). There was a significant association between whether or not a person was aware of JE and whether or not they practiced at least one mosquito avoiding practices ($p=0.03$). There was also an association between whether or not a person was literate and the use of mosquito bite prevention technique ($p=0.01$). However, there was no association between the use of at least one mosquito bite prevention technique and training on pig farming ($p=0.24$) or household income ($p=0.32$) (table 3).

Table 4: Use of mosquito avoiding practices by 400 pig farmers in four different districts in Nepal

https://doi.org/10.1371/journal.pone.0085399.t004

	Kathmandu (n = 100)	Morang (n = 100)	Rupandehi (n = 100)	Kapilvastu (n = 100)	p value
Use window screen	11	8	42	6	<0.001
Use repellants	25	8	18	4	<0.001
Use mosquito net	41	51	88	38	<0.001
Improve drainage	38	71	65	6	<0.001
Use mosquito coil	69	49	68	50	0.001
Stay indoors at dawn/dusk	39	22	42	42	0.007
Wear clothes that fully covers the body	40	22	32	4	<0.001

NT indicates associations were not tested because of the magnitude of difference.

doi:10.1371/journal.pone.0085399.t004

Logistic regression failed to produce model with significant predictive value (Cox and Snell $R^2 = 0.197$). A final model had five variables literacy, gender, time period of raising pigs, mosquitoes avoiding practices and knowledge about pig disease associated with awareness of JE. All of these variables were significant in the final model except mosquitoes practice (Wald=2.06; p=0.151) (table 5).

Table 5: Final logistic regression model farmer knowledge of Japanese Encephalitis and predictors (Cox and Snell; $R^2 = 0.197$)

https://doi.org/10.1371/journal.pone.0085399.t005

Variables	B	SE	Wald	Sig	Exp (B)
Constant	-0.248	0.311	0.634	0.426	0.781
Literacy	-0.639	0.275	5.39	0.020	0.528
Gender	0.773	0.263	8.64	0.003	2.166
Time period of pig raining	0.738	0.278	7.076	0.008	2.09
Mosquitoes avoiding practice	-0.416	0.289	2.06	0.151	0.660
Knowledge about pig disease	-1.608	0.287	31.336	0.000	0.200

doi:10.1371/journal.pone.0085399.t005

DISCUSSION

Each of the four study districts had recent histories of JE. In the 5 year period from 2007 to 2011, there were 157 JE human cases reported in

Kathmandu, 89 cases reported in Morang, 16 cases reported in Rupandehi and 16 cases reported in Kapilvastu district [18]. JE risk factors were common across farms and farm families in all four districts but there were district level differences in knowledge and practices related to on-farm JE risk reduction. A previous survey in Nepal identified three JE control priorities;(i) information, education and communication strengthening to increase awareness of individuals and communities;(ii) behavioral changes to increase prevention practices and (iii) environmental interventions to reduce risk factors [19]. We found that Kapilvastu district lagged behind in all three of these priority areas followed by the Morang district. Farmers in Kapilivastu not only had the lowest level of awareness of JE, but also had the lowest proportion of farmers who were (i) literate, (ii) had more than 3 years pig farming experience, (iii) had higher incomes, (iv) knew that people could acquire diseases from pigs and (v) used methods for mosquito bite avoidance. In all regions, women were less likely to have heard of JE than men. Literacy rate may be an important cause of these differences. Districts where farmers had a higher level of literacy (Kathmandu 61% and Rupandehi77%) had a higher rate of JE awareness than the districts with lower farmer literacy rates (Morang 45% and Kapilvastu 22%). The overall literacy rate in our sample population (51%) was lower than the national literacy rate of 65.9%. The literacy rate for women is less (57.4%) than that of men (75.1%) in Nepal [20]. Efforts to change farmer JE prevention behaviours will need to take into account the implications of low literacy rates when designing education programs.

Mosquito control would seem a critical target for Nepal not only due to the inconsistent use of control measures found in this study but also due to the presence of multiple vector-borne diseases in Nepal such as malaria and dengue fever. Community-based educational interventions have been shown elsewhere to affect understanding and involvement in mosquito control and vectorborne disease prevention [21], [22]. Work on community education for vector control to eliminate lymphatic filariasis in southern India, using pre-post surveys in exposed and control villages, found that an 87% reduction in mosquito density could be achieved for a per capita cost of $0.32 [23]. Programs targeting mosquito control have resulted in declines in JE elsewhere. For example, in Assam, India, a sharp reduction in JE sero-conversion rates in people and pigs was achieved when insecticide treated nets were used in both people and pigs [24]. Similarly, a population based case-control study in China found that use of insecticide treated nets was associated with significant reduction in JE cases [25].There were significant differences in use of various mosquito control techniques in different districts in our study, often related to socio-economic factors. For example, the use of window screen was higher in Rupandehi because the pig farmers more often had permanent houses with windows compared to pig farmers of Kathmandu, Morang and Kapilvastu districts who often lived in homes without windows. Higher literacy rates (p=0.01) and being aware of JE (p=0.03) were associated with the use of at least 1 mosquito control practice but income (p=0.32) and training on pig farming (p=0.24) were not.

A number of countries have achieved tremendous reductions in the number of human JE cases by vaccination of pigs and people along with environmental changes like separation of houses and pigpens [26]. Although the Nepal government had JE vaccination programs for people in our study districts, there was virtually no uptake in the pig farming community. Only one family had been vaccinated and this was the same family where JE had taken the life of a family member. Pig farmers were not opposed to immunization as they had vaccinated their pigs for diseases other than JE and supplemental questions founds that many had their children vaccinated for standard childhood vaccine preventable diseases. Reasons for lack of pig vaccination may be as easy to explain because of the lack of readily available commercial vaccine and lack of perception that JE caused illness in the pigs. Data on reasons for the exceedingly low uptake of JE vaccine in Nepalese pig farming families are lacking. A survey of mothers on demand side barriers to childhood vaccination in the Terai area of Nepal found lack of knowledge, misconception about immunization, lack of access to health services, heavy household work related to poverty, lack of permission from family to visit a health facility and religion, caste and gender factors were associated with lower vaccine uptake [27]. Public concern over adverse reactions led to refusal of the JE vaccination and a consequent decrease in coverage rate in Korea [28]. Further study is required to determine why pig farmers, who appear to be a high risk occupation group, have such low levels of vaccine uptake in a country where JE vaccine is free for people.

Income was not found to be a significant predictor of JE knowledge or practices. Although there were statistical differences in the proportion of homes with incomes more than 10000 NRS per month, all of the farm household could be considered to be very poor and well below the national average income of 202,374 NRS per year [29]. There is evidence for other vector-borne diseases that socio-economic status affects the uptake of prevention and treatment interventions [30]. It may be possible that the study population was so far below an income threshold that it obscured the effect of household income on JE prevention and control. There was a significant association between income and having had training on pig farming. It is not clear if the training increased income or income provided funds to attend training sessions. Having training or not had no bearing on awareness of JE likely because training was focused on pig husbandry and management rather than zoonotic disease prevention and control. Seventy three percent of pig farmers learned about pig diseases through their own experiences or from other pig farmers whom they trusted, suggesting that social networks may be key to disseminating information on JE. Field veterinarians were also a trusted source, especially about immunization and are therefore another key conduit of information into the community. Except for Kathmandu where the pig farming community was a major source of information, media was the main source of information on JE and thus should be targeted for public education campaigns. Media campaigns will need to use means for knowledge mobilization that can target the proportion of farmers and farm household members who are illiterate as well as be sensitive to gender differences in use of media sources.

There were three key reasons to prioritize pig farming families for JE education and control. First, this study showed that this occupational group lived and worked in close proximity to key JE risk factors like pigs, rice fields, ducks, wild birds, mosquitoes and standing water. The major JE vector *Culex tritaeniorhynchus* breeds predominantly in rice fields and open sunlit temporary and permanent habitats with vegetation and they have average flight range of 1.5 km [31]. We found the pig farms in all study areas were located within this flight range from rice fields and the standing water sources. Second, there was a low rate of use of JE prevention for families or pigs. Third, pig production is increasing in Nepal and is expected to grow. Add to these factors the spectre of an expanding JE range associated with climate change and land use changes and there is ample reason to conclude that pig farming families should be priority targets of JE control campaigns. Validation of this assumption will require case-control or similar studies to determine if pig farmers are indeed at a higher risk for JE disease in Nepal.

The design and delivery of future pig farm families JE education on prevention and control will need to take into consideration, not only how pig farmers differ from other members of society in terms of their exposure risks and capacities to understand, access and apply JE control actions but also how to tailor programs to differences in socio-economic variables across districts. This study re-enforces the need to attack root causes of people's personal disease prevention behaviours, such as literacy, when aiming to have wide impacts from public health or agriculture extension and education programs.

ACKNOWLEDGMENTS

The authors would like to thank the pig farmers of Nepal for their full and enthusiastic cooperation in this study.

AUTHOR CONTRIBUTIONS

Conceived and designed the experiments: SD DDJ AA DKP CS. Analyzed the data: SD DKP. Wrote the paper: SD DKP DDJ CS. Data collection via field interviews, including pre-testing: MS MD YS.

REFERENCES

1. Umenai T, Krzysko R, Bektimirov TA, Assaad FA (1985) Japanese encephalitis: current worldwide status. . Bull. WHO. 63: 625–31.
2. Rayamajhi A, Ansari I, Ledger E, Bista KP, Impoinvil DE, et al. (2011) Clinical and prognostic features among children with acute encephalitis syndrome in Nepal; a retrospective study. . BMC Infectious Diseases. 11: 294.
3. van den Hurk AF, Ritchie SA, Meckenzie JS (2009) Ecological and geographical expansion of Japanese encephalitis virus. . Annu Rev Entomol. 54: 17–35.
4. Erlanger TE, Weiss S, Keiser J, Utzinger J, Wiedenmayer K (2009) Past, present and future of Japanese encephalitis. Emer Infect Dis15 (1), 1–7.
5. LeBeaud AD (2008) Why arboviruses can be neglected tropical diseases? PLoS Negl Trop Dis 2(6): e247 Doi:https://doi. org/10. 1371/ journal. pntd. 0000247.
6. LeBeaud AD, Bashir F, King CH (2011) Measuring the burden of arboviral diseases: the spectrum of morbidity and mortality from four prevalent infections. . Pop Health Metrics. 9: 1.
7. Joshi D (1983) Incidence of Japanese encephalitis in children: 1978, 1979, and 1980 outbreaks. . J Nep Paediatr Soc. 2: 18–25.
8. Bista MB, Shrestha JM (2005) Epidemiological situation of Japanese encephalitis in Nepal. . J Nep Med Assoc. 44: 51–56.
9. Pant D (2009) Epidemiology of Japanese encephalitis in Nepal. J. Nepal Paediatr Soc. 29: 35–37.
10. Parajuli MB, Joshi DD, Pradhan SP, Chamling M, Joshi AB (1992) Incidence of Japanese encephalitis during 1989 in Nepal. J. Nepal Med. Assoc. 30: 7–14.
11. Igarashi A (2002) Control of Japanese encephalitis in Japan: immunization of humans and animals, and vector control. Curr Top Microbiol Immunol 267: 139–152.
12. SEARO/WHO (2002) Regional committee. Provisional agenda item 10. 2. Fifty-fifth session. Prevention and control of dengue, Japanese encephalitis and kalaazar in SEA region. Available at http://209. 61. 208. 233/LinkFiles/RC_55_7. pdf. Accessed 2013 Dec 6.

13. MOAD: (2012) Statistical information on Nepalese agriculture 2011/2012 (2069/069). Government of Nepal, Ministry of Agriculture Development. Available: http://www. moad. gov. np/downloadfile/ yearbook2012_1363677455. pdf. Accessed 2013 Dec 6.
14. Solomon T, Dung NM, Kneen R, Gainsborough M, Vaughn DW, et al. (2000) Japanese encephalitis. . J. Neurol. Neurosurg Psychiatr. 68: 405–415.
15. Dhakal S, Stephen C, Ale A, Joshi DD (2012) Knowledge and Practices of Pig Farmers Regarding Japanese encephalitis in Kathmandu, Nepal. . Zoonoses and Public Health. 59: 568–574.
16. Klein RE, Weller SC, Zeissig R, Richards FO, Ruebush TK (1995) Knowledge, beliefs, and practices in relation to malaria transmission and vector control in Guatemala. Am J Trop Med Hyg 52 (5): 383–388.
17. van Benthem BH, Khantikul N, Panart K, Kessels PJ, Somboon P, et al. (2002) Knowledge and use of prevention measures related to dengue in northern Thailand. . Trop Med Int Health. 7 (11): 993–1000.
18. WHO-IPD (2012) Records of JE cases (unpublished) obtained from World Health Organization, Immunization preventable diseases section, Nepal.
19. Houston R, Chhetry D (2003) Nepal: analysis of baseline survey data on Japanese encephalitis, kala-azar and malaria. Environmental health project activity report 121. 2003. Prepared for the Office of Health, Infectious Diseases and Nutrition, Bureau for Global Health, U. S. Agency for International Development, under EHP project 27052/E. X. NE5. ME. EO.
20. CBS (2011) National population and housing census 2011 (National report). Government of Nepal, National Planning Commission Secretariat, Central Bureau of Statistics.
21. Winch PJ, Leontsini E, Rigau-Perez JG, Ruiz-Perez M, Clark GG, et al. (2002) Community based dengue prevention programs in Puerto Rico: impact on knowledge, behavior, and residential mosquito infestation. Am J Trop Med Hyg 67 (4): 363–370.
22. Yasuoka J, Mangione TW, Spielman A, Levins R (2006) Impact of education on knowledge, agricultural practices, and community actions for mosquito control and mosquito-borne disease prevention in rice ecosystems in Sri Lanka. Am J Trop Med Hyg 74 (6): 1034–1042.

23. Nandha B, Krishnamoorthy K (2012) Impact of education campaign on community based vector control in hastening the process of elimination of lymphatic filariasis in Tamail Nadu, South India. Health Educ. Res. 27 (4): 585–594.
24. Dutta PS, Khan A, Khan AM, Bohora J, Sarmah CK, et al. (2011) The effect of insecticide- treated nets (ITMNs) on Japanese encephalitis virus seroconversion in pigs and humans. Am J Trop Med Hyg (84) 466–472.
25. Luo D, Yao R, Song J, Huo H, Wang Z (1994) The effect of DDT spraying and bed nets impregnated with pyrethroid insecticide on the incidence of Japanese encephalitis virus infection. Trans R Soc Trop Med Hyg 88: 629–631.
26. Obara M, Yamauchi T, Watanabe M, Haseqawa S, Ueda Y, et al. (2011) Continuity and change of Japanese encephalitis virus in Toyoma Prefecture, Japan. Am J Trop Med Hyg 84(5): 695–708.
27. Subedi G (2012) Demand side barriers to immunization services in Terai districts of Nepal. Nepal Population J 17 (16): 91–104.
28. Sohn YM (2000) Japanese Encephalitis Immunization in South Korea: Past, Present, and Future. Emerg Infect Dis. [serial on the Internet]. [cited Oct, 2013]. Available: http://wwwnc. cdc. gov/eid/ article/6/1/00-0103. htm. Accessed 2013 Dec 6.
29. CBS (2011b) Nepal living standard survey 2010/11. Statistical report volume two. Central Bureau of statistics. National planning commission secretariat. 2011b, Government of Nepal. November 2011.
30. Worrall E, Basu S, Hanson K (2005) Is malaria a disease of poverty? A review of the literature. . Trop Med Interna Health. 10 (10): 1047–059.
31. Henrich TJ, Hutchaleelaha S, Jiwariyavej V, Barbazan P, Nitatpattana N, et al. (2003) Geographic dynamics of viral encephalitis in Thailand. . Microbes Infect. 5: 603–611.

SECTION 4:
COMPANION ANIMALS

CHAPTER 21

A Survey of Canine Filarial Diseases of Veterinary and Public Health Significance in India

Puteri Azaziah Megat Abd Rani[1], Peter J Irwin[2] , Mukulesh Gatne[3] , Glen T Coleman[1] , Linda M McInnes[2] , Rebecca J Traub[1]

[1] School of Veterinary Science, The University of Queensland, Queensland 4072, Australia

[2] School of Veterinary and Biomedical Science, Murdoch University, Western Australia 6150, Australia

[3] Bombay Veterinary College, Maharastra Animal and Fisheries Sciences University, Parel, Mumbai 400012, India

Citation: Megat Abd Rani, P.A., Irwin, P.J., Gatne, M. et al. A survey of canine filarial diseases of veterinary and public health significance in India. Parasites Vectors 3, 30 (2010). https://doi.org/10.1186/1756-3305-3-30.

ABSTRACT

Background

Dirofilaria spp., *Acanthocheilonema* spp. and *Brugia* spp. have all been reported in Indian dogs. In previous studies, diagnosis was made by morphological identification only. This is the first geographically stratified cross-sectional study in India to determine the prevalence and geographical distribution of canine filarial species of veterinary and public health importance, using a combination of conventional and molecular diagnostic techniques.

Results

A total of 139 from 525 dogs (26.5%; 95% CI 22.7, 30.3) were positive for microfilariae. The most common species of canine filaria identified in this study was *A. reconditum* (9.3%) followed by *D. repens* (6.7%) and *D. immitis* (1.5%). Three out of 525 dogs were found to have mixed infections on PCR. The morphological and molecular evidence on the sequence of the 18S gene and phylogenetic analysis of the ITS-2 region provided strong evidence that the canine microfilariae discovered in the Himalayan city of Ladakh belong to a novel species of *Acanthocheilonema*. Two dogs in Ladakh were also found to have mixed infections of the novel species described above and a unique microfilaria which morphologically resembled *Microfilaria auquieri* Foley, 1921.

Conclusions

At least six species of filarial nematode are now known to infect dogs in India, two of which were reported for the first time in this study. The study also confirms and extends the geographical distribution of canine heartworm (*D. immitis*) which overlaps with *D. repens*, emphasising the importance for veterinary clinicians and diagnostic laboratories to utilise immunodiagnostic tests that will not cross-react between those two filarial species. From a public health viewpoint, the distribution and prevalences of these nematodes warrant an appropriate prophylaxis to be administered to dogs.

BACKGROUND

Filariasis in dogs is caused by several species of filariids. *Dirofilaria immitis*, the most pathogenic canine filarid is responsible for heartworm disease in dogs. Both *D. repens* and *Acanthocheilonema* spp. develop into adult worms in the subcutaneous tissue resulting in skin nodules. Adults of *Brugia* spp. are usually recovered from the mandibular, retropharyngeal and axillary lymphatics. Most infections *with D. repens*, *Acanthocheilonema* spp. and *Brugia* spp. are of minimal veterinary clinical significance, however all canine filariae have the potential to infect humans and remain important from a public health perspective [1, 2].

Diagnostic methods for filarial infections include isolation of adult worms followed by morphological identification, morphological observation of circulating microfilariae by stained blood smears, direct wet smears, modified Knott's technique and the Wylie's filtration technique [2, 3]. Histochemical or immuno-histochemical staining of circulating microfilariae has also been performed [4–6]. Detection of circulating antigen with commercial test kits is currently available and widely used for *D. immitis* [7, 8]. Molecular diagnostic approaches are also increasingly utilised for research and surveillance purposes [9, 10].

In India, *Dirofilaria* spp., *Acanthocheilonema* spp. and *Brugia* spp., have all been reported in dogs [6, 11–13]. Previous reports on filarial infections in India are summarised in Table 1. Based on these limited number of studies, it is currently accepted that *D. immitis* is geographically restricted to India›s north-east and *D. repens* to India›s south, with an overlapping area centrally. There are no reports of *D. immitis* occurring elsewhere in India, despite anecdotal evidence to suggest its occurrence in Delhi (Sharma, personal communication, July 2008, Delhi). This accepted view is questionable as competent mosquito vectors for *D. immitis*, belonging to the genera *Culex, Aedes* and *Anopheles*, that also happen to act as vectors for *D. repens*, are present all over India [14]. Moreover, a case of human pulmonary dirofilariasis due to *D. immitis* was reported in Mumbai in 1989 [15], which casts further doubt on its currently accepted geographical distribution. Recently, 16 out of 75 microfilaraemic dogs were shown to harbour *B. malayi* by researchers at the Kerala Agricultural University›s College of Veterinary and Animal Sciences using morphological and immunodiagnostic criteria [13]. However, since *B. ceylonensis* is endemic in Sri Lanka this finding needs to be confirmed using PCR as the microfilariae cannot be differentiated morphologically from *B. malayi* and it is likely the ELISA test cross-reacts among *Brugia* spp. [16].

Table 1: Previous reported prevalences of canine filarial species in different location in India

	Northeast India			Southern India	
	Mizoram n = 240	Orissa n = 7	Kolkata n = 3200	Kerala n = 160	Karnataka n = 400
Dirofilaria immitis	34%	57%	3%	0%	0%
Dirofilaria repens	0%	14%	0%	7%	21%

In previous reported studies, microfilarial identification relied on morphological assessment only. Despite the availability of published measurements of various microfilariae, the inadequacy of morphological diagnosis was demonstrated by Rishniw and colleagues (2006) [9] when microfilariae initially identified as *A. reconditum* were later determined to be *D. immitis* by molecular methods. Morphological identification of microfilariae not only requires experienced personnel but it may be difficult to detect multiple infections with more than one species of filarial worms [2].

A geographically stratified cross-sectional study was undertaken to determine the prevalence and geographical distribution of canine filarial species of veterinary and public health importance in India using a combination of conventional and molecular diagnostic techniques.

METHODS

Study Site

India features a wide range of climatic zones, ranging from montane (cold, wet alpine regions) and semi-arid regions to the wet tropics. The ecology of vectors of medical and veterinary importance (and therefore the diseases they transmit) is highly dependent on climate [17]. The study was stratified to include four climatic zones of India. The locations of the places sampled represent a unique climatic condition of their own, based on information produced by The World Meteorological Organization [18]. Ladakh in India's far north (3000 m altitude), experiences a temperature that rarely exceeds 27°C in summer, while in winter temperatures drop to minus 20°C. Mumbai's climate can be described as tropical with a high level of humidity. The mean average temperature for Mumbai ranges from 16°C during winter to 30°C in summer. The climate of Delhi is a monsoon-influenced humid subtropical climate with average temperatures range from 7°C during its dry winter to 39°C in summer. Sikkim's climate can be described as subtropical highland with mild temperatures range from 25°C in summer to 4°C in winter [19].

Figure 1: Political map of India. Areas outlined in red rectangles indicate sampling locations.

To facilitate the fieldwork, collaborations were established with Vets Beyond Borders, Jeevaashram, Krishnaashram, Bombay Veterinary College and In Defence of Animals India. These organisations allowed us access to stray and refuge dogs through their Animal Birth Control (ABC) and rabies vaccination programs. In these programs stray dogs are impounded, vaccinated, surgically neutered and released back to their original location [20]. The purpose of this program is to stabilise the street dog population and to help control the spread of rabies. Capillary and whole blood samples were collected from 525 dogs from four different cities (Figure 1), namely Gangtok and Jorethang in Sikkim (n = 101), Ladakh (n = 100) in Jammu and Kashmir, Delhi (n = 162) and Mumbai (n = 162). Blood samples were subjected to normal-thin and buffy-coat smears, air-dried and fixed in 100% ethanol and later stained with Giemsa for microscopic screening. Whole blood samples were also applied onto QIAcard FTA® Four Spots (Qiagen) for molecular-based screening later.

Microscopic Examination of Blood Films

Stained blood films were examined under ×200 and ×400 objective lens for microfilariae. The microfilariae were measured using Olympus BH-2 microscope (Japan) calibrated eye micrometer and photographed using Olympus DP12 digital microscope camera (Japan).

DNA Extraction of Adult D. Immitis and Blood from QIAcard FTA®

DNA from an adult worm of *D. immitis* (courtesy of Murdoch University) was extracted using the tissue sample protocol of the MasterPure DNA purification kit (Epicentre) according to the manufacturer's instructions and utilised as a positive control for this study.

Approximately 500 μl of blood from each animal was applied on QIAcard FTA® Four Spots (Qiagen) which were cut into two cm strips (vertically) and allowed to air dry. A modified DNA purification protocol for tissue samples using the MasterPure DNA purification kit (Epicentre) was used to extract DNA from FTA cards. A 2 cm^2 piece of dried FTA card impregnated with a blood sample was cut into small pieces using a sterile scalpel blade on a clean microscope slide and placed in a 1.5 ml microcentrifuge tube. One μl of Proteinase K, 150 μl tissue lysis buffer and 150 μl phosphate buffer saline were added to the FTA card sample and incubated overnight at 65°C. The lysed sample was then cooled on ice for 5 minutes, 175 μl of MPC protein precipitation reagent added and the sample centrifuged at 6, 800 g for 10 minutes. Following centrifugation the supernatant was transferred into a clean 1.5 ml tube and 500 μl of isopropanol added. The sample was then inverted 30-40 times to promote DNA precipitation, centrifuged at 6,800 g for 10 minute and the supernatant removed carefully so as to not dislodge the DNA pellet. DNA pellet was washed twice with 70% ethanol. The sample was air dried for 5 minute to remove any remaining ethanol before the DNA pellet was resuspended in 20 μl of TE buffer.

PCR Assays and DNA Sequencing

A single-step multiplex PCR targeted at amplifying the internal transcribed spacer-2 region of ribosomal DNA developed by Rishniw and colleagues (2006) [9] was utilised for molecular screening of canine filarial species in blood. Pan-filarial primers, forward: DIDR-F1 5'-AGT GCG AAT TGC AGA CGC ATT GAG-3' and reverse: DIDR-R1 5'-AGC GGG TAA TCA CGA CTG AGT TGA-3' were utilized to amplify and differentiate *D. immitis*, *D. repens*, *B. malayi*, *B. pahangi*, *A. reconditum* and *A. dracunculoides*. The PCR assay was carried out in a final volume of 25 μl containing 1 × PCR buffer, 1.5 mM $MgCl_2$, 200 μM dNTPs, 0.5 μM of each primer, 1 U of Taq polymerase and 1 μl of DNA template. The PCR procedure was executed according to Rishniw et al. (2006). The PCR products were run on a 2% agarose gel in 1 × TAE buffer at 100V and visualised using Geldoc (Biorad). The anticipated product sizes of each species of microfilaria are listed in Table 2.

Table 2: Primer sequences used to amplify PCR products from filarial and canine blood samples [9]

Primer pair	Primer sequence	Gene target	Product origin	Product size (based-pairs)
DIDR-F1	AGT GCG AAT TGC AGA CGC ATT GAG	5.8S-ITS2-28S	*D. immitis*	542
DIDR-R1	AGC GGG TAA TCA CGA CTG AGT TGA		*D. repens*	484
			B. malayi	615

			B. pahangi	664
			A. reconditum	578
			A. dracunculoides	584

Product size and species of filarial nematodes amplified are also reported.

PCR products from 30% of all positive samples were purified using Qiagen spin columns (Qiagen). When a multiple-band product was obtained, target bands were excised and purified with Qiaquick Gel Extraction kit (Qiagen) prior to DNA sequencing. DNA sequencing was performed using an ABI 3130xl Genetic Analyzer (Applied Biosystems) with Big Dye 3.0 chemistry, after which sequences were edited and assembled using Finch TV (Geospiza Inc.).

Primer design for 18S gene and amplification

A nested PCR was designed to amplify a partial region of the 18S rDNA of canine filarial species. Sequences of the near complete 18S rDNA of *B. malayi* [GenBank: AF036588], *Wuchereria bancrofti* [GenBank: AF227234] *D. immitis* [GenBank: AF036638] and *Dipetalonema* sp. [GenBank: DQ531723.1] were aligned using Clustal W http://align.genome.jp/ and an external and an internal set of primers (Table 3) were designed to amplify an approximately 800 bp and 700 bp product, respectively.

Table 3: External and internal primer sets for the amplification of a partial region of the 18S gene of most filarial species

Primer name	Sequence (5'-3')
PAFilariaF1 (external)	GGTGAAACCGCGAACGGCTC
PAFilariaR1 (external)	CCGTCCCTCTTAACCATTATC
PAFilariaF2 (internal)	CTATAATGTACTTGATGTTGATTATC
PAFilariaR2 (internal)	CCATGCACGAGTATTCAGGCA

The PCR assay was carried out in a volume of 25 μl containing 1 × PCR buffer (Qiagen), 1.5 mM $MgCl_2$, 200 μM of each dNTP, 0.5 μM of each of the forward and reverse primers, 1 U of Taq DNA polymerase (Qiagen) and 1 μl of extracted DNA. Five microlitres of Q-solution (Qiagen) was also added to optimise the PCR. The PCR conditions of both primary and secondary PCR are as follows: an initial activation step at 94°C for 2 min was followed by 35 cycles of amplification (94°C for 30 s, 56°C for 30 s and 72°C for 30 s) followed by a final extension step of 72°C for 7 min. The template for the secondary PCR amplification consisted of 1 μl of amplicon from the primary amplification.

Phylogenetic Analyses

Neighbour joining analyses were conducted with Tamura-Nei parameter distance estimates and trees constructed using Mega 4.1 software. Bootstrap analyses were conducted using 1000 replicates.

RESULTS

Out of the 525 dogs examined for circulating microfilaria, 27 (5.1%; 95% CI 3.3, 7.0) were positive using microscopic screening and 139 (26.5%; 95% CI 22.7, 30.3) confirmed for at least one filarial species by PCR. The most common species of canine filarial parasite identified in this study was *A. reconditum* followed by *D. repens* and *D. immitis*. The prevalence and geographical distribution of canine filarial species using both morphology and multiplex PCR are summarised in Table 4.

Table 4: The prevalences of canine filarial species in different cities in India by PCR and microscopy (in parentheses)

	Delhi n = 162	**Mumbai n = 162**	**Sikkim n = 101**	**Ladakh n = 100**	**Overall preva-lence n = 527**
Dirofilaria immitis	4.3% (0%)	0% (0%)	1% (0%)	0% (0%)	**1.5% (0%)**
Dirofilaria repens	4.9% (0%)	16.7% (8%)	0% (0%)	0% (0%)	**6.6% (2.2%)**
Acanthocheilonema reconditum	22.2% (0.6%)	4.3% (1.2%)	5.9% (0%)	0% (0%)	**9.3% (0.6%)**
***Novel* spp.**	0% (0%)	0% (0%)	0% (0%)	48% (13%)	**9.1% (2.5%)**
Microfilaria auquieri	0% (0%)	0% (0%)	0% (0%)	0% (2%)	**0% (0.4%)**

Morphology Identification

All microfilariae found in this study except for those identified from dogs in Ladakh could be identified based on previously described and published morphological criteria. In Ladakh, 13 dogs were found to harbour unsheathed microfilariae (Figure 2) with an average length of 165 μm (range 130 μm to 180 μm), that did not match the length of previously described canine microfilariae (Table 5). Two dogs in Ladakh were also found to have mixed infections of the unidentified microfilaria described above and another microfilaria of an unusual appearance (Figure 3) which was identified morphologically and presumed as *Microfilaria (Mf.) auquieri*, a species previously described by Foley [21] and Rioche [22] in Algeria, North Africa. Further analyses with regard to the genetic identity of the unidentified canine microfilariae in Ladakh were conducted using the 18S and ITS-2 genes.

Table 5: Measurements of microfilaria recorded by various author[a]

Filarial species	Special features of microfilaria	Microfilaria	
		Length (μm)	**Width (μm)**
Dirofilaria immitis	Unsheathed, tapered head, relatively straight tail	218 - 329	5.4 - 6.2
Dirofilaria repens		283 - 360	7.1 - 8.3
Acanthocheilonema reconditum	Unsheathed, round curved body, cephalic hook, blunt anterior end	250 - 270	4 - 4.5
Acanthocheilonema dracunculoides		195 - 230	Not available
Cercopithifilaria grassi		567	Not available
Microfilaria auquieri	Unsheathed	58 - 102	Not available
Microfilaria ochmanni	Sheathed	320	Not available
Brugia malayi	Sheathed, cephalic space: 6.3 - 6.7 μm	254 - 234	5.99-7.99
Brugia pahangi	Sheathed, cephalic space: 6.4 μm	200 - 189	4 - 5
Brugia ceylonensis	Sheathed, blunt tail, cephalic space: 6.3 - 6.7 μm	220 - 275	Not available

[a]Data obtained from [2, 6, 21, 53, 54]

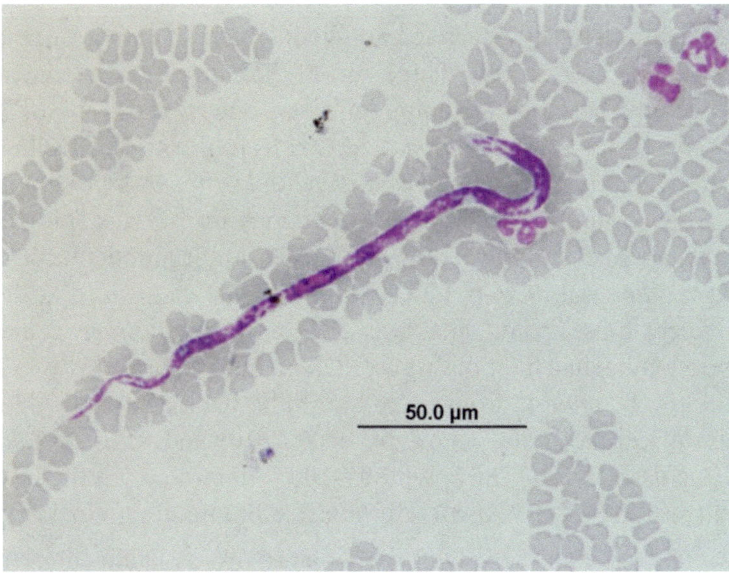

Figure 2: Unidentified microfilaria observed in Giemsa blood smears of dogs from Ladakh.

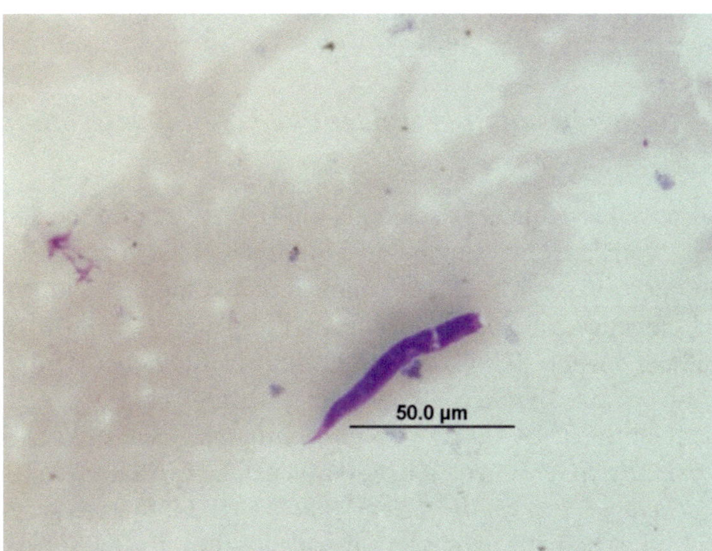

Figure 3: Microfilaria observed in Giemsa blood smears of dogs from Ladakh, India which conform to the morphological descriptions of *Microfilaria auquieri* Foley, 1921.

Single-step Multiplex PCR

Out of 525 samples, 139 produced amplicons corresponding to *D. immitis, D. repens* or *A. reconditum*. Three out of 527 dogs were found to have mixed infections based on the amplicon sizes; one from the city of Mumbai with *D. repens* and *A. reconditum*, and two from the city of Delhi with *D. immitis* and *A. reconditum*, and *D. immitis* with *D. repens*. Two samples from Ladakh which found to have mixed infections by microscopic screening with presumably *Mf. auquieri*, only produced a single amplicon corresponds to *A. reconditum*. All PCR products amplified and sequenced at the ITS-2 region, except those from Ladakh as well as *A. reconditum* isolates from Delhi and Mumbai had their multiplex PCR results confirmed with published sequences for *D. immitis* from Mizoram, India [GenBank: EU087699], *D. repens* from Kerala, India [GenBank: FJ717410] and *A. reconditum* from Taiwan [GenBank: AF217801] with 99-100% homology on the basic local alignment search tool, BLASTn http://blast.ncbi.nlm.nih.gov/Blast.cgi.

Acanthocheilonema reconditum isolates from Delhi and Mumbai displayed 95% and 90% homology to the published ITS-2 sequence for *A. reconditum* from Taiwan [GenBank: AF217801]. Sequences for the ITS-2 region of *Acanthocheilonema* isolates from each city; Delhi, Mumbai and Ladakh were submitted to GenBank under accession numbers GU593976, GU593978 and GU593978, respectively.

Genetic characterisation of unidentified canine microfilaria species

Clear and readable sequences spanning a 441 bp region of the SSU rDNA gene were obtained from a single Ladakhi isolate of the unidentified species of canine microfilaria. These were aligned and compared with previously published sequences of closely related filariid species *B. malayi* [GenBank: AF036588], *Loa loa* [GenBank DQ094173], *W. bancrofti* [GenBank: AY843436], *A. vitae* [GenBank: DQ094171], *Onchocerca cervicalis* [GenBank: DQ094174], *Setaria digitata* [GenBank: DQ094175] and *D. immitis* [GenBank: AF182647] using Clustal W (BioEdit v 7.0.5.3). *Thelazia lacrymalis* [GenBank: DQ503458] was used as an outgroup [23]. Although bootstrap support was low for the differentiation of all genera of filariid, the unidentified microfilariae isolated from dogs in Ladakh were distinctly placed within the same clade as *A. vitae* (Figure 4).

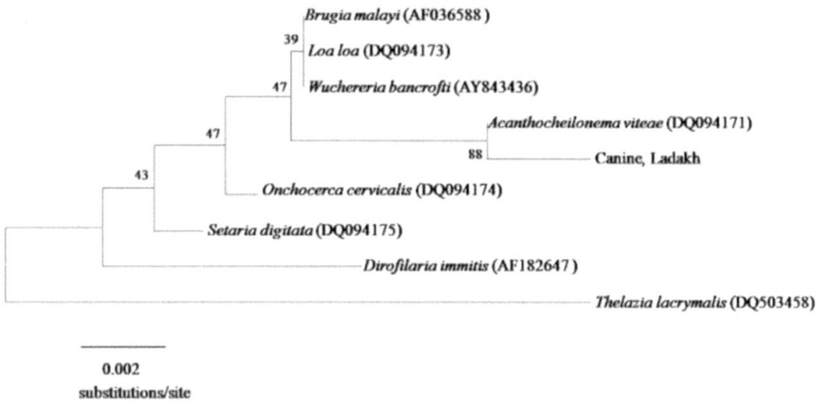

Figure 4: Phylogenetic placement of the unidentified species of microfilaria from Ladakhi dogs based on partial SSU rDNA gene sequences. Bootstrap values at nodes indicate percentagecalculated in 1000 replicates. *Thelazia lacrymalis* was used as an outgroup.

ITS-2 sequences of *Acanthocheilonema* were found to have identical sequence homology within the same geographical area. Sequences of six of the unidentified microfilaria isolates from Ladakh, two isolates of *A. reconditum* from Mumbai and Delhi and a single isolate from Sikkim, together with GenBank reference sequences of canine filarial species *A. reconditum* [GenBank: AF217801], *A. dracunculoides* [GenBank: DQ018785] were compared and aligned with a using Clustal W (BioEdit v 7.0.5.3).

The topology of the unrooted phenogram recognises five major groups within the genus *Acanthocheilonema*, which encompassed *A. reconditum* from Taiwan/Sikkim, *A. reconditum* from Delhi, *A. reconditum* from Mumbai, *Acanthocheilonema* isolates from Ladakh and *A. dracunculoides* as separate groups. Isolates from Mumbai formed a sister group to *A. reconditum* isolates from Taiwan, Sikkim and Delhi. There was very strong bootstrap placement for all six *Acanthocheilonema* isolates from Ladakh into a distinct group from all isolates of *A. reconditum* as well as *A. dracunculoides* (Figure 5).

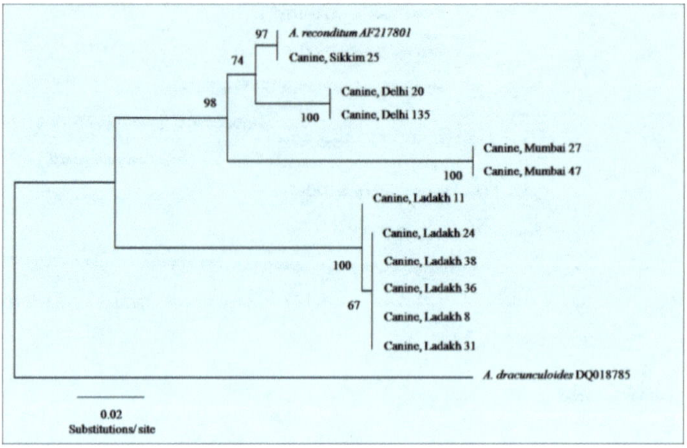

Figure 5: An unrooted phenogram of the ITS-2 region of the unidentified species of microfilariae from Ladakh using neighbour-joining analysis with the Tamura-Nei model. Bootstrap values at nodes indicatepercentage calculated in 1000 replicates.

Genetic distances of the ITS-2 region between *A. reconditum* and *Acanthocheilonema* isolates from Ladakh (11%) and Mumbai (9%) were similar to that between *A. reconditum* and *A. dranunculoides* (19%) (Table 6), whereas those between *A. reconditum* and isolates from Delhi were significantly less (3%).

Table 6: Distance matrix showing the nucleotide difference among ITS-2 gene sequences for microfilaria isolated from dogs from Delhi, Mumbai, Sikkim and Leh with reference sequences from GenBank for *A. reconditum* [GenBank: AF217801] and *A. dracunculoides* [GenBank: DQ018785]

A. dracunculoides _ **DQ018785**					
A. reconditum _ **AF217801**	0.19				
Canine, Delhi isolate (A.r)	0.20	0.03			
Canine, Ladakh isolate	0.22	0.11	0.13		
Canine, Sikkim isolate (A.r)	0.19	0.00	0.03	0.11	
Canine, Mumbai isolate (A.r)	0.26	0.09	0.09	0.17	0.09

Isolates from the study designated 'A.r' in parentheses are those that morphologically resembled A. reconditum.

DISCUSSION

This is the first comprehensive study that has utilised a combination of conventional and molecular techniques to determine the distribution and occurrence of canine filarial species in India. Despite anecdotal accounts of heartworm being present in Delhi, this study is the first to confirm its presence in that city, together with *D. repens*. This study is in agreement with previous reports [6, 11, 24] demonstrating a low prevalence of microfilaraemia overall, and that the most common filarial species found in India are *A. reconditum* and *D. repens*. Despite the relatively limited geographical locations used in this study, our results also support the aforementioned hypothesis that canine heartworm is primarily confined to tropical and sub-tropical areas of northern India, but seems to be absent towards the south. This hypothesis is debatable since competent mosquito vectors for *D. immitis* are present throughout central and southern India. For example, *Aedes albopictus* [25–27], a competent vector for *D. immitis* is present in Maharastra, Karnataka and Pondicherry [28], and heartworm is yet to be reported in dogs from these areas where climate would support larval development within the mosquito vector. It is noteworthy however that the prevalence of all filarial species determined in the current study may still an underestimate of its true value due to false negative test results. The high prevalence of *D. immitis* by necropsy that was reported in eastern and central India cannot be disregarded [29, 30]. Authors of these studies also mentioned that more than 30% of dogs positive for *D. immitis* had occult infections, which is in agreement with studies from Australia in the 1980s [31]. Several types of occult filarial infection in dogs have been documented: pre-patent infection, naturally occurring unisexual infection and immune-mediated clearance of microfilariae, a similar situation that exists also in human filariasis [3, 32–34]. PCR has been shown to detect occult infections of *Loa loa* in humans [35, 36], however a recent study by Duscher (2009) has shown that a minimum parasitaemia level of 6 ± 0.43 microfilariae per 100 μl of blood on the FTA cards is need for detection of *D. repens* microfilariae by PCR detection [37]. The authors explain that it is hard to distribute large extracellular metazoan stages such as microfilariae on a filter paper, and this can therefore lead to false negative results for low parasitaemia infections. Furthermore, macrocyclic lactones such as ivermectin are known to possess microfilaricidal effects. Almost half the refuge dogs from Mumbai and Delhi sampled in this study had a history of receiving treatment with ivermectin for the control of gastrointestinal and ectoparasites. A recent study by Bazzocchi and colleagues [38] confirmed that with prolonged treatment of ivermectin, a significant decreased in circulating microfilariemia in dogs (less than 100 microfilariae/ml) occurred.

The most common filarial species found in dogs from this study are *A. reconditum* and *D. repens*. Most infections with *A. reconditum* and *D. repens* do not contribute to any clinical illness in dogs, but this is not the case in humans [1, 2]. *Dirofilaria repens* causing subcutaneous nodules and sub-conjunctival infections in humans is now considered as a re-emerging zoonoses and it often leads to misdiagnosis of malignant tumours in endemic areas [1, 24]. Although *D. repens* is endemic to a wide geographical area spanning Europe and Asia, human cases, often involving nodules in organs such as lungs, male genitals and female breast are most commonly reported from Italy and Sri Lanka [39]. With regard to *Acanthocheilonema*, only one human case has been reported in Australia, which involved the eye; the recovered worm morphological features were consistent with an unfertilised adult female *A. reconditum* [40]. *Dirofilaria immitis* infection in humans is rarely reported, but associated with pulmonary lesions or radiological coin lesions of the lung. The significance of *D. immitis* infection in human is the confusion and invariable radiological misdiagnosis of a primary or metastatic lung tumour, which usually leads to thoracotomy for open lung biopsy or wedge resection of the lung to obtain the correct diagnosis [41, 42]. Sporadic reports of the immature heartworm in unusual locations such as the eye, mesentery, cerebral artery, spermatic cord and liver also exist [43–47]. To date, twelve cases of human subcutaneous dirofilariasis due to *D. repens* has been reported in southern India [24] and a single case of human pulmonary dirofilariasis due to *D. immitis* in Mumbai [15].

All microfilariae found in this study except for those identified from dogs in Ladakh could be identified based on previously described and published morphological criteria. Morphologic dimensions of the microfilariae genetically characterised as *A. reconditum* from Delhi, Mumbai and Sikkim matched previously documented measurements for *A. reconditum*. This was supported by molecular evidence to show that the isolates clustered within the *A. reconditum* group on phylogenetic analysis. Genetic distances of the ITS-2 region between *A. reconditum* and isolates from Delhi were within the range expected for intra-species variation of different geographical populations. However genetic distances of the ITS-2 region between *A.*

reconditum and isolates from Mumbai were within the range expected for separate species of the same genera. Different species of filarial nematodes are reported to vary in their degree of intra-species variation based on the ITS-2 gene [48], making this comparative analysis difficult to confirm. Future molecular epidemiological studies comparing the intra-species variation of *Acanthocheilonema* spp. in different geographical locations at multiple loci is necessary to confirm this hypothesis. This finding is in contrast to the situation discovered with the microfilariae found to infect nearly half the dogs in Ladakh. Morphologically, the length of these latter microfilariae did not match any of the documented or known species of canine filariae. Molecular evidence based on the phylogenetic analysis of the 18S and ITS-2 regions provided strong evidence to show that the canine microfilariae found in Ladakh belong to a novel species of *Acanthocheilonema*. Genetic distances between this novel species and *A. reconditum* were within the range expected for separate species of the same genera. We propose that in the interim, this new species of microfilaria be designated *Acanthocheilonema ladakhii* until further and more detailed morphological, molecular and clinic-pathological studies are undertaken to describe and name this novel species of canine filarial parasite and ascertain its veterinary significance.

This study also presents the first report of what we presume to be *Mf. auquieri* in India and this serendipitous re-encounter with this filarial species after 50 years is of great parasitological interest. Although previously reported in stray dogs of Algeria, there is still much to be discovered about this enigmatic parasite. The adult form of *Mf. auquieri* is not yet known as attempts to search for the adult parasite failed despite a systematic post-mortem by the early French researchers [22]. It is interesting to note also that almost all dogs in Ladakh, and a small proportion in Delhi, were found infested with a species of blood-sucking fly (authors' personal observation), identified as *Hippobosca longipennis* syn. *capensis* Fabricius, 1805 [49], which is known to be an intermediate host and vector of *A. dracunculoides* [50]. Our findings in Ladakh have prompted us to speculate whether this same *Hippobosca* fly is also the vector/intermediate host for *Mf. auquieri* or whether it is just mere co-incidence that this fly also happens to be common in Algeria where *Mf. auquieri* was first reported [49].

As a side note the authors would like to add that the application of blood samples to filter paper-based technology, such as the FTA cards used in this study, allow for the rapid and safe dispatch of samples to diagnostic facilities capable of PCR-based diagnosis, with the added advantage of providing an archival potential. However, using this technique in humid regions can result in fungal contamination of the blood-impregnated filter paper as we have experienced during this study. A careful drying process of the papers is therefore recommended to prevent this problem from occurring.

CONCLUSION

At least six species of filarial parasite are now known to infect dogs in India, two of which were reported for the first time in this study, namely *Mf. auquieri* and a novel species of *Acanthocheilonema*, both of which were discovered in the Himalayan city of Ladakh. The study also confirms and extends the known geographical distribution of canine heartworm in India. The distribution of heartworm in India extends from the Pakistani border in the west [51], Delhi and Sikkim in the north to the Burmese border in the east and Orissa to the south. The public health importance of *D. repens* also suggests that appropriate prophylaxis be administered to dogs throughout central and west India from as far south as Kerala to as far north as Delhi. From a diagnostic viewpoint, it is important to utilise an immunodiagnostic test that will not cross-react [8] between *D. immitis* and *D. repens* in areas where these two filarial species co-exist, for example, Delhi.

ACKNOWLEDGEMENTS

Financial support for this study was provided by Bayer Animal Health. We gratefully thank our collaborators; Vets Beyond Borders, Jeevaashram, Krishnaashram and In Defense of Animal for their help with the fieldwork. Special thanks to David Spratt and Odile Bain for their very insightful information about Mf. auquieri, Jenny Seddon for her help with construction and interpretation of the phylogenetic results, Brian Bynon and Mark Roper for their help with slides staining process and Myat Kyaw-tanner for her help with molecular methods. PhD scholarship support for Puteri Azaziah Megat Abd Rani is provided by The Ministry of Higher Education, Malaysia. Publication of this thematic series has been sponsored by Bayer Animal Health GmbH

REFERENCES

1. Macpherson CN, Meslin FX, Wandeler AI: Dogs, Zoonoses and Public Health. 2000, Anthony Rowe Limited, Eastbourne
2. Irwin PJ, Jefferies R: Arthropod-transmitted diseases of companion animals in Southeast Asia. Trends Parasitol. 2004, 20: 27-34. 10.1016/j.pt.2003.11.004.
3. Grieve RB, Lok JB, Glickman LT: Epidemiology of canine heartworm infection. Epidemiol Rev. 1983, 5: 220-246.
4. Kuecks RW, Slocombe JOD: Rapid histochemical differentiation of *Dirofilaria immitis* and *Dipetalonema reconditum*. Proceedings of the Heartworm Symposium; 23-24 February 1980; Dallas, Texas. Edited by: Otto GF. 1980, 48-50.
5. Peribanez MA, Lucientes J, Arce S, Morales M, Castillo JA, Gracia MJ: Histochemical differentiation of *Dirofilaria immitis*, *Dirofilaria repens* and *Acanthocheilonema dracunculoides* microfilariae by staining with a commercial kit, Leucognost-SP (R). Vet Parasitol. 2001, 102: 173-175. 10.1016/S0304-4017(01)00516-7.
6. Ananda KJ, D'Souza PE, Jagannath MS: Methods for identification of microfilaria of *Dirofilaria repens* and *Dipetalonema reconditum*. J Vet Parasitol. 2006, 20: 45-47.
7. Atkins CE: Comparison of results of three commercial heartworm antigen test kits in dogs with low heartworm burdens. J Am Vet Med Assoc. 2003, 222: 1221-1223. 10.2460/javma.2003.222.1221.
8. Ranjbar-Bahadori S, Eslami A, Bokaic S: Evaluation of different methods for diagnosis of *Dirofilaria immitis*. Pakistan J Bio Sci. 2007, 10: 1938-1940. 10.3923/pjbs.2007.1938.1940.
9. Rishniw M, Barr SC, Simpson KW, Frongillo MF, Franz M, Dominguez Alpizar JL: Discrimination between six species of canine microfilariae by a single polymerase chain reaction. Vet Parasitol. 2006, 135: 303-314. 10.1016/j.vetpar.2005.10.013.
10. Casiraghi M, Bazzocchi C, Mortarino M, Ottina E, Genchi C: A simple molecular method for discriminating common filarial nematodes of dogs (*Canis familiaris*). Vet Parasitol. 2006, 141: 368-372. 10.1016/j.vetpar.2006.06.006.
11. Gogoi AR: Filarids of animals in India. J Vet Parasitol. 2002, 16: 131-138.

12. Dam T, Das P: The importance of dirofilariasis in India. Internet J Parasitic Dis. 2006, 1:
13. Human filariasis parasite found in dogs. [http://www.hindu.com/2008/12/06/stories/2008120654580400.htm]
14. Sabu L, Subramanian H: Prevalence and vector potentiality of mosquitoes from Thissur, Kerala. J Vet Parasitol. 2007, 21: 165-167.
15. Badhe BP, Sane SY: Human pulmonary dirofilariasis in India: a case report. J Trop Med Hyg. 1989, 92: 425-426.
16. Supali T, Rahmah N, Djuardi Y, Sartono E, Rückert P, Fischer P: Detection of filaria-specific IgG4 antibodies using *Brugia* Rapid test in individuals from an area highly endemic for *Brugia timori*. Acta Trop. 2004, 90: 255-261. 10.1016/j.actatropica.2004.02.001.
17. Patz JA, Campbell-Lendrum D, Holloway T, Foley JA: Impact of regional climate change on human health. Nature. 2005, 438: 310-317. 10.1038/nature04188.
18. Current weather forecast and climatological information - India. [http://www.worldweather.org/066/m066.htm]
19. Current weather information - Gangtok. [http://imdsikkim.gov.in/]
20. Menezes R: Rabies in India. Can Med Assoc J. 2008, 178: 564-10.1503/cmaj.071488.
21. Foley H: Microfilaires du chien dans le Sud-Oranais. Ann Inst Pasteur (Paris). 1921, 35: 212-217.
22. Rioche M: Presence de *Dipetalonema dracunculoides* (Cobbold, 1870) chez le chien dans la region d'Alger. Arch Inst Pasteur Algér. 1960, 38: 386-398.
23. Casiraghi M, Anderson TJC, Bandi C, Bazzocchi C, Genchi C: A phylogenetic analysis of filarial nematodes: comparison with the phylogeny of *Wolbachia* endosymbionts. Parasitology. 2001, 122: 93-103. 10.1017/S0031182000007149.
24. Sabu L, Devada K, Subramanian H: Dirofilariosis in dogs & humans in Kerala. Indian J Med Res. 2005, 121: 691-693.
25. Lai C-H, Tung K-C, Ooi H-K, Wang J-S: Susceptibility of mosquitoes in central Taiwan to natural infections of *Dirofilaria immitis*. Med Vet Entomol. 2001, 15: 64-67. 10.1046/j.1365-2915.2001.00280.x.
26. Cancrini G, Frangipane di Regalbono A, Ricci I, Tessarin C, Gabrielli S, Pietrobelli M: *Aedes albopictus* is a natural vector of *Dirofilaria*

immitis in Italy. Vet Parasitol. 2003, 118: 195-202. 10.1016/j.vetpar.2003.10.011.

27. Tiawsirisup S, Kaewthamasorn M: The potential for *Aedes albopictus* (Skuse) (Diptera:Culicidae) to be a competent vector for canine heartworm, *Dirofilaria immitis* (Leidy). Southeast Asian J Trop Med Public Health. 2007, 38: 208-214.
28. Kumar NP, Rajavel AR, Natarajan R, Jambulingam P: DNA barcodes can distinguish species of Indian mosquitoes (Diptera: Culicidae). J Med Entomol. 2009, 44: 1-7.
29. Patnaik MM: On filarial nematodes in domestic animals in Orissa. Indian Vet J. 1989, 66: 573-574.
30. Borthakur SK, Sarmah K, Rajkhowa TK, Das MR, Rahman S: *Dirofilaria immitis* infection in dog. J Vet Parasitol. 2006, 20: 167-169.
31. Atwell RB: Basic pathophysiology and epidemiology in Australia. Heartworm symposium: Refresher course for veterinarians; 5 - 7 July 1988; Leura, NSW. 1988, 36-43.
32. Weil GJ, Powers KG, Parbuoni EL, Line BR, Furrow RD, Ottesen EA: *Dirofilaria immitis* .6. Antimicrofilarial immunity in experimental filariasis. Am J Trop Med Hyg. 1982, 31: 477-485.
33. Elsadr WM, Aikawa M, Greene BM: *In vitro* immune-mechanisms associated with clearance of microfilariae of *Dirofilaria immitis*. J Immunol. 1983, 130: 428-434.
34. Simón F, Morchón R, Gonzáles-Miguel J, Marcos-Atxutegi C, Siles-Lucas M: What is new about animal and human dirofilariosis?. Trends Parasitol. 2009, 25: 404-409. 10.1016/j.pt.2009.06.003.
35. Touré FS, Bain O, Nerrienet E, Millet P, Wahl G, Toure Y, Doumbo O, Nicolas L, Georges AJ, McReynolds LA, Egwang TG: Detection ofLoa loa-Specific DNA in Blood from Occult-Infected Individuals. Exp Parasitol. 1997, 86: 163-170. 10.1006/expr.1997.4168.
36. Touré FS, Mavoungou E, Kassambara L, Williams T, Wahl G, Millet P, Egwang TG: Human occult loiasis: field evaluation of a nested polymerase chain reaction assay for the detection of occult infection. Trop Med Int Health. 1998, 3: 505-511. 10.1046/j.1365-3156.1998.00260.x.
37. Duscher G, Peschke R, Wille-Piazzai W, Joachim A: Parasites on paper - The use of FTA Elute® for the detection of *Dirofilaria*

repens microfilariae in canine blood. Vet Parasitol. 2009, 161: 349-351. 10.1016/j.vetpar.2009.01.007.

38. Bazzocchi C, Mortarino M, Grandi G, Kramer LH, Genchi C, Bandi C, Genchi M, Sacchi L, McCall JW: Combined ivermectin and doxycycline treatment has microfilaricidal and adulticidal activity against *Dirofilaria immitis* in experimentally infected dogs. Int J Parasitol. 2008, 38: 1401-1410. 10.1016/j.ijpara.2008.03.002.
39. Pampiglione S, Rivasi F: Human dirofilariasis due to *Dirofilaria* (*Nochtiella*) *repens*: an update of world literature from 1995 to 2000. Parassitologia. 2000, 42: 231-254.
40. Huynh T, Thean J, Maini R: *Dipetalonema reconditum* in the human eye. Br J Ophthalmol. 2001, 85: 1391-10.1136/bjo.85.11.1384i.
41. Theis JH: Public health aspects of dirofilariasis in the United States. Vet Parasitol. 2005, 133: 157-180. 10.1016/j.vetpar.2005.04.007.
42. Foroulis CN, Khaldi L, Desimonas N, Kalafati G: Pulmonary dirofilariasis mimicking lung tumor with chest wall and mediastinal invasion. Thorac Cardiovasc Surg. 2005, 53: 173-175. 10.1055/s-2004-830567.
43. Dobson C, Welch JS: Dirofilariasis as a cause of eosinophilic meningitis in man diagnosed by immunofluorescence and arthus hypersensitivity. Trans R Soc Trop Med Hyg. 1974, 68: 223-228. 10.1016/0035-9203(74)90118-7.
44. Moorhouse DE: *Dirofilaria immitis*: A cause of human intra-ocular infection. Infection. 1978, 6: 192-193. 10.1007/BF01641909.
45. Tada I, Sakaguchi Y, Eto K: *Dirofilaria* in the Abdominal Cavity of a Man in Japan. Am J Trop Med Hyg. 1979, 28: 988-
46. Theis JH: Case report: Unusual location of *Dirofilaria immitis* in a 28-year-old man necessitates orchiectomy. Am J Trop Med Hyg. 2001, 64: 317-322.
47. Kim MK, Kim CH, Yeom BW, Park SH, Choi SY, Choi JS: The first human case of hepatic dirofilariasis. J Korean Med Sci. 2002, 17: 686-690.
48. Areekit S, Singhaphan P, Khucareontaworn S, Kanjanavas P, Sriyaphai T, Pakpitchareon A, Khawsak P, Chansiri K: Intraspecies variation of *Brugia* spp. in cat reservoirs using complete ITS sequences. Parasitol Res. 2009, 104: 1465-1469. 10.1007/s00436-009-1352-x.

49. Bequaert J: Notes on Hippoboscidae 2. The subfamily Hippoboscinae. Psyche (Stuttg). 1930, XXXVII: 303-326.
50. Nelson GS: *Dipetalonema drancunculoides* (Cobbold, 1870), from the dog in Kenya: wth a note on its development in the louse-fly, *Hippobosca longipennis*. J Helminthol. 1963, 37: 235-240. 10.1017/S0022149X00003825.
51. Wolfe MS, Aslamkhan M, Sharif M, E P: *Acanthocheilonema dracunculoides* (Cobbold, 1870) in dogs in Lahore, West Pakistan. J Helminthol. 1971, 45: 171-176.
52. Chakravarty A, Chaudhuri MN: Studies on canine filariasis in West Bengal. Indian J Anim Health. 1983, 22: 151-155.
53. Lapage G: Veterinary Helminthology and Entomology. 1956, London: Bailliere, Tindall and Cox, 4
54. Chakravarty A, Chaudhuri MN: Developmental stages of *Brugia pahangi* in the final host. J Parasitol. 1962, 48: 693-706. 10.2307/3275258.

Section: 5 Animal Models

CHAPTER 22

Guinea Pig Model for Evaluating the Potential Public Health Risk of Swine and Avian Influenza Viruses

Yipeng Sun[1]., Yuhai Bi[1,3], Juan Pu[1] , Yanxin Hu[1] , Jingjing Wang[1] , Huijie Gao[1] , Linqing Liu[1] , Qi Xu[1] , Yuanyuan Tan[1] , Mengda Liu[1] , Xin Guo[1] , Hanchun Yang[1] , Jinhua Liu[1,2]

[1] Key Laboratory of Zoonosis of Ministry of Agriculture, College of Veterinary Medicine, China Agricultural University, Beijing, China

[2] The Shandong Animal Disease Control Center, Jinan, Shandong, China

[3] CAS Key Laboratory of Pathogenic Microbiology and Immunology Institute of Microbiology, Chinese Academy of Sciences, Beijing, China

Citation: Sun Y, Bi Y, Pu J, Hu Y, Wang J, Gao H, et al. (2010) Guinea Pig Model for Evaluating the Potential Public Health Risk of Swine and Avian Influenza Viruses. PLoS ONE 5(11): e15537. https://doi.org/10.1371/journal.pone.0015537.

ABSTRACT

Background

The influenza viruses circulating in animals sporadically transmit to humans and pose pandemic threats. Animal models to evaluate the potential public health risk potential of these viruses are needed.

Methodology/Principal Findings

We investigated the guinea pig as a mammalian model for the study of the replication and transmission characteristics of selected swine H1N1, H1N2, H3N2 and avian H9N2 influenza viruses, compared to those of pandemic (H1N1) 2009 and seasonal human H1N1, H3N2 influenza viruses. The swine and avian influenza viruses investigated were restricted to the respiratory system of guinea pigs and shed at high titers in nasal tracts without prior adaptation, similar to human strains. None of the swine and avian influenza viruses showed transmissibility among guinea pigs; in contrast, pandemic (H1N1) 2009 virus transmitted from infected guinea pigs to all animals and seasonal human influenza viruses could also horizontally transmit in guinea pigs. The analysis of the receptor distribution in the guinea pig respiratory tissues by lectin histochemistry indicated that both SAα2,3-Gal and SAα2,6-Gal receptors widely presented in the nasal tract and the trachea, while SAα2,3-Gal receptor was the main receptor in the lung.

Conclusions/Significance

We propose that the guinea pig could serve as a useful mammalian model to evaluate the potential public health threat of swine and avian influenza viruses.

INTRODUCTION

Five influenza pandemics have resulted in significant morbidity and mortality around the world. Genetic analyses revealed that these pandemic strains were entirely or partially derived from the viruses of animal origin, including swine and avian strains [1]. So far, animal influenza viruses have repeatedly transmitted to humans with increasing genetic diversity, such as avian H5N1 and H9N2 influenza viruses, and swine H1N1, H1N2, H3N2 influenza viruses [2], [3], which served as an important reminder that another influenza pandemic is highly likely. Among these swine and avian influenza viruses, in addition to highly pathogenic H5N1 avian influenza viruses, many low pathogenic viruses are also high on the list of candidates that could potentially cause the next human pandemic. The efficiency of transmission is a key factor in determining the severity of influenza epidemics. Despite evidence for limited human-to-human transmission of some animal influenza viruses, they have yet to exhibit sustained transmission among humans [4], [5]. However, with typical human flu-like illness in humans [6], [7] that may be unrecognized, these less pathogenic animal viruses can develop the ability of human-to-human transmission. Thus, a mammalian disease model is needed to evaluate the epidemic potential of these viruses.

Mice are a common mammalian model used in influenza research but are not ideal for transmission studies and even the 1918 pandemic strain could not transmit between mice [8]. Ferrets, however, are susceptible to influenza infection and develop similar symptoms to humans and effectively transmit between individuals. Nevertheless, the ferret model presents several practical disadvantages [9]. Recently, guinea pigs have been shown to be an acceptable alternative mammalian model for the study of human influenza A virus transmission [10]–[12]. Guinea pigs are an attractive model host due to their high susceptibility to and ability to transmit low-passage human isolates (in contrast to mice), and based on several practical considerations such as size and cost, make them considerably more convenient for research purposes than ferrets [8]. Additionally, guinea pigs were also proposed as a useful model to understand the virus-host interactions in influenza A virus infection [13] because they possess similar innervations in the airway to humans [14] and also have human-like bronchus-associated lymphoid tissue in lungs [15].

Host receptor expression is one of the important factors influence the infection of the respiratory tracts by influenza viruses. The HA protein of influenza viruses initiates infection by binding sialic acid (SA) which are bound to glycans through an α2,3 or α2,6 linkage [16]. In general, human influenza viruses have a binding preference for SAα2,6-Gal and avian influenza viruses preferentially binding SAα2,3-Gal [17], [18]. The distribution of these two forms of sialic acid compounds is known to be cell and species specific. Thus, host receptor distribution is an important factor in consideration when evaluating an animal model.

Thus far, influenza studies using the guinea pig model have been performed with 1918 pandemic [19], 2009 pandemic H1N1 [11], seasonal human H1N1 and H3N2 [10], [13], and avian H5N1 influenza viruses [15], [20], [21], but little is known about other swine and avian influenza virus infections, particularly those isolated in recent years. To evaluate the applicability of the guinea pig model for swine and avian influenza viruses, we describe the infectivity and transmissibility in guinea pigs of the subtypes of viruses that repeatedly transmit to humans (swine H1N1, H1N2, H3N2 and avian H9N2 influenza viruses) compared with 2009 pandemic and human seasonal influenza viruses. Additionally, the expression of SAα2,3-Gal and SAα2,6-Gal linked receptors in guinea pig respiratory tissues were determined.

RESULTS

Replication of Human, Swine and Avian Influenza Viruses in Guinea Pigs

To evaluate the replication of human, swine and avian influenza viruses in guinea pigs, groups of 15 animals were anesthetized and intranasally inoculated with 10^6 50% egg infective dose (EID_{50}) of each virus listed in Table 1. On days 2, 4, 6, and 8 post-inoculation (p.i.), three animals from each group were anesthetized and the tracheas, lungs, hearts, livers, spleens, kidneys, brains, and colons were collected for virus titration. Nasal washes from the remaining three animals were collected on days 2, 4, 6, and 8 p.i., and titrated by EID_{50} assay. The study indicated that, similar to human influenza viruses, all swine and avian influenza viruses used in this study replicated efficiently in the nasal passages of guinea pigs, with peak virus titers in nasal washings ranging from $10^{4.4}$ to $10^{6.6}$ EID_{50}/mL. The virus titer of BJ/317/09 (H1N1) was lower than those of other pandemic 2009 (H1N1) virus reported previously [22], possibly due to strain specificity. Most viruses grew to peak titers on day 2 p.i. in the nasal passages of inoculated guinea pigs and dropped to undetectable levels by day 8 p.i.. However, Sw/GD/211/06 (H3N2), Sw/GD/968/06 (H3N2) and Qa/HK/G1/97 (H9N2) strains possessed different shedding kinetics and were cleared more slowly from the guinea pig nasal passages than other viruses (Fig. 1 A–D). The nasal virus titers of these three virus-infected animals remained high (>$10^{5.0}$ EID_{50}/mL) on day 6 p.i.. In particular, the Qa/HK/G1/97 (H9N2) virus still replicated well in nasal passages on day 8 p.i. with mean titers of $10^{5.3}$ EID_{50}/mL.

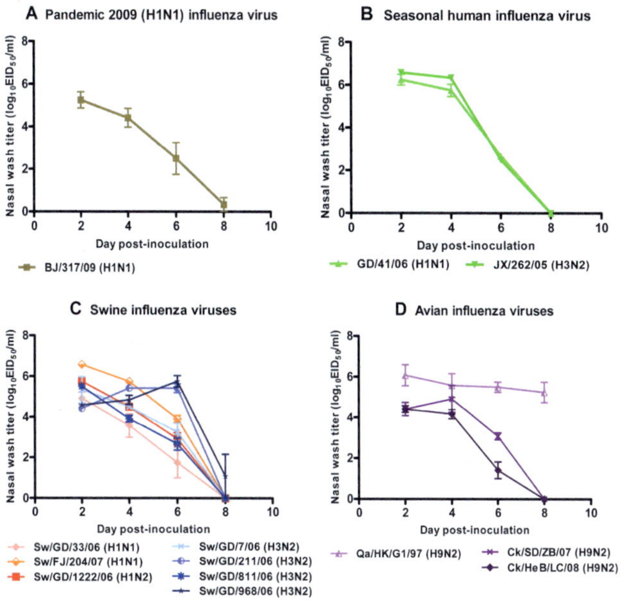

Figure 1: Growth kinetics of pandemic H1N1.

(A), seasonal human (B), swine (C), and avian (D) influenza viruses in the nasal passages of guinea pigs. Groups of three guinea pigs were intranasally inoculated with 10^6 EID_{50} of virus. Nasal wash titers (mean ± SD) are plotted on days 2, 4, 6, and 8 p.i. Nasal wash titers of different strains are represented by different colors. The lower limit of detection was $10^{1.5}$ EID_{50}/mL.

https://doi.org/10.1371/journal.pone.0015537.g001

Compared to nasal passages, tracheas and lungs had much lower virus titers and viruses were cleared more quickly from these organs (Table 2). Two human seasonal influenza viruses were not detected in the lower respiratory tract, while the 2009 pandemic H1N1 virus could replicate in the lung. For animal influenza viruses, only the Sw/FJ/204/07 (H1N1), Sw/GD/1222/06 (H1N2), Qa/HK/G1/97 (H9N2) and Ck/SD/1/06 (H9N2) viruses could be detected in the lung. All of these viruses could not be detected in tracheas and lungs on day 8 p.i..

Table 1: Virus, clinical signs, virus replication, transmission, and seroconversion of the guinea pigs

https://doi.org/10.1371/journal.pone.0015537.t001

Virus name	Abbreviation	Host	Subtype	Phylogenetic lineage of HA gene	Inoculated guinea pigs			Contact guinea pigs		Transmission
					Sneezing (day of onset)	Peak mean nasal wash titer±SD (day)[b]	Seroconversion (HI titer range)[c]	Virus detected in nasal wash[d]	Seroconversion (HI titer range)[c]	
A/Beijing/317/09	BJ/317/09	Human	H1N1	Novel swine-origin	3/3[a](2)	5.3±0.7 (2)	3/3 (160–320)	3/3	3/3 (40–160)	Efficient
A/Guangdong/41/06	GD/41/06	Human	H1N1	Human Seasonal	1/3 (3)	6.3±0.4 (2)	3/3 (320)	1/3	2/3 (80–160)	Inefficient
A/Jiangxi/262/05	JX/262/05	Human	H3N2	Human Seasonal	0/3	6.6±0.1 (2)	3/3 (160–320)	2/3	1/3 (80)	Inefficient
A/swine/Guangdong/33/06	Sw/GD/33/06	Swine	H1N1	Classical swine	0/3	4.9±0.5 (2)	3/3 (40)	0/3	0/3	None
A/swine/Fujian/204/07	Sw/FJ/204/07	Swine	H1N1	European avian-like swine	0/3	6.6±0.1 (2)	3/3 (40–80)	0/3	0/3	None
A/swine/Guangdong/1222/06	Sw/GD/1222/06	Swine	H1N2	North American triple	0/3	5.8±0.3 (2)	3/3 (80)	0/3	0/3	None
A/swine/Guangdong/7/06	Sw/GD/7/06	Swine	H3N2	Recent human-like	0/3	5.3±1.1 (2)	3/3 (40)	0/3	0/3	None
A/swine/Guangdong/211/06	Sw/GD/211/06	Swine	H3N2	Intermediate human-like	0/3	5.4±0.4 (6)	3/3 (40–80)	0/3	0/3	None
A/swine/Guangdong/811/06	Sw/GD/811/06	Swine	H3N2	Intermediate human-like	0/3	5.5±0.3 (2)	3/3 (40–80)	0/3	0/3	None
A/swine/Guangdong/968/06	Sw/GD/968/06	Swine	H3N2	Intermediate human-like	0/3	5.8±0.5 (6)	3/3 (40–80)	0/3	0/3	None
A/quail/Hong Kong/G1/97	Qa/HK/G1/97	Quail	H9N2	G1-like	0/3	6.1±0.5 (2)	3/3 (80)	0/3	0/3	None
A/chicken/Shandong/ZB/07	Ck/SD/ZB/07	Chicken	H9N2	Ck/Bei-like	0/3	4.9±1.2 (4)	3/3 (40–80)	0/3	0/3	None
A/chicken/Hebei/LC/08	Ck/HeB/LC/08	Chicken	H9N2	Ck/Bei-like	0/3	4.4±0.3 (2)	3/3 (40–80)	0/3	0/3	None

[a]Number of inoculated guinea pigs/total number.
[b]Peak nasal wash titers are expressed as the mean $log_{10}EID_{50}/mL \pm SD$.
[c]Serum was collected on day 21 p.i., and homologous strains were used with chicken RBCs in HI assays.
[d]The lower limit of detection was $10^{1.5}$ EID_{50}/mL.
doi:10.1371/journal.pone.0015537.t001

Table 2: Replication of influenza viruses in guinea pigs

https://doi.org/10.1371/journal.pone.0015537.t002

Virus (Subtype)	No. of positive guinea pigs/total no. of guinea pigs on dpi (mean virus titers [$log_{10}EID_{50}/gram$]±SD)[a]					
	2		4		6	
	Trachea	Lung	Trachea	Lung	Trachea	Lung
BJ/317/09 (H1N1)	3/3(3.4±0.1)	3/3(3.7±0.1)	3/3(3.3±0.1)	2/3(2.1±1.6)	1/3(3.3)	0/3
GD/41/06 (H1N1)	3/3(3.5±0.3)	0/3	3/3(1.8±1.3)	0/3	0/3	0/3
JX/262/05 (H3N2)	3/3(3.6±0.3)	0/3	3/3(3.3±0.1)	0/3	1/3(<)	0/3
Sw/GD/33/06 (H1N1)	3/3(2.6±1.4)	0/3	3/3(2.4±1.2)	0/3	1/3(<)	0/3
Sw/FJ/204/07 (H1N1)	2/3(3.4±0.2)	2/3(3.4±0.2)	3/3(3.4±0.2)	2/3(3.3±0.1)	1/3(3)	1/3(3.3)
Sw/GD/1222/06 (H1N2)	2/3(3.4±0.2)	3/3(3.7±0.1)	1/3(3.0)	3/3(3.2±0.1)	1/3(<)	3/3(1.7±1.2)
Sw/GD/7/06 (H3N2)	2/3(3.5±0.4)	0/3	1/3(3.5)	0/3	1/3(3.0)	0/3
Sw/GD/211/06 (H3N2)	0/3	0/3	1/3(4.3)	0/3	1/3(4.0)	0/3
Sw/GD/811/06 (H3N2)	0/3	0/3	0/3	0/3	0/3	0/3
Sw/GD/968/06 (H3N2)	1/3(3.5)	0/3	2/3(2.3±1.8)	0/3	0/3	0/3
Qa/HK/G1/97 (H9N2)	3/3(4.0±0.7)	3/3(4.3±0.7)	0/3	0/3	0/3	0/3
Ck/SD/ZB/07 (H9N2)	1/3(3.3)	3/3(3.8±0.5)	1/3(3.3)	3/3(3.4±0.3)	0/3	0/3
Ck/HeB/LC/08 (H9N2)	1/3(<)	0/3	0/3	0/3	0/3	0/3

[a]<, virus was detected in the undiluted samples. The lower limit of detection was $10^{1.5}$ EID_{50}/mL of tissue homogenate.
doi:10.1371/journal.pone.0015537.t002

It is interesting that the European avian-like swine H1N1 (Sw/FJ/204/07) and North American triple H1N2 (Sw/GD/1222/06) viruses, that were proposed to be the progenitors of 2009 pandemic H1N1 influenza viruses, both replicated more efficiently and distributed in more tissues than other swine influenza viruses, similar to the 2009 pandemic H1N1 virus (BJ/317/09). No virus was detected in other organs, including heart, liver, spleen, kidney, brain and colon during the observation period. Taken together, the replication of tested influenza viruses was restricted to the respiratory system of guinea pigs and virus replication in the upper respiratory tract was higher than in the lower respiratory tracts. Similar to human strains, swine and avian influenza viruses replicated efficiently in nasal tracts without prior adaptation.

Histopathology of Human, Swine, and Avian Influenza Viruses in Guinea Pigs

To compare the histopathological changes in human, swine, and avian influenza viruses in guinea pigs, the nasal, trachea, and lung specimens from infected animals on day 4 p.i. were fixed in 10% neutral buffered formalin and processed for routine histology. Nasal lesions in infected guinea pig were characterized by phlebectasia and congestion in the submucosal capillaries and veins, dropout of the mucous epithelium and erythrocytes adhering to the surface of the mucosa. The tracheal mucous membrane showed edema and thickening and dropout of tracheal epithelial cells. The infected lungs showed bronchiolitis with inflammatory cellular infiltrate and dropout of mucous epithelial cells in the bronchioles, peribronchiolitis was also observed with edema and inflammatory cells infiltrate around bronchioles and blood vessels. The results suggested that the tested viruses showed varying levels of pathological lesions. In particular, the histopathological changes in guinea pigs infected with Qa/HK/G1/97 were more severe than those caused by other viruses. In this group, the epithelial cells of the tracheal mucosa were severely desquamated resulting in exposure of the lamina propria. Severe bronchiolitis, peribronchiolitis, and bronchopneumonia were also observed with alveolar lumens and bronchioles flooded with edema fluid mixed with fibrin, erythrocytes, and inflammatory cells. For the tissues in virus was not detected, lesions were mild or absent.

Transmission of Human, Swine, and avian Influenza Viruses in Guinea Pigs

We next tested the propensity of the viruses to be transmitted by direct contact. Briefly, groups of three animals were intranasally inoculated with 10^6 EID_{50} of tested virus and housed in a cage. At 24 h p.i., three naïve guinea pigs were placed in the same cage for each virus. On days 2, 4, 6, and 8 p.i., the nasal washings of the three inoculated animals and three contact animals were collected and titered by EID_{50} assay. Evidence of transmission was based on the detection of virus in the nasal washing and seroconversion at the end of the three-week observation period. All of the viruses replicated efficiently in inoculated animals, as for the replication study; however, the transmissibility of these viruses in guinea pigs was variable. As shown in Fig. 2 and Table 1, transmission of BJ/317/09 (H1N1) virus occurred with 100% efficiency: all three exposed animals were detected to be positive on day 3 post-exposure (corresponding to day 4 p.i.). The two seasonal human influenza viruses transmitted less efficiently with only one of three and two of three exposed animals becoming infected, respectively. However, no virus shedding or seroconversion was detected in any direct contact animal of swine and avian influenza viruses. These results demonstrate that although the swine and avian influenza viruses replicated effectively in the inoculated guinea pigs, they lacked successful direct contact transmissibility. As no contact transmission occurred for swine and avian influenza viruses, we did not test the propensity of influenza viruses to be transmitted *via* the aerosol route.

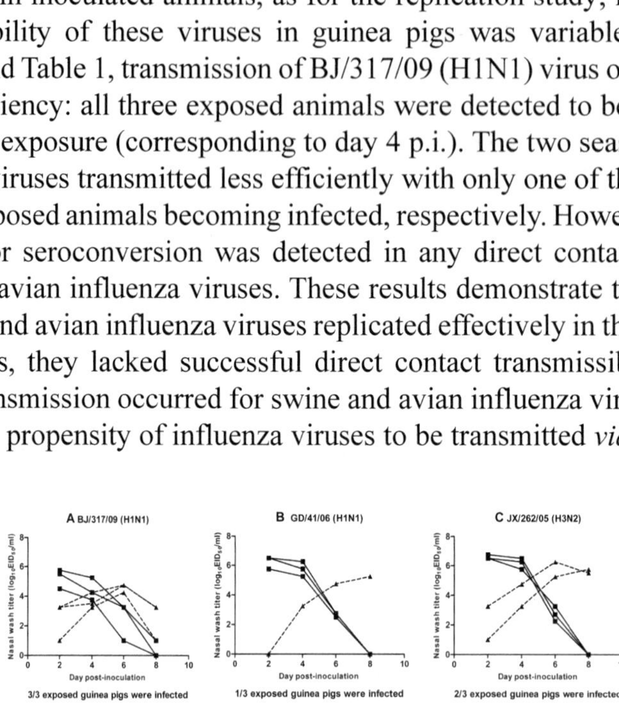

Figure 2: Contact transmission of pandemic H1N1 BJ/317/09, seasonal human H1N1 GD/41/06, and seasonal human H3N2 Jiangxi/262/05 in guinea pigs.

Three guinea pigs were intranasally inoculated with 10^6 EID_{50} of BJ/317/09 (A), GD/41/06 (B) or JX/262/05 (C) viruses and housed in a cage for each virus. At 24 h p.i., three naïve guinea pigs were placed in the same cage for each virus. Nasal washing titers are plotted as a function of time post-inoculation. Titers of intranasally inoculated animals are represented by dashed lines and filled squares; titers of exposed guinea pigs are shown with solid lines and filled triangles.

https://doi.org/10.1371/journal.pone.0015537.g002

Of note, we observed that guinea pigs in the inoculated group and contact group of BJ/317/09 (H1N1) infections began to sneeze on days 3 and 4 p.i., respectively. Some guinea pigs in the GD/41/06 (H1N1) infected group also sneezed during the observation period. Except for sneezing, no other clinical abnormalities were observed in any of the animals during the observation period.

Receptor Distribution in the Respiratory System of Guinea Pigs

We examined the receptor binding specificity in the respiratory tract of guinea pigs by lectin histochemistry assay. Biotinylated *Maackia amurensis* lectin II (MAA II), and *Sambucus nigra* agglutinin (SNA) were used to detect SAα2,3-Gal and SAα2,6-Gal linked receptors, respectively. In the nasal tract, both the SAα2,3-Gal and SAα2,6-Gal receptors were widely expressed by stratified squamous epithelial cells, vascular endothelial cells and the epithelial cells of the serous gland (Fig. 3 A, B). In the trachea, both of SAα2,3-Gal and SAα2,6-Gal receptors were widely expressed by epithelial cells of the tracheal mucosa (Fig. 3 D, E). In the lungs, SAα2,3-Gal receptor was mainly expressed by the alveolar cells and the vascular endothelial cells and was also expressed by the bronchial mucosa epithelial cells, however, SAα2,6-Gal receptor was only rarely expressed in the alveolar cells (Fig. 3 G, H). Taken together, the data suggested that SAα2,3-Gal and SAα2,6-Gal receptors were present in the nasal and tracheal areas of the guinea pig. In contrast, SAα2,3-Gal was the dominant receptor type in the lung.

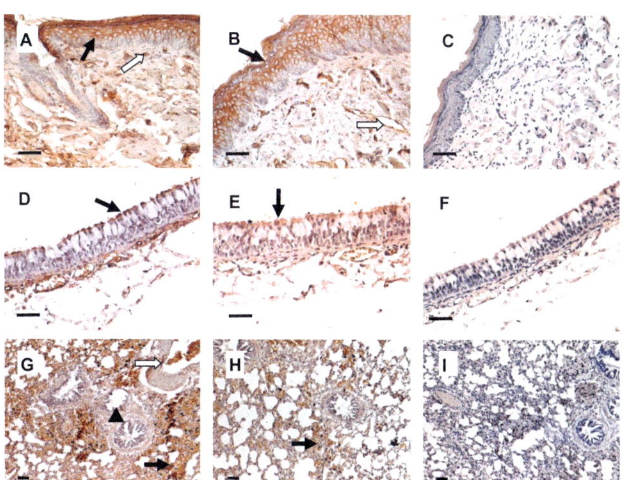

Figure 3: Receptor distribution in the respiratory system of guinea pigs.

Tissue sections were stained with MAA II (specific for binding the α-2,3-linked sialic acid) (A, D and G) or SNA (specific for binding the α-2,6 linked sialic acid) (B, E and H). C, F, and I, neuraminidase pre-treatment was applied prior to lectin staining. After pretreatment with neuraminidase, neither MAA II nor SNA showed a positive reaction. This confirms the specificity of both lectin stainings. A–C, nasal mucosa. (A) MAA staining is visible on stratified squamous epithelial cells (↑) and vascular endothelial cells (⇧). (B) SNA staining is identified on stratified squamous epithelial cells (↑) and vascular endothelial cells (⇧). D–F, trachea. (D) MAA II staining is visible on almost all epithelial cells (↑). (E) SNA staining is visible on almost all epithelial cells (↑). G–I, lung. (*G*) MAA II staining is positive on alveolar cells (↑), vascular endothelial cells (⇧) and bronchial mucosa epithelial cells (▲). (H) Except for some alveolar cells (↑), SNA staining is negative. C, F and I, following pre-treatment with neuraminidase, neither MAAII nor SNA shows positive reaction. This confirms the specificity of both lectin stains. Scale bar, 50 μm.

https://doi.org/10.1371/journal.pone.0015537.g003

DISCUSSION

In the present study, we investigated the guinea pig as a mammalian model for the study of the replication and transmission characteristics of selected swine H1N1, H1N2, H3N2 and avian H9N2 influenza viruses. We compared these characteristics to those of pandemic (H1N1) 2009, and seasonal human H1N1 and H3N2 influenza viruses. Our results showed that all the tested swine and avian influenza viruses replicated effectively in guinea pig nasal tracts without prior adaptation and the replication of these viruses was restricted to the respiratory system. The replication characteristics of swine and avian influenza viruses were similar to those of pandemic (H1N1) 2009 and seasonal human H1N1, H3N2 influenza viruses in addition to those of other influenza viruses reported by previous studies [13], [19], [21]. Moreover, some swine and avian viruses (namely, Sw/GD/211/06 (H3N2), Sw/GD/968/06 (H3N2) and Qa/HK/G1/97 (H9N2) viruses) were cleared more slowly from the guinea pig nasal passages than pandemic (H1N1) 2009 and seasonal human viruses. Histopathological analyses indicated that no significantly different pathological changes between human, swine and avian influenza viruses upon infection of the guinea pig. Of note, however, the histopathology after Qa/HK/G1/97 (H9N2) infection was the most severe of the viruses tested. Pandemic (H1N1) 2009 influenza viruses possess effective human-to-human transmission ability and rapidly spread around the world. Similar to the transmissibility in humans, the transmission results in the present study in guinea pigs indicated that the 2009 pandemic virus can be passed from infected guinea pigs to all the uninfected contacts. Steel *et al.* has also demonstrate that transmission of a pandemic (H1N1) 2009 strain occurred with 100% efficiency in guinea pigs [11]. Two human seasonal viruses also showed transmission between guinea pigs. Additionally, Lamb *et al.* reported that during the 1918 epidemic, a parallel outbreak of pneumonial disease among their laboratory guinea pigs emerged [23]. Although sporadic instances of human infection with swine and avian influenza viruses occur, human-to-human transmission is rare. The transmission phenotype of swine and avian influenza viruses in guinea pigs was similar to those found in the human population. Tested swine and avian influenza viruses could replicate effectively in the respiratory system of guinea pigs, however, none could be transmitted from infected guinea pigs to contact animals. Taken together, the experimental transmission of human, swine and avian influenza viruses in guinea pigs correlated well with the transmissibility of these viruses in humans. These studies show the utility of the guinea pig model for not only the evaluation of the transmissibility of pandemic influenza strains, human seasonal viruses and H5N1 influenza viruses, but could also be used for low pathogenic swine and avian influenza viruses to help recognize the strains that possess potential human-to-human transmission ability.

The genetic composition of the 2009 pandemic H1N1 viruses probably resulted from the reassortment of recent North American H3N2 and H1N2 swine viruses (i.e., avian/human/swine "triple" reassortant viruses) with Eurasian avian-like H1N1 swine viruses [24]. In the present study, the shedding kinetics and tissue tropism of Sw/GD/1222/06 (H1N2, North American triple) and Sw/FJ/204/07 (H1N1, European avian-like swine) were similar to BJ/317/09 (H1N1, pandemic) and support the proposed origin of pandemic H1N1 influenza virus. However, the transmission phenotypes of these swine influenza viruses were significantly different from those of the 2009 pandemic strain. The 2009 pandemic influenza virus was able to transmit efficiently from inoculated guinea pigs to all exposed animals, while no transmission was observed for the swine influenza viruses. Thus, further studies are needed to determine how swine viruses that lack human-to-human transmissibility acquired efficient transmissibility resulting in current pandemic strains.

Many studies have demonstrated that the guinea pig was a good animal model for influenza transmission research, however, the reason for the susceptibility of guinea pigs to influenza infection is not yet clear. Tang *et al.* proposed that the nasal epithelial cells of guinea pigs support influenza virus growth along with the excessive nasal mucus secretions and these factors may contribute to the susceptibility of guinea pigs to droplet spread [13]. In the present study, sneezing was often observed in all the guinea pigs infected by the 2009 pandemic H1N1 influenza virus and this strain had the most efficient transmissibility in this study. Sneezing is a clinical sign of humans infected with influenza virus and is usually observed in ferrets infected with influenza virus, and is considered an important ability of a virus for the efficient respiratory droplet transmission to elicit symptoms that promote expulsion of virus from the host [25]. Although previous studies have not reported such clinical signs in influenza virus infected guinea pigs, sneezing in guinea pigs is usually measured in studies for antigen-induced rhinitis [26], [27]. Here, we propose that sneezing is perhaps another reason that guinea pigs should be considered an ideal influenza transmission model host.

Host receptor expression is one of the factors important for infection of host cells by influenza viruses. We examined the receptor specificity of the guinea pig respiratory system. The results obtained in this study by histochemistry were similar to those of Gao *et al.* found by fluorescence [21]. We found a widespread presence of both SAα2,6-Gal (SNA) and SAα2,3-Gal (MAA II) receptors in the nasal tract of guinea pigs, which suggests that these organs are potential targets for avian, swine and human influenza viruses. The results of animal experiments performed in this study, in addition to previous studies, demonstrated that the human, swine and avian influenza viruses replicated efficiently in this area of guinea pigs. Furthermore, virus replication at this site is likely to result in viruses that can be readily shed to infect other animals leading to the efficient transmission of influenza viruses in guinea pigs. The SAα2,6-Gal receptor is rarely detected in the upper respiratory tract of BALB/c mice, while SAα2,6-Gal is widespread on epithelial cells of the upper airways of humans [18], [28]. Therefore, compared with mice, guinea pigs that express both SAα2,3-Gal and SAα2,6-Gal in the upper respiratory tract reflect a similar anatomy to that of humans.

In summary, we found the high susceptibility of guinea pigs to swine and avian influenza viruses, the correlation of transmissibility of these animal influenza viruses in guinea pigs with those in humans and the similar receptor distribution in the respiratory system of guinea pigs to those of humans. Additionally, guinea pigs possess many other human-like characteristics, including airway innervation [14], bronchus-associated lymphoid tissue in the lung [15] and response to various mediators [29]. Thus, we suggested that guinea pig could be used as an appropriate animal model for evaluating the potential threat of swine and avian viruses for humans. Animal influenza viruses are repeatedly introduced into humans and frequent reassortment and evolution often occurs. To this end, the guinea pig model offers a useful animal model to assess circulating strains that could challenge human public health and an important tool for pandemic preparedness worldwide.

MATERIALS AND METHODS

Ethics Statement

All animal research was approved by the Beijing Association for Science and Technology, the approve ID is SYXK (Beijing) 2007-0023, and complied with the guidelines of Beijing laboratory animal welfare and ethical of Beijing Administration Committee of Laboratory Animals.

Viruses

The 2009 pandemic H1N1 virus, human seasonal H1N1 and H3N2, swine H1N1, H1N2 and H3N2, and avian H9N2 influenza viruses were used in this study (Table 1). Human seasonal H1N1 GD/41/06 and H3N2 JX/262/05 viruses were kindly provided by Dr. G. F. Gao (Institute of Microbiology, Chinese Academy of Sciences). The avian H9N2 influenza Qa/HK/G1/97 virus was kindly provided by Dr. H. Kida (Hokkaido University Graduate School of Veterinary Medicine). The pandemic (H1N1) 2009 BJ/317/09 viruses, swine and other avian influenza viruses were isolated by our laboratory in China, as previously described [30], [31].

Replication of Viruses in Guinea Pigs

Hartley strain female SPF/VAF guinea pigs weighing 300–350 g that were seronegative for influenza virus were used in these studies and were obtained from Vital River Laboratory Animal Technology. Zoletil 100 (tiletamine-zolazepam; Virbac S.A., Garros, France) (10–15 mg/kg) were used to anesthetize animals by intramuscular injection.

To investigate the replication of human, swine and avian influenza viruses in the guinea pig model, groups of 15 animals were anesthetized and inoculated intranasally with 10^6 EID_{50} of tested virus in a 300 μL volume (150 μL on each side). On days 2, 4, 6, and 8 post-inoculation (p.i.), three animals from each group were euthanized and tracheas, lungs, hearts, livers, spleens, kidneys, brains, and colons were collected for virus titration in eggs. Nasal washes from the three remaining animals were collected on days 2, 4, 6, and 8 p.i., and titrated by EID_{50} assay. Nasal washing was performed by instilling a total of 1 mL of PBS containing antibiotics (Gibco) into the nostrils and collecting liquid runoff into a sterile Petri dish. These three animals were also observed for two weeks for body weight and temperature, signs of disease, and tested for seroconversion on day 21 p.i..

Contact Transmission in Guinea Pigs

Groups of three animals were intranasally inoculated with 10^6 EID_{50} of tested virus and housed in a cage. At 24 h p.i., three naïve guinea pigs were placed in the same cage and co-housed in this way for a total of 20 days. On days 2, 4, 6, and 8 p.i., the nasal washes of the three inoculated animals and three contact animals were collected and titered for EID_{50} assay. All of the animals were humanely euthanized on day 21 p.i. and tested for seroconversion.

Histopathology

On day 4 p.i., the nasal, trachea and lung specimens from inoculated guinea pigs were fixed in 10% neutral buffered formalin, routinely processed, and embedded in paraffin. Four-micrometer sections were stained with hematoxylin and eosin (HE).

Detection of SAα2,3-Gal and SAα2,6-Gal Receptors in Guinea Pig Respiratory Tissues

Biotinylated *Maackia amurensis* lectin II (MAA II), which is specific for α-2,3-linked sialic acid, and biotinylated *Sambucus Nigra* (Elderberry) bark lectin (SNA), specific for binding the α-2,6 linked sialic acid, were purchased from Vector Laboratories (Burlingame, CA, USA).

Two female Hartley SPF/VAF guinea pigs were euthanized and the nasal, trachea and lung tissues were fixed in 10% (w/v) neutral-buffered formalin for 24–48 h, and 4-μm thick paraffin-embedded tissue sections were made. Sections were stained with biotinylated sialic acid-specific lectins (MAA II or SNA) and the method of lectin histochemistry was performed as described by Yao *et al.* [32]. Briefly, paraffin-embedded serial sections were de-paraffinized and immersed in 3% hydrogen peroxide to eliminate endogenous peroxidase activity. A solution of 5% bovine serum albumin was used as the blocking agent to avoid non-specific staining. Sections were next incubated with SNA (0.625 g/mL) and MAA II (1.25 g/mL) at 4°C overnight. Serial sections of the same tissue were also incubated with PBS instead of lectin as negative controls. Optimal contrast between the specific labeling and the background was obtained with the use of a SABC kit (Dako, Carpinteria, CA, USA). Biotinylated lectin binding was visualized using a DAB (3,3′-diaminobenzidine-tetrahydrochloride) substrate chromogen kit (Zymed Labs, San Francisco, CA, USA) and slides were counterstained with hematoxylin.

Neuraminidase Pre-treatment

To confirm the specificity of the lectin stains, neuraminidase pre-treatment was performed. Because neuraminidase digests both α-2,3-linked and α-2,6-linked sialic acid residues, negative lectin staining after neuramidase pre-treatment would indicate that the lectin used was specific for detection of α-2,3-linked and α-2,6-linked sialic acids. Neuraminidase pre-treatment was performed as described by Yao *et al.* [32]. Briefly, the paraffin-embedded tissue sections were de-paraffinized and immersed in 3% hydrogen peroxide to eliminate endogenous peroxidase activity. Slides were covered with 12.5 U/μL neuraminidase solution (NEB, Ipswich, MA, USA) and incubated for 24 h at 37°C. Slides were washed three times with PBS and 5% BSA was used to block non-specific straining. Lectin staining was subsequently performed as described above. Additional negative controls were performed by using slides incubated with PBS instead of neuraminidase.

ACKNOWLEDGMENTS

We are indebted to the anonymous reviewers for their valuable suggestions. We thank Dr. G. F. Gao for providing the GD/41/06 and H3N2 JX/262/05 viruses and Dr. H. Kida for providing the Qa/HK/G1/97 virus. We also thank Honglei Sun, Jinming Tang and Qi Wang for excellent technical assistance.

AUTHOR CONTRIBUTIONS

Conceived and designed the experiments: YPS YHB JP JHL. Performed the experiments: YPS YHB JP YXH HJG LQL QX YYT MDL. Analyzed the data: YPS YHB JP YXH JJW XG HCY JHL. Contributed reagents/materials/analysis tools: JP JHL. Wrote the paper: YPS YHB JP YXH JJW JHL.

REFERENCES

1. Neumann G, Noda T, Kawaoka Y (2009) Emergence and pandemic potential of swine-origin H1N1 influenza virus. Nature 459: 931–939.
2. Peiris JS, Poon LL, Guan Y (2009) Emergence of a novel swine-origin influenza A virus (S-OIV) H1N1 virus in humans. J Clin Virol 45: 169–173.
3. Malik Peiris JS (2009) Avian influenza viruses in humans. Rev Sci Tech 28: 161–173.
4. Myers KP, Olsen CW, Gray GC (2007) Cases of swine influenza in humans: a review of the literature. Clin Infect Dis 44: 1084–1088.
5. Shinde V, Bridges CB, Uyeki TM, Shu B, Balish A, et al. (2009) Triple-reassortant swine influenza A (H1) in humans in the United States, 2005–2009. N Engl J Med 360: 2616–2625.
6. Butt KM, Smith GJ, Chen H, Zhang LJ, Leung YH, et al. (2005) Human infection with an avian H9N2 influenza A virus in Hong Kong in 2003. J Clin Microbiol 43: 5760–5767.
7. Bastien N, Antonishyn NA, Brandt K, Wong CE, Chokani K, et al. Human infection with a triple-reassortant swine influenza A(H1N1) virus containing the hemagglutinin and neuraminidase genes of seasonal influenza virus. J Infect Dis 201: 1178–1182.
8. Lowen AC, Mubareka S, Tumpey TM, Garcia-Sastre A, Palese P (2006) The guinea pig as a transmission model for human influenza viruses. Proc Natl Acad Sci U S A 103: 9988–9992.
9. Maher JA, DeStefano J (2004) The ferret: an animal model to study influenza virus. Lab Anim (NY) 33: 50–53.
10. Lowen AC, Steel J, Mubareka S, Palese P (2008) High temperature (30 degrees C) blocks aerosol but not contact transmission of influenza virus. J Virol 82: 5650–5652.
11. Steel J, Staeheli P, Mubareka S, Garcia-Sastre A, Palese P, et al. Transmission of pandemic H1N1 influenza virus and impact of prior exposure to seasonal strains or interferon treatment. J Virol 84: 21–26.
12. Bouvier NM, Lowen AC, Palese P (2008) Oseltamivir-resistant influenza A viruses are transmitted efficiently among guinea pigs by direct contact but not by aerosol. J Virol 82: 10052–10058.
13. Tang X, Chong KT (2009) Histopathology and growth kinetics of influenza viruses (H1N1 and H3N2) in the upper and lower airways of guinea pigs. J Gen Virol 90: 386–391.

14. Wang L, Chitano P, Murphy TM (2005) Length oscillation induces force potentiation in infant guinea pig airway smooth muscle. Am J Physiol Lung Cell Mol Physiol 289: L909–915.
15. Kwon YK, Lipatov AS, Swayne DE (2009) Bronchointerstitial pneumonia in guinea pigs following inoculation with H5N1 high pathogenicity avian influenza virus. Vet Pathol 46: 138–141.
16. Skehel JJ, Wiley DC (2000) Receptor binding and membrane fusion in virus entry: the influenza hemagglutinin. Annu Rev Biochem 69: 531–569.
17. Rogers GN, Paulson JC (1983) Receptor determinants of human and animal influenza virus isolates: differences in receptor specificity of the H3 hemagglutinin based on species of origin. Virology 127: 361–373.
18. Ibricevic A, Pekosz A, Walter MJ, Newby C, Battaile JT, et al. (2006) Influenza virus receptor specificity and cell tropism in mouse and human airway epithelial cells. J Virol 80: 7469–7480.
19. Van Hoeven N, Belser JA, Szretter KJ, Zeng H, Staeheli P, et al. (2009) Pathogenesis of 1918 pandemic and H5N1 influenza virus infections in a guinea pig model: antiviral potential of exogenous alpha interferon to reduce virus shedding. J Virol 83: 2851–2861.
20. Steel J, Lowen AC, Mubareka S, Palese P (2009) Transmission of influenza virus in a mammalian host is increased by PB2 amino acids 627K or 627E/701N. PLoS Pathog 5: e1000252.
21. Gao Y, Zhang Y, Shinya K, Deng G, Jiang Y, et al. (2009) Identification of amino acids in HA and PB2 critical for the transmission of H5N1 avian influenza viruses in a mammalian host. PLoS Pathog 5: e1000709.
22. Steel J, Staeheli P, Mubareka S, Garcia-Sastre A, Palese P, et al. (2010) Transmission of pandemic H1N1 influenza virus and impact of prior exposure to seasonal strains or interferon treatment. J Virol 84: 21–26.
23. Lamb FH, Brannin EB (1919) The epidemic respitatory infection at Camp Cody, N. M. J Am Med Assoc 72: 1056–1062.
24. Dawood FS, Jain S, Finelli L, Shaw MW, Lindstrom S, et al. (2009) Emergence of a novel swine-origin influenza A (H1N1) virus in humans. N Engl J Med 360: 2605–2615.
25. Maines TR, Chen LM, Matsuoka Y, Chen H, Rowe T, et al. (2006) Lack of transmission of H5N1 avian-human reassortant influenza viruses in a ferret model. Proc Natl Acad Sci U S A 103: 12121–12126.

26. Shaoqing Y, Ruxin Z, Yinjian C, Jianqiu C, Zhiqiang Y, et al. (2009) Down-regulation of endogenous hydrogen sulphide pathway in nasal mucosa of allergic rhinitis in guinea pigs. Allergol Immunopathol (Madr) 37: 180–187.
27. Shaoqing Y, Ruxin Z, Yinjian C, Jianqiu C, Chunsheng Z, et al. (2008) Possible contribution of endogenous carbon monoxide to the development of allergic rhinitis in guinea pigs. J Inflamm (Lond) 5: 23.
28. Shinya K, Ebina M, Yamada S, Ono M, Kasai N, et al. (2006) Avian flu: influenza virus receptors in the human airway. Nature 440: 435–436.
29. Ressmeyer AR, Larsson AK, Vollmer E, Dahlen SE, Uhlig S, et al. (2006) Characterisation of guinea pig precision-cut lung slices: comparison with human tissues. Eur Respir J 28: 603–611.
30. Bi Y, Fu G, Chen J, Peng J, Sun Y, et al. Novel swine influenza virus reassortants in pigs, China. Emerg Infect Dis 16: 1162–1164.
31. Sun YPJ, Jiang Z, Guan T, Xia Y, et al. (2010) Genotypic evolution and antigenic drift of H9N2 influenza viruses in China from 1994 to 2008. Vet Microbiol.
32. Yao L, Korteweg C, Hsueh W, Gu J (2008) Avian influenza receptor expression in H5N1-infected and noninfected human tissues. FASEB J 22: 733–740.

Section 6: Wild Animals

CHAPTER 23

A One Health Message about Bats Increases Intentions to Follow Public Health Guidance on Bat Rabies

Hang Lu[1], Katherine A. McComas[1] , Danielle E. Buttke[2] , Sungjong Roh[3] , Margaret A. Wild[2]

[1] Department of Communication, Cornell University, Ithaca, New York, United States of America

[2] Wildlife Health Branch, Biological Resources Division, National Park Service, Fort Collins, Colorado, United States of America

[3] Lee Kong Chian School of Business, Singapore Management University, Singapore, Singapore

Citation: Lu H, McComas KA, Buttke DE, Roh S, Wild MA (2016) A One Health Message about Bats Increases Intentions to Follow Public Health Guidance on Bat Rabies. PLoS ONE 11(5): e0156205. https://doi. org/10. 1371/journal. pone. 0156205.

ABSTRACT

Since 1960, bat rabies variants have become the greatest source of human rabies deaths in the United States. Improving rabies awareness and preventing human exposure to rabid bats remains a national public health priority today. Concurrently, conservation of bats and the ecosystem benefits they provide is of increasing importance due to declining populations of many bat species. This study used a visitor-intercept experiment (N = 521) in two U. S. national parks where human and bat interactions occur on an occasional basis to examine the relative persuasiveness of four messages differing in the provision of benefit and uncertainty information on intentions to adopt a rabies exposure prevention behavior. We found that acknowledging benefits of bats in a risk message led to greater intentions to adopt the recommended rabies exposure prevention behavior without unnecessarily stigmatizing bats. These results signify the importance of communicating benefits of bats in bat rabies prevention messages to benefit both human and wildlife health.

INTRODUCTION

With the highest case fatality rate of any infectious disease of humans, rabies remains an important public health concern in the United States [1, 2]. Reduction in rabies in domestic cats and dogs thanks to successful rabies vaccination campaigns since 1980 has meant that, today, more than 90 percent of reported rabid animals in the United States are wildlife [3]. Further, most human deaths caused by rabies in the past several decades have been associated with exposures to rabid bats [4].

Since vaccination of bats against rabies is not currently feasible, secondary intervention methods, such as public health education and exposure prevention, are the current means of preventing human infection with bat rabies [3]. Accordingly, improving rabies awareness among health care providers and the general public remains a high priority [1, 5] in that educational messages about how to prevent, recognize, and treat potential contact with rabid bats are crucial in reducing human deaths [1].

Concurrently, the conservation of bats and the ecosystem benefits they provide is of increasing importance due to declining populations of many species [6]. It is therefore imperative that wildlife-associated disease messages use an evidence-based approach to ensure that they do not produce unintended consequences that might diminish public support for wildlife [7]. Risk messages often center on negative information that is inherently fear arousing [8]. While this strategy is commonly used in public health messaging, its effectiveness for public health outcomes associated with wildlife disease is less known, and some research suggests it might even have negative impacts on wildlife conservation [7].

An approach that seeks to reformulate public or wildlife health initiatives under a larger framework is One Health. One Health focuses on the interdependence of human, domestic animal, wildlife, and ecosystem health and advocates an approach that enhances the health of all species [9]. In the context of disease messaging, a One Health approach would be exemplified by messages that convey both negative (e. g. , risks associated with bat rabies) and positive (e. g. , ecological benefits of bats) information [10]. However, because a One Health approach to risk messaging has received little evaluation, further research is warranted to test messages that seek to motivate wildlife-associated disease prevention behaviors while at the same time sustain general support for wildlife [7, 10]. In particular, the current study seeks to examine a One Health approach to messaging about bats and rabies to examine the effects of such messages on important public and wildlife health outcomes.

In this regard, we first examine the relative persuasiveness of simultaneous communication of both risk and benefit information (i. e. , the risk-benefit message), versus communication of mere risk information (i. e. , the risk-only message), in relation to intentions to adopt a bat rabies exposure prevention behavior in a U. S. national park. Studies in the wider risk communication literature have found inconsistent results regarding the usefulness of risk-benefit messages [11]. On the one hand, risk-benefit messages may provide more balanced information for individuals to make informed decisions on a health risk issue [12]. On the other hand, messages incorporating both risks and benefits may seem incongruent to audiences and decrease the credibility of or distract audiences from the prevention messages [13]. These inconsistencies further warrant the investigation of risk-benefit message efficacy in communicating bat rabies public health guidance.

Second, we examine the influence of the level of uncertainty associated with the risk [14]. Uncertainty can relate to perceptions of the temporal distance of a risk. For example, the probability of exposure to a health risk may be perceived higher when similar risks are prominent more recently than in the distant past [15, 16]. Testing the impacts of different levels of risk uncertainty conveyed in a bat rabies-related message has important practical considerations. As risk uncertainty levels differ across real-world situations, it is necessary to examine if messages capturing these variations will lead to different outcomes. For instance, certain levels of uncertainty may be more likely to result in unnecessary fear and strong denial of the message [17].

We conducted a field experiment in two U. S. national parks. We assessed the impact of bat rabies risk messages on park visitors' intentions to adopt appropriate rabies prevention behaviors. We manipulated messages in terms of the benefits of bats and the risk uncertainty associated with bat rabies. Our objectives were to determine if: 1) messages including the benefits of bats impacted public health prevention intent, and 2) risk uncertainty influenced public health prevention intent.

MATERIALS AND METHODS

Prior to the start of this study, the Office of Research Integrity and Assurance at Cornell University reviewed and approved this research including the consent procedure, as did the National Park Service, in consultation with the U. S. Department of Interior's Office of Management and Budget.

Trained research assistants approached visitors in the park and asked for their participation in the study if they were older than 18. If a group was encountered, the assistants asked the individual within the group who had the next birthday to serve as the respondent. Before starting the study, participants were informed that completion of the survey implied their consent to participate in this study. Therefore, the completion of the survey was regarded as written consent. We did not ask for participants' names to ensure anonymity and avoid any identifiable association between participants and their responses. We recruited a total of 521 visitor participants from two national parks (258 from Mammoth Cave National Park in Kentucky and 263 from Great Smoky Mountains National Park in Tennessee) in late July 2014. These parks were selected based on two criteria. First, visitors in these parks could get in close proximity with bat populations. Second, these parks have had previous bat-related public health issues, such as bat rabies and visitor-bat contact.

Design

Each participant was randomly assigned to either a control (no message) condition or one of four message conditions as part of a 2 (risk-benefit vs. risk-only) x 2 (low uncertainty vs. high uncertainty) between-subjects factorial design. In the message conditions, participants read one of four printed leaflets at the beginning of the experiment. The leaflets, which were designed to resemble flyers usually seen in a park and checked for accuracy by wildlife and public health professionals, mentioned the park's history of rabid bats, described the negative consequences of exposure to bat rabies, and recommended telling a park ranger right away should visitors encounter a bat that is behaving strangely. To distinguish between risk-benefit and risk-only messages, the risk-benefit messages included an additional description of the ecological benefits of bats and the One Health tenet, "Our health is linked to the health of all species and the environment. " We manipulated the high versus low risk uncertainty by varying the temporal distance of a rabid bat incident. The high uncertainty messages stated, "This park has had rabid bats in the past," whereas the low uncertainty messages stated, "A rabid bat was recently found in this area of the park. " After participants finished reading the leaflets, they received a printed questionnaire to complete. Those in the control condition completed the questionnaire without having read any leaflets.

Measures

The primary outcome variable of interest was the likelihood of participants telling a park ranger about a strangely behaving bat should they see one (where 1 = "very unlikely" and 5 = "very likely"). This behavior, which benefits not only the individual adopting this behavior but also other park visitors, was promoted as the bat rabies exposure prevention behavior in the leaflets.

In addition, we measured participants' beliefs about bats using six items. Participants were asked to indicate the extent to which they agreed or disagreed that bats were attractive, vulnerable, worthless (reverse-coded), harmless, native and frightening (reverse-coded) on a 5-point scale (where 1 = "strongly disagree" and 5 = "strongly agree"). Responses to these six items were averaged to create a belief composite scale.

For participants in the message conditions, we measured their evaluation of the messages using three items. Participants were asked to indicate the extent to which they agreed or disagreed that the message they read was clear, informative, or persuasive on a 5-point scale (where 1 = "strongly disagree" and 5 = "strongly agree"). These items were included to explore the underlying mechanism for potential message effects.

Analysis

We first examined if the experimental randomization was successful by testing for differences in participants' characteristics across conditions with the χ^2 test and analysis of variance (ANOVA). Next, we used a two-way ANOVA with benefit frame (risk-benefit vs. risk-only), uncertainty frame (high vs. low), and their interaction as independent variables, and the likelihood of telling a park ranger about a strangely behaving bat and beliefs about bats as dependent variables to examine the main and interaction effects of message frames. In addition, to compare each message condition with the control condition, we regressed the behavioral intention and belief variables respectively by using ordinary least squares regression on the dummy variables for the five conditions (control condition as the reference group). We also tested whether message effects on the behavioral intention were mediated respectively by perceived clarity, persuasiveness, or informativeness of the messages. We ran mediation analysis using the PROCESS macro (5000 bootstrap resamples, 95% bias corrected CI) [18].

RESULTS

The sample was 50. 1% female, 89. 8% White, 4. 2% Asian, 1. 7% African American, with a mean age of 44. 78 years (SD = 14. 34). Forty-five percent of the sample had completed less than a college education, and 14. 0% had completed only high school or less. Seventeen percent of the sample reported annual household incomes under $35,000. Eighty-one percent of the sample had visited a national park site before this particular visit. Three percent of the sample reported having seen a rabid bat, and eight percent reported being unaware that some bats carry rabies. The total sample in each condition was as follows: risk-benefit/low uncertainty (n = 113), risk-benefit/high uncertainty (n = 109), risk-only/low uncertainty (n = 99), risk-only/high uncertainty (n = 103), and control (n = 97).

First, we confirmed that the experimental randomization was successful by finding no significant differences across conditions in participants' age, gender, race, education, annual household income, experience with rabid bats, awareness of bat rabies, and visits to national park sites (Table 1).

Table 1: Demographic characteristics of participants in a visitor-intercept experiment to assess responses to messages about bat rabies exposure prevention

https://doi. org/10. 1371/journal. pone. 0156205. t001

	Test of randomization across 5 groups
Age (mean)	F(4, 508) = .67; p = .62
Female	Pearson $\chi^2(4)$ = 4.61; p = .33
White race	Pearson $\chi^2(4)$ = 4.62; p = .33
Education < Bachelor's degree	Pearson $\chi^2(4)$ = 1.20; p = .88
Household income <75, 000	Pearson $\chi^2(4)$ = .56; p = .97
Having seen a rabid bat	Pearson $\chi^2(4)$ = 2.16; p = .71
Not aware that some bats carry rabies	Pearson $\chi^2(4)$ = 5.68; p = .22
Having visited a national park site before	Pearson $\chi^2(4)$ = 3.01; p = .56

doi:10.1371/journal.pone.0156205.t001

The results of our two-way ANOVAs indicate a significant main effect of the benefit frame on participants' likelihood of telling a park ranger should they encounter a strangely behaving bat (M = 4. 31, SD = . 97), F(1, 419) = 4. 62, p < . 05, η^2 = . 01. Particularly, participants who read the risk-benefit messages (M = 4. 41, SD = . 90) showed stronger intentions to adopt the recommended behavior than those who read the risk-only messages (M = 4. 21, SD = 1. 04). We found no other significant main or interaction effects for this dependent variable or the beliefs about bats scale (Cronbach α = . 72; M = 3. 61, SD = . 62).

Table 2: Message effects on intention to tell a park ranger about a strangely behaving bat and beliefs about bats

https://doi. org/10. 1371/journal. pone. 0156205. t002

Message	Behavioral Intention, B(SE)	Beliefs about Bats, B(SE)
Risk-benefit/low uncertainty (Ref = control)	.28* (.14)	-.12 (.09)
Risk-benefit/high uncertainty (Ref = control)	.34* (.14)	-.08 (.09)
Risk-only/low uncertainty (Ref = control)	.18 (.14)	-.05 (.09)
Risk-only/high uncertainty (Ref = control)	.03 (.14)	-.17 (.09)
Intercept	4.10*** (.10)	3.70*** (.06)

*, **, *** Significant at the p = .05, p = .01, and p = .001 level, respectively.

doi:10.1371/journal.pone.0156205.t002

Comparing each of our message conditions with the control condition, we found that the risk-benefit/low uncertainty and the risk-benefit/high uncertainty messages increased intentions to tell a park ranger about a strangely behaving bat; further, none of the messages elicited significantly more negative beliefs about bats (Table 2).

We also examined the underlying mechanism accounting for the main effect of the benefit frame. Our mediation analyses found that the perceived informativeness ($M = 4.39$, $SD = .68$) rather than perceived persuasiveness ($M = 3.91$, $SD = .88$) or clarity ($M = 4.41$, $SD = .63$) of the messages fully mediated the relationship between benefit frame and intentions to tell a park ranger about a strangely behaving bat when controlling for uncertainty frame and their interaction term (Fig 1). More specifically, adding benefit information to the risk-only messages increased the perceived informativeness of the messages and thus further increased the likelihood of participants adopting the recommended behavior in the messages.

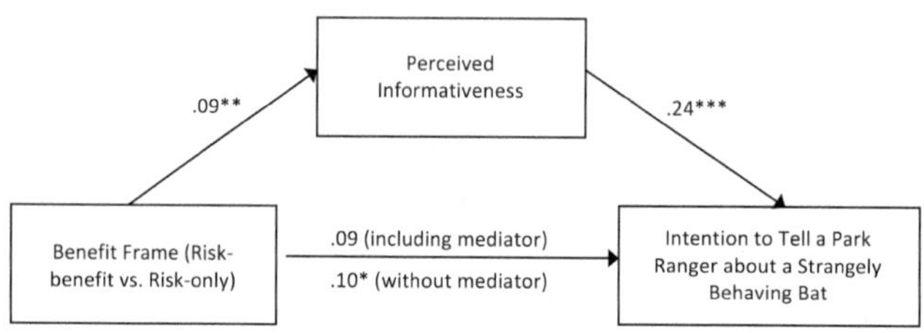

Figure 1: Mediation analysis investigating the relationship between benefit frame, perceived informativeness, and intention to tell a park ranger about a strangely behaving bat.

Note: Total indirect effect = . 02 (95% confidence interval = . 01, . 05). The uncertainty frame and the interaction term between the benefit frame and the uncertainty frame are not shown to reduce visual clutter. *$p < .05$; **$p < .01$; ***$p < .001$.

https://doi. org/10. 1371/journal. pone. 0156205. g001

DISCUSSION

Encroachment of human development on wildlife habitat and habitat loss have led to increased interactions between wildlife and humans. However, close-range interactions can sometimes have negative consequences for both wildlife and human health [19]. As bat rabies variants have become the greatest source of human rabies deaths in the United States since 1960, public health communication efforts focused on preventing human contact with rabid bats comprise a crucial component of effective rabies prevention and control programs [1]. This is true in general settings as well as specific ones, such as U. S. national parks.

Our results of a field experiment examining bat rabies risk messaging in two national parks showed that messages including both risks and benefits of bats were more effective in promoting the specific public health recommendation than messages mentioning only risks from bats. Our comparison between message conditions and the control (no message) condition further confirmed that while the risk-only messages were not superior to the control condition, the risk-benefit messages *increased* participants' intention to inform a park ranger of a strangely behaving bat. In addition, none of the four messages negatively influenced beliefs about bats relative to the control condition. Taken together, these findings suggest that the risk-benefit messages were more effective messaging strategies. These messages also reflect the One Health approach, which seeks to promote prevention behaviors to avoid wildlife-associated diseases while also maintain tolerance and support for wildlife [7, 10].

Why were risk-benefit messages more persuasive than risk-only messages? Our mediation analysis indicates that it was because participants perceived the risk-benefit messages as more informative. We believe that this perceived informativeness was not the by-product of any additional information in the message because the risk-only message conditions were no more efficacious than the control, no message condition. On the contrary, this perceived informativeness appears related specifically to the concept of One Health and the ecological benefits of bats provided in the risk-benefit messages. It is possible that because messages informed participants of contributions bats make to our environment, participants considered it more worthwhile to contact the park ranger should they encounter a strangely behaving bat. That is, such action would not only protect humans against rabies but also lower the chance of bats getting rabies or other infectious diseases from their own species.

We found no differential effects between the high versus low risk uncertainty messages on intentions to prevent bat rabies exposure. This finding has important practical implications. When warning visitors about potential bat rabies risks in the park, messages often mention the park's history with such diseases. Our manipulation of uncertainty reflected the temporal distance or acuteness of the disease. The lack of difference between uncertainty frames indicates that these messages can be used in parks with recent rabid bat cases as well as those with a remote history of rabid bats.

Put simply, general risk messages may be just as effective as specific risk messages with respect to when rabies was present in the park.

In viewing the results, we should consider study limitations. First, this experiment used a convenience sample of visitors in two national parks; therefore, the results should not be generalized to a broader public. However, we opted to test our messages in parks because large populations of both bats and people are found in many of the national parks, increasing the chance that people will encounter bats during their visits [5]. Additionally, it should also be noted that the primary goal of this study was to assess causal inference rather than representativeness. Future research seeking to confirm the external validity of our results should investigate similar messages with careful adaptions to other settings using a representative sample of the general population.

Second, we manipulated different levels of risk uncertainty by varying the temporal distance of past rabies incidence. While we could use other methods to accomplish similar manipulations (e. g. , by varying the likelihood of a risk occurring), our chosen approach was relatively easy to incorporate and more accessible in our current messaging context; it has also been shown to influence perceived uncertainty in other contexts [15, 20]. It is important to note that the lack of a significant effect does not indicate that messages varying in risk uncertainty might have no effects in this context. Rather, it only suggests that the current manipulation had little impact. Future research can explore other ways of manipulating risk uncertainty in similar contexts to further examine its relevance.

Third, we used a single-item measure of behavioral intention, which may be less reliable than a multi-item measure. In addition, behavioral intention cannot be equated with actual behaviors, and thus the findings should be interpreted with extra caution when it comes to anticipating actual behaviors. Furthermore, we measured behavioral intention immediately after participants read the message, when the message effects might be strongest. Future research should explore the long-term (e. g. , one hour or one day later) effects of these messages and their duration. Finally, we should point out that, while not uncommon in message testing studies, these findings should be interpreted with caution as they represent relatively small effects. Even so, it is noteworthy that relatively minor changes in the message wording showed effects, given the inability to control for all the competing messages visitors were exposed to in these parks.

Indeed, any risk communication effort in an uncontrolled setting, including national parks, can be quite challenging due to contextual factors introduced by the natural environment. Rather than abandon such messages to chance, we argue that these challenges further highlight the need for carefully designed messages that do not produce unintended effects. Public health risk messages understandably emphasize risks, but an overemphasis on risk can lead to unintended consequences that might conflict with and attenuate communication efforts [7]. This is especially the case when it comes to communication about wildlife-associated diseases, which by heightening awareness about risks could serve to decrease public support for wildlife. Our study suggests that a One Health approach that incorporates benefit information about bats alongside risk information may be more effective than risk-only messages. Therefore, incorporating risk-benefit messages into an overall approach may be a promising strategy for wildlife and public health professionals to communicate about wildlife-associated diseases, such as rabies.

ACKNOWLEDGMENTS

We would like to acknowledge the assistance of Graham Dixon, Joseph Steinhardt, and employees at Great Smoky Mountains National Park and Mammoth Cave National Park in data collection and thank Daniel Decker and Kirsten Leong for their guidance in preparation for this research.

AUTHOR CONTRIBUTIONS

Conceived and designed the experiments: KM DB SR MW. Performed the experiments: KM DB MW. Analyzed the data: HL KM SR. Contributed reagents/materials/analysis tools: HL KM DB SR MW. Wrote the paper: HL KM DB SR MW.

REFERENCES

1. Taylor LH, Costa P, Briggs DJ. Public Health Management of Humans at Risk. 2013:543–73.
2. Patyk K, Turmelle A, Blanton JD, Rupprecht CE. Trends in national surveillance data for bat rabies in the United States: 2001–2009. Vector Borne Zoonotic Dis. 2012;12(8):666–73. pmid:22607069
3. Dyer JL, Wallace R, Orciari L, Hightower D, Yager P, Blanton JD. Rabies surveillance in the United States during 2012. Javma-J Am Vet Med A. 2013;243(6):805–15.
4. De Serres G, Dallaire F, Cote M, Skowronski DM. Bat rabies in the united states and Canada from 1950 through 2007: Human cases with and without bat contact. Clin Infect Dis. 2008;46(9):1329–37. pmid:18419432
5. Zelepsky J, Harrison TM. Surveillance of rabies prevalence and bite protocols in captive mammals in American zoos. J Zoo Wildl Med. 2010;41(3):474–9. pmid:20945646
6. Kunz TH, Braun de Torrez E, Bauer D, Lobova T, Fleming TH. Ecosystem services provided by bats. Ann N Y Acad Sci. 2011;1223:1–38. pmid:21449963
7. Decker DJ, Siemer WF, Evensen DT, Stedman RC, McComas KA, Wild MA, et al. Public perceptions of wildlife-associated disease: risk communication matters. Human-Wildlife Interactions. 2012;6(1):112–22.
8. Witte K, Meyer G, Martell D. Effective Health Risk Messages: A Step-By-Step Guide: Sage Publications; 2001.
9. Decker DJ, Evensen DTN, Siemer WF, Leong KM, Riley SJ, Wild MA, et al. Understanding Risk Perceptions to Enhance Communication about Human-Wildlife Interactions and the Impacts of Zoonotic Disease. Ilar J. 2010;51(3):255–61. pmid:21131726
10. Buttke DE, Decker DJ, Wild MA. The role of one health in wildlife conservation: a challenge and opportunity. J Wildl Dis. 2015;51(1):1–8. pmid:25375941
11. Verbeke W, Liu R. The impacts of information about the risks and benefits of pork consumption on Chinese consumers' perceptions towards, and intention to eat, pork. Meat Sci. 2014;98(4):766–72. pmid:25119515

12. Vander Stichele RH, Vandierendonck A, De Vooght G, Reynvoet B, Lammertyn J. Impact of benefit messages in patient package inserts on subjective drug perception. Drug Inf J. 2002;36(1):201–8.
13. Regan Á, McConnon Á, Kuttschreuter M, Rutsaert P, Shan L, Pieniak Z, et al. The impact of communicating conflicting risk and benefit messages: An experimental study on red meat information. Food Quality and Preference. 2014;38:107–14.
14. Brashers DE. Communication and uncertainty management. Journal of Communication. 2001;51(3):477–97.
15. Wakslak CJ. The where and when of likely and unlikely events. Organizational Behavior and Human Decision Processes. 2012;117(1):150–7.
16. Baucells M, Heukamp FH. Probability and Time Trade-Off. Manage Sci. 2012;58(4):831–42.
17. Witte K. Putting the Fear Back into Fear Appeals—the Extended Parallel Process Model. Communication Monographs. 1992;59(4):329–49.
18. Hayes AF. Introduction to mediation, moderation, and conditional process analysis: A regression-based approach: Guilford Press; 2013.
19. Reynolds PC, Braithwaite D. Towards a conceptual framework for wildlife tourism. Tourism Manage. 2001;22(1):31–42.
20. Maglio SJ, Trope Y, Liberman N. The Common Currency of Psychological Distance. Curr Dir Psychol Sci. 2013;22(4):278–82.

CHAPTER 24

Urbanized White Ibises (Eudocimus albus) as Carriers of Salmonella Enterica of Significance to Public Health and Wildlife

Sonia M. Hernandez[1,2], Catharine N. Welch[1] , Valerie E. Peters[3] , Erin K. Lipp[4] , Shannon Curry[1] , Michael J. Yabsley[1,2], Susan Sanchez[5] , Andrea Presotto[6] , Peter GernerSmidt[7] , Kelley B. Hise[7] , Elizabeth Hammond[8] , Whitney M. Kistler[2] , Marguerite Madden[6] , April L. Conway[1], Tiffany Kwan[9] , John J. Maurer[9]

[1] Warnell School of Forestry and Natural Resources, University of Georgia, Athens, Georgia, United States of America

[2] Southeastern Cooperative Wildlife Disease Study, Department of Population Health, College of Veterinary Medicine, University of Georgia, Athens, Georgia, United States of America

---

Citation: Hernandez SM, Welch CN, Peters VE, Lipp EK, Curry S, Yabsley MJ, et al. (2016) Urbanized White Ibises (Eudocimus albus) as Carriers of Salmonella enterica of Significance to Public Health and Wildlife. PLoS ONE 11(10): e0164402. https://doi.org/10.1371/journal.pone.0164402.

[3] Institute for Environment and Sustainability, Department of Zoology, Miami University, Columbia, Ohio, United States of America

[4] Department of Environmental Health Science, University of Georgia, Athens, Georgia, United States of America

[5] Department of Infectious Diseases, College of Veterinary Medicine, University of Georgia, Athens, Georgia, United States of America

[6] Department of Geography, University of Georgia, Athens, Georgia, United States of America

[7] Centers for Disease Control and Prevention, U.S. Department of Health and Human Services, Atlanta, Georgia, United States of America

[8] Lion Country Safari Park, Loxahatchee, Florida, United States of America

[9] Poultry Diagnostic and Research Center, Department of Population Health, College of Veterinary Medicine, University of Georgia, Athens, Georgia, United States of America

ABSTRACT

Worldwide, *Salmonella* spp. is a significant cause of disease for both humans and wildlife, with wild birds adapted to urban environments having different opportunities for pathogen exposure, infection, and transmission compared to their natural conspecifics. Food provisioning by people may influence these factors, especially when high-density mixed species flocks aggregate. White Ibises (*Eudocimus albus*), an iconic Everglades species in decline in Florida, are becoming increasingly common in urbanized areas of south Florida where most are hand-fed. We examined the prevalence of *Salmonella* shedding by ibises to determine the role of landscape characteristics where ibis forage and their behavior, on shedding rates. We also compared *Salmonella* isolated from ibises to human isolates to better understand non-foodborne human salmonellosis. From 2010–2013, 13% (n = 261) adult/subadult ibises and 35% (n = 72) nestlings sampled were shedding *Salmonella*. The prevalence of *Salmonella* shedding by ibises significantly decreased as the percent of Palustrine emergent wetlands and herbaceous grasslands increased, and increased as the proportion of open-developed land types (e.g. parks, lawns, golf courses) increased, suggesting that natural ecosystem land cover types supported birds with a lower prevalence of infection. A high diversity of *Salmonella* serotypes (n = 24) and strain types (43 PFGE types) were shed by ibises, of which 33% of the serotypes ranked in the top 20 of high significance for people in the years of the study. Importantly, 44% of the *Salmonella* Pulsed-Field Gel Electrophoresis patterns for ibis isolates (n = 43) matched profiles in the CDC PulseNet USA database. Of these, 20% came from Florida in the same three years we sampled ibis. Importantly, there was a negative relationship between the amount of Palustrine emergent wetland and the number of *Salmonella* isolates from ibises that matched human cases in the PulseNet database ($p = 0.056$). Together, our results indicate that ibises are good indicators of salmonellae strains circulating in their environment and they have both the potential and opportunity to transmit salmonellae to people. Finally, they may act as salmonellae carriers to natural environments where other more highly-susceptible groups (nestlings) may be detrimentally affected.

INTRODUCTION

The genus *Salmonella* has a worldwide distribution and is one of the most common causes of intestinal diseases for both people and animals [1]. Despite major public education efforts and improvements to food hygiene practices, salmonellosis remains a significant source of enteric disease for people in the United States and worldwide. *Salmonella enterica* accounts for 1.2 million illnesses and 450 deaths in the US each year [2]. While most *Salmonella*-associated illnesses had been associated with the consumption of contaminated beef and poultry products [2, 3], there is an increasing frequency of illnesses associated with produce [4–14]. Despite public health education about food handling practices, the incidence of salmonellosis in people has remained the same in the last 20 years, and there has been a shift in serotype distribution from foodborne-associated strains towards environmentally-acquired strains, the source of which is not always known [3, 15, 16].

Although salmonellae are important human pathogens, they can also cause small to large-scale mortalities in wildlife species. For example, throughout the world, *Salmonella* outbreaks caused by *S.* Typhimurium have been reported in passerine birds associated with bird feeders, and to a lesser extent, colony-nesting birds (e.g., egrets and herons) [17–25]. There is still much to learn about the epidemiology and impact of salmonellosis on wild birds as the prevalence of avian salmonellosis appear to be on the rise, in some cases causing catastrophic outbreaks, some of which have been implicated with population-level effects [25, 26, 27]. The increase in outbreaks over the last 25 years has caused some to consider it as an emergent disease that may be directly related to anthropogenic activities such as backyard feeding and the use of contaminated habitats [23, 24, 28, 29].

In general, the level and duration of infection and *Salmonella* shedding in wild birds and potential to develop clinical disease is probably similar to domestic poultry but there are some important epidemiological differences. *Salmonella* prevalence of chicks of some colonial nesting species such as herons and egrets is higher compared to adult birds [25]. The immature intestinal flora related to both age and diet, may explain this difference where the infection rate for adults is relatively low (1–2%), but this may also be due to differences in habitat use. For example, adult herring gulls can be transient carriers of *Salmonella* and may serve in its transport while foraging between contaminated and pristine environments [30]. Several studies illustrate a direct correlation between the prevalence of *Salmonella* shedding in wild birds and their proximity to anthropogenic habitats [21, 31–35]. Urbanized birds likely have unique opportunities for enteric pathogen exposure, infection, and transmission due to various factors such as the quality of food and water they consume, the species with which they have direct contact, their consistent aggregations in high numbers in small areas and comparatively sedentary lifestyles [36]. Urbanized species with a potential to be carriers of *Salmonella* should be of special interest because of their frequent contact with the public and their potential to contaminate environments used by people.

Human *Salmonella* cases have been linked to direct or indirect contact with wild birds in several independent reports [17, 18, 29, 35, 37–42]. The majority of those reports relied on post-outbreak surveys and few utilize molecular typing to investigate similarities between human and avian isolates [37]. Therefore, studies that elucidate the phylogenetic, spatial and temporal relationships between human and bird infections and contribute to understanding of the underlying ecological interactions ecological interactions associated with non-foodborne cases of salmonellosis in humans.

The American white ibis (*Eudocimus albus*) is a gregarious member of the Pelecaniformes that forms large nesting colonies in natural wetlands. Ibises feed by constantly probing the substrate with their beaks for aquatic invertebrates, frogs and fish in wetlands, and roost and nest in mixed-species flocks over water bodies [43]. Their reproductive success and population size is highly dependent on food availability and the hydrology of the Florida Everglades Ecosystem, where most ibis breed [44, 45]. In South Florida, the utilization of urban habitats by white ibis has been steadily increasing since the late 1990s [46]. It is likely that a suite of factors, including natural wetland loss, and a constant source of food and water in urban parks promoted through rapid urban expansion in the region is responsible for this new distribution [47, 48]. Ibises are now abundant in neighborhood parks, golf courses and other artificial wetlands, where they have become sedentary, habituated to consuming food provided by people and where they regularly interact directly and indirectly with humans and other urbanized avian species (e.g, various gull species, semi-domesticated and wild ducks such as Muscovy ducks (*Cairina moschata*), and mallards (*Anas platyrhynchos*) (Hernandez, pers observation). In Australia, a related species of ibis (*Threskiornis molucca*) has become similarly habituated to anthropogenic sources of food and has been shown to be a carrier of *Salmonella enterica* [32]. Thus, we consider the American white ibis a good model species for understanding the role of urbanization on *Salmonella* transmission between wildlife and humans. Specifically, we aimed to determine whether the prevalence of *Salmonella* infection of urbanized white ibises in South Florida was related to either ibis behavior or landscape characteristics, and the relationship between *Salmonella* isolated from urbanized white ibises and human salmonellosis cases. Finally, although urbanized ibises primarily forage in urbanized environments during their non-reproductive period (Hernandez, unpublished data), they disperse long distances to natural areas to breed; thus, we sought to understand the role ibises might play in the dissemination of urban-associated salmonellae to those areas.

MATERIALS AND METHODS

Study Sites

We collected fecal samples from both adult and subadult birds at 17 public access sites in Palm Beach County, Florida, USA, (26°43'N, 80°2'W) with fresh water bodies from 2010–2013. We defined subadults (and use the term juveniles synonymously) as birds that had left the nest and were less than three years of age, which were readily distinguished from adults by feather coloration. Sites were chosen based on accessibility and a repeated record of more than 10 ibises per site during daily visits. Two sites were within natural areas, whereas the remaining 15 were urban parks, zoos or neighborhoods. We also collected feces from nestlings at three ibis rookeries located within Broward (n = 1) and Miami-Dade (n = 2) counties (Fig 1).

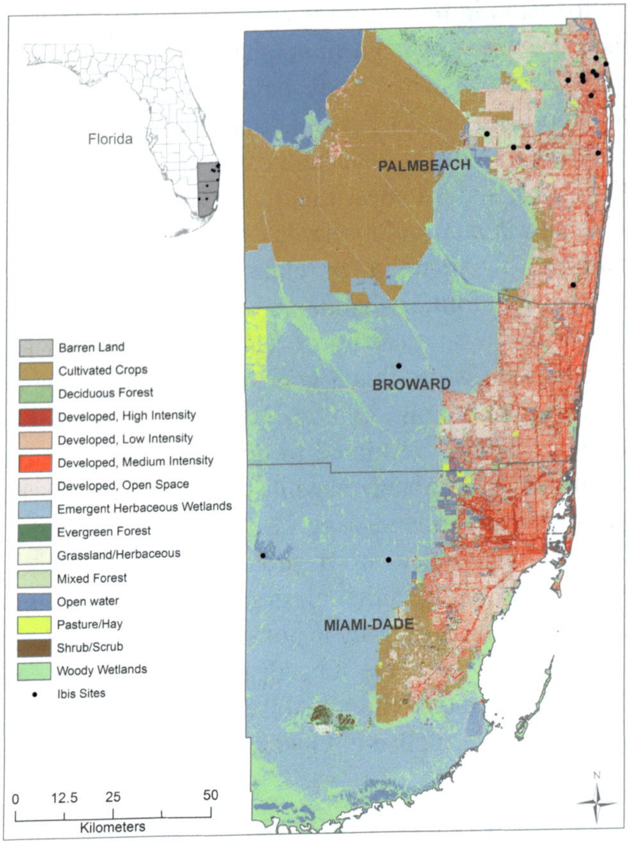

Figure 1: Map of Sites.

Fecal samples were collected from adult/subadult and nestling white ibises in Palm Beach, Broward and Miami-Dade counties, FL. The land cover for a 1 km radius surrounding each sampling site was determined using land cover maps.

https://doi.org/10.1371/journal.pone.0164402.g001

Using ArcGIS version 10.1, the land cover types were determined in a 1 km radius buffer around each site. Within each 1 km buffer, we quantified the percentages of each land cover class to examine the development of land and to classify sites by level of urbanization in each area. The publically-available land cover classes were developed by the Department of Commerce (DOC), National Oceanic and Atmospheric Administration (NOAA), National Ocean Service (NOS), Coastal Services Center (CSC) (http://gis.ncdc.noaa.gov/geoportal/catalog/search/resource/details.page?id=gov.noaa.

ncdc:C00814). The NOAA classification originally contained 25 classes, and the accuracy of classification for this data set is 86.5% and 85.1%. Initially, the studied areas represented 18 land cover types ranging from urban areas, agriculture, forested areas, and wetlands (Table 1). However, because some land cover types were <1% of the total land cover for that site, we only included 12 land cover classes in the analysis and excluded Cultivated/Crops, Pasture/Grasslands and Bare Land.

Table 1: Percent land cover of selected land use categories within 1km radius circles surrounding sampling site

https://doi.org/10.1371/journal.pone.0164402.t001

	Burns	Donald Ross	Garden	Juno	San Marco	Lion Safari	Loxa-hatchee	PGA	Prosperity	Royal Palm	San Matera	Zoo	Seasons	Horse shoe	ICP	Kissimmee	Dreher
Developed, High intensity	15.79	2.67	7.65	5.90	17.94	0.46	0.43	3.47	16.34	2.75	14.76	8.48	18.23	4.24	4.41	0.00	7.05
Developed, medium intensity	44.20	31.44	36.20	26.05	29.24	2.87	13.56	29.18	44.51	30.84	43.31	46.00	36.00	25.34	49.16	0.14	46.29
Developed, low intensity	22.27	26.89	19.18	10.15	19.69	11.69	32.16	25.97	20.29	22.61	24.71	21.41	31.70	40.64	19.81	0.40	22.24
Developed, open	9.29	5.82	4.56	10.89	4.33	6.31	5.68	20.55	3.01	7.62	4.01	5.93	3.87	6.56	9.09	0.00	6.08
Herbaceous grassland	0.00	0.37	0.03	0.11	2.35	4.50	2.87	0.37	0.03	2.75	0.00	0.03	0.00	0.06	1.06	19.32	0.03
Evergreen forest	0.00	0.00	0.00	0.00	0.03	0.69	0.49	0.03	0.09	0.03	0.00	0.03	0.00	0.03	0.03	0.11	0.03
Scrub	0.11	0.29	0.14	0.54	0.17	0.63	0.89	0.43	9.86	0.69	0.11	0.09	0.06	0.03	0.09	7.11	0.09
Forest, palustrine	5.93	20.81	22.84	17.03	8.37	63.57	34.48	12.35	0.92	15.85	9.26	10.52	8.46	20.84	4.82	20.04	10.32
Scrub, palustrine	0.66	2.35	1.49	2.01	1.32	5.68	3.15	0.57	1.00	2.04	0.63	1.18	0.43	0.72	1.75	32.04	1.26
Emergent wetland, palustrine	0.89	3.75	3.41	1.03	5.22	2.64	2.69	2.35	0.63	3.84	0.77	1.32	1.03	0.89	4.16	20.15	1.32
Bare land	0.14	0.09	3.41	0.66	5.07	0.23	2.49	0.00	0.00	8.05	0.00	0.00	0.00	0.00	0.03	0.00	0.00
Open water	0.26	4.67	0.72	24.39	5.50	0.03	0.00	3.90	3.24	2.04	2.29	4.61	0.20	0.69	4.64	0.03	4.70

doi:10.1371/journal.pone.0164402.t001

For clarification, we defined the four developed land cover types as follows: (1) Developed High Intensity- contained little or no vegetation, including heavily built-up urban centers as well as large constructed surfaces in suburban and rural areas-large buildings such as multiple family housing, hangars, and large barns, interstate highways, and runways, impervious surfaces accounted for 80–100 percent of the total cover (site range = 0 to 18%, Table 1); (2) Developed Medium Intensity- contained substantial amounts of constructed surface mixed with substantial amounts of vegetated surface, includes small buildings such as single family housing, farm outbuildings, and large sheds, impervious surfaces account for 50–79 percent of the total cover (site range = 0.1 to 49%, Table 1); (3) Developed Low Intensity- contained constructed surface mixed with vegetated surface, this class includes features seen in class 2, with the addition of streets and roads with associated trees and grasses, impervious surfaces account for

21–49 percent of the total cover (site range = 0.4 to 40%); and (4) Open Developed- included areas with a mixture of some constructed materials, but mostly vegetation in the form of lawn grasses, this subclass included parks, lawns, athletic fields, golf courses, and natural grasses occurring around airports and industrial sites, impervious surfaces accounted for less than 20 percent of total cover (site range = 0 to 20%).

Ibis Behavior

Ibis behavior may be directly related to pathogen transmission among themselves and between ibis and people. For example, ibises that beg for food get very close to people and defecate in very close proximity to people or on structures used by people. Therefore, ibis flocks of adults/subadults were scored by their behavior based on direct observations on a habituation scale of 1–4, where 1 were the least habituated to humans and 4 were the most (Table 2). We based our scoring system on observations of ibis with the general public. Highly habituated flocks responded to people by moving towards them immediately and even tolerated hand feeding. This score was assigned by observing the flock for 30 min prior to any sampling or manipulation. Highly habituated flocks tended to be larger, were more consistently present at a particular urban site and had more opportunities for direct contact with people. In contrast, flocks considered not to be habituated at all did not tolerate the presence of people and flew away when people got within 10 m of the flock. regardless of whether people offered food or not, varied in numbers of individuals and were not found predictably at a particular site. Flock density could be related to behavior and was calculated by counting the number of individuals in the approximate area actively utilized by the flock.

Table 2: Habituation behavioral scores assigned to white ibises foraging in urban parks in south Florida

https://doi.org/10.1371/journal.pone.0164402.t002

Score	Definition
1	Birds do not allow humans within 10 m and fly off if humans get within that distance
2	Birds allow humans within 10 m. but do not associate humans as a source of food, consume food provided by people or beg for food
3	Birds will allow humans within 10 m and will consume food provided by humans but do not allow humans within 10 ft. or beg for food
4	Birds allow humans within 3 m. and beg for food

doi:10.1371/journal.pone.0164402.t002

Fecal Sample Collection and *Salmonella* Isolation

The collection of feces from ibises was achieved in one of two ways. In some cases, individual animals (adults/subadults) were observed until they defecated, at which point a sample from the center of the fecal pellet was immediately collected from the ground. Care was taken not to resample individuals by separating one individual from the rest of the flock with a food reward until it defecated. In some instances, as part of a larger project, all nestlings and some adult or subadult birds were individually captured and physically restrained for other procedures, during which time they defecated on a clean, water repellent cloth placed across the holder's lap. Adult ibises were captured with leg lassos. Briefly, a leg lasso with a slip knot was constructed out of nylon monofilament and hidden in the substrate (i.e. grass). One person would hold the line approximately 6 m away, while a second person would bait a bird close with food, and encourage it to step into the lasso, at which point the line holder would swiftly tighten the lasso, capturing the bird by the leg. Both people would immediately physically restrain it and place it in a pillow case until it was processed. Juvenile ibis were captured prior to their ability to fly by visiting known nests, picking them, placing them in a pillow case and lowering them to the ground where they were immediately processed. In both cases, birds would defecate during handling and approximately 1 g of feces was collected with a sterile swab and immersed in 10 ml of dulcitol selenite broth (Becton, Dickinson and Company, Franklin Lakes, NJ), maintained at room temperature and shipped within 24 h to The Athens Diagnostic Laboratory at the University of Georgia. *Salmonella* enrichment and identification was performed as previously described [49]. Briefly, we performed a double selenite broth enrichment, followed by subculture on XLD and Brilliant Green agar plates (Remel). We selected 5–10 colonies from either selective plate that appeared to be *Salmonella* spp based on morphology. The number of colonies selected was determined based on the numbers of *Salmonella* suspect colonies present. Each colony was then serogrouped. One colony representative from each serogroup was selected and confirmed as *Salmonella* through the following biochemical tests: triple sugar iron (TSI), MIO, citrate, malonate, and phenylalanine. When isolates were confirmed, they were forwarded to the National Veterinary Service Laboratory (NVSL) at Ames, Iowa, for definitive serotyping. No animals were sacrificed as part of this research. All animal handling procedures were approved by the University of Georgia's Institutional Animal Care and Use Committee (IACUC; A2011 08–018). This work required permits from both the Florida Fish and Wildlife Conservation Commission (LSSC-11-00119F) and the U. S. Fish and Wildlife Service (MB779238-0). Both permits required prior approval by an IACUC. We obtained a Special Use Permit from Arthur R. Marshall Loxahatchee National Wildlife Refuge (B15_09). In addition, for urban parks we obtained permission from the Palm Beach County Parks and Recreations Department. For all sampling on private lands, we obtained permission from the relevant authority. All permits were obtained for the specific procedures of this specific study.

Molecular typing of Salmonella isolates by pulsed-field gel electrophoresis

Ibis isolates were strain typed by Pulsed-Field Gel Electrophoresis (PFGE) to determine their genetic relatedness [49] of these isolates to themselves, to archived bird isolates (both wild and domestic), to isolates from other animal sources, to water isolates, and to human isolates deposited in the CDC PulseNet national database which contains >50,000 entries [50]. A master database of *Salmonella* PFGE patterns was generated in BioNumerics (Applied Maths; Austin, TX). This database consisted of 1,047 total PFGE entries for *Salmonella* isolated from water (n = 400) and various animal species (n = 674) and consisted primarily of *S. enterica subsp. enterica* (97%) isolates representing 58 serotypes. Comparisons were made between PFGE patterns using Dice coefficient and unweighted-pair group method using average linkage (UPGMA) clustering [51].

Statistical Analysis

All data analyses were performed using R version 3.0.2 [52]. For the samples from adult/subadult ibises, generalized linear mixed models (GLMM) in the 'lme4' package of R [53]were used to analyze if *Salmonella* prevalence was influenced by the following fixed effects: (1) ibis flock density, (2) the behavioral variable 'habituation score', (3) sampling year, and (4) season. All models included site and sampling period as random variables, and used a binomial distribution. A likelihood ratio test was used to compare the GLMM models with and without the fixed effect, i.e. predictor variable, to test for the variable's significance. In addition to adults and subadults, we also sampled nestlings for *Salmonella*, but in a more limited fashion, i.e. all nestlings were tested during the same season (April/May), and the majority of nestlings were sampled in 2013. Nestlings also did not flock and we were unable to rate habituation scores for adult birds at the rookeries; therefore, we did not test these fixed effects against all three age classes. We were able, however, to test whether there was a difference in *Salmonella* prevalence rates among the three age classes sampled: adults, subadults and nestlings using a GLMM with a binomial distribution. For those predictor variables showing a statistically significant relationship with *Salmonella* prevalence, we calculated the marginal R^2, i.e. the R^2 value for only the fixed effects in the model using the package MuMIn in R [54].

To determine the effect of land use/land cover on *Salmonella* prevalence in adult/subadult birds, GLMMs were also used. All models included site and sampling period as random variables, and used a binomial distribution. Land use/land cover categories tested as fixed effects included all twelve of the aforementioned. In addition, due to the lack of predictive power for many of the land cover variables when analyzed separately, we also analyzed the effect of the landscape composition on *Salmonella* prevalence by combining the 12 land cover types into a principal components analysis (PCA), using the function prcomp in R [55]. PCA collapsed the land cover variables into three PCs that explained 73% of the landscape composition (PC1 = 41%, PC2 = 20%, PC3 = 12%). PCA1 loaded negatively on the percentage of land in the four developed categories. PCA2 loaded negatively on the amount of forested areas (Palustrine forest and Evergreen forest) in a landscape (96%). PCA3 loaded strongly and positively on the amount of open water and the amount of open developed land cover. We used GLMMs with a binomial distribution and likelihood ratio tests to evaluate the relationship between the principal components metrics and two response variables: (1) the number of different serotypes and (2) *Salmonella* prevalence. For landscape variables showing a statistically significant relationship with *Salmonella* prevalence, we also calculated the marginal R^2. We had an *a priori* expectation that there would be a significant difference in *Salmonella* prevalence between adult/subadult and nestlings; therefore, landscape analyses for prevalence did not include the sites from which nestlings were sampled.

To examine whether *Salmonella* prevalence displayed a spatial pattern among the sampled sites (excluding rookeries), we performed a Moran's I test using a Euclidean distance matrix and the variable *Salmonella* prevalence, using the package spdep in R [56]. To test for patterns of serotype composition by site, we performed a distance based redundancy ordination using the function capscale in the vegan package of R [57]. We examined the various landscape and bird predictor variables (e.g. habituation score) in the ordination. Only sites with at least one ibis testing positive for *Salmonella* were used in this analysis. As we were unable to rate habituation scores for nestlings at the rookeries, these sites were also excluded from this analysis.

To evaluate the potential effects of land use/land cover on the number of isolates from white ibises that matched human cases in the PulseNet database we again used GLMMS with a binomial distribution and likelihood ratio tests. In this analysis we included isolates from nestlings, as we had no *a priori* reason to expect differences among age classes. Land use/land cover categories tested as fixed effects included all twelve of the aforementioned. Marginal R^2 values were calculated for all significant land use/land cover predictors.

RESULTS

Salmonella Prevalence in Ibis

From January 2010 to July 2013 we collected and tested 261 fecal samples from birds at urban sites and 72 samples from nesting sites. We isolated *Salmonella* from 55 individual birds (33 adult/subadult birds, and 22 nestlings). The mean prevalence for birds (adult/subadult) by site was 13% (range 0–50%) and the mean prevalence for nestlings was 35% (range 6–50%) (Table 3). *Salmonella* prevalence was highest for nestlings, second highest for juvenile or subadult birds, and lowest in adult ibis (likelihood ratio test = 16.99, $P < 0.001$).

Combining all three years of the study, habituation score was not a significant predictor of *Salmonella* prevalence (likelihood ratio test = 0.17, $P = 0.68$). The prevalence in 2010 was significantly lower than in 2012 (likelihood ratio test = 8.01, $P = 0.005$), and although there was a trend towards a higher prevalence in 2013, the difference was not statistically significant. Although the majority of adult and subadult birds were sampled from December to March in both years, a few birds were sampled outside of this time period. We tested for an effect of season on *Salmonella* prevalence for adult/subadult birds and found no effect (likelihood ratio test = 0.20, $P = 0.90$). All nestlings were sampled in the same season, April/May. There was no relationship between prevalence and flock density (i.e. the number of white ibis per m^2) (likelihood ratio test = 1.89, $P = 0.17$); however, for the density calculation we were not able to consider habitat quality.

Table 3: Sampling sites, numbers of birds sampled and overall prevalence of *Salmonella* **infection o**f white ibises in south Florida

https://doi.org/10.1371/journal.pone.0164402.t003

Site	Dates Sampled	N	*Salmonella* prevalence %
Juno Beach urban park	1/2010; 3/2012; 12/2012	45	16
Horseshoe Acres neighborhood	1/2010	9	0
Dreher urban park	1/2010; 3/2012	12	17
Prosperity Oaks retirement community	3/2010; 3/2012; 12/2012	26	15
Garden Lakes business park	3/2010	15	7
Burns Rd urban park	3/2010	5	40
San Matera neighborhood	3/2010	15	13
San Marco Villas neighborhood	3/2010	19	0
Royal Palm urban park	3/2010	10	0
PGA golf community	3/2010	5	40
Donald Ross Rd empty lot	3/2010	10	20
Seasons 52 restaurant	3/2010	10	0
Indian Creek urban park	12/2012	10	10
Lion Country Safari park	3/2010; 3/2012; 12/2012	29	14
Palm Beach Zoo	7/2010; 3/2012	13	8
Loxahatchee National Wildlife Refuge	3/2010	13	0
Kissimmee Prairie Preserve State Park	3/2013; 7/2013	15	27
Tamiami West Colony	4/2012; 4/2013	60	33
Alley North Colony	4/2014	15	6
Ibis Colony	4/2013	28	50
6th Bridge Colony	4/2012; 5/2014	4	50

doi:10.1371/journal.pone.0164402.t003

Land Cover and *Salmonella* Prevalence in Ibis

Landscape composition metrics determined from PCA were not significantly related to either *Salmonella* prevalence or serotype diversity (numbers of *Salmonella* serotypes isolated; PC1: prevalence, z = 1.04, $P = 0.29$; serotype diversity, z = -0.77, $P = 0.44$). No spatial pattern was found in *Salmonella* prevalence among the sampled sites (Moran's I: observed = -0.04, expected = -0.06, $P = 0.90$). However, several land cover variables, when examined individually, showed a statistically significant relationship with the prevalence of *Salmonella* infection. As the percent cover of Palustrine wetlands and herbaceous grasslands in a 1km buffer surrounding the sampling site increased, the prevalence of *Salmonella* decreased (wetlands: LRT = 5.41, P = 0.02, grassland: LRT = 8.45, P = 0.004, Fig 2a, 2b and 2c). Palustrine wetlands are non-tidal wetlands dominated by trees, shrubs, and persistent emergent vegetation and low salinity. Within a 1 km radius around each capture point the amount cover of Palustrine emergent wetland ranged from 0.6% to 20% (Table 1). Conversely, the only developed land use type that was related to *Salmonella* prevalence, was the least intensive of the developed cover types, Open Developed, which was positively related to prevalence (Fig 2c, LRT = 5.51, P = 0.02).

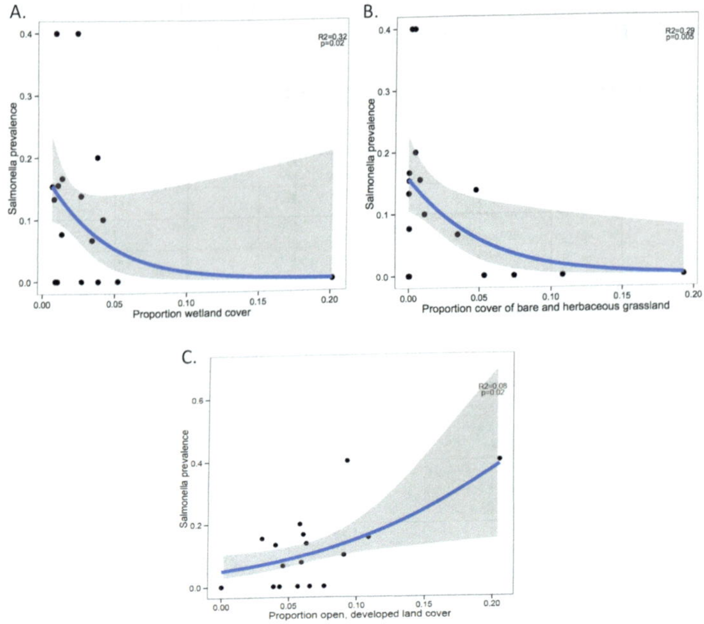

Figure 2: ***Salmonella*** **prevalence and habitat type.**

Salmonella prevalence in white ibises is negatively influenced by cover of emergent wetlands (a) herbaceous grasslands (b) and positively influenced by open, developed land (c) in Palm Beach County, Florida.

https://doi.org/10.1371/journal.pone.0164402.g002

Linkage between *Salmonella* Strains in Ibises and Cases of Human Salmonellosis

A high diversity of *Salmonella* serotypes (24 serotypes) and strain types (43 PFGE types) were isolated from white ibises (Table 4). These 43 *Salmonella* strain types belonged to one of 28 clusters identified by PFGE (≥75% similarity; Fig 3). We used rarefaction analysis to determine our sampling effort of serotype diversity in white ibis, which predicted that if another 32 *Salmonella*-infected ibis were tested, between 0 and 22 new serotypes would be found. Thirty-three percent of the ibis isolates were serotypes that ranked in the top 20 *Salmonella* serotypes associated with human cases in the U.S., as reported by the CDC for years 2010–2012 (*S.* Anatum, *S.* Bareilly, *S.* Braenderup, *S.* Javiana, *S.* Muenchen, *S.* Newport, *S.*

Saint Paul, *S.* Typhimurium) [16]. Fifteen percent of our isolates were serotypes most frequently reported in the state of Florida for the years of this study (*S.* Javiana, IV 50:Z4,Z23:- (formerly *S.* Flint), *S.* Newport, *S.* Saintpaul, and *S.* Typhimurium).

Table 4: ***Salmonella*** **serotype and strain diversity isolated from American** white ibises in South Florida

https://doi.org/10.1371/journal.pone.0164402.t004

Site [a]	Date	Source	Age [d]	Isolate	Serotype	PFGE Type	PulseNet	Matches [h]	Cases in FL [h]
Kissimmee Prairie Park	07/13	Ibis	Juvenile	KPP100	Rubislaw	Rb2	No Matches		
Kissimmee Prairie Park	07/13	Ibis	Juvenile	KPP102	Rubislaw	Rb7	No Matches		
Kissimmee Prairie Park	07/13	Ibis	Juvenile	KPP103	Miami	Mi1	TEAX01.0045	1	1
Kissimmee Prairie Park	07/13	Ibis	Juvenile	KPP104	Flint	Ft2	No Matches		
Indian Creek Park	12/12	Ibis	Adult	ICP301	Bareilly (10)[e]	Ba1	JAPX01.0064	8	1
Donald Ross Lot	03/10	Ibis	Juvenile	DR001	Baildon	Bl1	TDEX01.0001[g]	107	1
Donald Ross Lot	03/10	Ibis	Juvenile	DR002	Midway	Md1	No Matches	0	0
Juno Beach	01/10	Ibis	Adult	JB003	Rubislaw	Rb9	JLPX01.0002[g]	89	38
Juno Beach	01/10	Ibis	Adult	JB001	Rubislaw	Rb3	No Matches		
Juno Beach	01/10	Ibis	Adult	JB002	Rubislaw	Rb9	JLPX01.0002[g]	89	38
Juno Beach	01/10	Ibis	Adult	JB016	Rubislaw	Rb1	JLPX01.0059[g]	86	24
Juno Beach	03/12	Ibis	Adult	JB053	Anatum (17)[e]	An2	JAGX01.0001	104	3
Juno Beach	12/12	Ibis	Adult	JB303	San Diego	Sd3	No Matches		
Juno Beach	12/12	Ibis	Adult	JB308	Newport (3)[e]	Np1	JJPX01.0025	184	0
San Matera	03/10	Ibis	Adult	SM004	Typhimurium (2)[e]	Tm6	JPXX01.0946	96	0
San Matera	03/10	Ibis	Adult	SM009	Reading	Rd1	JLGX01.0012[g]	7	0
Garden Lake	03/10	Ibis	Adult	GL011	Rubislaw	Rb1	JLPX01.0059[g]	86	24
Prosperity Oaks	03/12	Ibis	Adult	PPO 080	Florida	Fl1	No Matches		
Prosperity Oaks	03/12	Ibis	Juvenile	PPO 082	Florida	Fl2	No Matches		
Prosperity Oaks	03/12	Ibis	Unknown	PPO 085	Bareilly (10[e]	Ba1	JAPX01.0064	8	1
Prosperity Oaks	12/12	Ibis	Adult	PPO 306	Muenchen (8)[e]	Mu2	No Matches		
PGA National	03/10	Ibis	Adult	PGA001	Braenderup (11)[e]	Br2	JBPX01.0008[g]	3	1
PGA National	03/10	Ibis	Adult	PGA004	Saint Paul (8)[e]	NT	ND		
Burns Road Park	03/10	Ibis	Adult	BR001	San Diego	Sd2	No Matches		
Burns Road Park	03/10	Ibis	Adult	BR007	San Diego	Sd2	No Matches		
Lion Country Safari	03/10	Ibis	Adult	LCS017	Newport (3)[e]	Np5	JJPX01.0714[g]	0	0
Lion Country Safari	12/12	Ibis	Adult	LCS040	Anatum (17)[e]	An2	JAGX01.0001	104	3
Lion Country Safari	12/12	Ibis	Adult	LCS300	Bareilly (10)[e]	Ba1	JAPX01.0064	8	1
Lion Country Safari	12/12	Ibis	Adult	LCS304	Newport (3)[e]	Np2	JJPX01.0880	2	0
Palm Beach Zoo	03/12	Ibis	Adult	PBZ060.1	Anatum (17)[e]	An2	JAGX01.0001	104	3
Palm Beach Zoo	03/12	Ibis	Adult	PBZ060.2	Baildon	Bl1	TDEX01.0001	107	1
Dreher Park	03/12	Ibis	Adult	DP071	Rubislaw	Rb1	JLPX01.0059[g]	86	24
Dreher Park	03/12	Ibis	Juvenile	DP077	Schwarzengrund	Sw3	JM6X01.0143	1	0
Ibis Colony, Greater Everglades [b]	04/13	Ibis	Nestling	IC003	Rubislaw	Rb6	No Matches		
Ibis Colony, Greater Everglades [b]	04/13	Ibis	Nestling	IC008	Tallahassee	Tl1	No Matches		
Ibis Colony, Greater Everglades [b]	04/13	Ibis	Nestling	IC009	Litchfield	Li1	No Matches		
Ibis Colony, Greater Everglades [b]	04/13	Ibis	Nestling	IC010	Gaminara	Ga1	No Matches		
Ibis Colony, Greater Everglades [b]	04/13	Ibis	Nestling	IC012	Litchfield	Li4	No Matches		
Ibis Colony, Greater Everglades [b]	03/13	Ibis	Nestling	IC016	Rubislaw	Rb5	JLPX01.0273[g]	3	0
Ibis Colony, Greater Everglades [b]	03/13	Ibis	Nestling	IC019	Hartford	Hf1	JHAX01.0003[g]	0	0
Ibis Colony, Greater Everglades [b]	03/13	Ibis	Nestling	IC020	Hartford	Hf3	No Matches		
Ibis Colony, Greater Everglades [b]	03/13	Ibis	Nestling	IC022	Flint	Ft1	TDHX01.0023	1	1
Ibis Colony, Greater Everglades [b]	05/13	Ibis	Nestling	IC023	Hartford	Hf2	JHAX01.0080[g]	0	0
Ibis Colony, Greater Everglades [b]	05/13	Ibis	Nestling	IC024	Litchfield	Li5	No Matches		
Ibis Colony, Greater Everglades [b]	05/13	Ibis	Nestling	IC026	Anatum (17)[f]	An3	JAGX01.0033	15	1
Ibis Colony, Greater Everglades [b]	05/13	Ibis	Nestling	IC027	Javiana (4)[f]	Jv4	No Matches		
Ibis Colony, Greater Everglades [b]	05/13	Ibis	Nestling	IC030	Litchfield	Li3	No Matches		
Tamiami W Colony [c]	03/13	Ibis	Nestling	TW004A	Typhimurium (2)[f]	Tm1	No Matches		
Tamiami W Colony [c]	03/13	Ibis	Nestling	TW010	Pensacola	Pn1	No Matches		
Tamiami W Colony [c]	04/12	Ibis	Nestling	TW034	Litchfield	Li2	No Matches		
Tamiami W Colony [c]	04/12	Ibis	Nestling	TW039	Muenchen(8)[e]	Mu1	JJ6X01.0426[g]	1	0
Tamiami W Colony [c]	04/13	Ibis	Nestling	TW162-2	Rubislaw	Rb8	No Matches		
Tamiami W Colony [c]	04/13	Ibis	Nestling	TW229B	Arechavelata	Ar1	AREX01.0029	0	0
Tamiami W Colony [c]	04/13	Ibis	Nestling	TW229C1	Braenderup(11)[f]	Br4	No Matches		
Tamiami W Colony [c]	04/13	Ibis	Nestling	TW231	Poona	Po1	JL6X01.0678[g]	1	1

[a] Sites are ordered in table corresponding to location from north (top) to south (bottom).

[b] Miami, Dade County.

[c] Ibis colony located in Broward County.

[d] Juveniles were identified by plumage color and were defined as birds < 3 years of age. Adults: >3 years of age. Nestlings were defined as birds that remained in the nest up to 25 days of age.

[e] Ranking of *Salmonella* serotype associated with human cases reporting for year of isolation in the U.S.

[f] *Salmonella* serotype in numerical ranking, in human cases for last year reported in in 2012.

[g] PFGE pattern not associated with any foodborne outbreaks for year of isolation.

[h] Matching entries for PFGE pattern for year of isolation.

doi:10.1371/journal.pone.0164402.t004

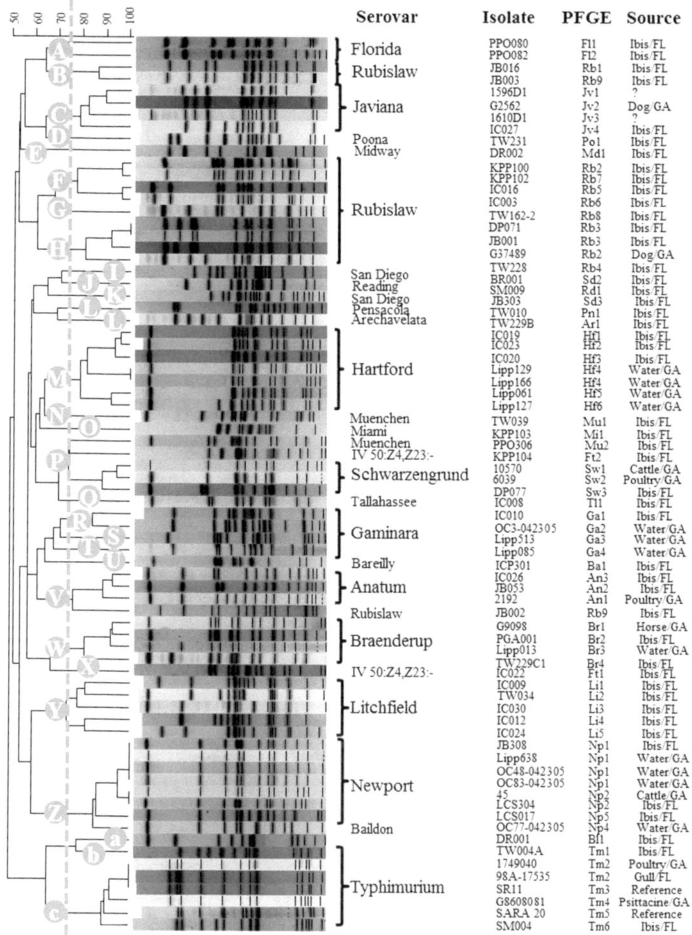

Figure 3: Cluter analysis of Salmonella.

Cluster analysis of *Salmonella* isolated from ibises by pulsed-field gel electrophoresis (PFGE) patterns generated with the restriction enzyme *Xba*I. Tiff images of *Salmonella* PFGE patterns were compared using DNA pattern recognition software, BioNumerics (Applied Maths, Austin, TX). Level of similarity was calculated using the band-based Dice similarity coefficient, and clustering of samples was performed using the unweighted pair-group method with arithmetic averaging (UPGMA). *Salmonella* PFGE patterns, generated in this study, were compared to a BioNumerics database of PFGE entries of *Salmonella* isolates from water and various animal species. PFGE patterns generated for ibis isolates were compared against an in-house database of *Salmonella* PFGE profiles for isolates from water or other animal species. A dendrogram was gen-

erated of PFGE patterns including those from isolates with the same or similar pattern (≥75% similarity). Clusters with >75% similarities were labeled A-Z, a-c on dendrogram tree on the left.

https://doi.org/10.1371/journal.pone.0164402.g003

Salmonella serotypes Anatum, Braenderup, Hartford, Newport, Schwarzengrund, and Typhimurium from ibises had identical or highly similar PFGE patterns to archived isolates from water or other animal species [49, 58–60] (Fig 3). However, the PFGE patterns for *Salmonella* serotypes *S.* Bareilly, *S.* Muenchen, and *S.* Rubislaw isolated from ibises were distinct (<75% similarity) from PFGE profiles generated for the same serotypes isolated from other wildlife and water sources (data not shown).

Additionally, forty-four percent of the *Salmonella* PFGE patterns for white ibis isolates (n = 43), matched profiles in the CDC PulseNet USA database. Several of these PulseNet patterns have not been associated with any foodborne outbreaks suggesting an environmental source (Table 4). Eighty percent of ibis isolates belonging to one of the top 20 *Salmonella* serotypes reported to the CDC (n = 15), had matching PFGE pattern with PulseNet. Twenty percent of entries reported to PulseNet, with matching patterns (n = 43), had reported human cases in Florida for the year we isolated the strain from ibis (n = 72, years 2010–2013). Florida had the most cases of any state associated with *S.* Rubislaw PulseNet patterns JLPX01.0002 and JLPX01.0059, and accounted for 86% human cases in Florida (n = 72). All *Salmonella* Rubislaw, isolated from ibises, produced unique PFGE patterns compared to PFGE patterns for *S.* Rubislaw isolated in Georgia [1]. Three of the eight white ibis *S.* Rubislaw PFGE patterns matched PFGE patterns for S. Rubislaw in PulseNet database of human cases. A non-random association was also observed in the distribution of *Salmonella* strain types from white ibises that matched human cases when examined by land cover type. There was a negative, but not statistically significant, relationship between Palustrine emergent wetland and the number of *Salmonella* isolates from white ibises that matched human cases in the PulseNet data base ($p = 0.055$; Fig 4). Furthermore, there was spatial and temporal overlap observed for *Salmonella* PFGE types isolated from ibises and humans in Florida (Table 5).

Table 5: ***Salmonella*** **PFGE patterns from American white ibis isolates that matched CDC** Pulsenet database and human cases reported in Florida

https://doi.org/10.1371/journal.pone.0164402.t005

Serotype	PulseNet Pattern	Date of Isolation		Florida County	
		Ibis	Human	Ibis	Human
Anatum	JAGX01.0001	03/2012;12/2012	02/24/2014	Palm Beach	Lee
			05/22/2012	Palm Beach	Broward
			11/20/2012	Palm Beach	Indian River
	JAGX01.0033	05/2013	12/28/2012	Dade	Hillsborough
			06/23/2013	Dade	Putnam
Baildon	TDEX01.0001	03/2012	08/07/2010	Palm Beach	Broward
Bareilly	JAPX01.0064	12/2012	08/20/2012	Palm Beach	Broward
Braenderup	JBPX01.0008	03/2010	02/12/2010	Palm Beach	Unknown
IV 50:Z4,Z23:- (formerly Flint)	TDHX01.0023	03/2013	10/31/2013	Greater Everglades	Volusia
Lomalinda	TDEX01.0001	3/2010	08/07/2010	Palm Beach	Broward
Miami	TEAX01.0045	07/2013	06/08/2013	Okeechobee	Pasco
Rubislaw	JLPX01.0059	01/2010; 03/2010	01/14/2010	Palm Beach	Lee
			02/16/2010		Palm Beach
			03/13/2010		Broward
			04/11/2010		Pinellas
			04/24/2010		Palm Beach
			05/08/2010		Orange
			05/29/2010		Orange
			06/12/2010		Palm Beach
			06/16/2010		Broward
			06/27/2010		Broward
			07/08/2010		Broward
			07/13/2010		Orange
			07/14/2010		Hillsborough
			07/18/2010		Palm Beach
			07/30/2010		Broward
			08/02/2010		Orange
			08/19/2010		Brevard
			08/19/2010		Sarasota
			08/23/2010		Broward
			08/28/2010		Hillsborough
			08/31/2010		Pinellas
			09/17/2010		Brevard
			09/21/2010		Hillsborough
			10/08/2010		Seminole
			10/09/2010		Seminole
			10/09/2010		Lake
			11/11/2010		Hillsborough
			11/15/2010		Hillsborough
			12/31/2010		Hillsborough
		3/2012	02/14/2012	Palm Beach	Lee
			02/23/2012		Palm Beach
			03/30/2012		Broward
			04/24/2012		Unknown
			06/07/2012		Hillsborough
			06/09/2012		Pinellas
			06/30/2012		Hillsborough
			07/01/2012		Pinellas
			08/09/2012		Polk
			08/18/2012		Lee
			08/19/2012		Lake
			08/21/2012		Hillsborough
			09/03/2012		Broward
			09/10/2012		Hillsborough
			09/14/2012		Palm Beach
			10/03/2012		Sarasota
			10/06/2012		Broward
			10/15/2012		Polk
			10/20/2012		Pasco
			11/01/2012		Lee
			11/04/2012		Palm Beach
			11/23/2012		Broward
			12/12/2012		Unknown
			12/28/2012		Palm Beach

doi:10.1371/journal.pone.0164402.t005

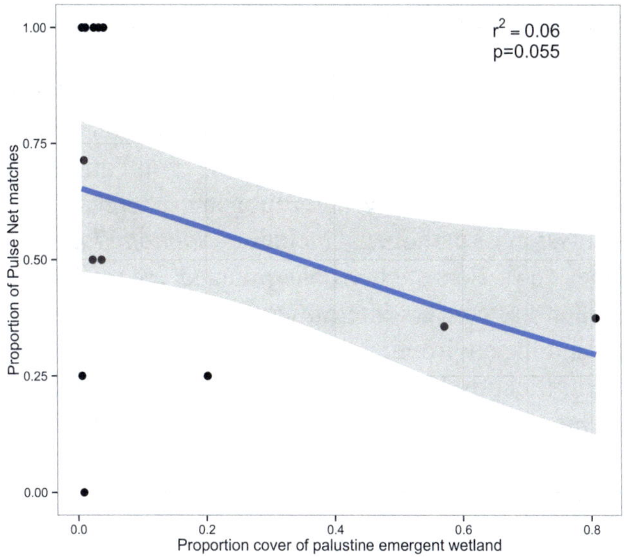

Figure 4: ***Salmonella*** **matches and emergent wetland habitat type.**

Salmonella isolates from white ibises in Florida that matched human isolates in the PulseNet database were negatively, and marginally statistically significantly related with the land cover type Palustrine emergent wetland.

https://doi.org/10.1371/journal.pone.0164402.g004

DISCUSSION

Previous reports have documented isolated and limited outbreaks of salmonellosis of colony-nesting birds, particularly during die-offs of nestlings [19, 39, 61–63]; however, this is the first report on the prevalence of *Salmonella* spp. in healthy, American white ibises. We found that the prevalence of *Salmonella* in ibises decreased as the percent of Palustrine wetlands and herbaceous grasslands increased, suggesting that land cover types that are closer to natural ecosystems support birds with a lower prevalence of carriage. White ibises are more highly dispersed in these habitats and wetlands/herbaceous grasslands provide high quality foraging habitat. Conversely, the prevalence of *Salmonella* shedding increased with the Open Developed land cover type. Ibises, as well as other peridomestic species known to carry *Salmonella*, seem to be highly attracted to open, developed areas because they are an optimal mixture of anthropogenic materials with high levels of vegetation that promote both foraging in grasses (particularly recently irrigated lawns and golf courses) and receiving handouts in parks and fields [64]. Urban parks in particular, where white ibises congregate in search of food provisions, promote the prolonged use of these sites by birds at high densities of individuals, likely resulting in cycles of *Salmonella* infection within those populations. Congregating around food sources, especially anthropogenic sources, is a well-known risk for transmission of pathogens, including *Salmonella*, among numerous wildlife species [65]. Persistently contaminated environments are a result of both birds that are carriers intermittently excreting salmonellae into the environment, and the environmental persistence of *Salmonella* spp [23, 66]. We found a significantly higher in prevalence in 2012, when compared to 2010, which may be attributed to hydrological conditions and its effect on the foraging behavioral patterns of white ibis. Years of unusually high or low precipitation create situations in natural wetlands that promote the utilization of anthropogenic food sources by white ibises [67] and the year 2012 was among the driest of our sampling period and one that would be consistent with increased urban foraging. While we did not find a positive association between the habituation score (as an indicator of dependency to urban habitats) and *Salmonella* prevalence, it is important to note that our comparison is based a small sample of animals that were not habituated and that behavior among birds may be site-specific.

The high diversity of *Salmonella* serotypes and strains found in white ibises suggests a lack of host-specificity and indicates that white ibises are likely to acquire *Salmonella* from their environment. Similarly, field and experimental studies of adult herring gulls (*Larus smithonianus*) and other urbanized birds demonstrate rapid elimination of *Salmonella* from the intestines, and a clear relationship between human/domestic animal waste and *Salmonella* colonization. This suggests that like these birds, ibis may be transiently colonized by *Salmonella* and may simply serve as a mechanical transport mechanism for the movement of salmonellae ingested from contaminated environments (for example, see Cizek et al.1994 [68], Reche et al 2003 [34] but for a review, see Tizard, 2004 [24]). Of the most frequently isolated serotypes from ibises, *S.* Litchfield, *S.* Rubislaw and *S.* Miami were specifically correlated with wetland land cover types—serotypes that are not among the most frequently reported in humans in Florida or US-wide.

Twenty-five of the PFGE patterns detected in ibis matched human clinical cases in the PulseNet database. Of those, 72% (n = 18) matched human cases from the state of Florida. There was a negative but only marginally statistically significant relationship between the number of isolates that matched human cases and emergent wetlands. At first glance, it might appear that age influenced this result, as nestlings made up a large proportion of the isolates from that land cover type and had the highest prevalence of infection; however, the nestlings we sampled were still fed solely by their parents who are foraging in nearby available habitat, confirming that the isolates from ibises in natural environments are dissimilar from human isolates.

The Florida Department of Health reports that the majority of the 5,000–6,000 human salmonellosis cases reported every year are sporadic in nature, or not associated with a food outbreak, and the most significant sources of these infections have not been identified to date (Florida Dept of Health, 2012 [69]). In urban parks in Palm Beach County, people actively feed white ibises at very close range (birds take food from people's hands) and it is common to see people, especially children and the elderly, leaning or sitting on surfaces contaminated with white ibis feces. Specifically in Florida, the state that houses the largest breeding colonies of white ibises in the US, there are over 5,000 cases of salmonellosis every year, of which nearly half are children less than 5 years of age for which consequences of infection can be severe (Florida Bureau of Environmental Public Health Medicine, Division of Environmental Health web 2011 [70]). Ibises have always been a common sight in southern Florida, but their adaption to urban habitats due to wetland loss and the increasing urbanization in Palm Beach County may increase the interaction between ibises and people [71]. Recently, urbanized ibis flocks have been noted in other urban centers in south and central Florida (Hernandez, personal observations). In Australia, highly urbanized ibises of a different species but with similar ecology are described as a nuisance and considered a threat public health threat, in part due to their carriage with *Salmonella* [32, 47]

These results indicate that ibises are likely acquiring salmonellae from environmental sources, may be good indicators of salmonellae circulating in their environment, and have the potential and opportunity to transmit salmonellae to people. Future testing of environmental samples is recommended. Additionally, our preliminary work of radio tagged ibises suggests that ibises tagged at urban sites move among several urban foraging sites on a weekly basis, and exhibit high site fidelity to urban habitats during the non-breeding season (using urban parks on a daily basis), yet fly long distances to return to natural areas to roost at night, such as the Loxahatchee National Wildlife Refuge or back and forth to nest and acquire food for nestlings between March-July in the Greater Everglades Ecosystem (Hernandez, unpublished data) where they can disseminate salmonellae acquired in urban areas. It is at these remote breeding areas that nestling mortality is likely to occur but least likely to be detected. Further research is needed to elucidate the relationship between habitat type, and the rate and persistence of *Salmonella* infection, and strain types.

ACKNOWLEDGMENTS

We truly appreciate the cooperation of the veterinarians Drs. Mark Cunningham and Dan Wolf with the Florida Wildlife Conservation Commission, staff at the Lion Country Safari, the Palm Beach Zoo, Kissimmee State Park, and the Solid Waste Authority, and the towns of Juno Beach and Palm Beach for permission and help in accessing urban ibises. Similarly, we appreciate the participation of several private landowners that allowed sampling. We thank Henry Adams for technical assistance with this manuscript. We sincerely appreciate the help of various graduate and undergraduate students who participated in the collection of samples and handling of ibises including Shaylin Duncan, Karen Christ, Henry Adams, Sarah Coker, Sebastian Ortiz and others.

AUTHOR CONTRIBUTIONS

1. **Conceptualization:** SMH MJY SS.
2. **Data curation:** SMH JJM SS.
3. **Formal analysis:** VEP.
4. **Funding acquisition:** SMH EKL.
5. **Investigation:** SMH MJY CNW SC PGS JJM SS AP KBH EH WMK MM ALC TK.
6. **Methodology:** SMH CNW.
7. **Project administration:** SMH.
8. **Resources:** JJM EKL SS SMH PGS AP MM.
9. **Software:** VEP.
10. **Supervision:** SMH.
11. **Validation:** JJM.
12. **Visualization:** VEP JJM.
13. **Writing – original draft:** SMH JJM EKL VEP.
14. **Writing – review & editing:** SMH JJM EKL SS VEP CNW SC.

REFERENCES

1. Thorns CJ. Bacterial food-borne zoonoses. Revue scientifique et technique-International Office of Epizootics. 2000;19: 226–239. pmid:11189717
2. Scallan E, Hoekstra RM, Angulo FJ, Tauxe RV, Widdowson MA, Roy SL, et al. Foodborne illness acquired in the United States—major pathogens. Emerg Infect Dis. 2011 Jan;17(1):7–15 pmid:21192848
3. Vugia D, Cronquist A, Hadler J, Tobin-D'angelo M, Blythe D, Smith K, et al. Preliminary FoodNet data on the incidence of infection with pathogens transmitted commonly through food—10 states, United States, 2005. MMWR. Morb & Mort Weekly Rep. 2006;55: 392–395.
4. CDC. Multistate outbreaks of *Salmonella* serotype Poona infections associated with eating Cantaloupe from Mexico—United States and Canada. MMWR. Morb & Mort Weekly Rep. 2002;51:1044–1047.
5. CDC. Outbreak of *Salmonella* serotype Enteritidis infections associated with raw almonds. 2004;53:484–487.
6. CDC. Multistate outbreaks of *Salmonella* infections associated with Raw Tomatoes eaten in Restraurants—United States, 2005–2006. MMWR. Morb. Mort. Weekly Rep. 2007b;56:909–911.
7. CDC. *Salmonella* Annual Summary, 2006. Atlanta: US Department of Health and Human Services, CDC. 2008
8. CDC. Outbreak of *Salmonella* serotype Saintpaul infections associated with eating alfalfa sprouts—United States, 2009. MMWR. Morb Mort Weekly Rep. 2009;58:500–503.
9. CDC. Preliminary FoodNet data on the incidence of infection with pathogens transmitted commonly through food—10 states, 2010. MMWR. Morb Mort Weekly Rep. 2010;59(14);418–422.
10. Greene SK, Daly ER, Talbot EA, Demma LJ, Holzbauer S, Patel NJ, et al. Recurrent multistate outbreak of Salmonella Newport associated with tomatoes from contaminated fields, 2005. Epidemiol Infect. 2008 Feb 1;136(02):157–65.
11. Hedberg CW, Angulo FJ, White KE, Langkop CW, Schell WL, Stobierski MG, et al. Outbreaks of salmonellosis associated with eating uncooked tomatoes: implications for public health. Epidemiol Infect. 1999;122:385–393. pmid:10459640

12. Lynch M, Painter J, Woodruff R, Braden C. Surveillance for foodborne-disease outbreaks—United States, 1998–2002. MMWR. Morb Mort Weekly Rep. 2006;55:1–34.
13. Mohle-Boetani JC, Farrar J, Bradley P, Barak JD, Miller M, Mandrell R, et al. *Salmonella* infections associated with mung bean sprouts: epidemiological and environmental investigations. Epidemiol Infect. 2009;137:357–366. pmid:18294429
14. Proctor ME, Hamacher M, Tortorello ML, Archer JR, Davis JP. Multistate outbreak of *Salmonella* serotype Muenchen infections associated with alfalfa sprouts grown from seeds pretreated with calcium hypochlorite. J Clin Microbiol. 2001;39:3461–3465. pmid:11574556
15. Olsen SJ, Bishop R, Brenner FW, Roels TH, Bean N, Tauxe RV, et al. The Changing Epidemiology of Salmonella: Trends in Serotypes Isolated from Humans in the United States, 1987–1997. J Infect Dis. 2001;183: 753–761. pmid:11181152
16. CDC. National *Salmonella* Surveillance Annual Report, 2012. Atlanta: US Department of Health and Human Services, CDC. 2014.
17. Tauni MA, Österlund A. Outbreak of Salmonella typhimurium in cats and humans associated with infection in wild birds. J Small Anim Prac. 2000;41: 339–341.
18. Alley MR, Connolly JH, Fenwick DG, Mackereth GF, Leyland MJ, Rogers LE, et al. An epidemic of salmonellosis caused by Salmonella Typhimurium DT160 in wild birds and humans in New Zealand. NZ Vet J. 2002;50: 170–176.
19. Refsum T, Vikoren T, Handeland K, Kapperud G, Holstad G. Epidemiologic and pathologic aspects of Salmonella typhimurium infection in passerine birds in Norway. J Wild Dis. 2003;39: 64–72.
20. Pennycott TW, Park A, Mather HA. Isolation of different serotypes of Salmonella enterica from wild birds in Great Britain between 1995 and 2003. Vet Rec. 2006;158:817–820. pmid:16782854
21. Une Y, Sanbe A, Suzuki S, Niwa T, Kawakami K, Kurosawa R, et al. *Salmonella enterica* serotype Typhimurium infection causing mortality in eurasian tree sparrows (*Passer montanus*) in Hokkaido. Jpn. J Infect Dis. 2008;61:166–167. pmid:18362416
22. Fukui D, Takahashi K, Kubo M, Une Y, Kato Y, Izumiya H. Mass mortality of eurasian tree sparrows (Passer montanus) from Salmonella typhimurium DT40 in Japan, winter 2008–2009. J Wild Dis. 2014;50(3), 2014, pp. 484–495

23. Friend M, Franson JC. Salmonellosis. Field manual of wildlife diseases. General field procedures and diseases of birds. No. ITR-1999-001. Geological Survey Madison, WI Biological Resources Div; 1999.
24. Tizard I. Salmonellosis in wild birds. Seminars in Avian and Exotic Pet Medicine. 2004;13:50–66
25. Daoust PY, Prescott JF. Salmonellosis. In: Hunter DB, Thomas NJ, editors. Infectious diseases of wild birds. Ames: Blackwell Publishing; 2007;270–88.
26. Daoust PY, Busby DG, Ferns L, Goltz J, McBurney S, Poppe C, et al. Salmonellosis in songbirds in the Canadian Atlantic provinces during winter-summer 1997–98. Can Vet J. 2000;41:54–59. pmid:10642873
27. Hall AJ, Saito EK. Avian wildlife mortality events due to salmonellosis in the United States, 1985–2004. J Wildl Dis. 2008 Jul;44:585–593. pmid:18689643
28. Lawson B, Howard T, Kirkwood JK, Macgregor SK, Perkins M, Robinson RA, et al. Epidemiology of Salmonellosis in Garden Birds in England and Wales, 1993 to 2003. Ecohealth. 2010;7(3): 294–306 pmid:20945078
29. Lawson B, de Pinna E, Horton RA, Macgregor ES, John SK, Chantrey J, et al. Epidemiological Evidence That Garden Birds Are a Source of Human Salmonellosis in England and Wales. PloS one. 2014 Feb 1;9(2):e88968 pmid:24586464
30. Girdwood RW, Fricker CR, Munro D, Shedden CB, Monaghan P. The incidence and significance of salmonella carriage by gulls (Larus spp.) in Scotland. J Hyg. 1985 Oct 1;95(02):229–41.
31. Daisuke F. Infectious Diseases Associated with Relation between Humans and Wildlife—Consideration on Wild Bird Mortality and Bird-Feeding. Jpn J Zoo & Wild Med. 2013;18(2):41–48
32. Epstein JH, McKee J, Shaw P, Hicks V, Micalizzi G, Daszak P, et al. The Australian white ibis (Threskiornis molucca) as a reservoir of zoonotic and livestock pathogens. EcoHealth. 2006 Dec 1;3(4):290–8.
33. Pedersen K, Clark L, Andelt WF, Salman MD. Prevalence of shiga toxin-producing *Escherichia coli* and *Salmonella enterica* in rock pigeons captured in Fort Collins, Colorado. J Wildl Dis. 2006;42:46–55. pmid:16699148
34. Reche MP, Jiménez PA, Alvarez F, García de los Rios JE, Rojas AM, de Pedro P. Incidence of salmonellae in captive and wild free-living

raptorial birds in central Spain. J Vet Med B Infect Dis. Vet. Public Health. 2003;50:42–44. pmid:12710500

35. Refsum T, Heir E, Kapperud G, Vardund T, Holstad G. Molecular epidemiology of Salmonella enterica serovar typhimurium isolates determined by pulsed-field gel electrophoresis: comparison of isolates from avian wildlife, domestic animals, and the environment in Norway. Appl & Environ Microbiol. 2002;68: 5600–5606.
36. Bradley CA, Altizer S. Urbanization and the ecology of wildlife diseases. Trends Eco & Evo. 2007;22: 95–102.
37. Hernandez SM, Keel K, Sanchez S, Trees E, Gerner-Smidt P, Adams JK, et al. Epidemiology of a *Salmonella enterica* subsp. *enterica* serovar Typhimurium strain associated with a songbird outbreak. Appl Enviro Microbiol. 2012;78:7290–7298.
38. Hauser E, Hühn S, Junker E, Jaber M, Schroeter A, Helmuth R, et al. Characterization of a phenotypic monophasic variant belonging to *Salmonella enterica subsp. enterica* serovar Typhimurium from wild birds and its possible transmission to cats and humans. Berl Munch Tierarztl Wochenschr. 2012;122:169–177.
39. Kapperud G, Stenwig H, Lassen J. Epidemiology of Salmonella typhimurium O:4–12 infection in Norway: evidence of transmission from an avian wildlife reservoir. Am J Epi. 1998;147: 774–782.
40. Nesse LL, Refsum T, Heir E, Nordby K, Vardund T, Holstad G. Molecular epidemiology of *Salmonella spp*. isolates from gulls, fish-meal factories, feed factories, animals and humans in Norway based on pulsed-field gel electrophoresis. Epidemiol Infect 2005;133:53–58. pmid:15724711
41. Palmgren H, Aspán A, Broman T, Bengtsson K, Blomquist L, Bergström S, et al. *Salmonella* in Black-headed gulls (*Larus ridibundus*); prevalence, genotypes and influence on *Salmonella* epidemiology. Epidemiol Infect 2006;134:635–644. pmid:16238820
42. Penfold JB, Amery HC, Peet PJ. Gastroenteritis associated with wild birds in a hospital kitchen. Brit Med J. 1979;29: 802.
43. Kushlan JA. Feeding Ecology and Prey Selection in the White Ibis. The Condor. 1979;81: 376–389.
44. Bildstein KL, Post W, Johnston J, Frederick P. Freshwater Wetlands, Rainfall, and the Breeding Ecology of White Ibises in Coastal South Carolina. The Wilson Bulletin. 1990;102: 84–98.

45. Gawlik DE. The Effects of Prey Availability on the Numerical Response of Wading Birds. Eco Mono. 2002;72: 329–346.
46. Kushlan JA. Supplemental Information for the White Ibis: Biological Status Review Report. Tallahassee: Florida Fish and Wildlife Conservation Commission. 2011; pp. 3.
47. Martin JM, French K, Ross GA, Major RE. Foraging distances and habitat preferences of a recent urban coloniser: The Australian white ibis. Lan & Urb Plan. 2011;102: 65–72.
48. Tiner RW. Wetlands of the United States: current status and recent trends. United Fish and Wildlife Service. 1984. Accessed: http://repositories. tdl. org/tamug-ir/bitstream/handle/1969. 3/21208/3386-Wetlands%20 of%20the%20United%20States-Current%20Status%20and%20 Recent%20Trends. pdf?sequence=1
49. Maurer JJ, Martin G, Hernandez SM, Cheng Y, Gerner-Smidt P, Hise KB, et al. Diversity and persistence of Salmonella enterica strains in rural landscapes in the Southeastern United States. PloS one. 2015;10(7).
50. Gerner-Smidt P, Hise K, Kincaid J, Hunter S, Rolando S, Hyytia-Trees E, et. al. PulseNet USA: a five-year update. Food Pathog Dis. 2006;3:9–19.
51. Zamperini K, Soni V, Waltman D, Sanchez S, Theriault EC, Bray J, et al. Molecular characterization reveals Salmonella enterica serovar 4,[5],12:i:- from poultry is a variant Typhimurium serovarserotype. Avian Dis. 2007;51:958–964. pmid:18251408
52. R Core Team. R: A language and environment for statistical computing. R Foundation for Statistical Computing. 2016; Available: http://www. R-project. org/
53. Bates D, Maechler M, Bolker B, Walker S, Christensen RHB, Singmann H, et al. Linear mixed-effects models using 'Eigen' and S4, ver 1. 1–8. 2015; Available at http://lme4. r-forge. r-project. org/
54. Bartoń K. Model selection and model averaging based on information criteria (AICc and alike), ver 1. 15. 6. 2016; Available at https://cran. r-project. org/web/packages/MuMIn/MuMIn. pdf
55. Mardia KV, Kent JT, Bibby JM. Multivariate Analysis. 1st ed. London: Academic Press; 1979

56. Bivand R. Spatial Dependence: Weighting Schemes, Statistics and Models, ver 0. 6–5. 2016; Available: https://cran. r-project. org/web/packages/spdep/spdep. pdf
57. Oksanen J, Guillaume Blanchet F, Friendly M, Kindt R, Legendre P, McGlinn D, et al. Community Ecology package. Ver 2. 4–0. Available at https://cran. r-project. org/web/packages/vegan/vegan. pdf
58. Liljebjelke KA, Hofacre CL, Liu T, White DG, Ayers S, Young S, et al. Vertical and horizontal transmission of Salmonella within integrated broiler production system. Food Path & Dis. 2005 Mar 1;2(1):90–102.
59. Meinersmann RJ, Berrang ME, Jackson CR, Fedorka-Cray P, Ladely S, Little E, et al. Salmonella, Campylobacter and Enterococcus spp. : Their antimicrobial resistance profiles and their spatial relationships in a synoptic study of the Upper Oconee River basin. Micro Eco. 2008 Apr 1;55(3):444–52
60. Haley BJ, Cole DJ, Lipp EK. Distribution, diversity, and seasonality of waterborne salmonellae in a rural watershed. Appl & Environ Microbiol. 2009 Mar 1;75(5):1248–55.
61. Phalen DN, Drew ML, Simpson B, Roset K, Dubose K, Mora M. Salmonella enterica subsp. Enterica in Cattle Egret (*Bubulcus ibis*) chicks from central Texas: prevalence, serotypes, pathogenicity, and epizootic potential. J of Wild Dis. 2010;46: 379–389.
62. Velarde R, Porrero MC, Serrano E, Marco I, García M, Téllez S, et al. Septicemic Salmonellosis aused by salmonella hessarek in wintering and migrating Song Thrushes (Turdus philomelos) in Spain. J of Wild Dis. 2012;48: 113–121.
63. Locke LN, Ohlendorf HM, Shillinger RB, Jareed T. Salmonellosis in a captive heron colony. J Wild Dis. 1974 Apr;10(2):143–5.
64. Oliver AJ, Hong-Wa C, Devonshire J, Olea KR, Rivas GF, Gahl MK. Avifauna richness enhanced in large, isolated urban parks. Landscape and Urban Planning. 2011 Sept 1;102(4): 215–225.
65. Miller S, Zieger U, Ganser C, Satterlee SA, Bankovich B, Amadi V. Influence of land use and climate on salmonella carrier status in the small Indian mongoose (Herpestes auropunctatus) in Grenada, West Indies. J Wild Dis. 2015 Jan;51(1):60–8.
66. Wales A, Davies RH. Environmental Aspects of Salmonella. In: Barrow PA, Methner U, editors. Salmonella in Domestic Animals. Oxfordshire: CABI International; 2013. pp 399–426.

67. Dorn NJ, Cook MI, Herring G, Boyle B, Nelson J, Gawlik DE. Aquatic prey switching and urban foraging by the White Ibis Eudocimus albus are determined by wetland hydrological conditions. Ibis. 2011;153, 323–335.
68. Cizek A, Literak I, Hejlicek K, Treml F, Smola J. Salmonella contamination of the environment and its incidence in wild birds. J Vet Med B. 1994;41(5): 320–327
69. Dept Florida. Health 2012.
70. Florida Bureau of Enviro 2011.
71. Mitsch WJ, Gosselink JG. Wetlands. 4th ed. Hoboken: John Wiley & Sons 2007.

CHAPTER 25

Live Exotic Animals Legally and Illegally Imported via the Main Dutch Airport and Considerations for Public Health

Annika van Roon[1], Miriam Maas[1], Daniela Toale[1], Nedzib Tafro[2], Joke van der Giessen[1]

[1] Centre for Infectious Disease Control, National Institute for Public Health and the Environment (RIVM), Bilthoven, the Netherlands

[2] Border inspection, Netherlands Food and Consumer Product Safety Authority (NVWA), Amsterdam, The Netherlands

Citation: van Roon A, Maas M, Toale D, Tafro N, van der Giessen J (2019) Live exotic animals legally and illegally imported via the main Dutch airport and considerations for public health. PLoS ONE 14(7): e0220122. https://doi.org/10.1371/journal.pone.0220122.

ABSTRACT

The trade in live animals and animal products is considered one of the major drivers of zoonotic disease emergence. Schiphol airport in the Netherlands is one of the largest European airports and is considered a main hub for legal and illegal import of exotic animals. However, so far there is little information about what pathogens these imported animals might carry with them. Therefore, this study aimed to assess the zoonotic risks of exotic animals imported into the Netherlands through Schiphol airport in 2013 and 2014. Based on a previous list of highly prioritised emerging zoonoses for the Netherlands (EmZoo list), WAHID and Promed databases, literature and expert opinions, a list of 143 potentially relevant zoonotic pathogens was compiled. In a step-wise selection process eighteen pathogen-host combinations that may pose a public health risk by the import of exotic animals via Schiphol airport were identified and these were assessed by expert elicitation. The five pathogens with the highest combined scores were *Salmonella* spp., Crimean-Congo haemorrhagic fever virus, West Nile virus, *Yersinia pestis* and arenaviruses, but overall, the public health risk of the introduction of these exotic pathogens into the Netherlands via the legal import of exotic animals was considered low. However, the vast majority of imported exotic animals were imported by trade companies, increasing the risk for specific groups such as retail and hobbyists/pet owners. It is expected that the risk of introduction of exotic zoonotic pathogens via illegal import is substantial due to the unknown health status. Due to changing trade patterns combined with changing epidemiological situation in the world and changing epidemiological features of pathogens, this risk assessment needs regular updating. The results could give directions for further adjusting of health requirements and risk based additional testing of imported exotic animals.

INTRODUCTION

Trade of exotic animals has existed since prehistory [1]. Due to the tremendous expansion of the human population in the last decades, economic growth and globalization, the demand for exotic animals increased and likewise the scale and extent of the exotic animal trade enlarged [2],[3]. The trade in domestic or exotic animals (products) has been indicated as one of the major drivers of zoonotic disease emergence [4],[5].These zoonotic pathogens have represented 75% of all emerging infectious diseases in the past decade [6]. The majority (60%) of these emerging infectious diseases are due to pathogens with wildlife origins [7], [8]. A similar percentage was obtained, when emerging zoonotic pathogens relevant for the Netherlands were ranked using multicriteria decision analyses: of the 86 pathogens ranked, more than 60% originated from wildlife [9]. Many zoonotic pathogens are present in third countries (i.e. non-European Union member states), and the international trade of exotic animals is one of the risk factors to introduce these pathogens [5]. There have been various examples in a wide range of species where the legal and illegal trade in live exotic animals raised public health concerns [10], [11], [12], [13]. Most of the trade in exotic animals is legal, but a significant part is not [14]. It is thought that the global illegal wildlife trade is worth at least 19 billion US dollars per year, which makes it the fourth largest illegal business after narcotics, counterfeiting and human trafficking [15],[16].

For import of live animals into the EU, health certificates or trade documents, issued by the competent Veterinary Authority of the exporting country, are required. The requirements are related to the country of origin and animal species. For example the international legal trade in many exotic animals, especially in mammals and birds, is only possible between registered premises as zoological gardens and bird parks. Regulations on the trade in endangered wild animals are defined in the EU Wildlife Trade Regulations [17] and the Convention on International Trade in Endangered Species of wild flora and fauna (CITES) [18].

Few risk assessments have been performed regarding animal imports and even fewer focus on exotic animal imports and the zoonotic health risks to humans. All studies emphasized the need for more research as the lack of data hampers risk assessment studies [2], [19], [20]. To better understand the risk of the global movement of exotic animals, knowledge is required about species of animals imported, volumes moved and trade routes [21].

Schiphol airport in the Netherlands is one of the largest European airports and is considered one of the main hubs for both legal and illegal import of exotic animals [22]. However, there are no studies published with numbers and species of exotic animals imported into the Netherlands and what pathogens these animals might carry. Therefore, in a collaborative project, an inventory was made of the numbers and species of exotic animals legally imported into the Netherlands in 2013 and 2014 at Schiphol airport, and the risk of introduction of pathogens relevant for livestock and public health were assessed. The inventory and the risk for livestock are described elsewhere [23]. In the current study, we aimed to assess the zoonotic risks of live exotic animals, defined as non-native animals that were not fish or invertebrate classes of animals, imported into the Netherlands. This was done by ranking animal host-pathogen combinations based on literature and expert opinion. Furthermore, an inventory was made of the numbers and species of exotic animals seized at Schiphol airport between January 2012 and December 2014 with the aim to assess the illegal import of exotic animals via Schiphol airport and to identify differences between legal and illegal trade.

MATERIAL AND METHODS

Hazard Identification

Import of exotic animals into the Netherlands

Numbers and animal species of exotic animals legally imported into the Netherlands between 2013–2014 were previously reported [23]. Briefly, the database contained 1404 consignments (i.e. shipment of animals subdivided according to species), representing a total number of 490,750 exotic animals. A little less than half of these animals (43%) was destined for the Netherlands, a small number (4%) was destined for other EU countries and the rest (53%) was in transit to third countries. The majority of the animals imported in the Netherlands were reptiles (93.8%), followed by amphibians (5.8%), birds (0.06%) and mammals (0.4%). The animals originated from 25 different countries, with most animals from the USA (78.8%), Vietnam (5.1%), Indonesia (3.5%) and Tanzania (3.1%). Zoonotic pathogens of concern in exotic animals legally imported into the Netherlands.

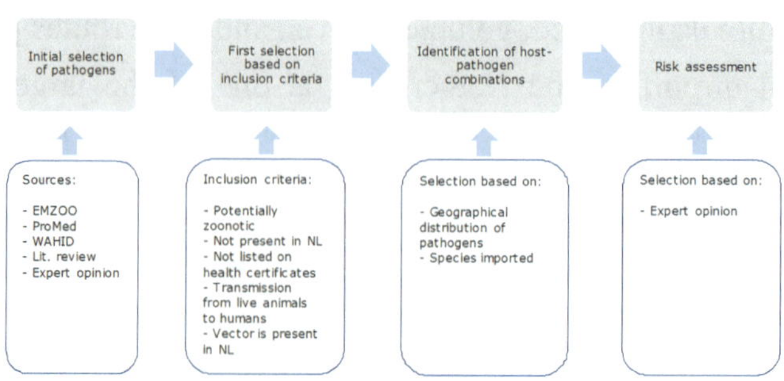

Figure 1: Stepwise selection of zoonotic pathogens of concern in exotic animals legally imported into the Netherlands.

https://doi.org/10.1371/journal.pone.0220122.g001

Pathogen-host combinations that may pose a public health risk by the import of exotic animals were identified following a stepwise selection process (Fig 1).

1. Initial selection of pathogens

To start the risk assessment, a longlist was compiled of pathogens that may pose a public health risk associated with the import of exotic animals. Multiple sources were used to compile this longlist. First, 86 zoonotic pathogens were included that had previously been selected and ranked as potential threats to public health in the Netherlands, hereafter referred to as "the EmZoo list" [9]. In addition, for the ten countries that exported the largest total number of animals to the Netherlands, a search was done on ProMed-mail and the WAHID database, using search terms "[country name] AND zoonosis". The list was supplemented with pathogens from literature studies on pathogens in exotic animals and by expert opinion.

2. First selection based on inclusion criteria

As a second step, pathogens on the longlist were included in the risk assessment when these pathogens complied to the following inclusion criteria:

- Zoonotic potential of the pathogen is proven (i.e. human cases have been described).
- Pathogen is absent in the Netherlands OR is present but more virulent genotypes do exist in other parts of the world.
- Pathogen is not listed on the health certificates of the imported animals and thus not already tested for/controlled.
- The pathogen can be directly transmitted from live animals to humans or via a vector or intermediate host
- The vector or intermediate host is present in the Netherlands for those pathogens that need a vector or intermediate host for transmission.

3. Identification of host-pathogen combinations

This step consisted of two parts: a check whether the pathogens were present in the list of countries that exported to the Netherlands, and a check whether the animal species that were imported in the Netherlands are susceptible to the pathogens. Pathogens were included in the risk assessment when, based on literature, they are present in a country that exported animals to the Netherlands In 2013 or 2014, and when the imported animals are susceptible to infection with these pathogens.

4. Risk assessment

Information was collected about the transmission route of the pathogen via animals to humans, the numbers of relevant animal species that were imported from the exporting countries, whether animals were imported into the Netherlands or in transit, and the disease the pathogen causes in humans. Then the pathogens were assessed by expert elicitation at the National Institute for Public Health and the Environment (RIVM). Experts of various disciplines including virology, vector-borne zoonoses, parasitology and wildlife-born zoonoses were asked to score the public health risk of each pathogen relative to the other pathogens. A score of 1 represents the lowest public health risk among all pathogens considered and 4 the highest public health risk among all pathogens considered. The involved experts were asked to give one score ranging from 1 to 4 for each pathogen while taking into account: the probability of introduction of the infectious pathogen, the probability of transmission of the pathogen to humans in the Netherlands and the impact of human disease. The scores given by the different experts are presented as a range per pathogen e.g. a pathogen receives score 2 from two experts and score 3 from the other two experts, leading to a final score of 2–3 for this pathogen. One expert scored only nine of the 18 pathogens as the other pathogens were outside her field of expertise.

Illegal Import of Live Animals into the Netherlands

Reports about seizures of live illegally imported animals at Schiphol airport between January 2012 and December 2014 were received from the Netherlands Food and Consumer Product Safety Authority (NVWA). These reports consisted of e-mail correspondence between NVWA staff members and customs about seizures of illegally imported animals or animal products, photographs of the seized animals and receipts for health checks and quarantine required for the animals. Only seizure reports that met the following criteria were included: (1) the seizure concerned live non-native animals that were not fish or invertebrate classes of animals, (2) there was no evidence that the animals were accidentally imported, e.g. a bat free-roaming in an airplane, (3) animals were seized because they were illegally imported and not because of welfare issues. The reports were documented in a database including animal species, country of origin, results of diagnostic tests that were performed, transport method and presence/absence of health certificates. Descriptive analysis was performed using Microsoft Excel 2010.

RESULTS

Zoonotic Pathogens of Concern in Exotic Animals Legally Imported into the Netherlands

Based on the first step as described above, a list of 143 potentially relevant zoonotic pathogens was compiled. In the subsequent selection steps, this list was reduced to a final list of 18 pathogens that are considered potentially relevant pathogens for public health of the legal import of exotic animals into the Netherlands (Fig 2).

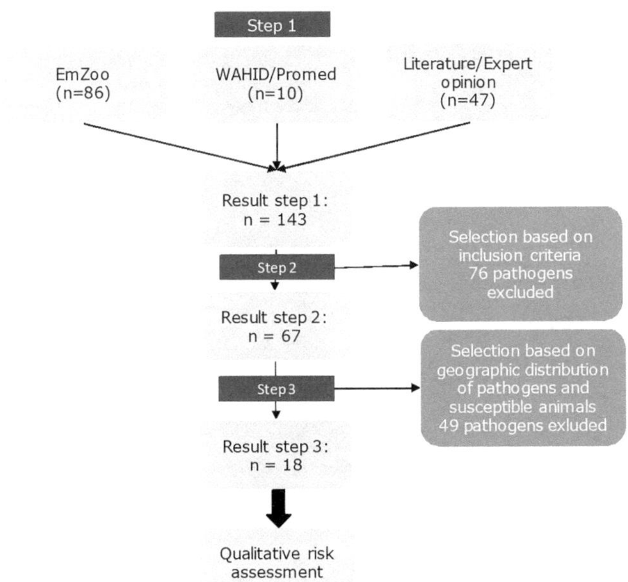

Figure 2: Flow chart presenting the selection steps of potentially relevant pathogens.

https://doi.org/10.1371/journal.pone.0220122.g002

Risk Assessment

The remaining 18 pathogens were scored relative to each other by expert elicitation. The information that was used for scoring the pathogens. The range of the expert opinions of each pathogen was determined and subsequently the pathogens were ranked according to this score Table 1. In case of equal concluding scores, the pathogens were ranked in alphabetical order. None of the pathogens were scored at the highest level, and the five pathogens with the highest combined scores were *Salmonella* spp., Crimean-Congo haemorrhagic fever virus, West Nile virus, *Yersinia pestis* and arenaviruses.

Table 1: Eighteen potentially relevant pathogens scored relative to each other and ranked by expert opinion

https://doi.org/10.1371/journal.pone.0220122.t001

Pathogens	Animal classes[a]	Country of origin[b]	Range[c]	
			Min	Max
Salmonella spp. [d]	Birds, reptiles, amphibians	Worldwide	2	4
Crimean-Congo haemorrhagic fever virus	Birds, mammals, reptiles	China, Egypt, Kenya, South Africa, Tanzania, UAE	2	3
West Nile virus	Birds	Brazil, Canada, South Africa, Tanzania, USA	2	3
Yersinia pestis[e]	Mammals	Argentina, Egypt	2	3
Arenaviruses[e]	Mammals	Argentina	1	3
Eastern equine encephalitis virus	Birds, reptiles	Brazil, Canada, Columbia, Ecuador, USA	2	2
Japanese encephalitis virus	Birds	Philippines, Singapore	2	2
Western equine encephalitis virus	Birds, reptiles	Argentina, Canada, Guyana, Peru, Suriname, USA	2	2
Rocio virus	Birds	Brazil	1	2
Ross river virus	Mammals	Indonesia	1	2
Saint Louis encephalitis virus	Birds	Argentina, Brazil, Canada, Cuba, Peru, Suriname, USA	1	2
Sindbis virus[e]	Birds	Tanzania	1	2
Venezuelan equine encephalitis virus[e]	Birds	Peru, USA	1	2
Echinococcus granulosus[e]	Mammals	Chile	1	2
Rickettsia africae/ African tick bite fever[e]	Mammals	South Africa	1	1
Rickettsia typhi[e]	Mammals	Mexico	1	1
Rickettsia rickettsii[e]	Mammals	Argentina	1	1
T-cell lymphotropic virus 1/HTLV-1	Mammals	Ghana, Peru, Tanzania, USA	1	1

[a] Animal classes that were imported into the Netherlands in 2013–2014 that could play a role in the transmission of the pathogen.

[b] Countries where pathogen is present and where animals were imported from into the Netherlands in 2013–2014

[c] A scale of 1 to 4 was used, where 1 represents the lowest risk and 4 the highest risk

[d] Only scored by two of the experts as the pathogen was outside the field of expertise of one expert and the other expert indicated that there was not enough information to decide on the risk

[e] Only scored by three of the experts as the pathogen was outside the field of expertise of one expert

https://doi.org/10.1371/journal.pone.0220122.t001

Illegal Import of Exotic Animals into the Netherlands 2012–2014

Two hundred forty one seizures of animals or animal products that were imported or in transit at Schiphol airport were identified as illegal. These seizure reports contained information about numbers of animals, species of animals, country of origin, results of diagnostic tests, transportation method, the presence or absence of health certificates or (wrong) import papers, whether it concerned CITES listed species and destination of animals after seizure. Not all seizure reports contained all information. Of the 241 seizures, only 38 met the inclusion criteria and were considered illegally imported live animals that were non-domesticated and not fish or invertebrate classes of animals Table 2. The 38 seizure reports consisted of 58 consignments and 514 individual animals.

Table 2: Number of consignments (C) and individual animals (I) per class and country of origin that were attempted to be illegally imported into the Netherlands or were in transit through Schiphol airport between the years 2012–2014

https://doi.org/10.1371/journal.pone.0220122.t002

	Amphibia C (I)	Aves C (I)	Reptilia C (I)	Total C (I)
Canada		1 (3)		1 (3)
China			5 (16)	5 (16)
Curacao		1 (5)	2 (6)	3 (11)
Guyana			2 (203)	2 (203)
Indonesia		3 (22)		3 (22)
Iran		1 (2)		1 (2)
Iraq		1 (2)		1 (2)
Mexico		5 (76)		5 (76)
Morocco		3 (8)		3 (8)
Netherlands Antilles			1 (7)	1 (7)
Panama			1 (1)	1 (1)
Suriname		24 (127)		24 (127)
Tunisia			1 (1)	1 (1)
Turkey		2 (15)		2 (15)
USA	4 (19)			4 (19)
Unknown			1 (1)	1 (1)
Total	4 (19)	41 (260)	13 (235)	58 (514)

C: Number of consignments (i.e. shipment of animals subdivided according to species)
I: Number of individual animals

https://doi.org/10.1371/journal.pone.0220122.t002

Most consignments of the illegal import originated from Suriname (41%), all concerning birds, followed by five consignments of reptiles from China (9%) and five consignments of birds from Mexico (9%). When analysing the number of individual animals, the top three of countries for illegally imported animals are Guyana, Suriname and Mexico. In contrast to these illegal imports, in the same period, most consignments of legally imported animals originated from Indonesia, Suriname and the USA. On the individual animal level, the top three countries of origin for legally imported animals were Peru, the USA and Vietnam [23].

The reported animals were exotic amphibians, birds and reptiles. Of the 58 consignments, 11 consignments (19%) concerned CITES-listed species. The majority of the consignments concerned birds (71%), followed by reptiles (22%) and amphibians (7%). In contrast to legal import, in the same period, 1404 consignments of exotic animals with in total 490,750 individual animals were legally imported into the Netherlands or in transit to other countries. The consignments of legally imported animals concerned reptiles (65%), amphibians (18%), birds (11%) and mammals (5%) [23].

According to European legislation, all birds that are confiscated should be tested for Avian Influenza and Newcastle Disease. Ninety-five percent of the confiscated birds tested negative for Avian Influenza and of the remaining 5% the results were not documented. For confiscated reptiles and amphibians, there are no health requirements.

In this project, a rough estimate of the number of illegally imported exotic animals was made based on the expert opinion of an expert on wildlife crime and the expert opinion of an expert ranking member at EcoHealth Alliance. The first opinion was that the entire illegal trade can be extrapolated from the seizures, the number of seized animals is 10% of the entire illegal trade. The second opinion was that illegal trade is equivalent to the legal trade. This gives a broad estimate of 1,700 to 245,000 illegally imported and transited animals per year between 2013–2014.

DISCUSSION

This study aimed to assess the zoonotic risks of exotic animals imported into the Netherlands via Schiphol airport by literature and expert elicitation. Ideally, the risk factors for introduction and transmission of zoonotic pathogens would be systematically weighted and compared, but there were no data available to apply this method. The five pathogens that scored highest were: *Salmonella* spp., Crimean-Congo haemorrhagic fever virus, West Nile virus, *Yersinia pestis* and arenaviruses. However, the overall public health risk of these pathogens via the legal import of exotic animals was considered low reflected by the low scores in general.

Salmonella spp. was scored highest, although scoring *Salmonella* spp. was difficult, as over 2500 different Salmonella serotypes exist, which differ in virulence [24]. While some serotypes are host-specific, all mammals, birds, reptiles and amphibians can become infected with *Salmonella* spp. A large number of reptiles (468,621) as well as birds (2,938) and amphibians (17,292) were imported into (43%) or transited through (57%) the Netherlands in 2013–2014. Since exotic animals are not being tested for *Salmonella* spp. and a large number of animals are imported by trade companies (99.8%) [23] and mostly destined for the pet industry, the probability of exposure of humans to (rare and exotic) *Salmonella* spp. is high. Salmonellosis has been linked to pet turtles in the USA [25], to bearded dragons in Canada [26] and pet reptiles in Europe [27]. In Sweden, a rise in the number of reptile-associated *Salmonella* cases was observed since 1996 when the import regulations no longer required certificates stating that imported reptiles were free of *Salmonella* [28]. In the Netherlands, reptile-associated salmonellosis was studied during a 30-year period (1985–2014) and a significant annual increase of 19% was found in reptile-associated salmonellosis in humans [29]. Amphibians are also associated as a source of human *Salmonella* [30], but in our study amphibians were not assumed to be as relevant as reptiles as much lower numbers were imported. However, this number might be highly underestimated as many amphibians are imported along with, and classified as, live ornamental fish [23]. Additional risk is that many amphibians and also reptiles are often caught in the wild and exported. In order to assess the risk of amphibians properly, the reporting of imported amphibians should be improved.

Scoring CCHFV was complicated as it is still uncertain if the imported exotic animal species, although they are a potential reservoir, develop a viremia high enough for transmission of CCHFV and, thus, can play a role in the transmission of CCHFV. Therefore, predominantly the risk of transporting virus-infected larvae and nymphs when importing exotic animals should be considered. For an ongoing transmission concerning multiple persons, both competent hosts and a vector population are needed. The *Hyalomma* tick, the main vector of CCHFV, cannot (yet) establish in the Netherlands due to the cold and wet climate, but CCHFV has also been found in other tick species, including *Rhipicephalus sanguineus* [31],[32], which is incidentally found in the Netherlands. There were birds (403), testudines (1092) and rodents (217) imported or transited through the Netherlands in 2013–2014 that came from countries where CCHFV is present. A small part (39) of these testudines concerned *Testudo graeca* [23], which have been found before to carry *Hyalomma aegyptium* ticks with a high prevalence (30.2%) of CCHFV in Turkey and Syria [33]. As no animal health certificates are required for reptiles and there are no regulations about tick control, there is a risk of importing CCHFV into the Netherlands with infected ticks and potentially with viraemic exotic animals. Scores were also relatively high, because the human impact after exposure is high, although more information is needed to better assess the risk.

The risk of West Nile Virus (WNV) from imported exotic animals for public health was scored second highest together with CCHFV and *Yersinia pestis*. WNV can either be introduced into the Netherlands by the import of an infectious vector or by the import of an infected bird with a viremia high enough to infect endemic mosquitoes, since the competent vectors are present in the Netherlands. The probability that WNV will be introduced in the Netherlands by the import of an infectious vector is very low as consignments of live animals from third countries have to be sprayed with insecticidal aerosols against vectors [34]. When it concerns non-detected illegal import these animals are not sprayed. But even in the latter case, the chance of import of mosquitoes is considered low due to the fact that mosquitoes do not stay on the host after blood feeding. Nevertheless, exotic mosquitos are regularly found at the EU airports. In case an infectious vector is imported, it must be able to survive (long enough) in the Netherlands in order to further transmit WNV. The probability that WNV will be introduced in the Netherlands by the import of an infected bird with a viremia high enough to infect Dutch mosquitoes is also not very high as most birds are only viraemic for a few days [35]. However, in the Netherlands there is a range of suitable vectors present and there are still questions on the role of mammals, reptiles and amphibians in the lifecycle of WNV, complicating a risk assessment of this pathogen. Therefore, also keeping in mind the rapid spread of WNV reported in other countries, WNV is one of the top five pathogens on the priority list. When humans become infected with WNV, disease can be severe, although most infections are asymptomatic.

Yersinia pestis can be introduced in the Netherlands by infected rodents and a wide range of flea species. More than 200 species of rodents and lagomorphs have been implicated as reservoirs for *Yersinia pestis*, but the exact number of rodent species that are more than accidental reservoirs is uncertain [36]. As fleas, that are the main transmission route, are permanent ectoparasites, which usually stay on their host, the import of fleas with rodents is to be expected. Human plague is a very serious illness that can lead to death, but can be treated with commonly available antibiotics [37]. In 2013–2014, 226 rodents from Egypt and Argentina were imported into (217), or transited through (9), the Netherlands. In the USA, Egypt and Argentina the presence of sylvatic foci of the plague is suspected [38],[39]. On the general veterinary health certificate that accompanied these legally imported animals, it is stated that animals should be either born in captivity or have been held in captivity for at least six months. Thus, there is a chance that these animals were sourced from the wild, though this is not very likely.

The only arenavirus that was relevant for this risk assessment was Junin virus, as the other viruses are present in countries that are not in the database [23] and/or no rodents were imported from these countries in 2013/2014 or, in case of Lymphocytic Choriomeningitis Virus (LCMV), the virus is already present in the Netherlands. Junin virus can be introduced in the Netherlands via import of infected rodents, that can shed the virus for a lifetime without developing disease themselves. Whether rodent species in the Netherlands are susceptible to Junin virus and can be part of a natural cycle in the Netherlands, is unknown. In 2013–2014, only nine rodents from countries where Junin virus is present were transited through the Netherlands and these were not the main reservoir species *Calomys musculinus*. However, Junin virus has been isolated from several rodent species and also hares, making the association of Junin virus with a single rodent species less definite than for other arenaviruses [40],[41]. Although the risk of exposure is considered very low, human impact of Junin virus infection can be considerable.

In the Netherlands, import control and registrations of exotic animals is done by the Netherlands Food and Consumer Product Safety Authority (NVWA). Import data are automatically transferred to TRACES (TRAde Control and Expert System), a web application of the European Commission connecting official veterinary services. This application makes it possible to trace back all imports of animals and enables quick response to any alert for better protection of animal and public health. However, the records of shipments of exotic animals in this database often only included a rough classification of the type of animals imported i.e. mammal, reptile, amphibian, bird or fish. More detailed information about numbers of animals and sometimes species had to be extracted from the attached (often handwritten) health certificates. For reptiles and amphibians, these health certificates are very basic trade documents concerning species, number and origin of animals without any obligatory testing of animals. Under the current legislation, it is not possible to perform routine microbiological testing of exotic animals upon arrival at Schiphol airport. However, this could be of great value for improvement of import control regulations and the risk assessment would be strengthened by the microbiological testing of exotic animals for zoonotic pathogens that are not yet covered by animal health legislation. The current risk analysis only assessed the public health risk of imports of live exotic animals into the Netherlands in the period of 2013 and 2014. Differences in trade flows and the epidemiological situation in the world can influence the public health risks. These differences were already observed between 2013 and 2014, although no trends could be extracted from this short period [23]. In the USA, the number of annual wildlife shipments has doubled over a period of thirteen years [2]. Whether this holds true for the Netherlands is unknown, but it shows the importance of updating this risk assessment regularly as trade patterns of exotic animals can change and also epidemiological situation and/or new information on the susceptibility of exotic animals to zoonotic pathogens can become available.

To get an indication of the illegal import of live exotic animal species and trade routes into the Netherlands via Schiphol airport, only e-mail correspondence between NVWA staff members and customs about seizures of illegally imported animals or animal products was available. An inventory was made of the numbers and species of exotic animals seized at Schiphol airport between January 2012 and December 2014. In 2015, an online registration system became available for recording seizures of illegally imported animals via Schiphol Airport. This electronic database for NVWA staff members, in which they can enter all necessary information concerning seizures of illegal animals and animal products, is the first step towards building a reliable database for further analysis.

Including additional data on illegal imports in this risk assessment could increase the risk for public health of the total exotic animal trade, as pathogens that were now excluded because no relevant animal species were imported, may be included when assessing the risk of illegal imports as well. The health status of illegally imported animals is furthermore unknown and may pose a much higher risk for public health than legally imported animals. The estimated illegal trade range of 1,700 to 245,000 illegally and transited animals per year shows the difficulty of estimating the numbers of illegally imported animals.

CONCLUSION

So far, this is the first time the public health risk of legally imported live exotic animals into the Netherlands was assessed. This study showed that the overall public health risk of the introduction of exotic pathogens into the Netherlands via the legal import of exotic animals is considered low, which is supported by the lack of large scale outbreaks in the human population caused by exotic animals. However, due to changes in the epidemiological situation worldwide and changing trade patterns, this risk assessment needs regular updating. As this is very time-consuming, and not quantitative, an easily adaptable (quantitative) risk assessment tool should be developed, which can contribute to the early detection of high-risk trade flows of exotic animals in the future. The risk assessment tool can be strengthened by the incorporation of microbiological data. To start with, the large numbers of imported reptiles and amphibians should be risk based or at random sampled and tested for *Salmonella* spp. to enable better estimates of the risk of importing exotic animals infected with *Salmonella* spp. Illegally imported animals did not have a health check before exportation and are not accompanied by a health certificate; therefore, it is expected that the illegal imports pose a more substantial risk for public health than legally imported animals. Analysis of the seizure data of illegally imported animals via Schiphol Airport registered in the recently developed online registration system can contribute to risk-based checks of incoming flights and development of risk mitigation measures.

ACKNOWLEDGMENTS

We are grateful to William Karesh and Daan van Uhm for fruitful discussions on the trade of exotic animals. We also thank the discussions in the project on exotic animals with Clazien Vos and Manon Swanenburg from the Wageningen Bioveterinary Research Institute (WBVR) and Olaf Stenvers of the NVWA. We are also grateful to Arno Swart for discussions in the project on the methodology.

REFERENCES

1. Bush ER, Baker SE, MacDonald DW. Global trade in exotic pets 2006–2012. Conservation Biology 2014;28: 663–676. pmid:24661260
2. Smith KM, Zambrana-Torrelio C, White A, Asmussen M, Machalaba C, Kennedy S et al. Summarizing US Wildlife Trade with an Eye Toward Assessing the Risk of Infectious Disease Introduction. EcoHealth 2017: 1–11.
3. Nijman V. An overview of international wildlife trade from Southeast Asia. Biodiversity and conservation 2010;19: 1101–1114.
4. WHO. Report of the WHO/FAO/OIE joint consultation on emerging zoonotic diseases. World Health Organization, Geneva, Zwitserland. Available from: https://apps. who. int/iris/bitstream/handle/10665/68899/WHO_CDS_CPE_ZFK_2004. 9. pdf;jsessionid =77F3D288ECF1B84B6DA9160B0BE4A2D5?sequence=1 Cited: 9 March 2017.
5. Karesh WB, Cook RA, Bennett EL, Newcomb J. Wildlife trade and global disease emergence. Emerging infectious diseases 2005;11: 1000–1002. pmid:16022772
6. Taylor LH, Latham SM, Mark EJ. Risk factors for human disease emergence. Phil. Trans. R. Soc. Lond. B. 2001;356: 983–989.
7. Jones KE, Patel NG, Levy MA, Storeygard A, Balk D, Gittleman JL, et al. Global trends in emerging infectious diseases. Nature 2008;451: 990–993. pmid:18288193
8. Kuiken T, Leighton FA, Fouchier RAM, LeDuc JW, Peiris JSM, Schudel A, et al. Pathogen surveillance in animals. Science 2005;309: 1680–1681. pmid:16150997
9. Havelaar AH, Van Rosse F, Bucura C, Toetenel MA, Haagsma JA, Kurowicka D, et al. Prioritizing emerging zoonoses in the Netherlands. PloS one 2010;5: e13965. pmid:21085625
10. Van Borm S, Thomas I, Hanquet G, Lambrecht B, Boschmans M, Dupont G, et al. Highly pathogenic H5N1 influenza virus in smuggled Thai eagles, Belgium. Emerging infectious diseases 2005;11: 702–705. pmid:15890123
11. CDC. Multistate outbreak of monkeypox—Illinois, Indiana, and Wisconsin, 2003. Morbidity and mortality weekly report 2003;52: 537. pmid:12803191

12. Unger F, Eisenberg T, Prenger-Berninghoff E, Leidner U, Ludwig ML, Rothe M, et al. Imported reptiles as a risk factor for the global distribution of Escherichia coli harbouring the colistin resistance gene mcr-1. International Journal of Antimicrobial Agents 2017;49: 122–123. pmid:27916289
13. Avashia SB, Petersen JM, Lindley CM, Schriefer ME, Gage KL, Cetron M, et al. First reported prairie dog–to-human tularemia transmission, Texas, 2002. Emerging infectious diseases 2004;10: 483. pmid:15109417
14. Broad S, Mulliken T, Roe D. The nature and extent of legal and illegal trade in wildlife. In: Oldfield S (Ed.). The trade in wildlife: regulation for conservation. Earthscan Publications; 2003. pp. 3–22.
15. Wyler LS, Sheikh PA, Pervaze A. International illegal trade in wildlife: threats and US policy. CRS Report for Congress, Congressional Research Service, US Congress, Washington DC 2013. Available from: https://apps. dtic. mil/dtic/tr/fulltext/u2/a486486. pdf Cited: 9 March 2017.
16. Europol. EU organised crime threat assessment. Available from: www. europol. europa. eu/sites/default/files/publications/octa_2011_1. pdf Cited: 9 March 2017.
17. European Commission. The European Union and Trade in Wild Fauna and Flora. Available from: http://ec. europa. eu/environment/cites/legislation_en. htm Cited: 17 July 2018.
18. CITES. Discover CITES. Available from: https://www. cites. org/eng/disc/what. php Cited 17 July 2018.
19. Pavlin BI, Schloegel LM, Daszak P. Risk of importing zoonotic diseases through wildlife trade, United States. Emerging infectious diseases 2009;15: 1721. pmid:19891857
20. Bueno I, Smith KM, Sampedro F, Machalaba CC, Karesh WB, Travis DA. Risk prioritization tool to identify the public health risks of wildlife trade: the case of rodents from Latin America. Zoonoses and public health 2016;63: 281–293. pmid:26414429
21. Fèvre EM, Bronsvoort BMDC, Hamilton KA, Cleaveland S. Animal movements and the spread of infectious diseases. Trends in microbiology 2006;14: 125–131. pmid:16460942
22. Van Uhm DP. Uncovering the Illegal Wildlife Trade: inside the world of poachers, smugglers and traders [doctoral dissertation]. Utrecht University; 2016.

23. De Vos CJ, Swanenburg M, Tafro N, van Roon A, Stenvers OF, Elbers AR. Animal health risk of legally imported exotic animals into the Netherlands in the period 2013–2014. Microbial Risk Analysis 2017;6: 9–20.

24. CFSPH. CDC Salmonellosis technical information website. 2013. Available from: http://www. cfsph. iastate. edu/Factsheets/pdfs/nontyphoidal_salmonellosis. pdf Cited: 09 March 2017.

25. CDC. CDC Salmonella technical information website. Available from: http://www. cdc. gov/salmonella/general/technical. html Cited 9 February 2017.

26. PHAC. Public Health Notice: Outbreak of Salmonella infections linked to contact with bearded dragons. Available from: http://www. phac-aspc. gc. ca/fs-sa/phn-asp/2014/salm-0527-eng. php Cited: 1 October 2015.

27. Bertrand S, Rimhanen-Finne R, Weill FX, Rabsch W, Thornton L, Perevoscikovs J, et al. Salmonella infections associated with reptiles: the current situation in Europe. Euro surveillance: bulletin Europeen sur les maladies transmissibles = European communicable disease bulletin 2008;13: 717–727.

28. De Jong B, Andersson Y, Ekdahl K. Effect of regulation and education on reptile-associated salmonellosis. Emerging infectious diseases 2005;11: 398–403. pmid:15757554

29. Mughini-Gras L, Heck M, van Pelt W. Increase in reptile-associated human salmonellosis and shift toward adulthood in the age groups at risk, the Netherlands, 1985 to 2014. Eurosurveillance 2016;21.

30. Zarecki SLM, Bennett SD, Hall J, Yaeger J, Lujan K, Adams-Cameron M, et al. US outbreak of human Salmonella infections associated with aquatic frogs, 2008–2011. Pediatrics 2013;131: 724–731. pmid:23478862

31. Papa A, Christova I, Papadimitriou E, Antoniadis A. Crimean-Congo hemorrhagic fever in Bulgaria. Emerging infectious diseases 2004;10: 1465–1467. pmid:15496250

32. Tahmasebi F, Ghiasi SM, Mostafavi E, Moradi M, Piazak N, Mozafari A, et al. Molecular epidemiology of Crimean-Congo hemorrhagic fever virus genome isolated from ticks of Hamadan province of Iran. Journal of vector borne diseases 2010;47: 211–216. pmid:21178213

33. Široký P, Bělohlávek T, Papoušek I, Jandzik D, Mikulíček P, Kubelová M, et al. Hidden threat of tortoise ticks: high prevalence of Crimean-Congo haemorrhagic fever virus in ticks Hyalomma aegyptium in the Middle East. Parasit. Vectors 2014;11: 101.
34. EUR-Lex. COMMISSION REGULATION (EU) No 206/2010 of 12 March 2010. Available from: http://eur-lex. europa. eu/legal-content/en/ALL/?uri=CELEX:32010R0206 Cited: 10 March 2017.
35. Pradier S, Lecollinet S, Leblond A. West Nile virus epidemiology and factors triggering change in its distribution in Europe. Rev. Sci. Tech 2012;31: 829–844. pmid:23520737
36. Dennis DT, Gage KL, Gratz N, Poland JD, Tikhomirov E. Plague manual: epidemiology distribution surveillance and control. Available from: https://apps. who. int/iris/bitstream/handle/10665/66010/whocdscsredc992Fa. pdf Cited 10 March 2017.
37. CDC. CDC Plague Ecology and Transmission. Available from: https://www. cdc. gov/plague/index. html Cited: 10 March 2017.
38. Dennis DT, Campbell GL. Plague and other Yersinia infections. Harrisons principles of internal medicine 2008;16: 921.
39. Lotfy WM. Plague in Egypt: Disease biology, history and contemporary analysis: A minireview. Journal of advanced research 2015;6: 549–554. pmid:26199744
40. Mackenzie R. Public health importance of rodents in South America. Bulletin of the World Health Organization 1972;47: 161. pmid:4539412
41. Mills JN, Ellis BA, McKee K Jr, Ksiazek TG, Oro J, Maiztegui JI, et al. Junin virus activity in rodents from endemic and nonendemic loci in central Argentina. The American journal of tropical medicine and hygiene 1991;44: 589–597. pmid:1650148

Index

C

D

E

Q

R

S

T

Printed in the United States
by Baker & Taylor Publisher Services